Medieval Cosmology

Medieval Cosmology

Theories of Infinity, Place, Time, Void, and the Plurality of Worlds

Pierre Duhem

Edited and Translated by Roger Ariew

The University of Chicago Press
Chicago and London

This is an abridged edition in English translation of Pierre Duhem's *Le Système du monde. Histoire des doctrines cosmologiques de Platon à Copernic*, originally published in ten volumes, in French, by Hermann, éditeurs des sciences et des arts, Paris.

The University of Chicago Press, Chicago 60637
The University of Chicago Press, Ltd., London
© 1985 by The University of Chicago
All rights reserved. Published 1985
Paperback edition 1987
Printed in the United States of America

96 95   6 5 4

Library of Congress Cataloging in Publication Data

Duhem, Pierre Maurice Marie, 1861-1916.
    Medieval cosmology.

    Bibliography: p.
    Includes Index.
    1. Cosmology, Medieval. I. Ariew, Roger.
II. Title.
BD495.5.D832513   1985   113'.09'02   85-8115

ISBN 0-226-16923-5 (paper)

Contents

Part II. Place

Part III. Time

Foreword

The selections here translated represent, for all their length, but a fraction of a well-nigh unique monument of individual scholarship, Pierre Duhem's *Le Système du monde*. A ten-volume, profusely documented analysis of cosmological theories from Plato to Copernicus, the *Système du monde* is in turn a minor part of a vast set of publications by one who excelled as a theoretical physicist no less than as philosopher and historian of science. The six years stretching from 1909 to 1916, during which Duhem wrote this monumental opus, should seem woefully inadequate even if they had been devoted exclusively to the task. However, during the same period, Duhem published his two-volume *Traité d'energétique*, a highly original thermodynamical synthesis of macrophysical phenomena. In addition, to speak only of his major publications during those years, he wrote the third volume of his Leonardo studies. These dealt with the sciences which Leonardo inherited from medieval sources and which, through Leonardo's mediation, guided Galileo's immediate predecessors and even the great Pisan himself.

Like any great accomplishment, the *Système du monde* was a work of faith. Long before Duhem received assurance in 1913 from the Académie des Sciences and the Ministère d'Instruction publique concerning financial support,* the first volume was ready for print and the manuscript of four other volumes was substantially completed. He also had to have faith in his own strength. Even Duhem's best friends, who knew that he hardly ever had to rewrite

---

*About these and other details mentioned in this introduction, ample documentation is given in my book, *Uneasy Genius: The Life and Work of Pierre Duhem* (Dordrecht: Martinus Nijhoff, 1984).

a phrase, let alone a page, were staggered at the magnitude of the project. Duhem had never been in the best of health. Would he last twenty more years? Would he endure the continued deprivation of that happiness he lost when his wife died in 1892 in the third year of their marriage? A second heart attack within two weeks felled him on September 14, 1916, at the age of fifty-six. Death came to Duhem in the house of his maternal ancestors in Cabrespine, a small village north of the famed medieval town of Carcassonne. He was then correcting the proofs of the fifth volume of the *Système du monde*, published posthumously in 1917.

Duhem left behind a large pile of manuscripts, obviously the continuation of the *Système du monde*. His daughter, Hélène, then twenty-four, did what seemed to be eminently sensible. She deposited the manuscripts with the Académie des Sciences where a committee of four was appointed to study the manuscripts and to arrange for their publication. Most interested among the four was Darboux, perpetual secretary of the Académie, a former teacher of Duhem and a good friend as well. But Darboux died in early 1917 and the manuscripts were shelved. Only in the 1930s did Hélène become fully aware of her mission to activate the long dormant project. Among those who gave her support was Mme Tannery, the able editor of the works of her late husband, Paul Tannery, a prominent historian of science and a friend of Duhem. At Mme Tannery's urging, *Isis* carried in 1937 an appeal which called for the publication of the manuscripts through international subscription. A year later, the French branch of the Société internationale d'histoire des sciences devoted its meeting to the various aspects of Duhem's work and issued a similar appeal in its organ, *Archeion*. Last, but not least, Hélène herself published a moving biography of her father in which she displayed time and time again something of his penetrating mind and brilliant pen.

All these appeals and efforts were made on behalf of a manuscript in publishable form. But because of financial difficulties and ideological opposition only a small part of what later became the sixth volume was typeset in 1938. Another batch of galleys followed ten years later, but once more work was disrupted. Finally, in 1953 financial support was obtained from the Centre Nationale de la recherche scientifique. In 1954 the sixth volume appeared together with a reprint of the first five volumes. The remaining four volumes were published between 1956 and 1959.

Had Duhem lived another year or two, he would have almost certainly completed the *Système du monde* with a discussion of

Copernicus's achievement. He had planned, after completing his magnum opus, to write, during a summer vacation in Cabrespine, a 300-page summary of it for the general public. It would have certainly become a classic. He also had hoped to leave the study of the history of science behind and to devote most of his energies to theoretical physics again. A series of communications on electromagnetic theory which he sent in 1916 to the Académie des Sciences, proved all too well that theoretical physics was his abiding interest. It was as a physicist that he became a corresponding member of the Académie in 1900 and a nonresident member in 1913, one of six on whom that status was first conferred.

Meanwhile in France, as well as abroad, perplexity grew over the fact that the professional career of a savant of such caliber was confined to provincial universities. Duhem first taught at Lille (1887-93), then in Rennes (1893-94), and finally in Bordeaux (1894-1916). Partly at fault were his teachers at the Ecole Normale, which he entered in 1882 at the head of his class, a position he never lost. Not that his teachers, especially Jules Tannery (vice-director of studies at the Ecole from 1884), overestimated young Duhem's abilities by letting him write a doctoral thesis while only a second-year student and not even an *agrégé*. Duhem wrote his thesis on his great insight, the thermodynamical potential, which not only permitted a broad generalization of thermodynamics, but, unlike Marcelin Berthelot's favorite idea, the principle of maximum work, also could cope with chemical reactions in which the energy transfer was very small. Duhem and his teachers gave too much credit to Berthelot's much professed commitment to the policy of judging everything and everybody on objective merit alone.

Duhem's thesis was rejected by the Sorbonne in June 1885. Patently nonscientific factors must have been at play. The verdict written by the head of the jury, Gabriel Lippmann, is hardly an example of scientific care. Lippmann himself may have felt threatened, as Duhem's thesis was far superior to the thermodynamics found in Lippmann's treatise on the subject. Lippmann, although a student of Gustav Kirchhoff, excelled far more in experimental physics (he was to receive the Nobel Prize for his discovery of color photography) than in theoretical work. But the deepest source of opposition was Berthelot, whose principle of maximum work was politely but thoroughly discredited by Duhem's thesis. Duhem was brilliant enough to present another thesis two years later and to defend it with flying colors before another jury. Once more he wrote on thermodynamics, though (to prevent

Lippmann and Berthelot from interfering directly) not in reference to chemical processes but to electromagnetic induction in its mathematical aspects. Duhem earned a doctorate, not in physics but in mathematics. The brilliance of his rejected thesis had already been proved through its publication by A. Hermann, a leading scientific publisher, who brought it out in an important series devoted to the latest advances in physics. No degree of brilliance, however, could outweigh the gravity of the professional, scientific wound inflicted on Berthelot's ego. Berthelot, who held one after another, sometimes simultaneously, the most powerful positions in French academic, educational, and political circles, could, well beyond his death in 1907, force the "establishment" to obey his verdict: "This young man shall never teach in Paris."

This is not the place to speculate on what would have happened had Duhem's career as professor of physics run its course in Paris, a center of the intellectual world. On account of his anti-atomism and antirelativism, Duhem is too often branded as a reactionary. Yet on both counts his stance seems rather prophetic in retrospect. As modern atomic physics progressed, less and less room was left for the mechanistic way of thinking about matter, Duhem's chief argument against atomism. As to relativity, the expansion of the universe and the 2.7^0 K cosmic background radiation stand out in bold defiance to the equivalence of all reference systems. His objections to relativity were in defense of a logic which forbids the turning of mathematical formalism into pseudo-epistemology. Had he lived a dozen or so more years, he would have combated on the same ground the Copenhagen interpretation of quantum mechanics. While embodying Mach's idea of economy of thought, Duhem's philosophy was the very opposite of Mach's sensationalism, a solipsist stance in ultimate analysis. Duhem was a realist, though not in the facile style of mechanists whom he spiritedly combated. The fallacy of mechanistic realism was in his eyes that of a pseudo-metaphysics which threatened true or realist metaphysics, because it expected metaphysics to deliver particular truths about the inner structure of matter and its interactions. While in this respect Duhem held metaphysics impotent, he saw metaphysics as the only source that could justify commonsense realism, the indispensable tie which connects physical theory to the physical world and gives meaning to the physicist's labor. Such is the foundation of Duhem's philosophy of science, which in no way was derivative of his Catholic faith, and which reveals, as time goes on, more and more depth regardless of efforts that try to discredit his writings as mere "preachment" or mere logicism.

The writings of Duhem, which especially create either admiration or resentment, relate to his researches on the history of science. Although the son of deeply devout Catholic parents and an alumnus of Collège Stanislas, a bastion of Catholic secondary education, young Duhem learned only one positive thing about the Middle Ages: its artistic splendor expressive of religious vision. Louis Cons, his much admired history teacher at Stanislas, was an agnostic layman and a Comtean. As late as 1903, when Duhem, at the request of the editor of *Revue générale des sciences pures et appliquées*, published a series of articles on the philosophical history of mechanics, he dismissed the Middle Ages in one line as an age sterile scientifically. The jump from the Greeks to Galileo was by then a hallowed cliché of intellectual history, especially for historians of science. To perceive the illusory character of that cliché, an interest, however vivid, in the history of mechanics was not enough. Equally insufficient was the interest in history which Comte's popularity inspired in the late nineteenth century. The decisive factor that alerted Duhem to the illusion seems to have been his bent on complete rigor, a principal feature of his physical theory. The specific direction which his historical research was to take was determined by his overriding interest in the notion of thermodynamic potential. He naturally sought its origin in the idea of virtual velocity, the basis of statics and dynamics. He did not suspect that his search would lead him deep into the Middle Ages as he published the first installment of his history of statics in the October 1903 issue of *Revue des questions scientifiques*. There too he made the same jump from the Greeks to Galileo without suspecting its illusory character.

Within a month or so he came across a cryptic reference in a late sixteenth-century book to an elusive author (Jordanus de Nemore). Although that reference had already been noted by several historians of science, only Duhem had the heroic perseverance and penetrating acumen to unfold the full story and to perceive its portent. The result was the unearthing of extensive medieval discussions about virtual velocity. Other equally startling discoveries followed in quick succession. The jump from the Greeks to Galileo soon appeared to Duhem as a monumental somersault.

That jump has been reinstated in its erstwhile repute, tellingly enough, through a heavy reliance on jumps, or rather mutations (genetic and mental—often in blissful disregard of the possibility that the two may be very different in nature), as the ultimate form of explanation. While courteous homage, if not mere lip service, is often paid to Duhem's greatness ("Duhem is the teacher of us all," was declared in 1961 in a prominent gathering of historians of science), his main conclusions are firmly opposed by many in that profession. Readers of these selections will not have to go beyond the very first page to see the cause of this baffling situation. Duhem would not, of course, be surprised by that new breed of studies on science in the Middle Ages in which every effort is made to keep out of sight the role—a very central one—which the Christian creed played throughout medieval centuries. Having been always sincerity incarnate, he would speak out candidly, fearless of the charge of lèse-majesté all too often leveled against those who dare to dissent from the "received view." Fond of logic, he would point out that the Christian creed made no sense whatever without its very first tenet: belief in the Maker of heaven and earth, of all things visible and invisible—the dogma of creation out of nothing. The logician in Duhem would be seconded by the historian. As he declared in the very first page of these selections and in many other pages of the Système du monde, it was the vivid awareness of medieval Christian thinkers of the implications of that tenet that provoked them to criticize Aristotelian physics and cosmology steeped in pantheism and, its chief corollary, eternalism.

Had a Buridan and an Oresme cavorted with pantheism, they would have assigned, in their commentaries on Aristotle, the motion of heavenly bodies to a desire expressive of their divine connaturality, and the prospect of the eventual formulation of the laws of motion would have been once more nipped in the bud. Instead, in a sound rebuttal of the Stagirite, they attributed the motion of those bodies, created by God as everything else, to another creative act, the imparting by God of an impetus to them, and kept by them undiminished owing to the absence of friction in heavenly spaces. Newton might have recognized in this his first law of motion and he might have appeared less original. But by 1700 the Age of Faith, which provided such scientific light, had long been branded as the Dark Ages. As a result, the interpretation of intellectual history, increasingly wary of the God of revelation, had to rely more and more on factors that were all too often not better than a deus ex machina to account for the rise of science.

Duhem saw the rise of science in "the resolute struggle which the University of Paris, then staunch defender of Catholic orthodoxy, waged against Peripatetic and Neoplatonic paganism," and he asked: "How could a Christian fail to thank God for that?" Duhem knew, of course, that whatever the proofs of the medieval and Christian origin of science, they were not proofs of Christianity as such. Nor did he expect conversions on the part of those reading his historical studies. His vast correspondence proves that he was on cordial terms with scholars of the most varied persuasions and backgrounds. He received from Protestant, Jewish, and agnostic scholars, French and foreign, admiring notes for having discovered— single-handedly—the existence of science during the Middle Ages. He did not expect a fair hearing from the type of scholar that took Condorcet's sketch of mankind's mental development for a creed. He knew that the specter of science germinating in the Middle Ages could only loom ominous for those who believed, and still do, that Revelation in general and Christianity in particular had to be thoroughly discredited and sidelined if science was to rise and remain true to itself. Those who find unpalatable Duhem's emphasis on the ultimate theological mainspring of science in the Middle Ages, will still find much enjoyment in this volume on account of its mind-boggling riches relating to the history of scientific ideas, logic, and philosophical analysis. Even Alexandre Koyré, the most articulate spokesman of opposition to Duhem's claim about the organic connection between medieval and Galilean science, admitted, as the *Système du monde* began to be printed in full, that it was "a work of unlikely richness and . . . an incomparable and indispensable source of information and tool of study."

The selections translated here are mostly confined to the posthumously published parts of the *Système du monde* which deal with the abstract foundations of cosmology. Infinity versus finitude, space, time, void, and the plurality of worlds are keenly analyzed topics in this twentieth century which saw the birth of the first truly scientific cosmologies by virtue of Einstein's general theory of relativity—cosmologies free of the contradictions that plagued the notion of an infinite Euclidean universe. In setting forth the views of medievals on such topics, Duhem loved to point out, with his astonishingly wide expertise, the often stunning similarity of some of those views to the dicta of such leading mathematicians among his contemporaries as Couturat, Cantor, and others.

For selecting text with much judiciousness, the translator deserves no small praise. Admiration for his work will only grow when the clarity and graceful flow of his phrases are considered. To render a master stylist like Duhem into another language calls for considerable skill. To present difficult medieval texts calls not only for linguistic expertise, but also for a keen perception of often elusive, but all important, philosophical, theological, and scientific nuances.

The translator also displays courage and independence of mind. At a time when the concepts of revolution and mutation have appropriated the status of ultimate intelligibility in most studies on the history of science, both courage and independence of mind were required to bring to the center stage a Duhem who never failed to hold high the truth of the slow and organic growth of ideas. In his masterful account of the origins of modern chemistry, a book written as he was publishing the fourth volume of the *Système du monde*, Duhem compared Lavoisier's "revolution" to the story of an acorn whose shell breaks at the proper moment so that the seed may take root and grow into a tree. Similar instructive statements occur in the *Système du monde* and in these selections. Had they been read and pondered, the recent historiography of science may have kept itself free of ideologies which Duhem would have been the first to see as utterly destructive to the idea of scientific progress. In that progress he believed, with a passionate devotion to the cause of science as a part of human understanding.

STANLEY L. JAKI

Preface

The Principle of Selection

As Professor Jaki points out, during 1956, on the occasion of the posthumous publication of volumes VI and VII of Pierre Duhem's *Le Système du monde, Histoire des doctrines cosmologique de Platon à Copernic*, Alexandre Koyré, who was then the leading historian of science and leader of the opposition to Duhem's historical views, still felt compelled to praise the *Système du monde* by describing it as an intellectual landmark. Koyré wrote that the *Système du monde* "is a work of permanent value whose richness of documentation is the fruit of labor so large that it confounds the mind; in spite of forty years of study and research, it remains a source of knowledge and an instrument of research which has not been replaced, and is therefore indispensable."[1] Twenty-five years later, Koyré's claim is still valid, and it can be extended. No other complete synthesis of cosmological doctrines from the Greeks to the modern age has been attempted, and no other synthesis of medieval cosmological doctrines has been accomplished. The medieval period is clearly Duhem's forte, and it is the period in which his contributions have been most impressive. Working in a vacuum of knowledge about medieval science, he is said to have single-handedly destroyed the myth of the "scientific night" of the Middle Ages.

But even if other such works existed, the greatness of the *Système du monde*, and the need for making it more widely available, would not be diminished. The *Système du monde* does not merely provide us with information about cosmological doctrines, it enables us to view a powerful thinker synthesizing historical data in order to use them as support for a philosophical point of view—which

is still at the forefront of the philosophy of science—and for a
physical theory. For Duhem, an understanding of the history of
science is crucial for doing science, because it is "the memory of
past attempts and their happy or unhappy fate that prevents science
from accepting hypotheses which have led more ancient theories
to their ruin."[2] That is why the *Système du monde* makes for such
interesting reading, and that is basically the rationale for this
translation of *Medieval Cosmology: Theories of Infinity, Place,
Time, Void and the Plurality of Worlds*. Of course, medieval
cosmology has great intrinsic interest, but beyond the mere
description of a particular content resides its analysis and the
interpreter's purpose in dealing with that particular content. In
the *Système du monde* the reader is not only treated to medieval
cosmology, but also to Duhem's analysis of medieval cosmology.
One might compare this with the metamorphosis of *Othello* in
Verdi's hands. Without doubt, *Othello* is a great Shakespearean
tragedy, but *Otello* is a great opera as well. Both can be great and
intrinsically interesting; the greatness of the one does not diminish
the greatness of the other.

Medieval Cosmology is a selection of translations from the
Système du monde, dealing with cosmology proper and restricted
to the Middle Ages. This selection represents the core of the *Système
du monde* (though, naturally, no single volume is going to represent
its ten volumes completely) since, as the subtitle of the *Système
du monde* indicates (*History of Cosmological Doctrines from Plato
to Copernicus*), cosmological doctrines are the focus of the work.
The principle of selection—to translate only properly cosmological
discussions of medieval texts in the *Système du monde*—provides
one with a unified volume in a number of ways and for a number
of reasons.

1. The *Système du monde* itself is divided into parts: Part V,
"Parisian Physics during the Fourteenth Century," which spans
volumes VI to IX, is divided topically, instead of temporally or
by author. This enables one to choose whole selections without
having to change Duhem's work to any great degree. The selections
from part V can be classified into three main groups: (1) cosmology
proper, including such topics as the infinitely small and the
infinitely large, place, time, void, and the plurality (or singularity)
of worlds; (2) physics—meaning mechanics, dynamics, and
kinematics—including such topics as the change of forms,
movement in the void, projectile movement, the accelerated fall
of weights, and the rotation of the world (or the earth); and (3)

other physical sciences, including astrology, the theory of tides, the equilibrium of the earth and of the seas, and the origins of geology. It becomes a simple matter to pick out the chapters of the first group as selections for *Medieval Cosmology* and supplement them as necessary for coherence and completeness.

It is no secret that Duhem's work was accomplished some sixty-five years ago, and that, in some details, it might have been superseded by the work of others building upon Duhem's great pioneering achievement. This is more true with respect to the second group of possible selections than with respect to the first or third group. Of the historians of medieval science who have labored after Duhem, the most respected have been Clagett and Maier; their careful scholarly work has been concerned primarily with physics. For example, Clagett's major work, *The Science of Mechanics in the Middle Ages*, deals with statics, kinematics, uniform acceleration, impetus theory, and free fall. In it, Clagett talks about cosmology only in chapter 10—which is 48 pages long—entitled "Mechanics and Cosmology." There Clagett states, "Obviously I have not attempted to give a complete historical treatment of each of the problems discussed, nor have I ranged over the large spectrum of cosmological problems revealed in Pierre Duhem's *Le Système du Monde*."[3] So Clagett himself refers his readers to Duhem for discussions of medieval cosmology. Moreover, Clagett does not refer his readers to Maier's work on cosmology, for Maier's work, which precedes Clagett's, is also primarily about physics—intension and remission of forms, impetus theory, the structure of matter, the cause of fall, acceleration, free fall in a vacuum, and so forth. In this respect, there is as strong an argument for a volume of Duhem on the fringes of science—astrology, theory of tides, equilibrium of the earth and seas, and origins of geology—as there is for Duhem on cosmology.

2. A second unity can be achieved using our principle of selection. The selections derived can be characterized as those dealing with the medieval commentaries on books III and IV of Aristotle's *Physics* (supplemented by book I of Aristotle's *De Caelo*). In books III and IV of the *Physics*, Aristotle also proceeds topically, so that cosmology proper may be defined roughly as the topics Aristotle discusses there, namely, movement (*Physics* III, 1-3), the infinite (*Physics* III, 4-8; *De Caelo* I, 5-7), place (*Physics* IV, 1-5), the void (*Physics* IV, 6-9), time (*Physics* IV, 10-14), and the singularity and eternity of the world (*De Caelo* I, 8-12). *Medieval Cosmology*

therefore closely follows Aristotle's cosmological topics of books III and IV of the *Physics* and book I of the *De Caelo*.

3. The third unity derived from our principle of selection is that, because of it, the reader is given the best set of selections by which to judge Duhem's bold historical thesis that the condemnations of 1277 (brought forth by Etienne Tempier, the Bishop of Paris, and the faculty of theology of the University of Paris) were instrumental in the birth of modern science—were its "birth certificate." Repeatedly in these selections, Duhem attributes the change in medieval cosmological doctrines before and after 1277 to the condemnations. For example, in the chapter on infinity, Duhem states: "From the start of the fourteenth century the grandiose edifice of Peripatetic physics was doomed to destruction. Christian faith had undermined all its essential principles. . . . This substitution [of a new science for the old science] . . . often sprang . . . from the desire not to incur the condemnations brought forth by ecclesiastic authority. . . . Among what was overthrown during the fourteenth century were the principles formulated by Aristotle on infinity in number and infinity in magnitude." In the chapter on place, Duhem asserts: "Until 1277, one proceeded by repairing Aristotle's system. . . . In 1277, the decrees brought forth by Etienne Tempier, Bishop of Paris, formulated a proposition contradicting Aristotelian philosophy with respect to the mobility of the ultimate sphere and of the whole universe. Afterwards there appeared theories of place that broke with Aristotelian tradition." In the chapter on time, Duhem states: "The memorable sixteenth-century discussions about the theories of place and movement received their starting point with Etienne Tempier's decision; they were inaugurated by John Duns Scotus who dared to assert the proposition that even if there exists no immobile term [as reference] a body could still move by local movement. . . . Duns Scotus added another assertion to the above, one that is no less contrary to the Stagirite's physics: Even if heaven stopped, time would continue to be and to measure the movements of the other bodies." In the chapter on void, Duhem summarizes his thesis by stating that: "In 1277, Etienne Tempier, Bishop of Paris, condemned the following two errors: The First Cause cannot make more than one world; God cannot move the heavens in a straight line, the reason being that He would then leave a void. The first of these condemnations denied what Peripatetic philosophy taught about the impossibility of infinite magnitude, both potential and in actuality. . . . The second of these condemnations upset the Peripatetic theory of place. . . . These

two condemnations also contributed toward ruining the Peripatetic theory of time. . . . Everything that Aristotle's *Physics* asserted about infinity, place, and time shattered when it was confronted by the power of the condemnations of Paris. . . . But we have not yet described all the consequences of the decisions brought forth by Etienne Tempier and his counsel. We shall now see how they required Scholasticism to deny Aristotle's objections against the possibility of void." And finally, with respect to the plurality of worlds, Duhem wrote: "The question about the plurality of worlds, like many other problems, seems to place in opposition the impossibilities decreed by Peripatetic physics and the creative omnipotence Christianity recognized in God. . . . Christianity . . . considered the assertion that a second world cannot be produced as an impious pretension of philosophers, placing a limit on God's power. This belief finds expression in the decree brought forth on March 7, 1277, by Etienne Tempier, Bishop of Paris, and his advisors." This last statement indicates straightforwardly why the thesis seems so compelling (as do all of the others, directly or indirectly). Ironically, the intended repressive condemnations removed the fetters placed by Aristotelian philosophy on the medieval philosophers' ability to conceive alternative cosmological schemes. Peripatetic philosophy denied the very *possibility* of infinity, the movement of the earth, atoms, the void, and a second world. For medieval theologians, to deny such a possibility would be to limit God's omnipotence; this limitation seemed impious and was condemned. Philosophers were then *required* to conceive what was formerly considered inconceivable. Given the formidable task, it does not seem that the labors proceeded too slowly or too laboriously. The results were often dramatically different from their starting points, and even if, in the end, some medievals came to validate their starting points, it was after putting their philosophy through extremely interesting and revealing meditations and analyses.

The Problem of Translation and the Critical Apparatus

Clearly, there is a serious problem when translating an early twentieth-century French text that makes frequent use of quotations and paraphrases of thirteenth-to-fifteenth-century Latin texts (not to mention older Greek, Roman, Arabic, and Jewish texts). In addition to the standard difficulties of translation, one has to deal with the text within the text, a text which itself has its own difficulties because of manuscript variations, new editions, and abbreviations (for example, the published editions of Albert of Saxony's *Commentary on the De Caelo* are so varied that entire questions disappear and are changed from one edition to another). Of course, one could always provide photoduplications of the originals, or edited texts, but the cost of publication would become prohibitive. Moreover, the text would become extremely cumbersome. For these reasons, and for readability, it is best to provide a translation and then to do as much as possible to indicate where the interested reader might find the original texts. This gets us back to translating a translation. Obviously the translator of a translation has a choice to make; he can attempt to translate the original text most faithfully and perhaps provide a jumbled text or a text that opposes the interpretation given to it in the main text, or he can attempt to translate the translation and perhaps provide a text which is more removed from the original. But between these two extremes, there are two compromises available. The translator can do both and provide an odd text in which a battle rages between the text and the footnotes (and thereby undercut his credibility with the translated material—we are essentially back to the path of providing the original text and variants). Finally the translator can attempt to reconcile the translated text and the original. This final alternative is a compromise that, if not successful, may be open to accusations at both ends—providing a jumbled text and providing a text removed from the original—but it has the virtue of attempting to be faithful to what is being translated while not being blind to the problems of translating a translation. In the present case, the choice is not difficult to make—really there is no choice. As already indicated, Duhem's analysis of the original texts is as important as the texts themselves, so the translator has decided to attempt to remain as faithful as possible to Duhem. For most interpreters, one would wish that they interpret faithfully; for a few, it is *their truth* which is valuable.

What I have attempted to do in the translation is to make Duhemian sense out of every passage. This does not mean that I have disregarded the original texts. I have made every effort to consult the original texts (I spent the summer of 1981 at the Bibliothèque Nationale looking at Duhem's sources) and to obtain every source available on this continent, not so much to "correct" Duhem, but to arrive at the most faithful interpretation of the text reconcilable with Duhem's writings. Another reason to track down Duhem's original sources and what is available here is to be able to indicate where the interested reader might find these documents.

Toward that end, I have included citations to available documents in the notes, indicating (in the corresponding bibliographical citation) whether the item is a microfilm, whether there is a standard edition of the various manuscripts and books, and whether there are English translations available. I believe that this is the most useful thing I could do for the interested reader. In keeping with this policy, I have reproduced Duhem's notes and bibliography, translating his French into English, but leaving the Latin as is. Whenever possible in the notes, I also give a second note—in square brackets—referring to an updated entry—also in square brackets—in the bibliography of works cited by Duhem. The updated entry immediately follows Duhem's entry in the same bibliography. Similarly, when Duhem refers in the text to a manuscript using its original Latin title, I have done the same; but when he refers to a manuscript by giving it a French title, I have translated the title into English.

Some Background on the Relevant Aristotelian Cosmological Doctrines

For background on ancient cosmology relevant to medieval cosmological discussion, the reader is referred to volume I of the *Système du monde*, especially chapters 4 and 5. But it might be useful here to summarize, as briefly as possible, the relevant Aristotelian cosmological doctrines, with an eye toward what the medievals might find particularly problematic. However, I will dispense with the summary of Aristotle's position on the singularity or plurality of worlds, since I include Duhem's section on this topic from volume I. (The reason for my inclusion of this section is that, unlike other sections of volume I, it is paired with a section on Simplicius and Averroes on the plurality of worlds.)

The Two Infinites

Aristotle's doctrine on infinity may be summarized as follows: There are two infinites to be considered, infinite by addition, and infinite by division; the infinitely large, and the infinitely small. But since, when we say that something is infinite, the "is" in that sentence means either what potentially is or what actually is, there are four possibilities, namely, potentially infinite by division and addition, and actually infinite by division and addition. Aristotle argues against the existence of actual infinities (thereby denying both the actual infinitely large and the actual infinitely small).

His doctrine about potential infinity is more complex. He wishes to affirm the existence of the potentially infinite by division in magnitude and number, while denying the potentially infinite by addition in magnitude, except in the case where one is adding a part determined by a ratio, instead of keeping the parts equal.

However, there is a problem with accepting the existence of the potentially infinite while denying the actually infinite, a problem recognized by Aristotle. Generally for Aristotle, what is potential will be actual—which seems to license the inference from the existence of the potentially infinite to the actually infinite: The phrase "potential existence" is ambiguous. Using Aristotle's example, when we speak of the potential existence of a statue, we mean that there will be an actual statue. It is not so with the infinite. There will not be an actual infinite. "The word 'is' has many senses, and we may say that the infinite 'is' in the sense in which we say 'it is day' or 'it is the games', because one thing after another is always coming into existence."[4] There are then two (or more) senses of "potential" according to Aristotle; one sense, which the potential infinite shares with the Olympic games and things whose being is not like that of substance, consists in a process of coming to be and passing away, a process which is finite at every stage, but always different. The Olympic games are potential both in the sense that their being consists in a process, and in the sense that they *may* occur. It is only in the latter sense that when a state is potential, there will be an actual state. Hence, Aristotle can affirm the existence of potential infinities such as the infinite in time, in the generations of man, in the division of magnitudes, and in numbers, while denying the existence of the actually infinite. However, Aristotle also wishes to deny the existence of the potentially infinite in magnitude by the addition of equal parts. He does so by asserting that "there is no infinite in the direction of increase. For the size which it can potentially be it can actually be. Hence, since no sensible magnitude is infinite, it is impossible to exceed every assigned magnitude; for if it were possible, there would be something bigger than the heavens."[5] Aristotle seems to be playing on the ambiguity of "potential" in this case; that will be a difficulty which the medievals will have to unravel.

Some of the medieval refinements with respect to Aristotle's doctrine on infinity will be concerned with the ambiguity of "potential" in potentially infinite—as in "the games are potential." A first attempt to bring out Aristotle's intention will be to distinguish between the infinite *in facto esse* and the infinite *in fieri*, the latter being a mixture of the infinite in actuality and potentiality. But soon the *logica moderna* will impart another distinction, one that will provide the standard medieval terminology for dealing with the problems of infinity. The logicians distinguished between categorematic terms and syncategorematic terms, or terms that have a signification by themselves, and terms that have no signification apart (cosignificative terms). Examples of the first kind are substantival names and verbs, and examples of the second kind are adjectives, adverbs, conjunctions, and prepositions. A list of the syncategorematic terms would commonly include: every, whole, both, of every sort, no, nothing, neither, but, alone, only, is, not, necessarily, contingently, begins, ceases, if, unless, but that, and infinitely many. One might call these words logical constants (or perhaps connectives, functions, or quantifiers) and distinguish them from predicative terms. The distinction is applied to infinity to yield both a categorematic and syncategorematic infinite. This is not unusual. A modern logician might consider existence both as a quantifier and as a predicate. "The phrase 'infinitely many' is both syncategorematic and categorematic, for it can indicate an infinite plurality belonging to its substance either absolutely or in respect to its predicate."[6] The distinction allows one to solve logical puzzles, since it may be true that something is infinite, taken syncategorematically, and false that something is infinite, taken categorematically. The distinction also enables one to ask separately whether the syncategorematic infinite exists and whether the categorematic infinite exists, without worrying about potentialities. Of course, medievals will take differing views with respect to the existence of various infinities, and will often disagree with Aristotle's doctrines. It is not difficult to see why this should be so, given that parts of Aristotle's doctrine about infinity are clearly in conflict with the conception of an absolutely omnipotent God who is a creator.

Other interesting issues connected with infinity discussed by medievals will include questions about the natural minimum of a substance, whether there is such a minimum or whether the division can go on indefinitely; questions about indivisibles, such as points, lines, and surfaces, and their ontological status; and questions about the notion of limits, of maximum and minimum.

Place

Aristotle's conception of place is extremely distant from our conceptions of place and space. He defines place as the boundary of the containing body in contact with the contained body, a contained body that can be moved by locomotion. But he immediately modifies this definition by asserting that place is the innermost motionless boundary of what contains. (To make matters worse, there even seems to be, in chapter 6 of Aristotle's *Categories*, if not a different theory of place, at least a different emphasis, or an earlier account of place.) There is clearly some tension between the two definitions of the *Physics*, and it gives rise to questions about whether place is mobile or not. There is also some difficulty with the ontological status of place, connected with the ontological status of surface. But the most interesting problem is the one with respect to the place of the ultimate sphere. If having a place must depend on being contained, the ultimate sphere will have no place since there is no body outside it to contain it. Aristotle recognizes this: "Heaven . . . is not anywhere as a whole, nor in any place, if at least, as we must suppose, no body contains it."[7] Moreover, heaven needs a place since it rotates, and movement involves change of place. Aristotle's resolution of this problem is a distinction between place *per se* and place *per accidens*. Place *per se* is the place that bodies capable of locomotion or growth must possess. Place *per accidens* is the place that some things possess indirectly, "through things conjoined with them, as the soul and the heaven. Heaven is, in a way, in place, for all its parts are; for on the orb, one part contains another."[8]

The medievals will bring forward new precisions to the definition of place. An important medieval distinction relative to these problems, one originally formulated by Aristotle but without any elaboration, is the distinction between place and the *ubi*, the former being the inner surface of the containing body and the latter the outer surface of the contained body (and variations thereof). Also thoroughly discussed is the important question about the place of the ultimate sphere. The medievals will formulate many variant solutions to the problem, and will add another, theologically inspired, of an immobile ultimate heaven beyond the mobile ultimate heaven.

Time

For Aristotle, time is the number of movement, that is, time is the enumeration of movement. There cannot be any time without there being some change. The link between time and movement is extremely close, given that we measure movement by time and time by movement. Consequently, there are as many times as there are movements, and these times are all able to serve as the definition of time. However, the choice of a movement to measure time is not an indifferent choice; the measure must be of the same kind as the object it serves to measure, but it must also play the role of principle with respect to the latter. Although Aristotle thinks that time has no reality independent of the movement it measures, he does not think that time has no reality independent of the measurer of the movement. Aristotle asserts that time is independent of soul: "Whether if soul did not exist time would exist or not, is a question that may fairly be asked; for if there cannot be some one to count there cannot be anything that can be counted, so that evidently there cannot be number; for number is either what has been, or what can be, counted. But if nothing but soul, or in soul reason, is qualified to count, there would not be time unless there were soul, but only that of which time is an attribute, i.e., if *movement* can exist without soul, and the before and after are attributes of movement, and time is these *qua* innumerable."[9]

The obvious questions raised by the medievals will deal with the subjectivity of time and its intimate connection with movement. Other topics discussed by the medievals will include the equivalence of movements for measuring time (whether there is an absolute clock), the asymmetry between the theories of time and place, and possible atomistic conceptions of time (time atomistic in reality and continuous in the mind, instead of vice versa).

Void

Aristotle denies the existence of the void. He does so with a number of different arguments which can be separated into two general kinds. First, there are arguments that conclude that the void is impossible, if it is thought to be a place with nothing in it, or a place deprived of body, distinct from the bodies that occupy it. But then if it is three-dimensional, it would be a body, and could not accept another body in the same place: "The void is thought to be place with nothing in it. The reason for that is that people take what exists to be body, and hold that while every body is in place, void is place in which there is no body, so that where there is no body there must be void. . . . If void is a sort of place deprived of body, when there is a void where will a body placed in it move to? It certainly cannot move into the whole of the void."[10]

The second type of argument concludes that movement is

impossible in the void. The argument is derived from Aristotle's principles of dynamics. A body moving by violence moves in proportion to the force exerted on it and in inverse proportion to the resistance of the medium in which it is situated. Since a void would provide no resistance, the body "would move with a speed beyond any ratio"[11]—which is impossible. Aristotle also adds the interesting observation that in the void, "no one could say why a thing once set in motion should stop anywhere; for why should it stop *here* rather than *there*? So that a thing will either be at rest or must be moved *ad infinitum*, unless something more powerful gets in its way."[12]

The medievals will distinguish between the void within the ultimate sphere and beyond it (a possibility taken seriously by Stoicism), and will make some progress in both realms. With respect to the void within the circumference of the world, they will investigate the argument based on the impossibility of movement in the void, that it must be instantaneous, and that there is no reason for a body to come to a rest in the void. As Duhem indicates, meditating on this topic might lead one to conceive the notion of mass. With respect to the void outside the ultimate sphere, the medievals will investigate the connection between its impossibility and the impossibility of plural worlds.

Acknowledgments

As with all enterprises of this sort, this project has benefited from the assistance of numerous institutions and individuals. I would like to take this occasion to thank a number of them, while apologizing for the many others that should have been acknowledged but will inevitably be left out. The Bibliothèque Nationale was extremely kind in allowing me to use its marvelous facilities and incomparable collections—although I am sure that Charles V might have been upset if he knew that a commoner would be allowed to use some books and manuscripts from his library. I must thank the University of Chicago's Humanities Collegiate Division and William Rainey Harper Fellow Program (Braxton Ross, then Master of the Humanities), and Virginia Polytechnic Institute and State University's (Virginia Tech) Department of Philosophy for giving me release time to work on the project; moreover, Virginia Tech's Center for Programs in the Humanities (Wilfred Jewkes, Director) was charitable enough to award me a stipend to travel to France, and Virginia Tech's Center for the Study of Science in Society

(Arthur Donovan, then Director of the Center) graciously provided me with two excellent typists (Bonnie Meredith and Becky Cox) to input my manuscript into a word processor. Virginia Tech's Center for Programs in the Humanities, Department of Philosophy, and College of Arts and Science (Henry Bauer, Dean) also collaborated to fund the preparation of the manuscript, whose camera-ready copy was produced in Virginia Tech's Print Shop. Moreover, I must thank Robert Ariew, Michael Crowe, Alan Donagan, Daniel Garber, and Robert Wengert for their invaluable advice and moral support. Finally, Susan Ariew has spent countless hours proofreading the translation and manuscript with me; no doubt, proofreading, like translation, is one of the labors which is never acknowledged sufficiently, since when it is done perfectly it is invisible, but given that it is never done perfectly, all one ever sees are the oversights.

ROGER ARIEW

I
The Two Infinites

1
Infinitely Small and Infinitely Large

Actual Infinity in Number and the Immortality of the Soul

From the start of the fourteenth century the grandiose edifice of Peripatetic physics was doomed to destruction. Christian faith had undermined all its essential principles; observational science, or at least the only observational science which was somewhat developed—astronomy—had rejected its consequences. The ancient monument was about to disappear; modern science was about to replace it. The collapse of Peripatetic physics did not occur suddenly; the construction of modern physics was not accomplished on an empty terrain where nothing was standing. The passage from one state to the other was made by a long series of partial transformations, each one pretending merely to retouch or to enlarge some part of the edifice without changing the whole. But when all these minor modifications were accomplished, man, encompassing at one glance the result of his lengthy labor, recognized with surprise that nothing remained of the old palace, and that a new palace stood in its place. Those who, during the sixteenth century, became aware of this substitution of one science for another were possessed by a strange delusion; they imagined that the substitution was sudden and that it was their doing. They proclaimed that Peripatetic physics, that dark den of error, just succumbed to their blows, and that they had built upon its ruins, as if by magic, the bright domain of truth. The men of subsequent centuries were either the dupes or the accomplices of the sincere delusion or vain error of these men. The physicists of the sixteenth century were celebrated as the creators to which the world owed its scientific renaissance; but they were often merely continuators, and sometimes plagiarists. During the substitution of the astronomy of eccentrics and epicycles for the

3

astronomy of homocentric spheres, experience acted alone; it had another important role to play in the destruction of Aristotle's dynamics and in the creation of the new dynamics. Except for these two occasions, experience contributed little to the substitution of the new ideas for the old ideas. This substitution was the result of philosophical discussions, and these discussions themselves often sprang from the wish to admit nothing as true that did not conform with Catholic orthodoxy, from the desire not to incur the condemnations brought forth by ecclesiastic authority. One can state that the excommunications delivered in Paris on March 7, 1277, by Bishop Etienne Tempier and the Doctors of Theology were the birth certificate of modern physics.[1] For example, among what was overthrown during the fourteenth century were the principles formulated by Aristotle on infinity in number and infinity in magnitude. New ideas concerning this subject were expressed, discussed, and formulated to the point of preparing and even beginning the creation of the infinitesimal calculus. This was due to a belief in two dogmas: the dogma of the personal immortality of human souls, and especially, the dogma of the creative omnipotence of God.

Let us recall that for Aristotle the whole theory of the infinitely large can be summarized in four assertions.[2]

1. The existence of an actual infinity of objects distinct from one another is contradictory.
2. A multitude of objects distinct from one another may be infinite potentially; that is, however large a finite [3] number of such objects, one can always add another, thus constituting a larger number.
3. The existence of an actual continuous infinite magnitude is contradictory.
4. The existence of a potential continuous infinity in magnitude is an impossibility; that is, when adding *actually existent* magnitudes of the same kind to one another, one cannot exceed a certain limit in size, for the world is limited. It is only in the imagination of the mathematicians that for a given magnitude one can always add another (in thought).

The first proposition of this doctrine was almost immediately contradicted by the belief in the individual survival of each human soul.

Aristotle believed that the world had no beginning and that it would have no end; he thought that it would eternally start again in its periodic existence. Even this doctrine seemed contradicted by the impossibility of an actual infinite multitude. When Aristotle was writing, had not the world already traveled through an infinite multitude of cycles? Had not the starry sphere and the sun accomplished an infinity of revolutions?

These infinite multitudes which had unfolded in time, these multitudes whose previous unities ceased to exist at the moment their present unity came into being, did not seem worthy to Aristotle of the name "actual infinity"; the name "potential infinity" for these multitudes was more to his liking. He was careful to state that if he allowed the notion of potential infinity, it was precisely in order to safeguard these three truths:[4]

1. Number can be indefinitely augmented by means of addition.
2. Continuous magnitude can be indefinitely subdivided.
3. Time had no beginning and will have no end.

Belief in the individual survival of the human soul gives the objection a more pressing formulation. Aristotle attributes no beginning to the human species in the same way that he attributes no beginning to the world. At each instant of time it is true that an infinity of men were born, lived, and died; but according to the belief in the individual survival of the human soul, the souls of each of these men actually subsist and remain actually distinct from one another. Hence, at each instant of time, the souls of the dead form an actual infinite multitude of distinct objects.

How would Aristotle have resolved this objection which he had not even mentioned? Doubtless he would have denied the individual survival of the human soul; either he would have considered it as destroyed with the body, in accordance with the opinion of Alexander of Aphrodisias, or he would have asserted, as Averroes did assert for him, that after death all souls unite themselves into a single intelligence, common to all humanity.

The objection that Aristotle had not even considered could not have worried the pagan philosophers who succeeded him either. The Stoics denied any belief in the immortality of the soul. The Neoplatonists thought that souls were limited in number and that for each cyclical revolution of the world a soul would repossess a body similar to the one which it had occupied during previous revolutions.

Jews, Christians, Mohammedans—all those who believed in the individual survival of the human soul and who at the same time denied the periodic reincarnation of souls—would have regarded the objection as without force. All of them agreed with the statement that the world did not have a past eternity, that it had a beginning, and there was a first man, and that therefore the souls of the dead were finite in number.

What did not, until then, embarrass anyone appeared, however, to be a serious difficulty when the philosophers attempted to reconcile, on the one hand, the Peripatetic metaphysical doctrine that neither the world nor the human species had a beginning, and, on the other hand, their religious belief that the human soul subsists after death, eternally distinct from other souls, and exempt from perpetual new beginnings—in other words, when Avicenna and his disciples first discovered that there is a contradiction between the impossibility of an actual infinity in number and the personal immortality of the human soul.

Avicenna allowed that the impossibility of an actual infinite multitude is not absolute; according to him, one has to restrict it to the multitude composed of objects, each occupying some place, a position. It does not extend to the multitude of souls stripped of all position.

This opinion of Ibn Sina [Avicenna] was conserved for us by Ibn Rushd [Averroes] who, of course, did not adopt it:

> The assumption that souls deprived of matter are numerically multiple is unknown in the opinions of the philosophers. In fact, for them it is matter that causes numerical multiplicity. . . . Besides, the impossibility of an actual infinity is a recognized principle of the philosophers, whether the objects in question are bodies or not. One finds no one who distinguishes between what has a position and what does not have a position, except Avicenna. Among the rest, I have found none who talk in this fashion. . . . For the philosophers deny all actual infinity, whether it is bodily or not; and the existence of such an infinite would allow that there may be an infinite which is larger than another infinite. Avicenna intends to flatter the common people by upholding that which they understand by soul, but his discourse is quite inadequate.[5]

In order to reconcile the assumed eternity of the world and of the human species with the individual survival of each soul, Avicenna required that an actual infinity be possible. But his philosophy required, on the other hand, that this infinite not always be possible. In fact, elsewhere he reasoned thus: a contingent thing, which by itself is merely possible, owes its existence to some cause; this cause itself can owe its existence to some other cause, but one cannot ascend indefinitely in the hierarchy of these causes, each of which engenders the existence of the lower cause. One must therefore arrive at a first cause which is not caused by any other.

The proof of the existence of the first cause therefore requires the major premise: an actual infinite series in the hierarchy of causes is impossible. The proof would flounder if one admits, without restriction, the possibility of an actual infinite multitude.

What would be the suitable restriction concerning this possibility? Must we except all series of causes? An exception so large and absolute would be unacceptable. In fact the world is eternal; there have been generations and corruptions from all eternity. From all eternity some air has been transformed into water, some water into air or earth. In each of these transformations, the corrupted substance was the material cause of the substance engendered. Hence one can state that each of the concrete substances existing in the world today is the result of a series of material causes, and that this series comprises an infinite multitude of causes.

Avicenna therefore does not extend the impossibility of an actual infinite multitude to all series of causes; he does not extend it, in particular, to series of causes such as material causes that cease to exist when their effect is produced. He retains the impossibility in the case of essential causes, causes conferring existence upon their effect and necessarily coexisting with their effect.

Such is the thought Ibn Sina [Avicenna] expresses on two occasions in his *Metaphysics*:

> When we demonstrate that within series causes are finite in number, we intend to refer to series of essential causes. We do not deny that before the essential causes there was an endless series of auxillary or preparatory causes. Better yet, it is necessary that it be so. . . .
>
> With regard to the essential causes of a thing, those by which a thing exists effectively, we have already shown that they must necessarily coexist with the thing; they must not be prior to it in such a way that it may be possible to suppress them without suppressing their effect. This possibility, in fact,

may be accorded only to causes which are neither essential
nor proximate. I do not deny that these causes which are
neither essential nor proximate can proceed to infinity; rather
I claim that they must.[6]

Elsewhere Avicenna considers a certain mass of air and the
mass of water issuing from the corruption of this air or engendering
it anew by its own corruption:

> Each one becomes, at its turn, the subject from which
> the other is engendered, for the first is corrupted into the
> second and the second into the first; assuredly, then, neither
> one is essentially prior to the other—it is prior only
> accidentally, a priority which refers to the individual and not
> to the species. Water is not naturally more worthy of being
> the principle of air than air being the principle of water; these
> two bodies are thus interchangeable in existence (*vicissitu-
> dinaria in esse*). This particular mass of water may be prior
> to this particular mass of air [or vice versa]. I do not deny
> that with these particular beings there can be no beginning
> and no end. But we do not claim that these particular beings
> are in need of a beginning, only that their species are. Neither
> do we claim that there is a need for an accidental beginning,
> only for an essential one. In fact, in the previous cases, the
> causes can precede [or succeed] one another to infinity, in
> the past and in the future. Moreover, we do not need to
> demonstrate that things are finite in number except those
> things which are essential causes.[7]

What Avicenna asserted in various places on the subject of
the possibility of an actual infinite multitude, al-Gazali compiled
and attempted to form into a system. Toward this end he
distinguished between things whose very nature ranks them within
a determined order, such as things each of which is the effect of
some preceding thing and the cause of some succeeding thing, and
things whose order is purely accidental, such as things whose order
may be changed without any change in their nature. The multitude
of the things whose hierarchy follows an essential order cannot,
in any case, be actually infinite. It is the same for things whose
hierarchy is not essentially ordered, but which follow among
themselves an order of location in such a way that they are located
on this side or that side of each other. It is not the same for things

that need not be ranked according to an order of nature or an order of location.

Being is divisible between finite being and infinite being.

There are four ways in which being is said to be infinite; of these four ways, two exist and two do not exist, as is recognized by reasoning.

It is said that the movement of heaven has neither beginning nor end; and that has already been demonstrated.

It is also said that there is an infinity of human souls separated from their bodies; and that is necessarily true if one rejects finitude for time and for the movement of heaven, which one has accomplished by rejecting that they have a beginning.

The third way is, for example, the way in which one says that there is an infinite body or an infinite space from the top to the bottom; but this is false.

The fourth occurs when one states that the series of causes is infinite, since a thing has a cause and this cause itself has a cause, and that one cannot attain, in this fashion, a first cause which has no other cause; but this is also false.

The sense of these assertions is that we cannot conceive an infinity of multiple existent things which have, by their nature, a certain order, or else which are on this side or that side of each other.

It will be thus for an infinite series of causes; necessarily, the order between the cause and the effect is an order of nature, for if one suppresses this order, there is no longer a cause.

It is the same for bodies or spaces which are conceived as able to be ordered; some of them, in fact, are necessarily to one side of the others, as soon as one begins from somewhere. But that is not an order of nature; it is only an order of location. The difference between these two orders is indicated in the treatise, *De prius et posterius*.

But for all multitudes where one of these orders cannot be found, infinity is not prohibited.

Thus it is not prohibited from the movement of heaven (though it has a certain order and a certain rule of progression) because all the parts of this movement do not exist simultaneously following the same disposition. In fact, when we state that the movement of heaven (had no beginning and) is without end we do not mean by that to suppress the finitude of movements existing (presently) only to suppress the set of all movements which have been, are, and will be.

In the same fashion, we grant that the human souls which, upon death, have been separated from their bodies, are infinite in number, because there is no order of nature between them such that if one suppressed it, they would cease to be souls; none of them, in fact, is the cause of the others. They exist simultaneously without being ordered by nature or by location (*sunt simul sine prius et posterius natura et situ*). In fact, one conceives the before and after between them only with reference to the time of their creation. But in their essences, by which they exist and by which they are souls, there is no hierarchy; they are all essentially equal (*sed sunt aequales in esse*). It is the contrary with respect to spaces, bodies, and causes and effects.[8]

Infinity, then, is not by itself repugnant to multitude; the repugnance toward being infinite is introduced into multitude when an order of nature or of location in space is introduced.

Why this distinction between multitudes that can be actually infinite and those that cannot be? Al-Gazali gives no reason. Probably he has no wish other than to reconcile the four propositions that play an essential role in Avicenna's philosophy:

1. The world is eternal.
2. Human souls are immortal and remain distinct one from the other after they have left their bodies.
3. The world is finite in size.
4. By ascending the series of causes, one necessarily arrives at a first cause that has not been caused.

But when al-Gazali abandoned the philosophy of Avicenna in order to return to the pure religious doctrine of the Koran, he became eager to deny the possibility of all actual infinity and to draw from it an argument against the eternity of the world: if the world had no beginning, the number of revolutions that the sun had already accomplished would be infinite, and the number of revolutions of Saturn would also be infinite; however, the first number would be to the second in a ratio inverse to the ratio between the periods of revolution of the two stellar bodies. Two infinite numbers would then have a determined ratio, which al-Gazali deems contradictory.[9]

The Peripatetic philosophers used against al-Gazali and the

Mutakallimun the doctrine set forth by Aristotle, without giving any clearer reason for it. A multitude cannot be infinite when all the objects forming this multitude exist simultaneously. It can be infinite, on the contrary, when it is a collection of objects existing one after the other. That is the thesis upheld by Averroes[10] and by Moses Maimonides.[11]

It seems that the second position taken by al-Gazali should have been immediately adopted by all Christian philosophers; the negation of actual infinity, even for a collection of objects existing successively, would require that the world had a beginning, and hence, that it was created. As a consequence, the souls of the dead would be finite in number; the individual survival of these souls would pose no difficulty.

But soon the influence of some philosophers, particularly Maimonides, would suggest an opinion which Saint Thomas Aquinas defended ardently: the world had a beginning—that is a dogma which faith teaches us—but reason is incapable of demonstrating it; the proofs one gives for it are clearly worthless.

Consequently, it is true that since the beginning of the world, heaven only accomplished a finite number of revolutions, but there is no contradiction in the world having no beginning, and in it accomplishing an infinity of revolutions—hence the necessity of upholding, with Averroes and Maimonides, Aristotle's thesis: an infinite multitude of objects existing one after the other is not contradictory.

But a difficulty remains. Faith teaches us that there was a first man; reason cannot prove it for us. The assumption that humanity had no beginning therefore cannot imply a contradiction. But according to this supposition, an infinite number of men have already lived; the souls of the dead would make up an infinite multitude of coexistent objects.

Neither Averroes nor Maimonides would be embarrassed by this difficulty suggested by Avicenna, al-Gazali, and the Mutakallimun, for neither one nor the other believed in the individual survival of the human soul. It would seem to be a serious objection for a Catholic theologian like Thomas Aquinas who wanted there to be no impossibility in the assumption of an eternal world.

In order to avoid this difficulty, Thomas Aquinas seems disposed to take up the position of al-Gazali, perhaps even to go further: "We have not yet demonstrated that God cannot make it be that there is an actual infinity. (*Adhuc non est demonstratum quod Deus non possit facere ut sint infinita actu.*)"[12]

This thought clearly dictates the correction that the Angelic Doctor, commenting on Aristotle's doctrine about infinity, brought to the subject. The Philosopher declared that no power is capable of producing by successive additions a magnitude that surpasses all quantities. Thomas takes care to introduce the following modification: there exists no such power *in nature (in natura)*.[13] He thereby safeguards God's creative power.

This prudent reservation is even more noticeable in the conclusion of one of the *Quaestiones disputatae*:

> If the infinite can exist in actuality according to the nature of things, or even if it cannot exist in this manner because of an impediment which is not itself the ground (*ratio*) of infinity, I state that God can create an actual infinity. But if actual existence is repugnant to infinity due to its own ground, then God would not be able to produce this existence, no more than He would be able to make it that man were not a rational animal. As for whether actual existence is repugnant to infinity because of its own ground, or not, it is a question which is incidental to this discussion. Hence, for now, I shall postpone it.[14]

This assertion cannot be invoked either for or against the possibility of an actual infinity. It sufficed, however, to authorize certain authors to believe themselves Thomists because they formally admitted the possibility of an infinite in actuality. Among these was the Dominican Graziadei of Ascoli. Let us present to our readers this mostly unknown philosopher whose opinions we often have cause to recall.

In 1484, the *Questions on Aristotle's Physics, Disputed at the University of Padua*, by Friar Graziadei of Ascoli, were printed in Venice.[15] In 1503 these *Disputed Questions* were again printed in Venice; but this time they were preceded by *Literal Questions* on Aristotle's *Physics*. In order to indicate his respect for the Angelic Doctor, Graziadei kept the order of his lecture the same as Thomas Aquinas's lecture on Aristotle's *Physics*. The colophon of the 1503 edition states that the *Questions* of Graziadei were printed following an exemplary manuscript recently discovered. It therefore suggests that the *Questions* were already old; but in which century were they composed? The colophon does not indicate this, and it is not easy to discover.

Fathers Quétif and Echard number many writings as those of Graziadei,[16] but they impart little information about their author. They tell us that, according to Leander Albertus, he flourished in 1341;[17] but they do not hide from us that Alonzo Fernandez had him living in 1480 and Altamura in 1314.

Any indication able to quell the indecision of these dates would obviously be welcome. Here is one that seems secure to us: when we study the theory of time, we see that Graziadei, as most Scholastics, examines the question, does time exist in the soul or outside the soul? We see that the answer he proposes, on two separate occasions, is completely in conformity, both in word and spirit, with the remarkable solution which Joannes Canonicus develops, after Francis Bleth, in his *Questions on the Physics*. It does not seem doubtful that Graziadei had before him the writings of Joannes Canonicus and that he summarized them faithfully.

And the *Questions* of the Canon can be dated approximately; they were drawn up while Gerard of Odon was general minister of the Franciscan order, that is, between 1329 and 1342.[18] The *Disputed Questions* of Graziadei and the subsequent *Literal Questions* could not have preceded the year 1340 by much. Could they have been written much later than this date? Nothing seems to indicate so. The most recent opinions discussed—and without ever naming their authors in any case—are those of Peter Aureol, William of Ockham, and Joannes Canonicus; the doctrines examined seem to have been those debated in Paris just before 1350. One might therefore think that the fourteenth century was not yet halfway finished when Graziadei of Ascoli succeeded in becoming a Doctor of Theology at Paris, according to the rules of his order. When Leandro Alberti told us that he flourished in 1341, he was not leading us astray. Graziadei, like others at the University of Paris at the time, believed that God could create an actual infinite magnitude; he further thought that this opinion was in conformity with that of Saint Thomas Aquinas.

> To the question, can God create an actual infinite magnitude, I only intend to respond in conformity with what I take to be Saint Thomas's position. If someone is able to demonstrate that Saint Thomas's opinion was contrary to what I am saying, I would be willing to retract it completely. In fact, Thomas stated, in his disputed question about the infinite, that if the actual existence of the infinite is not contrary to the nature of things and does not imply a contradiction, then according to him, God can make such

an infinite. Truly he does not declare that the actual existence of the infinite is not contrary to the nature of things; but it seems that this is not difficult to see.[19]

Graziadei hence applied himself to proving that the actual existence of an infinite magnitude is exempt from all contradiction, internal as well as external. He concluded that God can create such an infinite magnitude, and he believed that by formulating this conclusion he was following Thomas's intention; assuredly, he misjudged Thomas's intention.

Saint Thomas Aquinas had tested out the possibility of granting to God the power to produce an infinite multitude in actuality, and to create a magnitude which was infinite at least in power. But these possibilities were promptly suppressed by the tyrannical authority of Peripatetic philosophy.

In one of his quodlibetal discussions, Thomas maintained that God can create anything that does not imply a contradiction—and therefore an actual infinite if the existence of such an infinite is not contradictory.

> But the Commentator, in the sixth commentary on the fifth book of the *Metaphysics*, stated that neither the absolutely infinite nor the accidentally infinite can exist in actuality; the accidentally infinite can exist potentially, but the absolutely infinite cannot. According to Averroes, then, everything that would relate to actual existence is repugnant to infinity, and this side seems closer to the truth than the other side. (*Et sic, secundum eum, esse infinitum omnino repugnant ei quod est esse in actu; et hoc verius esse videtur.*)
>
> In natural reality, there cannot exist an unspecified thing, a thing that would be indifferent between diverse specific determinations. Doubtless, our intellect can conceive an animal that has not yet received the specific determination making it rational or making it devoid of rationality; but there cannot actually exist an animal which is neither rational nor devoid of rationality. In addition, according to the Philosopher, a thing cannot be within a genus without belonging to one of the species of that genus.
>
> And all quantity is specified by some quantitative determination. The species of numbers, for example, are two, three, etc.; the species of length are two cubits, three cubits, etc.; and in the same way other magnitudes are specified by some measures. It is therefore impossible to find a magnitude in actuality not delimited by its own terms.
>
> But the infinite attains a magnitude only by the

suppression of all terms; it is in virtue of this suppression that a magnitude is said to be infinite. It is therefore impossible for an actual infinite to exist.

Further, the Philosopher states in the third book of the *Physics* that the infinite is like some matter not yet specified and existing only under privation, that the infinite behaves as both part and container.

Consequently, in the same fashion that God cannot make a rational horse, he cannot make it be that something actual is infinite. (*Et ideo sicut Deus non potest facere equum rationale, ita non potest facere ens actu esse infinitum.*)[20]

We find again, in the *Summa Theologica*, the reasoning contained in the above passage:

Assuming from what has preceded that no creature is infinite in essence, it still remains to inquire whether any creature can be infinite in magnitude.

We must therefore observe that a body, which is a complete magnitude, can be considered in two ways: mathematically, in which case we consider its quantity only, and naturally, in which case we consider its matter and form.

That a natural body cannot be infinite in magnitude results from the substantial form of the body requiring a quantity between a determined maximum and minimum; in fact, the Philosopher states that a certain measure and reason is suitable for the magnitude of every natural being.

The same applies to a mathematical body. For if we imagine a mathematical body actually existing, we must imagine it under some form, because nothing is actual except by its form. Hence, since the form of quantity as such is a shape, such a body must have some shape. It would therefore be finite, for shape is confined by a term or boundary.[21]

If he rejects the possibility of an actual infinite in magnitude, does Thomas at least allow what Aristotle called potential infinite magnitude, that is, magnitude that grows with the addition of new parts beyond any limit one might assign it? The possibility of indefinite division of magnitude allowed by the Philosopher seems to have as counterpart the possibility of indefinite addition of magnitudes. Thomas Aquinas rejects this analogy. In order to do so he invokes an odd reason he borrows from Aristotle and Averroes. While seeking in what order of causes the infinite must be ranked, Aristotle wrote this somewhat enigmatic phrase: "It is evident that the infinite is a cause in the sense of matter, and that its essence is privation.''[22]

This brief mention attracted the attention of Averroes, who commented on it thus: "It is evident that matter is the cause of the infinite. If infinite is regarded as cause, it would be cause insofar as it is matter. In fact, the essence of the infinite is the privation of all end and matter is the cause of all privation."[23] Averroes also adds: "The essence of the infinite is only potential, and in this way it resembles the essence of matter, not form; in fact, the essence of the infinite resides in its potentiality, while the essence of form and limitation resides in its actuality. The finite therefore resembles form and the infinite matter."[24]

The above thoughts inspired in Thomas Aquinas the following odd argument:

> The infinite in quantity, as was shown above, belongs to matter (se tenet ex parte materiae). Now, by division of the whole we get closer to matter, since parts are as matter (se habent in ratione materiae); but by addition we get closer to the whole which is as form. Therefore the infinite is not found in the addition of magnitude, but only in division.[25]

Aristotle excluded potential infinite magnitude from the reality of things, the possibility to produce by successive addition a magnitude greater than any limit, because there can be no magnitude greater than the world. At least he allowed the mathematician to conceive it and to use it in his reasoning. Did Thomas Aquinas give the same latitude to the geometer? It is possible to doubt this because of the following: "A geometer does not need to assume that a given line is actually infinite; he needs to take some actually finite line, from which he can subtract whatever he finds necessary. This line he calls infinite."[26] It is possible to think that Thomas Aquinas did not understand how infinite lines enter into the geometer's reasoning.

From magnitude, let us pass to number. Can there exist an actual infinite multitude of coexisting objects? "Some, as Avicenna and al-Gazali, said that it is impossible for an absolutely infinite multitude to exist actually, but that an accidentally infinite multitude is not impossible."[27] This phrasing, and the explanation the *Summa Theologica* gives for it, do not do justice to the opinion of Avicenna and al-Gazali. The *Summa*, in any case, says nothing about what caused the opinion; it does not here refer to the difficulty one might have in reconciling the eternity of the world with the individual survival of the human soul. Further, Thomas Aquinas formally rejects the possibility of any actual infinite multitude, whether absolute or accidental. The only other infinite multitude he allows is potential infinite multitude—potential in the sense that Aristotle understood. However, there is another infinite multitude that he, imitating the Stagirite, admits as a possibility; the world might not have had a beginning; the number of revolutions actually completed by heaven, the sun, the fixed stars, and other stellar objects would be an infinite number. The Angelic Doctor is not repulsed by this conclusion as he was by the existence of an actual infinite magnitude:

> Movement and time are not actual as wholes, but successive, and hence, they have a potentiality mixed with actuality. But magnitude is an actual whole, and therefore the infinite in quantity refers to matter, and does not agree with the totality of magnitude; yet it agrees with the totality of time or movement; for to be in potentiality befits matter.[28]

We rediscover, then, in the *Summa Theologica*, Aristotle's whole doctrine of the infinite; perhaps it gained in narrowness that which its argumentation lost in clarity. The Thomists, following their master, allowed that the world can exist from all eternity. They then encountered the corollary: the multitude of the souls of the dead might be actually infinite. But they were wary of admitting this corollary; they did not follow the example of Avicenna and al-Gazali, an example that Thomas Aquinas might have been tempted to follow. They preferred the supposition that within the eternal world mankind had a beginning, or else that God had decreed the periodic reincarnation of human souls in limited numbers. Thinkers like Giles of Rome and Godfrey of Fontaines accepted these hypotheses rather than admit the possibility of an actual infinite multitude. In order to reconcile the eternity of the world and of mankind with the belief in the individual survival of human souls, Avicenna's Neoplatonism had attempted to enlarge the Peripatetic theory of the infinite. But the circle, momentarily broken, was repaired; it was broken again, but that time at another point and for different reasons. The ideas of Avicenna and al-Gazali continued to figure in the large list of arguments that Scholastic discussion had pitted itself for or against; but whether they were admitted or rejected, they were no longer ideas that commanded the allegiance of their readers.

Geometric Arguments against Infinite Divisibility

Before broadening itself, the theory of the infinite became more precise; this precision it owed above all to the study of the infinitely small, of the infinite divisibility of the continuum—questions about which almost all Scholastics, except for the rare atomists, such as Gerard of Odon, Nicholas Bonet, or Nicholas of Autrecourt, agreed among themselves and with Aristotle.

The opponents of the doctrine of Leucippus and Epicurus frequently attempted to deny atomism in the name of geometry; they liked to draw consequences from it that contradicted the teachings of the mathematicians. It seems that one should consider Roger Bacon as the innovator of this method; in his *Opus Majus*, which he addressed in 1266 or 1267 to Pope Clement IV, he argues as follows:

> If lines are composed of atoms, the diagonal of a square and its side will have the same ratio as the number of whole atoms making up these lengths; therefore these lengths are commensurable, contrary to what the mathematicians teach.[29]

The thought contained in the above passage was developed by John Duns Scotus in his commentary on the *Four Books of the Sentences of Peter Lombard*. Duns Scotus distinguishes between two versions of the theory he opposes. One affirms that the continuum is composed of *indivisibilia*, that is, discontinuous atoms separate one from the other; the other affirms that it is composed of *minima* continuously welded one to another. Scotus pits Roger Bacon's argument and other similar arguments against each of the two versions. Here is an example: Concentric circles are intersected by any radius from the center; hence the circles must contain the same number of atoms, and, consequently, they must all be equal.[30] The geometric refutations of the doctrine of indivisibles, proposed by Roger Bacon and Duns Scotus, were first propagated in the schools by the disciples of the Subtile Doctor. The Scotist, Joannes Canonicus, summarily takes up the arguments of his master in his *Questions on Aristotle's Physics*.[31] And William of Ockham, in one of his *Quodlibets*, summarizes the geometric demonstrations of Scotus.[32] Among the logicians who follow the tradition of Ockham, the most powerful that we have encountered is surely Gregory of Rimini. Gregory of Rimini was a hermit of Saint Augustine. On May 28, 1357, he was named general prior of the order that Giles of Rome had formerly honored; but he had barely taken over these functions when he died a year later, in 1358 at Vienne (Isère). We have a voluminous commentary by Gregory of Rimini on the first two books of the *Sentences*; happily this commentary, to which we will often refer, is dated—a datum too frequently omitted from manuscripts. Gregory's lessons were reported to have been given at Paris in 1344. And in his commentary on the second book of the *Sentences*, Gregory of Rimini attacks the doctrine that attempts to build a continuous magnitude by means of a limited number of indivisibles, by invoking the geometric impossibilities noted by Duns Scotus. He completes his treatise by stating, "One might construct many other mathematical arguments, but these should suffice."[33] Of course, some were not content to reproduce or summarize the arguments of the Subtile Doctor; they extended them by various means. Such was the work of the celebrated Thomas of Bradwardine who died in 1349, the moment when the sovereign pontiff ratified his election as archbishop of Canterbury. He composed a *Tractatus continui*, of which the only book surviving in manuscript is the first; a summary of it was published by Maximillian Curtze.[34]

In his work—whose complete publication would be most interesting—the illustrious mathematician and theologian of Oxford University attempts to refute various atomist sects by mathematical reasons. He attacks those who construct a finite continuum by means of a limited number of contiguous indivisible elements, and those who construct it out of a limited number of separate points. He also attacks those who view the continuum as composed of an actually existent infinity of points. These ingenious demonstrations, which oppose the propositions of geometry with the consequences of the division of a continuous magnitude into points or into indivisible elements, were also thought extremely important by John Buridan; but the role he attributed to them was, so to speak, the inverse of that which his predecessors attributed to them.[35] Bacon, Duns Scotus, and Ockham had faith in the absolute certainty of geometry; since the existence of indivisibles seems to imply conclusions contrary to geometry, then certainly, continuous magnitudes were not composed of indivisibles. The proposition, continuous magnitude is not composed of indivisibles, is not viewed by Buridan as a corollary whose truth is assured by the necessity of not contradicting geometry, the science of incontestable certitude; he sees in it a principle whose truth the geometer is obliged to admit in order to construct his science, a postulate which the geometer cannot do without, whose denial would compel him to renounce his work. Far from geometry's certainty guaranteeing the truth of the proposition, it is the truth of geometry that is subordinated to the correctness of the proposition; and the correctness of the proposition is not for geometry, but for physics or metaphysics to establish. Doubtless Buridan's notion was too profound since it does not appear to have been adopted by even his most faithful disciples. Albert of Saxony[36] and Marsilius of Inghen[37] argued against the hypothesis of indivisibles using Scotus's arguments; they did not hesitate to rely on geometry in order to refute the hypothesis.

Are Indivisibles Pure Abstractions?

Except for the few atomists like Gerard of Odon or Nicholas of Autrecourt, the Scholastics of the thirteenth and fourteenth centuries affirmed the proposition, a continuum is not actually and really composed of indivisibles. Is the result of this that indivisibles are pure abstractions? For example, a volume would therefore not be composed of planes piled up on top of one another. From this, does it result that, outside our mind, surfaces that delimit bodies are nothing at all? Are there not, on the contrary, some physical properties that must be attributed not to the whole body but to its surface? Must we not conclude that a surface has some reality itself, even though it cannot exist independently from the body it delimits? That is a question John Duns Scotus poses and examines. He formulates the first of two opposing hypotheses as follows:

> The indivisible is nothing more than the lack of continuity; the instant, for example, is formally nothing more than the lack of continuity; the point is nothing more than the absence of length, and the word point expresses nothing positive. . . . Hence the line is simply the privation of width, and the surface the privation of depth.[38]

Among the objections that the Subtile Doctor formulates against this hypothesis, here is the last, which is also the most important:

> Many sensible and corporeal qualities exist on the surface, as we can see; the surface is not, therefore, a simple privation. The premise is proven by the colors and shapes which are visible by themselves, and which are consequently positive things. Besides, shape follows the surface in an absolutely proper fashion . . . in such a way that it seems to be an accident

which manifests the surface; and it does not seem probable that the positive entity of such a quality which follows naturally and manifests the surface essentially implies in itself a privation, or that it has a privation for immediate support (*subjectum*).[39]

To attack the opinion that seems most probable to Duns Scotus was one of the favored tasks of William of Ockham. We have previously described the treatise in which a disciple of Ockham sketched his teacher's doctrines;[40] let us quote from this treatise:

According to the preceding principle he posits that one must not admit indivisibles such as those commonly conceded, such as points, lines, surfaces, and things of that kind. In fact, neither reason, nor experience, nor authority prohibit us from doing so.

He states that the texts authorized by Aristotle should be interpreted conditionally. When Aristotle asserts, for example, that the circle is a shape such that the lines from its center to its circumference are all equal, he states that it must be understood thus: the circle is a shape such that, if a point existed, the lines from this point to its circumference will be equal. That is how one ought to explain all the postulates and conclusions relative to indivisibles.[41]

In fact, Ockham affirms in several works that indivisibles, such as points, lines, and surfaces, are only pure negations, that they represent nothing positive. Sometimes he formulates the assertion as an incidental proposition when dealing with questions of physics; sometimes he takes it as the formal subject of his discussion. For example, the first two questions of his *De sacramento altaris* are about "whether points, lines, and surfaces are absolute things distinct from volumes (*quantitas*)."[42] The first of these two questions, which has as its object the refusal to admit any positive reality to points, is the longer discussion; it is a remarkable example of the penetrating discussions for which the mind of Ockham is marvelously well suited. Those who agree with the author do not think that points are indivisible real things that terminate a line. They demonstrate that "a point is not some positive and absolute thing really distinct from any volume (*quantitas*) and especially a line."[43]

They argue that formally, everything divisible is through its proper nature finite and terminated; that if it is continuous, it is so through its proper nature, without any other thing added to it; that from the point of view of causality, it is finite, terminated and continuous through God's agency and the other causes maintaining it, whatever those causes are. The line therefore is finite, terminated and continuous without anything else being added to it. If God, destroying all the other things, conserved this line, it would still be truly finite, terminated and continuous. And this indivisible thing is not admitted for any other reason [than to assure finitude, termination, and continuity to the line]. It therefore seems impossible, as well as superfluous, to admit that the point is such an indivisible thing.[44]

What is a point then? "Do not be duped by words" is the attitude Ockham brings to his philosophical problems; the force of his logic derives from his ability to escape abstract terms, and the subtlety by which he is able to penetrate their meaning. He demonstrates, with great finesse, that point does not designate a simple idea, but a complex set of diverse concepts; that this set, moreover, is not always the same; and that it changes according to the context in which it is used: "the name point will be equivalent to a line of such or such length or a line not farther lengthened or extended; or to some whole composed of a noun and verb with an intervening conjunction or adverb [or the pronoun *quae*]."[45] This ability to display the complex sense of an abstract term introduced into Ockham's language a precision and rigor entirely comparable to that in which modern mathematicians pride themselves. Here is an example: The adversaries of those who deny all reality to point formulate the following objection:

Let a perfectly spherical body be realized by Divine power, and be put in contact with a perfect plane; that is impossible. That a spherical body touches a plane implies a contradiction; if it touches it, it must be touching it by a divisible portion, since, in fact, it cannot touch it at an indivisible point [by hypothesis]. And, whichever way this portion is given, it will be spherical, since it is part of a spherical body.

One might state this otherwise, and perhaps better; one might say that the spherical body touches the plane by one of its divisible parts. But then one might say that this part is not spherical. I reply by denying this consequence, since it is not a result of the premises, unless it is given that the totality (*secundum se totam*) of the part touches the plane, in such a way that every part of this part touches the plane; then the reasoning would conclude that the body is not absolutely spherical. But I suppose that the body does not touch the plane by a part, every part of which touches the plane; it does not therefore touch the plane by a part which would have priority over any other tangent part. But whatever tangent part one names, one can still take a half which does not touch the plane in all its extension (*immediate*), and it is the same for half of this half, and so on to infinity.[46]

It is by this accuracy of analysis and precision of language that the nominalists succeeded in dispelling the paralogisms that accumulated around the notions of infinitely small and infinitely large. It is by similar proceedings that the mathematicians of the nineteenth century were able to disencumber infinitesimal analysis from the slippery reasonings and the ill-founded conclusions which disfigured it. One might easily count the Dominican, Durandus de Sancto Porciano, as a precursor of Ockhamism; but one might more easily view him as a philosopher who yielded to the influence of William of Ockham. In 1326, William of Ockham was summoned to Avignon in order to respond to the errors of which he was accused; it is possible that this summons ended his philosophical career. His flight to the side of Louis of Bavaria opened another career for him, that of writing polemics against the Holy See. Elsewhere, on March 13 of the same year, Durandus de Sancto Porciano was leaving the Episcopal See of Puy-en-Velay in order to occupy that of Meaux, which he occupied until his death on September 13, 1334. He was previously Dominican at Clermont and master of the sacred palace; on August 16, 1317, he was named bishop of Limoux, and on February 14, 1318, bishop of Puy. At the end of his *Commentary on the Sentences*, Durandus placed a brief conclusion starting with these words, "I began this work on the four books *Sentences* during my youth, but I completed it during my old age. (*Scripturam super quatuor Sententiarum libros juvenis inchoavi, sed senex complevi.*)"[47] Durandus therefore drew up his work while William of Ockham was giving his audacious and innovative lectures at Paris. Perhaps even he pursued his writing after the break between the Church and the fiery Franciscan. When one compares the teaching of Ockham and that of Durandus, one encounters some evident similarities; we would rightly deem the former the initiator and the latter the imitator. Such similarities are very apparent in the question interesting us.

Some think that the point is a positive and indivisible reality that actually terminates the line without belonging to the essence of the line.

But that cannot be true for even without any extrinsic termination, one must admit that the line is in itself finite; in fact, if the line were in itself infinite, an extrinsic termination could not impose finitude upon it. Moreover, if the point were an extrinsic termination, once the point was excluded, the line, taken in itself, would remain as long as before; it would be neither greater nor smaller. What then does the point being an extrinsic termination do for the finitude of the line? It does not seem that for the finitude of the line one must admit of such a termination; the thing remains completely hidden to us—it is ridiculous, in fact, to consider that the line would extend to infinity in both directions if the points terminating the line did not exist. All finite things are finite in themselves or by virtue of something which is intrinsic to them. . . . One must not think, then, that a point is a certain positive and indivisible nature terminating a line; the line is terminated by itself in that it has a certain extension and no more—such that the termination of the line indicates the privation of a further continuation. Since the privation of continuation and divisibility seems to be like something indivisible, we imagine for ourselves that the point is such an indivisible when no indivisible positive thing is intrinsic to the line or adjoins the line.

In the same way that the point is said to be the termination of the line, the line is said to be the termination of the surface, and the surface is said to be the termination of the body; hence one would have to imagine that each body would have a surface whose depth is indivisible, actually differing from the body, and containing it as an extrinsic termination—that is absurd.

Let us hold, then, that the line does not behave toward the point as a divisible thing towards a positive indivisible thing, but simply as it would towards a privation of further continuation; hence, one says that the line is actually terminated by two points insofar as it is deprived of any further extension.[48]

The above is the thought and almost the language of William of Ockham; at the same time it is the exact counterpart to the teaching of Duns Scotus. The school of Paris divided itself evenly between the opinion Duns Scotus had proposed and the opinion just defended by Ockham and Durandus. Walter Burley, Ockham's constant adversary, was one of those who held an anti-Ockham position. Burley's commentary on Aristotle's *Categories* reproduces Ockham's arguments against the reality of indivisibles (from the *Summa*

Logicae chapter, "On Quantity against the Moderns").[49] Burley then continues in this fashion:

> It must be known that some moderns, because of the reasonings above, deny that points, lines, and surfaces exist. . . . Against them one can show that, among the beings here, there is something that has length and width, but lacks depth, and something that has length, but no width, and finally something that is absolutely indivisible and that lacks any dimension.
> In order to prove these propositions, one must first assume that bodies are finite and terminated. . . . One also assumes that everything which is terminated has a final termination above which there is nothing belonging to it; one assumes, thirdly, that the final termination is indivisible, unlike the divisibility of the thing bounded by the termination.[50]

That is simply to assume what is in question. What William of Ockham and Durandus de Sancto Porciano deny is that "a body in order to be terminated needs a termination beyond itself." The above thesis "the contrary of which is upholded by the moderns" is taken up by Burley in his commentaries on Aristotle's *Physics*; there he reproduces almost textually Ockham's arguments as given in several places, and attempts to refute them in detail. But here, as in his commentary on the *Categories*, the whole force of the argument consists in viewing as certain and evident the assertions contrary to the following: "[The moderns might state that] corporeality has no termination if not of itself, that formally it is terminated and finite by itself; or else they state that any portions of the body above which the body does not extend is the termination of the body."[51] According to Burley, the contrary of what Ockham asserts is a certain and evident principle. Gregory of Rimini, on the other hand, follows exactly the doctrine formulated by William of Ockham and Durandus de Sancto Porciano with respect to this question. Here is the conclusion he attempts to demonstrate:

> There is no indivisible intrinsic to any magnitude. I say intrinsic because I do not wish to be involved in a quibble about the intellectual soul or the angel who is indivisible and yet is not in a body or in a place. . . . Hence, I do not here speak about such indivisibles, but about indivisibles of extension (*indivisibilia situalia*), such as things completely indivisible, which the people who conceive of such things call points; or things divisible only in one dimension, which they call lines; or things divisible in two dimensions which

they call surfaces. Since their opinion is extremely widespread, it is not necessary to explain it further.[52]

Let us cite a passage from the lengthy exposition Gregory develops to support his conclusion, a passage in which the spirit of Ockham may be discerned:

No experience, no reason forces us to admit such points— nor does any authority we are not allowed to reject. Therefore one must not assert that these points exist. The reasoning (*consequentia*) is clearly correct. As for the premise, I can demonstrate it.

First, let us refer to experience; it is certain that we have no experience of points. Even better, it is very difficult, or impossible, to imagine or to conceive these points.

There is, in any case, no real authority with respect to this subject.

Finally, that no reason forces us to admit points will be shown. In fact, if one could argue for this opinion (as those who assume such points actually do), one would argue from the finiteness of magnitude or from the continuity of magnitude, giving as support that all magnitude is finite or that all magnitude is continuous.

But I can prove that the first proposition does not require us to admit the existence of points. I ask, in fact, how it is that they conceive that a line is terminated by a point. Either they understand it thus: because of the point, the line has a fixed (*certae*) amount and extension; it has a final portion, a determined magnitude. If the point did not terminate the line, the line would not have a fixed extension; it would not have a final portion, but beyond all the parts of this magnitude, there would be another part of the same magnitude; it would be infinitely extended. Or else that is not what they mean when they assert that the point terminates the line; it is some other thing. But one does not perceive any evident other way of understanding that the line is terminated by the point, and nothing that we know makes us necessarily suppose another. . . . And, in the first case, it is not necessary to admit the existence of the point in order for the line to have an end; I can prove it thus: if it were necessary, and one had destroyed the point while conserving the magnitude without any other point being added or created in place of the first, without any other magnitude having been added to the first, then the first magnitude would have an infinite extension or would be infinitely continued. The reasoning is clearly conclusive, but the conclusion is false, for no magnitude can become larger by having had something removed from it. One must not therefore imagine that the point prevents magnitude from extending itself, and that if it were not prevented, it would extend its parts throughout the world; as if it were a cork shutting tight the opening in a barrel, preventing the wine from spilling into the whole house. . . .

Thus the point is not necessary for the line to have a termination. It is easy to prove that it would not be necessary in order that some magnitude be continuous either. In fact, whether some portion of a magnitude is continuous with another portion, or whether they together constitute a single magnitude, is neither less possible nor less easy to know than the continuity of the point itself with some portion of magnitude or with several such portions; and yet they do not suppose that in order to make up a single magnitude, a point needs the intermediary of another point. Consequently, two portions of magnitude would just as easily, or rather more easily, since they are of the same nature, constitute by themselves a single thing.[53]

How can one safeguard geometry while denying the actual existence of points and lines? That is what Gregory of Rimini tells us:

The words line, surface, and body can be taken in two different senses.

In the first sense, they signify real magnitudes, actually existing outside the mind.

In this first sense, what we call line, surface, and body is the same magnitude, but considered from different points of view (rationes). We call this magnitude line insofar as it is extended in a certain dimension or as it is shown by a difference in location; insofar as it is extended in two dimensions, we call it surface, and body when it is extended in three dimensions. Now all magnitude existing outside the mind is extended simultaneously in one dimension, two dimensions, and three dimensions; there are none who are extended in one or two dimensions only.

Therefore, if one understands the words in this sense, each line is, at the same time, surface and body; one can assert similar things about the surface and the body. . . .

The writers state, the line is a magnitude having extension in only one dimension. But after what has been stated, the exclusion here formulated is not to be understood as, the real thing which is a line has no extension in more than one dimension; it signifies that the definition of the line does not imply that this thing is extended in many dimensions, but only that it is extended in one dimension.

These words can be taken in a second sense, as signifying fictitious or imaginary magnitudes, or images of magnitudes the mind imagines in itself, not by any of its sensitive powers, but only through its intellect. In external reality, there is no area without depth; however, experience shows us that we can imagine and consider in ourselves an area without considering any depth, that is, consider a certain magnitude extended only in two dimensions; we can similarly consider a pure length stripped of any width. Moreover, we can consider a shape endowed with depth, that is, a magnitude extending in three dimensions, following three differences in location. The fictive magnitudes of this kind we name surfaces, lines, and bodies. . . .

The geometer does not assume that there is, outside the mind, real indivisibles of this kind; he allows only that they are imagined by the mind, and he defines them in the manner above. . . . If he acts thus, here is the reason: all magnitude which the geometer entertains has length, width, and depth; but some properties belong to such a magnitude as it is long— others as it is [long and] wide, and others as it is [long, wide, and] deep. Because he wants to give us a clear awareness of these various properties, he imagines a pure length, in order to be understood better; the proposition he demonstrates about this pure length must be understood as being proper to magnitude insofar as it is long, or in other words insofar as it is length. Similarly, he imagines a width, so that what he states is to be understood as belonging to magnitude insofar as it is wide. He also imagines points, but it is not as if he wishes to conceive, by means of these points, indivisible things existing outside our mind; when he imagines that there is a point, it is in order to conceive the negation of all magnitude. If he states, for example, that some lines converge on a point, he does not understand by that that an indivisible is situated between these lines, but only that these lines do not contain any space between them. If he states that a line is drawn from a certain point or is continued to a certain point, he simply understands that this line does not extend further. It is the same in all the other cases where he uses points. . . .

Let us hold as certain that no geometric truth requires the existence of points or indivisibles outside the mind; anyone who knows this science sees this clearly—moreover, the geometer uses such fictions so that one better understands the truths he wishes to demonstrate.[54]

No one has more rigorously or more completely expounded Ockham's doctrine of indivisibles than John Buridan.

> We allow that surfaces are to bodies what lines are to surfaces, and points are to lines; then if one does not admit indivisible points in a line, one should not admit lines indivisible in width in a surface, and surfaces indivisible in depth in a body.[55]

Once the above is posited, Buridan concentrates on showing, by means of an argument copied from Ockham, that one should not admit indivisible points in a line—that one should not, in particular, assume that a line is terminated by two points which are two indivisible things extrinsic to the line. He then formulates this conclusion:

> Points [and instants] are divisible things since they exist and are not indivisibles; they are then divisibles. As a result, lines are divisible in length, and surfaces are divisible in depth. As a result, surfaces are bodies, for anything divisible in length, width, and depth is a body; also the line is a surface and the point is a line. Consequently, it follows that points are bodies.[56]

This language should not surprise us; the person speaking is a physicist, and he is speaking about real things, things that have matter. Later we will hear the geometer speak; for now what is at stake are real points, physical points. The following conclusion is therefore formulated with respect to actual physical things, not geometric abstractions:

> Points are parts of lines . . . lines are parts of surfaces . . . surfaces are parts of bodies . . . surfaces are so named because they are the termination of bodies—lines because they are the termination of surfaces, and points because they are the termination of lines. If, for example, one said that there is a point in the middle of a line which is not the termination of a line, this point is still so named because it is the termination of lines which are part of the first, because it is the beginning of one of these lines and the end of the other.
> We therefore allege that points are the termination of lines, and lines are the termination of surfaces; we allege that surfaces are the termination of bodies, and not an extrinsic separate termination, for the body would then be terminated even though nothing existed outside it.
> In addition, we allege that the termination of a line is

not the line itself; in fact, there are different terminations in
a line, one at one end and one at the other. . . .

The following conclusion results from this: a point is
a line which is the intrinsic termination of another line; it
is not itself the line it terminates, but it is a line which is
a part of the line it terminates. One can say the same about
the line with respect to the surface and the surface with respect
to the body. . . .

One cannot find a portion of a continuous magnitude
which, in its totality, terminates this continuum. (I take the
expression "in its totality" syncategorematically.) This is
evident because no part of a continuum is, in its totality, the
first part—or the last part. Let us assume the contrary, that
some part of the continuum is in fact, in its totality, the first
part—or the last part; the result of the assumption would be
that every part of this part is also the first part of the
continuum—or is also the last part—and that it is also the
termination of the continuum, which is false: the part which
one supposes is the first part, let us divide it into two other
parts, A and B; one of them, A, will be before the other, B,
and B will not be the first.[57]

Buridan then asks the following question: Since the point, which
in reality is a body, is divisible as all bodies are, why is it commonly
asserted that the point is indivisible? He replies that the point is
not really indivisible, and that the proposition is not literally true,
but that one means to assert various truths by the proposition.
Among these truths are those that have reference to the bonds uniting
the present physical theory with geometry; we will refer to them
later. Others are purely physical; they follow from what has just
been stated:

One says that the point is indivisible because one calls
it point insofar as it is the termination of a line. And any
termination of a line is indivisible, not literally, but indivisible
insofar as it is not divisible into parts, each of which is still
the termination of this line.

In yet another fashion, we call the point indivisible
because it is called first part or last part. If one calls it first
part or last part, it is in fact in order to distinguish it from
or to count it among the other parts as being a certain part;
and if one says that it is one, it is in virtue of a reason of
nondivisibility. . . .

About this matter, there is need to remark that the point
is infinitely small (*in infinitum parvum*) because indefinitely

there is a termination, a first part or a last part of the continuum. If, for instance, one calls a third of line B the first part of the line, we can still call a tenth of the line the first part, and that is less than a third; and one can still call a hundredth of a line—which is smaller yet—the first part of the line, and a thousandth—which is smaller yet—and so forth to infinity. However small is the given portion, there is a first part which is smaller (*ideo quantumcunque parva parte data, adhuc minor est prima pars*).[58]

In order to see the infinitely small defined with such precision one would have to wait until the nineteenth century; moreover, our most rigorous algebraists have no definition for the infinitely small other than the one Buridan used. So far we have heard Buridan speak in a language that would surprise the geometer, a language in which the geometer might not recognize the indivisible points, lines without width, and surfaces without depth he was accustomed to consider; Buridan does indicate what, in his opinion, is the meaning one must attribute to these various indivisibles of geometry. He formulates several answers to the question: "Why does one say that the point is indivisible?" He places the following answer first, before the ones we have just cited.

By stating that the point is indivisible we do not mean to say that it is so, nor that the proposition is literally true; but we say that it is so in another fashion, following the imagination of the mathematicians, as if it were indivisible. Not that one believes that it is so, but that when we measure, we arrive at the same conclusions as if it were so. (*Hoc non dicitur quia sit ita vel quia sit verum de virtute sermonis, sed uno modo hoc dicitur secundum imaginationem mathematicorum, ac si esset punctum indivisibile; non quia debeant credere quod ita sit, sed quia in mensurando revertuntur eadem [conclusiones] sicut si ita esset.*)[59]

If one wishes to understand Buridan's thought completely, one should relate the previous passage to a passage from Buridan's *Metaphysics* that concerns the astronomy of eccentrics and epicycles. Here is what Buridan writes on this subject:

The Commentator would respond as follows concerning the authority of astronomers: this fashion of positing or imagining epicycles and eccentrics is useful for calculations, in order to know the locations of the planets, the relations holding between them and us; that is all astronomers demand. It is therefore possible for us to use such imagined things even though it is not this way in reality. (*Iste modus ponendi seu imaginandi epicyclos et eccentricos bene valet ad computationem et ad sciendum loca planetarum et habitudines eorum adinvicem et ad nos; et nihil plus petunt astrologi; ideo licet uti talibus imaginationibus quamvis non sit ita in re.*)[60]

The close similarity between the expressions used in the two passages reveals analogous thought. Buridan evidently considers that points, lines, and surfaces, the various indivisibles, are for geometers what the eccentrics and epicycles are for astronomers; they are pure fictions that have no reality beyond the mind. But by reasoning about these fictions one achieves results in conformity with measurements carried out on real bodies. Buridan therefore juxtaposes two conceptions of geometry, a geometry that considers points, lines, and surfaces as nothing but constructions of the mind, and a physical geometry in conformity with reality, that only treats bodies. Buridan also attempts to show how the propositions of the former geometry must be interpreted in the latter:

One can draw a straight line between any two points; that signifies that between any two bodies distant from one another, there is a straight corporeal dimension.

We also allow that there is an infinity of points in a line, because there is an infinity of parts of a line each having a first part and a last part. . . . But how is one to set forth the definition of circles and spheres, shapes in whose middle is a point from which all lines drawn from it to the circumference are equal? That is an easy question to answer according to the view that points and lines are fictions in one's imagination. But according to the truth, there is a part situated within the circle and within the sphere that one calls center, and around it there is an extreme part that one calls circumference. The part which is always in the middle is said to be the termination of all the radii. This central part is not purely and simply indivisible, but one calls it indivisible because it is not divisible into several parts, each being the termination of all the radii of the circle or sphere; one calls it indivisible because this central part or this termination is an infinitely small part; one calls it a point because this part is the termination of a line or of all radial lines.

That is why one says that the earth is the natural center of the world sphere; or else why one calls the infinitely small part, located within the center of the earth and equidistant from the circumference of the world, the center of the world.[61]

Buridan always seems to be thinking about the coexistence of the two geometries, the fictive geometry and the true geometry; he wastes no occasion to remind his reader of it. In his *Metaphysics*, for example, he has occasion to recall what the astronomers call the concentric sphere of the world; he writes on this subject, "One must know that in the world, the natural center is the earth itself; it is only in the imagination that one posits an indivisible center; however, let us imagine as the center of the world a point located in the middle of the earth."[62] Buridan owes the essential principles of this doctrine to William of Ockham; however, he diverges from the doctrine of the Venerable Inceptor in an important way, since this divergence deals with the real meaning of geometric propositions. According to Ockham these propositions have only conditional validity; they would be true if there existed indivisibles, points, lines, and surfaces. Since these indivisibles do not exist, the postulates and theorems of geometry seem to be stripped of all objects. The above conclusion is not of such a nature as to be pleasant to the mathematicians; Ockham doubtless did not worry about this since he seemed to have little esteem for mathematicians. To those who are surprised to see him carefully demonstrate that a continuous magnitude is not composed of indivisibles, and that such indivisibles are non-existent, he replies:

> No doubt a single means was sufficient to prove that there is no indivisible; however, other demonstrations would not be superfluous, and so especially in this case; for to prove that nothing is indivisible in these things here-below cannot be accomplished except through subtle reasons, which mathematicians and others less skilled in metaphysics and logic would not comprehend. But to prove that a continuum is not composed of indivisibles by various methods, even if there were such indivisibles, can be accomplished by reasons more apparent to the mathematician and others, whatever those reasons are.[63]

The theory of the non-existence of indivisibles does not run the risk of offending the mathematicians when it is given Buridan's interpretation; with it they are able to see how their reasonings derive their legitimacy. Because of the philosopher of Béthune, the theory of William of Ockham, Durandus de Sancto Porciano, and Gregory of Rimini attained a perfect form; nothing essential was then added to it. Albert of Saxony does no more than summarize Buridan's doctrine.[64] As for Marsilius of Inghen, his position, as demonstrated in a number of questions dealing with the infinite, changed profoundly with time. In his *Abbreviationes libri physicorum*, which seems to be a work from his youth, he reproduced Buridan's entire doctrine, although in a somewhat abbreviated manner.[65] But that is not what he does in his *Quaestiones* which indubitably were composed many years later. Marsilius answers the question "Are there indivisibles such as points in a continuum?" as follows:

> There are two opposed opinions answering this question.
> The first allows that there are no indivisible things in a continuous magnitude; even more, it allows that there is nothing indivisible in the world except intelligences, separated substances, and acts of intellect. . . .
> One should note that those who hold this opinion differ when it comes to saving the propositions of the mathematicians and Aristotle, for the former admit the existence of points.
> Some assert that all these propositions, the point exists, the line exists, length exists, etc., are false; for in all these propositions there is a term that signifies nothing. They therefore state that mathematical propositions and some of Aristotle's propositions in the sixth book of the *Physics* cannot but be understood conditionally.
> Others assert that it is not necessary that the proposition, the point exists, be true; but that it suffices for the mathematician's purposes that one can imagine that the point exists. In fact, the Commentator stated, in the third book of the *Metaphysics*, that Aristotle's mathematical propositions occur in the chapter about propositions able to be imagined.
> Others, in order to save the propositions of the mathematicians, suppose that the word point is a purely negative word. In the same fashion that the word volume represents a body and designates its extension at the same time, the word point represents a body without designating its extension; it represents a body as we conceive it under the notion, termination of distance or volume. In this fashion, almost anything in the world may be called point; it is in the same fashion that an astronomer, when he wishes to measure the distance between two stars, calls each one of the stars a point.
> The second opinion is contrary to the preceding opinion; it admits that there are indivisible parts such as points within any continuum. . . .
> Let us prove this last thesis.[66]

And Marsilius establishes the following conclusion:

> There is a point of prime matter, a point of substantial form, a point of quality, and a point composed of matter and form. The proof is as follows: In virtue of the previously established conclusion, there is a point; this point is then either a point of magnitude, or a point of matter, or a point of form, and so on. Moreover, whichever alternative is agreed upon, the others follow; in fact, if there exists a point of matter, there exists a point of form which is the form informing this matter, and if there exists a point of form, there exists a point of quality since there cannot be a substantial form not accompanied by qualities. Similarly, if there is a point of magnitude, there has to be a point of matter corresponding to this point of magnitude; magnitude cannot be separated from matter, nor, consequently, from the point of magnitude. Hence whatever alternative one admits, the others follow. . . .
>
> There are lines indivisible in width and depth. . . .
>
> There are surfaces with length and width, but indivisible in depth.
>
> In the same way that there exists a point composed of matter and form, there exists a line composed of matter and form, and a surface also.[67]

After having held (following his teacher, Buridan) that the geometric indivisibles—points, lines, and surfaces—are pure fictions, Marsilius of Inghen came to consider them as complete substances, having matter, form, and qualities. Neither Duns Scotus nor even Walter Burley had professed such a radical geometric realism.

The Natural Minimum of a Substance

No continuous magnitude is composed of indivisibles; therefore, however numerous and however small are the parts into which a magnitude has been divided, one can still divide these parts into as many portions as needed. Divided as they are about other matters, the masters we have just heard are united in asserting the preceding truth. Since the division of a magnitude can be undertaken without limits, is it possible for it to be continued indefinitely without altering its substance? Would the substance keep the same nature no matter how small the portions are into which it is reduced? That is the new question which greatly occupied Scholasticism. The masters who treated this question often utilized Aristotle's authority. In the first book of the *Physics*, when he

discussed the famous axiom of Anaxagoras and Empedocles—
everything subsides in everything else—the Stagirite wrote that
"flesh is quantitatively definite in respect both of greatness and
smallness."[68] The assertion does not seem to have interested Averroes
who merely repeated it as evident.[69] But Scholasticism quickly
latched onto it and developed the doctrine latent in it. Robert
Grosseteste commented very briefly on Aristotle's *Physics*, but in
his few pages were the seeds of many fertile ideas. In some old
editions, Grosseteste's *Summa* is attached to Aquinas's commentary
on the *Physics*; and, near the end of this work, in the sixth book,
one can read the following:

> There are doubts about whether continua are composed
> of indivisibles, but it seems that it must be so. . . . In fact,
> a natural body is composed of minima; therefore, it is a
> continuum composed of indivisibles. The premise is true
> because according to the Philosopher (in the *Physics*) one must
> admit a natural minimum body; the reasoning is clearly
> conclusive. Therefore, the conclusion is equally true.[70]

To this objection he replies thus:

> There are two ways in which one can conceive the
> minimum body. . . . It is in the second way that one can
> assign a minimum to a natural body; for insofar as it is a
> volume, a natural body is continuous and consequently
> infinitely divisible.[71]

Albertus Magnus limits himself to paraphrasing Aristotle's text:
"Flesh is limited and infinite in size, both large and small. Its
greatness cannot be so, that it no longer is flesh; its smallness also
cannot be so, that it no longer is flesh, for so small a size cannot
act as flesh."[72] In his *In libros physicorum Aristotelis expositio*,
Saint Thomas Aquinas allows an error to slip by him:

> There is no difficulty in finding infinite unequal parts
> in some finite thing, if we consider only the nature of quantity
> because, if the continuum is divided according to the same
> proportion, we can continue to infinity; for example, if we
> take a third part of the whole, and a third of the third, etc.
> However, the parts will not be equal in quantity. But if the
> division is made with equal parts, we cannot proceed to
> infinity; not even if we consider only the nature of quantity
> in the mathematical body (*sed si fiat divisio per partes aequales,
> non proceditur in infinitum, etiam si sola ratio quantitas in
> corpore mathematico consideretur*).[73]

How is it possible that Thomas Aquinas did not realize that one can divide a body into three parts, then each into three others, and so forth indefinitely? Some stories tell that the *Doctor Communis* was subject to distractions; assuredly, he must have been distracted when he was writing the passage we have just cited. So far we have not seen Aristotle's text give birth to anything but simple remarks; however, there is the seed of a fertile theory in this text. This theory made its appearance in the works of Saint Thomas Aquinas. Aristotle's principle is clearly enunciated in the *Summa Theologica*.

Every natural body has some determined substantial form. Therefore, since accidents follow upon the substantial form, it is necessary that determinate accidents should follow upon a determinate form; and among those accidents is quantity. So every natural body has a greater or smaller determinate quantity. (*Unde omne corpus naturale habet determinatam quantitatem in majus et in minus.*)[74]

Thomas Aquinas uses the principle in the *Summa* only in order to establish that a natural body cannot be infinitely large. Elsewhere he gives it the following, more precise formulation: the substantial form of a natural body imposes a limit on the smallness of the parts into which a body may be divided, and a limit on the greatness of the volume a body can attain by dilation; and if, by division or dilation, one of these limits has been transgressed, the substantial form is destroyed and another form takes its place. Water that one divides into overly small parcels or that one dilates too much changes into air. Here is a passage in which Thomas Aquinas formulates the corollaries of the principle enunciated in the *Summa*:

Though mathematical bodies may be divided to infinity, natural bodies can only be divided to a certain limit, for there is a corresponding quantity determined by nature for each form, as there is for all other accidents. Moreover, the rarefaction [of a natural body] cannot be continued to infinity, it can only be continued until a precise limit corresponding to the rarity of fire. Besides, one could rarefy water until it is no longer water, but air or fire; that is what can happen if one exceeds the bounds of rarity proper to water. Water cannot, in a natural fashion, occupy more space than air and fire (in order to surpass the rarity of air or fire) without losing its aqueous nature.[75]

Giles of Rome was almost always a faithful Thomist; he developed the theory of natural minimum, for which Thomas Aquinas clearly laid down the principle, to such an extent that the Scholastics often thought of him as the creator of the theory. Giles of Rome's theory is founded on an essential distinction: Magnitude can be conceived in three different ways.[76] First, one can conceive it as a pure magnitude, by abstracting it from the matter in which it subsists; that is *imagined magnitude*. Secondly, one can conceive it somewhat more concretely, as in matter, but without specifying the nature of this matter; that is *real magnitude*. Finally, one can conceive it in a more concrete fashion, as in some matter whose nature is specifically determined, so that it is the magnitude of a body of that kind, the magnitude of some quantity of water, for example; that is *natural magnitude*. Pure magnitude abstracted from all matter, magnitude such as the geometer conceives, is obviously infinitely divisible. It is the same for magnitude in matter whose nature remains undetermined. But it is otherwise with magnitude in matter whose nature is determined; this magnitude would not be able to be divided beyond some limit without resulting in a change in the nature within which it subsists. One should conceptualize Giles of Rome's theory as follows: one can imagine that one is subdividing indefinitely the volume of a cubic foot abstracted from all matter. One can equally conceive that one is dividing to infinity the matter occupying the volume of a cubic foot, but on the condition that one is not interested in whether the matter always retains a particular nature—if, for example, it remains water always. But if one takes a cubic foot of water and one requires it to remain water always, one cannot continue indefinitely with division; the matter one is dividing will cease to be water at some point, and will be transformed into some other substance.

> Strictly speaking, then, continuous division is not repugnant to natural magnitude purely and simply because it is natural; but it is repugnant to natural magnitude inasmuch as it exists under one or another species—that is why one can assign a minimum to flesh and a minimum to water. . . . If division to infinity is repugnant to natural things, that is because it is repugnant to the specific form. (*Divisio autem in infinitum, si repugnat rebus naturalibus per se loquendo, repugnat formae specie.*)[77]

Giles of Rome did not treat the natural minimum only in his

commentaries on Aristotle's *Physics*; he also discussed the existence of this minimum of nature, an existence that does not contradict the infinite division of the mathematical continuum, in one of his questions on *De generatione et corruptione*.[78] There Giles of Rome takes up Aquinas's strange assertion: one can infinitely divide a body into unequal parts whose magnitudes diminish in a geometric progression, but one cannot divide it to infinity in equal parts. But by equal parts Giles of Rome understands parts of a given magnitude, parts that would all be equal to a grain of sand, for instance. That is to replace an absurdity with a truism. Giles also finds occasion to affirm the essential principles of his theory of natural mimima in one of his quodlibetal discussions and in his *Theoremata de corpore Christi*[79]—principles, let us recall, that Saint Thomas Aquinas formulated: For a given substance, the minimum volume is numbered among the accidents that result necessarily from the substantial form; in the same fashion that, by its substantial form, water is necessarily humid, a volume of water is necessarily larger than a given lower limit. It would be contradictory and inconceivable that water is not humid; in the same fashion, it is contradictory and inconceivable that a volume of water be smaller than its natural minimum—infinite division is repugnant to the specific form. Therefore, one must conclude that God cannot create some water that is not humid, nor a mass of water smaller than its natural minimum; what He might create would not be water. The above conclusion appeared excessive to Richard of Middleton. Given a volume of fire, one can conceive that it be divided, and that the resulting portion be divided into smaller portions, and so forth without end. Each portion, no matter how small it is, would actually be a portion of fire—in each portion there would always be the specific matter of fire, and the specific form of fire. God can, in this way, indefinitely divide a volume of fire, and maintain the existence of the portions thus obtained, no matter how small they are. This division, no matter how far one pushes it, does not therefore alter the specific matter or the specific form of fire, but it can be pushed far enough to alter some of the properties or virtues of the fire.

> For instance, one can attain portions so small that they might no longer be kept in existence only by created forces, because the virtues of such particles would be too weakened. However, God can conserve such a particle. Only He can actually effect such a division; neither angel nor intellectual

soul can actually effect such a division, but they can conceive it in thought.[80]

Likewise an extremely small particle of fire, though remaining fire specifically, might not have enough virtue to engender its kind, to move itself, and to affect our senses; with respect to these various properties, fire is not divisible to infinity. The Scholastics who treated with some detail the question that occupies us were divided between the opinion of Giles of Rome and that of Richard of Middleton. John of Jandun appears to admit the opinion of Giles of Rome, but he introduces a modification to it. This modification is indicated in an extremely brief and unclear fashion by a few lines of his *Questions on Aristotle's Physics* devoted to this doctrine. If we completely understand these lines, they signify that no lower limit restricts the divisibility of a magnitude as long as the parts remain attached to the whole, hence as long as the division is conceived but not accomplished—but that the parts obtained by the division cannot be separated from the whole and subsist isolated if their magnitude does not surpass a certain minimum.[81] John of Jandun is more explicit in the *Questions* he composed on the *De substantia orbis* of Averroes. Among these questions, which were very popular during the Renaissance at the Averroist school of Padua, and which were so often printed, there is one in which the author examines "if each natural form is terminated by a maximum and a minimum." After having responded affirmatively to the question, John of Jandun examines some difficulties which might be given as objections. Every natural form is united to some matter; this matter, according to Averroes's doctrine, of which John of Jandun is the steadfast defender, necessarily and by itself possesses three dimensions—meaning that the matter is divisible to infinity following each of the three dimensions. Is it not necessarily the same with respect to the substance constituting this formed matter? Our Averroist responds as follows to the objection: A natural substance, fire for example, insofar as it is a quantity, as it occupies a certain volume, is divisible to infinity; insofar as it is a natural substance, it is no longer indefinitely divisible. If one pushes the division of the substance too far, its form is destroyed; fire, for example, so divided would be transformed into the element, air or water (depending upon which of them it happens to touch). But, it may be said, if one divides fire in this way, at the moment when the division attains the minimum of magnitude below which fire can no longer subsist, the entire mass of the fire being submitted to this division would change instantaneously into air or water; and that cannot be. That is not how one should understand the operation by which fire, when one pushes its division too far, is transformed into the element within which it is located. It must not be imagined that the parts produced by such a division can be transformed while they remain united among themselves; it is only when one separates each from the whole that the part takes the form of the element that contains it, and unites itself to it. "There is no minimum of magnitude for a continuous natural substance, as long as the parts remain united to the whole; there is no natural minimum for these parts except insofar as they are separated from the whole."[82] We see nothing more in the teaching of John of Jandun than the doctrine of Thomas Aquinas and Giles of Rome; the difference lies in that John of Jandun has attempted to avoid a possible faulty interpretation. Burley does not seem to have seen it thus; he appears to hold the opinions of Giles of Rome and John of Jandun as distinct opinions:

> One can state that division to infinity is repugnant to magnitude as it is realized in sensible matter, while nonsensible magnitude simply realized in prime matter is divisible to infinity. One can also conceive of another interpretation: magnitude realized in sensible matter is divisible to infinity as long as the division consists only in marking [in thought] the distinction between the various parts; but this magnitude realized in sensible matter is not divisible to infinity when the division is an actual division, separating the parts from one another.[83]

In any case, Burley neglects to tell us which is his own view. William of Ockham is clearly on the side of Richard of Middleton:

> A natural minimum cannot be brought about that is not able to be infinitely divided into smaller parts while conserving its natural form; it is evident, for example, that flesh cannot be made so small that it cannot be divided into smaller parts indefinitely by Divine power.
> To the Philosopher I reply that he must have understood this as follows: A natural minimum, a minimum of flesh, for example, can be brought about when it becomes incapable of subsisting by itself and of resisting the extrinsic corrupting agents such as cold and heat, etc.—in such a way that if there were a portion of flesh smaller than this minimum, it could not resist the extrinsic agents. Almost immediately, because of its lack of resistance, it would yield to those surrounding

it (*cederet in continuis*); the form of flesh would be destroyed, and some other new form would be introduced. But God could suspend the action of the extrinsic agents and protect the portion of flesh against destruction; He would then be able to subdivide it into smaller pieces to infinity so that there would be no end to the division, no minimum of flesh. Neither would any of these pieces yield to the surrounding elements since the action of the extrinsic agents was suspended and the flesh preserved from corruption.[84]

The doctrine of Richard of Middleton and Ockham undergoes a substantial modification with Buridan; the philosopher of Béthune allows that one can take from any corporeal substance a quantity so small that it would not be able to resist external agents *for a long time*. He introduces to this question a temporal consideration his predecessors did not consider.

One can bring about that something should be so small that such a body or a smaller body, isolated from bodies of the same kind, would not be able to be conserved in a natural fashion for any length or noticeable amount of time. This small body would tend constantly to be corrupted; it would be corrupted quickly by neighboring bodies. These neighboring bodies which are, in fact, of another species, would have a contrariety with it, and because of this contrariety, they would become agents of corruption toward it. And this body might be so small that its resistance might be too weak to resist for a noticeable amount of time.

We must note, however, that one cannot bring about a volume so small—that some flesh be so small—that it cannot be conserved by itself without any tendency to become corrupt by means of God's power, for as long as God wills it. But I have stated that this cannot happen naturally. Perhaps this is what Aristotle intended when he stated that natural beings are limited in greatness as in smallness.[85]

Buridan allows that for any substance one can take a quantity so small that it cannot be conserved *for a long time*. Can one take a quantity from any substance, so small that it cannot be conserved *for some given time, however short*? That is a new question; its answer does not result necessarily from what has been asserted previously. Buridan's predecessors responded affirmatively to the question; however, if one wishes to hold that any portion of a substance, no matter how small it is, requires some time to transform itself into another substance, one must answer negatively.

A more difficult subject of doubt is whether one can bring about that a natural body existing separately be so small that the separate existence of a smaller body of that kind is impossible. It is certain that such a body can exist by divine power; what one wishes to ask is whether it is possible by natural powers.

Some people assert in a probable fashion that one cannot bring about such a minimum; they state, for example, that one cannot bring about a minimum of flesh separate from other flesh such that a smaller amount of flesh separate from other flesh cannot exist.

They prove it thus: this flesh is still divisible, for it still has parts; if one were to separate these parts from one another, they would not be corrupted instantly by the effect of the separation. Each part, in fact, has some resistance that will not yield until the whole part is corrupted.

This reasoning may also be confirmed thus: we suppose that some small amount of fire is at the bottom of the sea. It is not possible that something else—some air or some water for example—be generated from it without its matter having been disposed of, by some prior alteration, to take up the form of air or of water; and it cannot be corrupted without something else being generated from it.

Also, other things being equal, a natural agent acts more forcefully on the parts closest to it than on the ones farthest; the external parts of the fire are then more forcefully and hence more quickly acted upon by the water containing them and corrupting them than the parts located in the middle. [The former are then already corrupted while] the latter remain fire, separate from all other fire; moreover the remaining part is smaller than the part that one has taken as the minimum, since the remaining part is smaller than the whole.[86]

The above theory is more completely opposed to the doctrine of Saint Thomas Aquinas and Giles of Rome than was the theory of Richard of Middleton and William of Ockham. No matter how small a portion of substance is, it can still be conserved separate from substances of the same kind, and without resorting to God's supernatural power; but the time during which it can be conserved will be as short as the portion is small. Albert of Saxony also takes up the question of the natural minimum. The responses he formulates reject the theory of Thomas Aquinas and Giles of Rome in a sharper fashion; they account for Buridan's modification, but they also introduce a new modification which accounts for the nature of the medium in which the portion at stake is located.

Let us consider homogeneous substances such as bone, flesh, etc.

First conclusion: one cannot assign a minimum to the matter from whose power one may draw out the form of flesh; that is evident, for the form of flesh is divisible and the portions of this form can be produced one after the other. Therefore, this form cannot be drawn from a portion of matter so small that one cannot draw out a smaller form of flesh from a smaller chunk of matter. Hence, in virtue of the definition of minimum, one cannot assign a minimum to the quantity from which flesh can be produced.

Second conclusion: one cannot assign a minimum to the matter from whose power one may draw out a form of flesh sufficient for the composite of this form and matter to be called flesh. Let us prove this: If one has a quantity of matter sufficient for this end, a smaller quantity will be sufficient also, for any part is sufficient; in fact, flesh is a homogenous substance, every part of which receives the name of the whole. Every part of flesh is flesh; that is how homogenous substances differ from heterogenous substances.

Third conclusion: One can assign a minimum to a quantity of flesh which, within a medium determined in a specific and detailed fashion, does not tend to become corrupted. That is evident. Let us suppose that the power [of conservation] of an amount of flesh is 2, and that this flesh is situated within a medium whose contrary power is 2; the flesh in this medium would remain without becoming corrupted, for its resistance is equal to the activity of the medium. But if one has a smaller amount of flesh, it would not be able to remain in this medium without becoming corrupted, for its power would be surpassed by that of the medium; the former flesh was therefore the minimum amount of flesh which would remain in such a medium without becoming corrupted.

Fourth conclusion: with reference to a medium designated absolutely [as opposed to a medium determined in a specific and detailed fashion] one cannot assign a minimum to a quantity of flesh capable of subsisting in this medium without becoming corrupt. Let us prove this: if one has a quantity of flesh which does not become corrupt in a given medium, one can always have a smaller quantity which does not become corrupt as long as one situates it in a less-active medium. For example, if some flesh of magnitude 2 were in a medium of power 2 and did not become corrupted, then flesh of magnitude 1 would not become corrupted in a medium of power 1.[87]

The change from the doctrine of Saint Thomas Aquinas and Giles of Rome to that of Albert of Saxony is astounding: the former was a pure doctrine of Peripatetic metaphysics—each form requires a minimum of matter; the latter is a problem of mechanistic chemistry—one seeks the minimum quantity below which a certain substance situated within a determined medium will dissolve. This problem reveals that the mind of the thinker preoccupied by it has interests similar to those of modern physicists; already, although in a timid and unpolished way, there is an attempt to express the reasoning in a mathematical language. The immediate successors of Albert of Saxony seem to have added nothing essential to his doctrine of the natural minimum. In his *Abbreviationes libri physicorum*, Marsilius of Inghen merely asserts that "the form of homogenous matter can be produced in infinitely small matter; for example, the form of fire can be produced in a portion of matter of any size, however small, for any part of fire is still fire."[88]

In his *Quaestiones super libros physicorum*, the future rector of Heidelberg develops Albert of Saxony's doctrine, but without adding anything of substance to it.[89]

Infinite Divisibility: Categorematic and Syncategorematic Infinites

In the previous section the divisibility of matter was considered from a physical point of view; in this section, it is considered from a logical point of view. Most of the Scholastics agreed with Aristotle that a magnitude cannot actually be divided into an infinite multitude of parts; with Aristotle they denied the existence of indivisibles. However they allowed that one can pursue the division of a continuous magnitude into smaller and smaller parts without end. This truth was formulated by Aristotle as "magnitude is potentially infinite by division."[90] However, it posed an extremely serious difficulty. According to Aristotle, what exists potentially will some day exist actually; what cannot exist actually at any moment, cannot exist potentially either. This principle, asserted many times by the Stagirite, dominates Peripatetic metaphysics. Moreover, Aristotle does not forget the principle when he considers the infinite; he invokes it when justifying the following: the actual existence of the infinitely large is impossible, therefore the infinitely large does not exist even potentially. In fact, "the size it can potentially be, it can actually be."[91] Following Aristotle, his faithful Commentator writes, "If a magnitude can potentially become greater than any given magnitude, it would then be in actuality greater than any given magnitude; it would then be an actually infinite magnitude."[92] But then there is a serious difficulty with respect to the infinite divisibility of a continuous magnitude; there is no reason not to apply to infinite division the principle used with respect to infinite addition. If what is potential can always be actually realized, then since magnitude is potentially infinite by division, it can be infinitely divided in actuality. Walter Burley justly comments that this reasoning is valid "if a certain magnitude can grow to infinity, it is possible that a certain magnitude be infinite in actuality; the following reasoning seems similarly conclusive: if it is possible to divide indefinitely a magnitude, it is possible that a magnitude be actually divided into infinity."[93] The principle posited by Aristotle and Averroes appears to lead one to the conclusion that a magnitude can actually be divided into an infinity of parts; however, their teaching formally denies it. This apparent contradiction forced the Scholastics to delve into the meaning of *potentially indefinitely divisible* to a greater extent than had Aristotle and his successors. They distinguished two types of potentiality. There are potentialities susceptible of being completely actualized—what is potential in this way can, at any given moment, be actual (*in facto esse*). There are also potentialities that can never be completely actualized—no matter how far one accomplishes its actualization, there is still some potentiality not actualized—what is potential in this way can never be conceived as actual; it is always becoming actual (*in fieri*). Roger Bacon appears to have been the first to have conceived this. (It does occur in Aristotle's work, though in a somewhat confused manner.) In his *Opus Tertium*, Roger Bacon expresses it with remarkable precision:

> A body's potential for division cannot be reduced to actuality, purely and completely. It is a potentiality that one can only reduce to actuality impurely and incompletely, where there is always a mixture with a potential for further actualization; it is actually reduced but in such a way that there remains the potential for another division. That is the potential of the continuum and that which constitutes infinite divisibility; when this potential is reduced by actual division, the possibility of another division is not excluded. Actually, it is required; in fact, the portion which is the result of division

is a magnitude; hence it is still divisible, and so forth to infinity.[94]

Those who maintain the possibility of actual division to infinity reject Bacon's conception by means of the following argument: If each particular proposition is possible at the same time as each other, then surely the universal proposition is possible. And it is possible that a line is actually divided at point A, at point B, and at point C; since it is possible that it is actually divided at point A and point B it must therefore be able to be actually divided at all its points. Here is Bacon's reply to this argument:

> Each particular proposition is possible in itself; it is *compossible* with all other actually given particular propositions; but it is incompatible with a particular proposition not actually given, given in the future. . . . One must therefore concede that division at point A does not preclude division at any other point given presently and in actuality; it does preclude division at some point not yet given. And the points of division cannot be given simultaneously; they are given successively, by a succession which extends to infinity. . . . The club of Hercules is therefore broken in this fashion. This has not been accomplished without effort, for the common people do not know these things; some worthy people do know these things, but they are few in number.[95]

Roger Bacon's response to those who hold actual division to infinity is not altogether right. The division of a line at point A is certainly compatible with the division at any given point; one sees no reason why it ceases to be compatible with a point that is not given, but is still in the future. That is not an impossibility one has yet encountered; the impossibility is not introduced until one considers the division at all points at the same time.

Richard of Middleton improved upon Bacon's doctrine:

> When one asserts that continuous magnitude is infinitely divisible, I reply that it is true as long as one understands it thus: it can be divided without end, but such that the number of parts thus formed is always finite. If one allows that it is divided in this fashion, there is no impossibility; the existence of an infinite *in facto esse* does not result, but only that of an infinite *in fieri* which is what one calls the mixture of infinite in actuality with potentiality.[96]

But Richard did not attempt to resolve Bacon's difficulty; he did not seem to be aware of the difficulty. However, Duns Scotus clearly perceived it, and the solution he proposed far surpassed Bacon's. At some instant (*nunc*) a continuum may be actually divided in *a* parts; it can be actually divided in *b* parts, or at the same time in *a* parts and in *b* parts. But it cannot be that at some instant, even an undetermined instant, the continuum can be actually divided in *a* parts, *b* parts, *c* parts . . ., *a*, *b*, and *c* being all the possible numbers. Each of these divisions may be actualized at some instant; it is the same with any group of such divisions, but they are not *all* compossible in the same *nunc*. The potentialities which are infinite in number cannot *all at the same time* be reduced to actuality. In order to make himself better understood, Duns Scotus exhibits an ingenious example of possibilities that can be realized individually or in conjunction with others, but cannot all be realized together: Socrates can carry 9 stones, and we have 10. Socrates can carry any stone or any group of stones consisting of 2, 3, . . ., 9 stones; but Socrates cannot carry all 10 stones at the same time.[97] Walter Burley, with his usual clarity, expounds upon the thoughts of Duns Scotus:

> When one states, a continuum can be actually divided into the parts it is divisible, I reply that it is not true. In fact, the continuum can be actually divided at each of its points; however it is impossible that it be actually divided simultaneously at each of its points. If one questions the truth of the proposition, the continuum can be actually divided at any point, I reply that the proposition has two senses: one of composition, and one of division. Given its sense of composition, the proposition is false for it signifies the possibility of the proposition, a continuum is actually divided at any point; and it is impossible (and false) that the continuum be actually and simultaneously divided at each of its points. But given its sense of division, it is true, for each of the singular propositions corresponding to it is true; for, in the sense of division it is a universal proposition, and in the sense of composition it is a singular proposition. . . .
>
> One may object that all singular propositions corresponding to the proposition at stake are possible and compossible. I admit that each of the singular propositions is possible, but I state, however, that the singular propositions cannot all be taken simultaneously. Although no single proposition is repugnant to another single proposition, a large number of these propositions taken together is incompatible with other singular propositions taken together; and if one were to take all singular propositions except for one, they

would repulse it. One can take any number of propositions, each of which would not repulse the other, but if we take all but one, they would repulse it. . . .

Thus singular propositions correspond to the proposition, a continuum is actually divided at any point. Each of the singular propositions is possible, and compossible with each other; however, none of the singular propositions is compossible with all the others taken together.[98]

Following Duns Scotus's example, Walter Burley, in order to resolve the difficulty with the infinite divisibility of continuous magnitude, resorts to a distinction; he distinguishes two senses of universal propositions: a sense of composition, and a sense of division. This distinction was commonplace; it was used at Paris as well as at Oxford to unravel sophistries, a scholarly exercise which was then very fashionable. The proposition, a line is divisible at all its points, is therefore true if it is understood according to division and false according to composition. The distinction no longer relies upon the notions of potentiality and actuality, and it is no longer prey to the objection facing Aristotle's doctrine; A line is divisible at all its points is true if division is understood as potential and false if it is understood as actual. But the logicians mark another distinction. In the proposition, a line is divisible to infinity, infinity may be taken categorematically (*cathegoreumatice*) or syncategorematically (*syncathegoreumatice*); in every proposition where infinite is pronounced, this distinction holds. We encounter the distinction between the categorematic and the syncategorematic infinite for the first time in a work which was the classical treatise on logic at the University of Paris during the fourteenth, fifteenth, and even into the sixteenth century. We are referring to the *Summulae Logicales* of Peter of Spain (*Petrus Hispanus*). This Peter of Spain is commonly identified, perhaps without sufficient evidence, with the Portuguese Pedro Juliani (1226-77) who became Pope John XXI. The *Summulae* of Peter of Spain are divided into seven treatises; of these seven treatises, the last, called the *Parva Logicalia* is the most extensive and the most original. The *Parva Logicalia* is also divided into seven treatises; the seventh of these treatises is what interests us:

Infinite is taken two ways; in one way it is taken categorematically, significatively as a general term, and thus it signifies the quantity of the thing which is subject or predicate, as when one says, the world is infinite. . . .

In another way it is taken syncategorematically, not insofar as it indicates the quantity of the things which is subject of predicate, but insofar as the subject is related to the predicate, and in this way there is distribution of the subject and [it is] a distributive sign.

We offer three rules concerning these distinctions.

The first rule is that infinite taken syncategorematically and placed in the subject causes the general term following to have confused supposition only, as in: infinite men run. In this case, men has confused but not movably confused supposition.

The second rule is that a proposition concerning the infinite, taken syncategorematically, is expounded by a copulative whose first part affirms the predicate of the subject taken according to some quantity, continuous or discrete, and whose second part denies that the predicate is in such a subject according to a determined quantity; as in: infinite men run, which is expounded thus: some men run and not so few that they will be no more than two or three, or thus: some men run and as many more as you wish.

The third rule is that a proposition concerning the infinite, taken categorematically or significatively, is expounded through a copulative whose first part asserts quantity of the subject and whose second part denies the terminus of that quantity; as in: a line is infinite, which is equivalent to: a line is long and does not have an end to its quantity. This is the case if infinite is in the predicate. But if it is in the subject, the first part asserts the predicate of the subject and the second part denies the terminus of that quantity; as in: some infinite body is white, which is equivalent to: some great body is white and the same body does not have an end to its quantity.[99]

This distinction between the categorematic and the syncategorematic senses of infinite is completely independent from the distinction between potentiality and actuality. Peter of Spain reminds us of this toward the end of the passage we have just cited:

Note that it is necessary to deny the terminus of that quantity according to the way that the infinite is spoken of, so that if the infinite in actuality is spoken of, the terminus of the actual quantity must be denied; and if the potential infinite is spoken of, whether according to addition or division, so that the infinite according to potential quantity and not according to actual quantity is spoken of, the terminus of potential quantity and not of actual quantity must be denied.[100]

It is not possible to assert more clearly that a categorematic infinite is not necessarily an actual infinite, that something might be only potentially infinite, which by itself does not render it into a syncategorematic infinite. Since categorematic and syncategorematic are two radically different senses of infinite, we should not be surprised that the same proposition containing infinite might be true or false according to whether one takes it categorematically or syncategorematically. This consequence of Peter of Spain's distinction has been noted by Walter Burley who gives the following example: The proposition, for any given magnitude there is an infinity of equal and separate parts, can be true or false; it is false if one takes it *cathegoreumatice*, understanding that one can distinguish between an infinity of equal parts and an infinity of parts equal to a previously given quantity. It is true if one takes it *syncathegoreumatice*, as affirming the possibility of finding in any given magnitude a growing number of parts whose size has not been previously determined.[101] Gregory of Rimini thoroughly examined Peter of Spain's distinction between the categorematic and the syncategorematic infinite. Here is what he wrote at the start of his inquiry concerning the infinite:

> The discussions of the opinions that some philosophers profess on this topic bring us to posit a distinction with respect to the word infinite which may be taken in two different senses; according to common terminology, it can be taken syncategorematically or categorematically.
> Concerning continuous quantities, the former sense is equivalent to the phrase, a quantity which cannot be so great that there is no greater (*non tantum quin majus*). Concerning collections of distinct objects, it is equivalent to the phrase, a multitude which cannot be so numerous that there is none more numerous (*non tot quin plura*).[102]

These definitions, modeled on what Peter of Spain asserted about the syncategorematic infinite, are not completely satisfactory for the subtle Augustinian, who proposes a different formulation to characterize the syncategorematic infinite:

> I think that it would be better to say, for a given finite quantity, however large, there is something larger, or for any given finite collection, however numerous, there is something more numerous (*quantocunque finito majus, vel quotcunque finitis plura*).
> If one wishes to take the infinite categorematically, referring to continuous quantities, one explains its sense by

the following phrase: A quantity so large that there is, and
can be, no larger; referring to distinct objects, one defines it
as: a multitude so considerable, that there can be no greater.[103]

Here again Gregory of Rimini does not seem disposed to accept
the manner of speaking then fashionable:

> The manner of expositing the notion of categorematic infinite
> does not seem suitable; according to the Philosopher, the
> ultimate heaven, or at least the universe, is a body so great
> that there is no, and can be no, greater. However, it is not
> an infinite body. Similarly, according to many modern doctors,
> there does exist a multitude more numerous than an infinite
> multitude.
>
> Hence others give a better definition of the [categorematic]
> infinite by stating, with reference to continuous quantities,
> that it is larger than one foot, two feet, three feet, and any
> given finite magnitude—with reference to a collection of
> distinct objects, by stating that it is more numerous than two,
> three, four, and any finite multitude.
>
> One can state that the infinite, taken in this sense, with
> respect to continuous magnitudes, can be defined by the
> following phrase, it is larger than any given finite quantity,
> however large (*majus quantocunque finito*). With respect to
> a multitude of distinct objects, it can be characterized by the
> phrase, it is more considerable than any finite multitude,
> however numerous (*plura quotcunque finitis*).[104]

Gregory therefore characterized the two senses of infinite by a simple
transposition of terms; he said *quantocunque finito majus* for
syncategorematic infinite and *majus quantocunque finito* for
categorematic infinite. This manner of speaking, which cannot be
translated, was broadly adopted. The proposition, *in infinitum
continuum est divisibile* signified: the continuum is infinitely
divisible in the syncategorematic sense of the word infinite. The
proposition, *continuum est divisibile in infinitum*, signified: the
continuum is infinitely divisible, taking the word infinite in its
categorematic sense. This extremely simple convention added much
clarity and conciseness to the discussions. Buridan knew and used
the convention; for example, we see him oppose the two
propositions, *infinita est linea gyrativa* and *linea gyrativa est
infinita*, to each other, taking the former syncategorematically and
the latter categorematically. The considerations he develops toward
this end enable us to recognize the origins of this convention; writing
about the latter proposition he states, "The word line is taken in

a determined sense, since there is no word before it which gives it a confused supposition."[105] We can recall that Peter of Spain sketched the following rule: "When the word infinite is applied to the subject, it causes the term which follows to have confused supposition." The medievals therefore owe the above convention to the teachings of the *Summulae*. Buridan seems to use this convention irregularly or intermittently; he even attacks those who pretend "that any word placed next to the predicate must be understood categorematically and not syncategorematically."[106] This convention, however, is scrupulously respected by Albert of Saxony and his successors. Thus, in his *Exposition* on the *De Caelo*, dated October 15, 1514, Agostino Nifo attributes to "the Peripatetic Albertilla,"[107] that is, Albert of Saxony, the invention of this manner of speaking. It was prior to him, though; we have just given evidence for this, and we can cite a formal testimony for this. The testimony in question concerns an anonymous disciple of Ockham whose work is conserved for us in a manuscript donated to the Sorbonne around 1350 by Henry Pistoris of Lewis. The author tells us that according to Ockham, "God can increase charity to infinity or, *in order to speak more clearly as a logician*, to infinitely increase charity (*Deus potest augmentare caritatem in infinitum vel, magis logice loquendo, potest in infinitum caritatem augmentare*); actually, however great is the given charity, the existence of a greater charity is not contradictory."[108] But let us return to Gregory of Rimini and to the distinction he formulates between the categorematic and syncategorematic infinites:

> The two senses of the word infinite differ notably. In fact, if the word infinite is on the side of the subject and is understood syncategorematically, it renders the proposition into a universal proposition, which does not occur if it is understood categorematically. Also it can happen that a proposition is true if understood one way and false if understood another. Let us assume that the world must remain perpetually as it is, for example, and that one asserts the proposition, an infinity of men will be dead. If infinite is taken syncategorematically, the proposition is true, for whatever finite number is given, the number of dead will be greater (*quotcunque finitis plures erunt praeteriti*). But if one takes the word infinite categorematically, the proposition is false, for the dead will not be more numerous than any finite number (*non plures quotcunque finitis*).[109]

John Buridan attaches no lesser importance than Gregory of

Rimini to the distinction between the categorematic and the syncategorematic infinites; he also attempts to render more precise the definitions given by Peter of Spain:

> This question about infinite divisibility presents numerous difficulties, the first of which is as follows: Since the word infinite can be taken categorematically and syncategorematically, the question must be formulated one way or another; and since the choice is arbitrary, many have prescribed their definitions according to their whims. Afterwards, one must speak in conformity with these definitions, since, as Aristotle said, the definition is the point of departure for any doctrine. It seems to me that if he had understood it categorematically, Aristotle would have defined infinite for magnitudes as what is extended without limit, or else, what is extended and not limited. . . .
>
> One should note that the word infinite taken categorematically has numerous properties . . . the first is that it is opposed to the finite as privation, in the same manner that the nonlimited is opposed to the limited and having no termination is opposed to having a termination. . . .[110]
>
> Let us now speak about the syncategorematic infinite and note that many definitions on the topic of the infinite taken syncategorematically have been given. One defines it at first for [continuous] magnitudes: it has some magnitude but is not so large that it cannot be larger (*aliquantum, et non tantum quin majus*); one defines it in an analogous manner for multitudes [of distinct objects]: it is numerous, but not so numerous that it cannot be more numerous (*aliquanta et non tanta quin plura*).[111]

Buridan is clearly no more satisfied with these definitions inspired by the *Summulae* than was Gregory of Rimini; like Gregory he seeks to replace them with more precise definitions:

> It seems to me that this definition which is more succinct and more precise is equivalent to the preceding definition: To state that B is infinite in magnitude, signifies that for any B, there is a larger B. (*Infinitum esse B secundum magnitudinem significat quod omni B est B majus.*) Thus B is infinite in length signifies that for any B corresponds a longer B; it is the same for infinite speed, infinite slowness, infinite smallness, etc. In any case, I understand one and the same thing for infinite in length, infinitely long, and of infinite length (*infinitum secundum longitudinem, infinite longum, infinitum longum*).[112]

We have already seen Buridan define the infinitely small, and we have admired the precision of his language (which modern algebra has done no more than recapture). However Buridan is not content with this precision and attempts to increase it. He notes the following consequence of the definition he has given the syncategorematic infinite:

> If there were an infinite body, infinite being taken categorematically . . . the proposition, there is an infinite body, where infinite is taken syncategorematically, would be false. . . . For there would be a certain body, that is a [categorematically] infinite body such that no body would be greater. . . . Nevertheless, if this [categorematic] infinite existed, the following proposition would be true: the finite body is infinite [syncategorematically] since, for every finite body there would correspond a larger finite body (*infinitum est corpus finitum, quia omni corpore finito esset majus corpus finitum*).[113]

The formula reminds one that the study of the syncategorematic infinite considers only finite magnitudes; it is the formula which Buridan utilizes when he wishes to announce an important proposition. Here is a good example:

> There can be an eternal or infinite movement and similarly, eternal time, at least in the future. . . . This conclusion is evident if one takes the words eternal and infinite syncategorematically. According to Aristotle one should state, there is no movement, no time of so great a duration that there is no movement, no time of longer duration. And [it is the same] according to the truth of our faith. Time and movement can endure perpetually and to infinity. Hence finite movement can be infinite, for there can be no finite movement so great that there could be no greater finite movement. (*Igitur infinitus potest esse motus finitus quia non potest, esse tantus finitus quin possit esse major finitus.*)[114]

In his definition of the syncategorematic infinite, John Buridan attained a precision no one has been able to surpass; his immediate successors were far behind him, and even behind Gregory of Rimini. Thus after having critiqued the two formulations of Peter of Spain, *non tantum quin majus, non tot quin plures*, Albert of Saxony was satisfied with the two definitions, one for the continuous magnitudes, and the other for collections of distinct objects: *Aliquantum, et quantumlibet majus; aliquot et quantumlibet plures vel plura.* Neither Gregory nor Buridan thought these formulations precise or clear enough. On the other hand, Albert of Saxony does not have to yield to his predecessors for the clearness with which he indicates the logical heterogeneity of the categorematic and syncategorematic infinites:

> If one formulates two similar propositions, but infinite is taken categorematically in one and syncategorematically in the other, these two propositions are radically heterogeneous (*impertinentes*); they do not result from one another; neither are they repugnant to one another.[115]

The truth of each has to be proven by itself, and without worrying about the truth of the other. "That is why the proposition, the continuum is divisible to infinity (*in infinitum continuum est divisibile*), does not depend (and vice versa) on the proposition, the continuum is infinitely divisible (*continuum est divisibile in infinitum*); for the former consists of a syncategorematic infinite, and the latter of a categorematic infinite."[116] The apparent antinomy that Roger Bacon's contemporaries found so difficult—what Bacon called the "club of Hercules" and boasted about having broken—has now been dispelled.

The Concept of Limit: Maximum and Minimum

If one takes infinite syncategorematically, a continuum is infinitely divisible; this truth is commonly admitted by the Scholastics of the fourteenth century who use it constantly. The example they most often invoke is the following: One divides the continuum into two equal parts; one of the halves thus obtained is then divided into two equal parts. One of the two fourths is divided into eighths, and so forth. One thus forms a series of parts whose magnitudes decrease in a geometric progression by one half. This is what was called during the Middle Ages, *dividing a continuum into proportional parts.* The well-known paralogism, "Achilles and the Tortoise," attributed to Zeno of Elea, led logicians to meditate upon this division which is forever pursued without ever being achieved. Here is what Giles of Rome asserted on this subject:

> There is a difficulty in the infinite division of time. If this division to infinity can be realized in actuality, then a speedy horse can never reach an ant. Let us assume that a horse moves by half a palm and stops, and then he moves

after that for half of the remaining palm and stops again, and so forth. Since the continuum is infinitely divisible, it can never complete the whole palm. Therefore, when one divides a continuum, if each part has its own separate existence and if these parts are produced by an actual division, the division of the continuum will never be achieved. . . . Thus time is infinitely divisible, but its parts are only potential.[117]

These remarks and similar remarks compelled the masters of Scholasticism to reflect on the case where a variable magnitude tends toward a limit without ever reaching it; soon their minds, sharpened by continual exercises in logic, were able to reason about these topics with a rigor that was rarely surpassed; for example, here is a passage from Walter Burley:

What we have just expounded upon proves the truth of the following proposition which is not known by many: Given any line, one can mark off segments whose lengths decrease proportionally, and one can also indicate a point which cannot be reached by a finite operation. That will occur if one takes as the first segment half the length to the extremity which cannot be reached by a finite operation; one takes as the second segment half the first segment, and so forth. On the other hand, every point before the extremity can be reached by a finite operation. That can easily be demonstrated geometrically, but for now we will not insist on its demonstration.[118]

According to Burley the proposition was not known by many. But soon the paralogisms of Zeno of Elea became an inexhaustible source of *sophismata*, whose resolutions were a favorite exercise of those at the University of Paris and, above all, those at Oxford University. Gregory of Rimini cites several of the *sophismata* discussed at Oxford by Henricus Hibernicus, Adam Goddam, and Clienton (or Clymeton) Lengley.[119] The resolution of every one of these consisted of the following remark, intimately connected with Burley's: "In any magnitude there is an infinity of proportional parts, infinity being taken syncategorematically; a result of this is that none of these parts is the last one."[120] A large portion of the sophisms generated by the division of a length into proportional parts was due in fact to the use of vicious locutions such as these: I take all the proportional parts of a continuum, I consider the ultimate proportional part formed in the division of a continuum, etc. Buridan excelled at showing that the locutions implied some error. He takes, for example, a cylindrical column divided into proportional parts by planes parallel to the base; he imagines that one draws a perpendicular to these planes, in the following way:

A first segment crosses the first proportional part and goes no further; a second segment crosses the second proportional part, without going further, and so forth. It is manifest that none of these rectilinear segments, nor any straight lines composed of these segments, goes beyond these proportional halves. . . . And it is manifest that no straight line can be drawn through all these proportional halves without being able to go beyond all these halves; in fact, if it were drawn through the whole column—until the final termination of the column which touches an external body—the line would go beyond all these halves.

A line drawn in this way can cross all these halves [taking all syncategorematically], but no line drawn this way can cross all these halves [taking all categorematically]. (*Per omnes est aliqua protensa, sed non est aliqua protensa per omnes.*) There is a line that can be drawn through 100 parts, there is one that can be drawn through 1000 parts, and so forth, for any given number; but it does not follow that there is a line drawn through an infinity of parts or through all the parts, for there are no parts that are an infinity of parts and all the parts (*quia nullae sunt infinitae et nulla sunt omnes*).[121]

It is true that if one takes infinity syncategorematically, there is an ultimate part to an infinity of parts; it is true that there is an ultimate part for all the proportional parts of a line. However, there is no ultimate proportional half in the sense that there is no other which comes after it; there is therefore an ultimate part, but it is not the last one of all the parts.[122]

In other words, in each of the successive states of the division into proportional parts, there is a finite number of parts among which there is an ultimate part; but a new state of division into proportional parts engenders a new part located after the one which was ultimate. One can still object to these considerations as follows: Can we not take an infinity of proportional parts of a line—all the proportional parts of a line? Is it not sufficient, in order to do this, to take the whole line? "Assuredly," replies Buridan, "when I take my book, I take an infinity of parts of my book, for I am taking three parts, 100 parts, 1000 parts, and so forth without end. But what is impossible, is that one takes an infinity of parts successively, counting one after the other."[123] The conciseness of Buridan's speech is recaptured in Albert of Saxony's writings; he also asserts that "there are no parts of a continuous magnitude which are all the proportional parts of this continuum; that is evident for, whichever parts are given, there are still others. There are therefore no parts which are all the parts (*quibuscunque datis, adhuc sunt plures; ergo nullae sunt omnes*).[124]

Such rigor of thought and precision of language must have been extremely useful to the Scholastics of Paris in their discussion of the problems about a variable magnitude tending toward a limit that cannot be attained. We will demonstrate this by summarizing the history of one that was well known in the schools. In order to recapture the origin of this problem, we must, as always, read Aristotle, and above all, the commentaries of Averroes. Aristotle observes that a man who can walk 100 stadia, can walk 2; if one asked how many can he walk, the reply would not be 2, but 100 stadia; "Power is of the maximum, and a thing said, with reference to the maximum, to be incapable of so much is also incapable of any greater amount. It is, for instance, clear that a person who cannot walk 1000 stadia will also be unable to walk 1001."[125] Averroes develops these remarks as follows:

> It is evident that the powers of things must be defined by their terminations; the terminations distinguish the powers of things that have differing powers. . . . A power, then, is defined by the termination of its action, and not by that which is before this termination. . . . On the other hand, incapacity is defined by the least it can do; in other words, the deficiency of power is defined by the minimum of its power, the inverse of that which was needed for the definition of power.[126]

After having noted Averroes's commentary, let us look at Aquinas's commentary:

> In the same way that one determines a power by the maximum one can accomplish, one determines what cannot be done by the least of the things that are impossible. That is how one characterizes incapacity. If, for example, the maximum number of stadia that can be traveled is 20, and the minimum number of stadia that cannot be traveled is 21, one should characterize one's incapacity by the latter number, and not by saying that one cannot accomplish 100 stadia or 1000 stadia.[127]

The doctrine Aquinas seems to profess in this passage raises lively and proper criticisms from John of Jandun. He states that, for any natural virtue, there is a maximum that it can accomplish; thus there is a maximum to the number of pounds that a man can carry. Some philosophers think that there is a minimum that this virtue cannot accomplish, and that this minimum is distinct from the preceding maximum. For example, let us assume that a man is capable of carrying any weight up to 100 pounds maximum. The weights he cannot carry admit of a minimum, and this minimum is not 100 pounds; it would be more than 100 pounds. John of Jandun can easily demonstrate that the maximum and the minimum cannot differ by any divisible magnitude. Let us in fact assume that they so differ, and let us take an intermediary weight, between the maximum and the minimum. The man can carry this weight, since it is less than the minimum of weight he cannot carry; however, this weight is more than the maximum that he can carry. The contradiction is manifest. In order for it to disappear, the maximum and the minimum would have to be separated only by an indivisible. The impossibility of indivisibles closes off this line of inquiry so that John of Jandun thinks himself authorized to formulate the following conclusion: "It is true that to each natural virtue corresponds a maximum that can be accomplished [. . .] it is not true that there corresponds a minimum that cannot be accomplished."[128] John of Jandun clearly demonstrated the contradiction within Saint Thomas Aquinas's theory, but his attempted solution to the problem was worthless. It was only a short time before a proper solution was proposed, as Buridan shows us. Buridan states that with respect to the limits of active and passive powers,

we commonly propose only probable conclusions. Here is the first conclusion: Let A be a power capable of lifting a large weight. We cannot assign a maximum weight to what A can lift. This conclusion can be proved by allowing that there is no action when the agent is equal or less than the resistance. . . . Suppose that A lifts weight B and that this weight is the maximum weight that A can lift (according to our opponent); then there would have to be some excess of A to B. Let us suspend a weight C to B such that the new resistance becomes equal to A's power; it is true that A cannot lift B and C together. But since C is divisible, we can remove half, and let the other half—called D—remain attached to B; A's power exceeds the resistance of B and D and consequently A can lift it. However, B and D is greater than B; B is therefore not the maximum weight that A can lift.
One can also reason thus: Let A be a power capable of

lifting weights, and B a weight whose resistance equals A's power. A cannot move B, but its power can move a smaller weight than B, for it will be greater than it by some amount; and one cannot give a weight smaller than B by an indivisible amount because a continuum is not composed of indivisibles. Hence, given any weight smaller than B, one can always give an intermediary weight larger than it and smaller than B; therefore, given any weight that A's power can lift, there is a larger weight that this power can lift. . . .[129]

There are other conclusions which are rightly deduced from the conclusions that have just been posited.

The first is as follows: One can assign a minimum to the weight that A cannot lift. It is certain, in fact, that the weight can be increased so that A can no longer lift it. It is therefore necessary that some weight mark the termination of this power; and one cannot understand that this power stops at such a weight, if it is not in one of these two ways: either his power can lift such a weight and cannot lift anything heavier—that would be the maximum weight that can be lifted (which we know to be impossible)—or his power cannot lift this weight, but can lift any lesser weight—which is our conclusion; this weight is the smallest weight which cannot be lifted, since any smaller weight can be lifted.[130]

Buridan took up again, in his *Quaestiones super libris de Caelo et Mundo*, the question, "Must a power be defined by the minimum that cannot be done?"—but this time with less precision. The disorder that can be observed there might indicate that we are dealing with *reportata* rather than the work of the master. The author indicates that the question is extremely difficult; he then formulates the opinions of Aristotle and the Commentator and makes the following observation:

One must assert with Aristotle that an active power must be determined by the maximum that can be done (*per maximum ad quod ipsa potest*). We understand by this that we know the magnitude of an active power when we know the maximum that it can produce, or at least, that we know the maximum below which it can do all (*maximum infra quod ipsa omne potest*). I posit this distinction because of a difficulty I will examine later.

Others admit that one cannot give a maximum that a power can produce, but the minimum among those it cannot produce; this minimum is not the maximum that the power can produce, but the action below which it can produce all.[131]

Those "others" who uphold this opinion are actually Buridan and those who, like Albert of Saxony, have understood him. Why this opinion is the right one and why the one formulated before is inadequate, the author of the *Quaestiones* does not yet indicate. He merely shows, in a somewhat scanty fashion, that "we know what is the strength of a power when we can distinguish it from any stronger power and any weaker power, and by noting the maximum it can produce or, at least, the maximum below which it can do all."[132] The author until now remained indifferent between these two modes of speaking; he makes his choice in the question following the long, thorny, confused twenty-first question: Can one assign a maximum to the action that a power can produce?[133]

> Many think that they can demonstrate the contrary with the help of this supposition: in order that a mover move a mobile, the strength of the mover must surpass the strength [resistance] of the mobile. One first asserts that there is no action if the former has the same or less strength than the latter. One assumes that however small the excess of motive strength over the resistance of the mobile, the mover can move the mobile, if there is not some other impediment or some more resistance.
>
> Let us allow that our opponent designates the largest sphere that the lunar motor can move. It is certain that the force of that mover surpasses the resistance of that mobile; and this force cannot surpass the resistance by some indivisible. It surpasses it by a divisible quantity. Hence, the mover which suffices to move this mobile can move a larger mobile.[134]

That is clearly the reasoning Buridan developed in his *Physics*, but some essential intermediary steps have disappeared. The author concludes:

> I think, however, that this reasoning is not conclusive. In fact, the mobile has no resistance to the movements that intelligences communicate to celestial bodies; there is actually a small inclination of the mobile toward the movement it is given. This reasoning is therefore conclusive when the mobile resists the agent, but where it does not resist it—in celestial movements, for example—the argument is not conclusive.[135]

One must therefore limit one's application "to those agents that the mobile resists. For those, one supposes as a principle that there is no action if the resistance is stronger or equal to the active virtue;

but there is an action if the active virtue is stronger than the resisting virtue."[136] In that case one can conclude "as is commonly done," that one can give the minimum resistance for which the mover cannot move the mobile. One can also conclude that it is impossible to assign a maximum weight that Socrates can lift, but it is possible to assign a minimum to the weight that he cannot lift; that is the weight equal to his strength. We arrive thus at the conclusion Buridan achieved in the *Physics*. But in obtaining it the philosopher of Béthune proceeded with greater clarity and rigor than in his *Quaestiones super libri de Caelo et Mundo*. Perhaps, as we have already stated, we should conclude that we do not have in these *Quaestiones* a text written by the master, but simply the composition of some disciple.[137] The solution of the problem that escaped Saint Thomas Aquinas and John of Jandun now follows logically; this solution Buridan does not claim as his own. On the contrary, he tells us that the conclusions he enunciates are commonly held. Obviously, the questions they resolved were commonly examined then, at Oxford as well as at Paris. Someone named Swineshead, a member of Merton College, was the leader of a disturbance provoked by the election of the Oxford chancellor in 1348.[138] This same Swineshead wrote a treatise on physics entitled *De primo motore* (manuscripts of which are conserved in English libraries); the Bibliothèque Nationale possesses portions of it, in poor condition and not very legible, collected in some philosophy notebooks.[139] In this treatise, Master Swineshead attempted to posit some distinctions relative to maxima and minima for active and passive powers; he considered the maximum to which a power can act (*maximum in quod potest*) and the maximum in which a power can no longer act (*maximum in quod non potest*).[140] The philosophy notebooks, to which we have just referred, conserved, in addition to the extended passages from *De primo motore*, the discussion of three questions which seem to be Swineshead's and are entitled *Tres dubia parisiensa*.[141] And, of these three *Doubts of Paris*, the last two are devoted to the question we are considering. The second doubt examines if there is a maximum weight that a man called Socrates (or by abbreviation, Sortes) can carry: *Utrum sit dare maximum pondus quod homo Sortes potest portare*.[142] From then on this becomes the formulation for the problem of the upper limit of a power. At Oxford as in Paris, the powers that lift weights, the mobiles that cross distances, are no longer called A and B, but Sortes and Plato. The humanists of the Renaissance poked fun at this custom; Agostino Nifo gave the title of "Socratizers" (*Sorticolae*) to the dialecticians of Paris and Oxford, whose logical precision he could never have rivaled. The third doubt of Paris begins as follows:

> With respect to the end or the termination of an active or passive power, one poses two distinctions; one is by the maximum of what the power can do, or by the minimum of what it cannot do; the other is by the maximum of what it cannot do or by the minimum of what it can do (*una per maximum in quod potentia potest vel minimum in quod non potest; alia per maximum in quod non potest vel minimum in quod potest*).[143]

The name John of Dumbleton first appeared on the registers of Oxford's Merton College in 1331. On September 27, 1332, John of Dumbleton was presented for the rectory of Rotherfield Peppart near Hensley, in the archdeaconry of Oxford; he resigned this office in 1334. He took part in some assemblies at Merton College between 1338 and 1339; in February 1340 (1341 according to present methods) he was named among the first fellows of Queen's College, according to the original statutes of this college. He was again at Merton College in 1344 and 1349.[144] We have a voluminous work by John of Dumbleton that was never printed, whose manuscripts are entitled *Summa logicae et naturalis philosophiae* or *Summa de logicis et naturalibus*; some even carry the inappropriate title *Summa de theologia major*. The notebooks holding the excerpts of *De primo motore* of Swineshead also hold excerpts of Dumbleton's *Summa*; in addition, the Bibliothèque Nationale owns a complete specimen of this work.[145] John of Dumbleton dedicated the first two chapters of the sixth part of this work to the questions that were the object of Swineshead's last two *Doubts of Paris*. In the first of these chapters, the author enumerates the various opinions one can hold about the limits of power; the ultimate reason—which he prefers—is formulated as follows: "The third thesis holds that any agent is determined by a natural action such that the agent cannot accomplish any greater action."[146] The second chapter attempts to justify this thesis: "One asks whether there is a maximum action that a man can be accomplishing."[147] Swineshead and Dumbleton's arguments on these topics were what the Oxford arguments almost always were during that period, a web of paradoxes and sophisms, intertwined in an inextricable way, knotted for the pleasure of unraveling them. Clarity disappeared and truth was often eclipsed in this game of complex dialectics. Thus one would be seeking in vain, in the writings of Swineshead and Dumbleton, for what Buridan perceived so well: one can assign an upper limit to the effects that a power is capable of producing such that its effects can approach it as much as one wishes, but can never attain it. We have already referred to the philosophy notebooks in which a Parisian student inserted selections and summaries of Dumbleton's *Summa*. In these notebooks, Dumbleton's thought is represented as follows:

> When one wishes to know the termination of powers, one conceives that a power is determined by its *maxima in quod sic* difficulty; it is the maximum difficulty which this power can overcome or the minimum difficulty which it cannot overcome.[148]

The expressions, *maximum in quod sic, minimum in quod non*, are not found in the writings of Swineshead, Dumbleton, or Buridan. On the other hand, they occur frequently in the writings of later authors; for example, they can be discovered in some questions attributed to Robert Holkot. The year in which Robert Holkot, the Dominican who wrote some *Questions* on the *Sentences*, died is given as 1349. Clearly we can attribute the writing of these questions to an earlier date. With Holkot's *Questions* on the *Sentences* are appended some *Determinations on Some Other Questions*. Are these *Determinations* also the work of the Dominican Doctor? Josse Bade, who edited them, tells us that "many suppose that these questions have been gathered together by some disciple of Holkot, or that he, while he was teaching gave them publicly; others believe that they were written by himself.[149] That is to say, the authenticity of these questions was doubtful and uncertain even during the period in which they were written. The first determination starts with an article in which the author distinguishes the *maximum in quod sic* from the *minimum in quod non*, the *minimum in quod sic*, from the *maximum in quod non*; understandably, the classic example of the limit between the weight that Socrates can carry and the weight he cannot carry is the first Holkot uses to illustrate his definitions. The logical discussion that the Dominican master delivers on these various notions is long and meticulous, but its tedious subtleties remind one of Swineshead and Dumbleton's discussions; one finds nothing there worthy of the notice of a modern mathematician, nothing of the justified rigor which we have met in John Buridan's analysis, and will be able to admire anew in Albert of Saxony's analysis.

The two propositions Albert of Saxony discussed are as follows:[150] Given an active power, there is a *maximum* resistance among the resistances it *can overcome (maximum in quod sic)*. Given an active power, there is a *minimum* resistance among the resistances it cannot overcome (*minimum in quod non*). But before discussing these two propositions, he established their meanings with a precision a modern mathematician might envy. By stating that a resistance is *maximum* among those a given power *can* overcome, he meant that the power can overcome *that resistance* and any lesser resistance, while it cannot overcome any greater resistance. When defining the meaning of the phrase, such resistance is a minimum among those a given power cannot overcome, Albert's predecessors were content with: the given power cannot overcome this resistance and any greater resistance, but it can overcome any lesser resistance. Our logician required a greater precision; he stated that the given power cannot overcome either the minimum resistance or a greater resistance, but that if one were to designate any resistance whatsoever, less than the minimum resistance, there can exist a resistance greater than the one designated that the power can overcome.[151] This precision enabled him to sidestep some objections; Albert thought that the effects of some powers might not only admit of an upper limit, but also a lower limit. Sight is the example Scholastics invoked to demonstrate the existence of such powers; we do not see what is too near nor what is too far. These definitions having been carefully posited, Albert formulates the following conclusions, which are also those of John Buridan: It is not true that there is a maximum among the resistances that a given power can overcome (*potentia activa non terminatur per maximum in quod sic*), but there is a minimum among the resistances it cannot overcome (*terminatur per minimum in quod non*).

Let A be an active power; one can give it an equal resistance and designate it by B. And this resistance is the minimum resistance among those A cannot overcome. A cannot overcome resistance B because it does not exceed it. But if we give ourselves a resistance smaller than B by any amount, we can find a resistance larger than it that A's power can overcome— a resistance smaller than B; we can find another resistance larger than it and smaller than A. Since the least excess is enough to determine movement, a resistance smaller than B being given, we can find a resistance larger than that and the active power of A can overcome it; hence, a consequence of the definition of *minimum in quod non* given above, B is the minimum resistance among those A cannot overcome.

We can state that one knows the magnitude of an active power by knowing what is the minimum resistance it cannot overcome. In fact, we know what is the strength of an active power if we can distinguish it from any stronger power or any weaker power; and that is what we know when we know the least resistance it cannot overcome, for in order to know this minimum, we have to know three things: we have to know first that the given power cannot overcome either such a resistance or any stronger resistance, and these two items of knowledge allow us to distinguish the given power from all other greater powers; we have to know, further, that if one gives any resistance, however smaller than this minimum, one can find a greater resistance than it which the given power can overcome, and this last item of knowledge suffices to distinguish it from any weaker power.[152]

The resistances that a power can overcome therefore make up a set of magnitudes which admit of an upper limit, but which cannot attain this limit (as in Walter Burley's example). The possibility to formulate propositions that are true or false according to whether we take them categorematically or syncategorematically logically follows from this.

It would not be logical to assert: Socrates has the power to carry any portion of this weight; therefore he will carry any portion of this weight. Let us consider a weight A, weighing 8 units and suppose that 8 units is Socrates' power. It is evident that Socrates has the power to carry any portion of weight A; however, it is impossible that he carry every part for he would then carry weight A itself. And that is false, for there can be no action when the power is equal to the resistance.[153]

In that case, then "the universal proposition is impossible, while each particular proposition is possible and compossible with each other. . . . One shifts from a *divided sense* which is true, to a *composite sense, which is false.*"[154] Here Albert speaks in the manner of someone who has read Duns Scotus. Instead of being given an active power and considering the various resistances it can overcome, one can, on the contrary, fix a resistance and consider all the power able to overcome it. The powers able to overcome this given resistance do not admit of a *minimum in quod sic*, but they admit of a *maximum in quod non*, whose magnitude can serve to characterize the given resistance. This proposition can be illuminated by an example:

> Let us suppose that Socrates' power to lift a weight is equal to the resistance of one pound, such that Socrates has precisely enough lifting strength as the pound has resistance. Socrates' strength is the maximum of the lifting power that cannot lift the pound, for no force smaller than Socrates' can lift a pound, and any force greater can lift it, so that Socrates has the greatest power among those which cannot lift a pound. Thus the active power equal to the resistance is the maximum power among those which the resistance does not give way; and the resistance equal to the active power is the minimum of the resistances that the power cannot overcome.[155]

Let us now bring together these last two definitions:[156] When power is equal to resistance, neither of the forces overpower the other. "They are like two men of equal strength attempting to pull each other; neither of them acts on the other, but each of them prohibits the action of the other." One of these two antagonistic forces counterbalancing each other, being augmented by as little as one wishes, would be sufficient for it to overpower the other. When Socrates carries on his head a stone whose resistance is equal to his power, if one were to augment Socrates' strength whatsoever, he would be able to lift the stone; if one augmented the weight of the stone, Socrates would fail. Thus Albert of Saxony, when considering the antagonism between a power and a resistance, divides the circumstances that can occur into two categories: on the one hand are the circumstances in which the action is carried out according to the power, and on the other hand are the circumstances in which the action is according to the resistance. The two categories are separated by a common limit, and the limiting circumstances belong to neither category; when they are realized, there is no action, neither according to power nor resistance—there is equilibrium. We should ponder about the process by which Dedekind and Jules Tannery introduced the notion of incommensurable numbers to arithmetic; we would not fail to recognize a striking analogy between these proceedings and those by which Albert of Saxony defined a power. Even more striking is the resemblance between the considerations just detailed and those which thermodynamics underwent in order to give a precise sense to the notion: *reversible modification*. Assuredly, our modern treatises of analysis have nothing to teach Walter Burley, John Buridan, and Albert of Saxony when it comes to the art of thinking rigorously and talking precisely about the notion of limit. It is probable that they would have nothing to teach Nicole Oresme either. Oresme's *Traité du Ciel et du Monde* was written nine years after Albert of Saxony's *Subtilissimae quaestiones in libros de Caelo et Mundo*, in 1377. Oresme holds the same opinions as Albert on almost everything; specifically, that is the case with the topic of how an active power ought to be defined. Here is what he concludes in chapter 29 of the first book, entitled "He solves the problem of the possible or impossible with respect to a given force."

> I say accordingly, that any force, with respect to its resistance, is measured precisely by the resistance which equals that force, so that it represents the smallest resistance of all those its power cannot overcome; but it can overcome any lesser resistance, except the one which will be stated very soon. . . .
>   Perhaps another force may be such that it cannot be measured, but requires a distance moderate in size; it is limited to the smallest distance it cannot reach—for the other distances are too great and too far—and the largest of those it cannot reach, because the objects are too small or too near, as in the case of the force of visible objects.[157]

Communications between the two universities, Paris and Oxford, were frequent; we learned from Swineshead that the masters of Oxford examined the *Doubts of Paris*. We should not be surprised to see contemporaries of Nicole Oresme and Albert of Saxony at Oxford profess doctrines similar to those taught at Paris. William Heytesbury is mentioned as a fellow of Merton College as early as 1330; he was bursar in 1338, and his name was on the examination lists of the college in 1338 and 1339. One finds a William Heightilbury—probably none other than Heytesbury—among the first students of Queen's College in 1340. There is no document pertaining to him from 1340 to 1371, but in 1371 one finds a William Heighterbury or Hetisbury, Doctor of Theology and chancellor of Oxford University. This Gulielmus Hentisberus composed various treatises devoted to developing the Logical methodology introduced by Peter of Spain's *Summulae Logicales* and, above all, to unraveling the sophisms that one can compose on various subjects.[159] William Heytesbury's writings had a greater influence on dialectical studies, first at Oxford, then at Paris and in the Italian schools. Among

Heytesbury's writings is a small treatise, *De sensu composito et diviso*. There the author distinguishes several modes of distinctions that one might wish to establish between these two senses of a single proposition. One of these modes proceeds from

> whether the terms are taken categorematically or syncateg-orematically; if one wishes to conclude a composite sense from a divided sense, the reasoning is faulty. For example, one cannot conclude that Socrates has a categorematic infinity of equal parts located one outside the other (*ergo Sortes habet infinitas partes aequales non communicantes*) from the fact that Socrates has a syncategorematic infinity of equal parts located one outside the other (*infinitas partes aequales non communicantes habet Sortes*).[160]

We see Heytesbury distinguish the categorematic and syncategorematic by the placement of the word infinite in the proposition, according to the Paris custom, that the Oxford logician is careful to recall. Among the many sophisms unraveled by the distinction between the composite and divided senses, our author cites the following, known from the time of Duns Scotus: Socrates can carry stone A; he can also carry stone B; therefore he can carry both A and B.[161] One of the opuscula of Heytesbury is entitled, *Regulae solvendi sophismata*. These *Rules for Unraveling Sophisms* are distributed among several small treatises, which the fifth, *De maximo et minimo*, concerns the limits of active and passive powers:

> Either active power has as termination the maximum of what it can do (*maximum in quod potest*), or the minimum of what it cannot do (*minimum in quod non poterit*). In fact, since Socrates' power is a finite power, one can assign either a maximum to what Socrates can carry or a minimum to what he cannot carry.[162]

Which of the two ways of defining Socrates' lifting strength is better? The latter—the minimum of what Socrates cannot carry. One can still assign "the maximum weight among those which can be carried by men stronger than Socrates."[163] Heytesbury therefore accepts the conclusion we have seen formulated by Buridan, Albert of Saxony, and Oresme. But one does not find in the *Regulae* of the Oxford logician the clear and conclusive reasoning of the Parisian masters; instead, one finds several pages of complex examples, bizarre sophisms, and thorny discussions. There is no more striking contrast revealing the differing mentalities of fourteenth century Oxford and Paris than the comparison of what Heytesbury, on the one hand, and Albert of Saxony, on the other, wrote about the maximum and the minimum. However, the clear and precise reasonings of the masters of the University of Paris can be recovered in a work attributed to the chancellor of Oxford. The final work in the collection of works by our logician is entitled: *Preclarissimi viri ac subtilissimi sophistae Gulielmi Hentisberi probationes profundissimae conclusionum in regulis positarum.* As this title clearly indicates, the subject matter of this treatise is to give demonstrations of numerous propositions asserted without sufficient proof in the treatise *De sensu composito et diviso* or the treatise *Regulae solvendi sophismata.* Moreover, the second of the *Probationes,* about the treatise *De maximo et minimo,* is intended to justify these two propositions:[164]

> One cannot assign a maximum to the weight Socrates can carry.
> One can assign a minimum to the weight Socrates cannot carry.

The reasonings developed in support of these two propositions are almost those Buridan used in similar circumstances. But it is possible to doubt that the *Probationes conclusionum* were written by William Heytesbury. The *Probationes* constitute a commentary on the *Regulae solvendi sophismata.* It would be surprising if Heytesbury commented on himself in this manner. But more importantly, one should note the radical difference between the manners of reasoning and writing the author would have utilized had he composed both the *Regulae* and the *Probationes.* The *Regulae* are an example of the disorganized, confused, sophistical argumentation fashionable at Oxford, and from which Heytesbury did not depart in his other works; the *Probationes* recall, on the other hand, the order, clarity, and rigor of the writings of Buridan and Albert of Saxony. Most of the time they borrow their reasoning and style from these masters. It seems better to regard the *Probationes conclusionum* as a commentary written by some Parisian master, some disciple of Albert of Saxony, on the *Regulae solvendi sophismata* of William Heytesbury. Be that as it may, the author of the *Probationes conclusionum* carefully preserved the opinion of John Buridan, Albert of Saxony, and Nicole Oresme with regard to the *maximum in quod sic* and the *minimum in quod non.* But in Paris this clear and correct opinion began to be forgotten—a manifest sign of the decadence the university in which these great masters taught underwent after their deaths. When he treats of physics *secundum nominalium viam,* Marsilius of Inghen almost always follows the order of questions formulated by Albert in his *Physica* or his *De Caelo*; but he willingly contradicts him in his conclusions, and almost always in an untoward fashion. He devotes three questions in his commentary on Aristotle's *Physics* to the study of the limits that terminate the effect of a power or a resistance; these are clearly inspired by Albert's two questions on this topic from his *De Caelo*. But Albert's precision and rigor are neglected by Marsilius of Inghen. After having defined the *maximum in quod sic* in the fashion of Albert of Saxony, Marsilius adds: "One can define the *minimum in quod non*, the *maximum in quod non*, and the *minimum in quod sic* in the same fashion." Then abandoning the distinctions Buridan and Albert so carefully drew, he formulates the following erroneous conclusion: "For any active power, there is a *maximum in quod sic* among the resistances it can overcome, and a *minimum in quod non* among those it cannot overcome';"[165] this maximum and minimum are one and the same resistance. In his *Abbreviationes libri physicorum*, Marsilius of Inghen, more faithful to the teachings of Albert of Saxony, borrowed from him his careful definition of the *minimum in quod non*; he extended this definition to the *maximum in quod non*, asserting that these definitions were better than those previously given. But the conclusions he formulated were the false conclusions he repeated in his *Quaestiones*.[166] Marsilius's teachers understood with admirable distinctness that a set of magnitudes can have as a limit a magnitude that does not belong to this set. This truth escapes their illustrious disciple.

2
Infinitely Large

The Scholastic Formulation of the Problem of the Infinitely Large

Any problem with the infinitely small is a problem with the infinitely large—the study of one infinite is not separate from the study of the other—that is a truth the masters of Scholasticism clearly perceived. They applied to the infinitely large the methods which allowed them to treat the infinitely small; even better, they often gave a single theory for the two problems. This analogy between the theory of the infinitely large and the infinitely small was not recognized by Peripatetic philosophy, however. For Aristotle, no infinite magnitude exists in actuality, for the universe is limited. It cannot exist potentially either; however great a quantity is realized, there exists a limit that cannot be surpassed, for no quantity can exceed the boundaries of the world. No power would therefore be able to realize a magnitude exceeding any given magnitude. The above reasoning is valid for a power that has to accept the world as it is, a world one considers as bounded, that cannot add any body, no matter how small, to the bodies already existing; it is valid, hence, for a power that cannot create. It is not valid for a power that can produce bodies without end, one that can forever move back the boundaries of the world. Aristotle did not admit any creative power; he thought that the world contained all existing matter, all matter that can exist, and that this matter is in limited quantity. He could then sustain, without contradiction, the denial of the potential infinitely large. But Scholastic Christianity could not tolerate the absolutism of this proposition; perhaps the power to produce a potential infinity is not given to the world, which cannot create, but it is surely not beyond God's omnipotence. In 1277, Etienne Tempier condemned this error: "That the first cause

73

cannot make several worlds. (*Quod prima causa non potest plures mundos facere.*)"[1] The effect of the condemnation was to deny that the world included all possible matter within its ultimate sphere, to deny the principle upon which Aristotle based his rejection of the potential infinitely large. That is what Walter Burley explains to us with his usual clarity:

> If one admits that the addition [of magnitudes] is accomplished not by new parts, but by the indefinite addition of preexisting parts, the conclusion of the Philosopher is logical. And that is how the Philosopher understands that the addition must be accomplished, for according to him, prime matter is not capable of being generated and corrupted. Similarly, for the Commentator, any portion of matter is eternal, for any quantity of matter is either part of the celestial matter, which is eternal, or part of prime matter and inseparable from it. A new quantity of matter would therefore not be able to be produced. Hence when one wishes to add a body to another or a magnitude to another, this addition cannot be accomplished by the generation of a new portion or a new magnitude; it can only be accomplished by the addition of a preexisting magnitude. If one wishes to pursue this addition indefinitely, one would have to remove a portion from a preexisting magnitude and add it to the magnitude being formed. That is the true intent of the Commentator. . . . It is a clear consequence of what has been stated that the theologians who assert that God can create a new quantity of matter, add it to a finite body, and continue this indefinitely, would not be making use of the following proposition of the Philosopher: if a magnitude is potential by the mere addition of preexisting parts and without the generation of new parts, then there is an equal magnitude to it in actuality. . . .
>
> Certain theologians allow that God can increase the volume of heaven, that He can, for example, make heaven be twice as large, three times as large, and so forth, indefinitely, such that, given any magnitude whatever, God can create a magnitude twice it. These theologians however would deny that God can create an actual infinite magnitude, for this latter proposition may hold a contradiction; in any case, the proposition, given any magnitude, God can make a magnitude twice it, and twice the product does not formally entail the proposition, God can make an actual infinite magnitude. One might state that any magnitude that can be conceived as potential can also exist in actuality—that it would be formed by the simultaneous addition of the parts which have been

created. I assert that this proposition is false; that is not how one ought to understand the noted proposition, but as it was stated above, that is, in the following way: if a magnitude can be conceived as potential by the simple addition of preexisting parts without the creation of new parts, an equal magnitude can exist in actuality. This remark allows one to respond easily to any difficulty that one can oppose to the growth of forms to infinity.[2]

This interesting passage by Burley not only demonstrates with extreme clarity the antagonism existing between the Peripatetic theory of the infinitely large and the Christian dogma of God's omnipotence, but it also exhibits the relations between the various theses that Scholastics examined. First, since God has the power to create new bodies, one cannot refuse Him the power to produce an infinite magnitude in the syncategorematic sense of the word. Moreover, one would not be licensed to conclude immediately from that, that God can create an infinitely large body in the categorematic sense of the word, invoking, in order to justify the conclusion, the Peripatetic adage, what can be conceived as existing potentially can be conceived as existing actually. When a creative power intervenes, this adage is no longer applicable. And, in order to know whether God has the power to produce an infinite magnitude in the syncategorematic sense of the word, one would have to examine whether such an infinite magnitude implies a contradiction, since God can make anything that implies no contradiction. The new form which the Christian dogma of God's creative power brought to the question of the infinite did not appear immediately after the decree of Etienne Tempier. Henry of Ghent, who took part in the deliberations preparatory to the decree, did not, at first, perceive all its consequences. The Solemn Doctor admits, however, in a steadfast and formal manner that, if He wishes, God can create a new body outside this world:

> I say that God can create a body or another world beyond the ultimate heaven, in the same way that he created the earth within the world or within heaven, and in the same way that he created the world itself, and the ultimate heaven.[3]
> The sun contains all its matter, that is all the matter capable of receiving the form of the sun or, at least, all the matter that has been made already; however, it does not contain all that which will be made or can be made by God. That is why God can make new matter capable of receiving the form of the sun, matter which is the same as that which now

exists under the form of the sun; moreover, if He so wishes, He can make a new sun.[4]

Despite these principles so distinctly affirmed, with respect to the infinitely large Henry of Ghent maintained the conclusions of Aristotle, of Averroes, and above all, of Saint Thomas Aquinas; in fact he was visibly inspired by the teaching of Saint Thomas. In one of his quodlibetal discussions, the Solemn Doctor responds to the following question: "Should one hold that there is an infinity of ideas or notions in God?"[5] The examination of this question requires another thus formulated: "According to the essence and nature of creatures, should one assume that they, by imitating divine perfection, can surpass one another in such a way that their degree of perfection increases to infinity?"[6] Henry of Ghent allows that the progress by which the degree of a perfection increases in intensity, by which perfection imitates divine perfection more and more, is accomplished by the addition of a new form to the preexisting form. This progress *per additionem ad formam* is what Richard of Middleton, Duns Scotus, and William of Ockham and his disciples will admit. Thomas Aquinas and Giles of Rome formally denied that the progress of a form is accomplished in this manner; Godfrey of Fontaines and Walter Burley afterward held that progress in perfection is accomplished by the destruction of the less perfect form and the generation of the more perfect form. According to Henry of Ghent, the progress of a perfection is likened to the increase of a magnitude: "As we can see, there is no difference with respect to this topic between the magnitude of a body and the degree of a perfection."[7] The question is referred to a more general question, "If the perfecting of a form, of which we have spoken, can proceed to infinity, then any increase by addition, considered *absolute* and *simpliciter*, can proceed to infinity."[8] In particular, the addition of one volume to another can proceed to infinity. The problem is thus related to another problem resolved by Aristotle, and our author clearly admits the Philosopher's solution. He admits that an infinite body cannot actually exist. He admits that the addition of permanent magnitudes to one another cannot proceed to infinity if there does not exist an infinite magnitude of the same kind in actuality. He is thus led to summarize his whole argument with this proposition: "If the increase of a form can proceed to infinity, one would have to allow that the [actual] existence of an infinite body is possible."[9] Henry of Ghent's whole argument, like Aristotle's, rests entirely on this axiom: The possibility to proceed to infinity by addition assumes the existence of the actual infinitely large. Elsewhere the Solemn Doctor, like the Philosopher, admits that the division of a magnitude can proceed to infinity; however he denies the existence and the actual possibility of the infinitely small. Our author tells us why this opposition between indefinite addition and indefinite division exists:

> The Commentator teaches that potency is the essence of matter and the infinite; on the other hand, form and the finite are in actuality. The finite is therefore similar to form and the infinite to matter. That is why if we admitted that magnitude can grow indefinitely, the existence of the actual infinite will result. On the contrary, when we admit that division can proceed to infinity, no impossibility results, and here is the cause: any diminution of a real thing goes toward nothingness, and the cause of this nothingness is matter; on the other hand, any addition goes toward being, and form is the cause of being; the infinite exists entirely by matter as the finite form.[10]

Henry of Ghent maintains the essential conclusions of Aristotle's teaching; like the Stagirite he denies the possibility of infinite magnitude in actuality, and he claims to conclude from it the impossibility of potential infinite magnitude, but during his argument he abandons Aristotle's basic reason for it. For the Philosopher, the impossibility of infinite magnitude, potentially or in actuality, follows from this assertion: there is, from all eternity, a certain quantity of matter, a quantity no creative act can increase. It is there, and not in the analogy of limitation with form and infinity with matter, that the reason for the difference between addition to infinity and division to infinity resides. From the moment this reason disappeared, from the moment that Christianity recognized God's power to create new matter from nothing, the doctrine professed by the Peripatetics on the subject of the infinitely large was destroyed at its foundations. Henry of Ghent did not see this. He taught that outside the limits of this world God can create a new world, or a new stone; he did not conclude that after this stone he can create another, and another again, and so forth without limit. He did not recognize that the proposition he formulated carried with it the possibility of at least a potential infinite magnitude. He fought against this possibility, but he was one of the last to do so.

The Possibility of the Syncategorematic Infinite

The first person who taught the possibility of the syncategorematic infinite was perhaps a Franciscan who in many other respects showed himself to be the faithful disciple of Henry of Ghent, that is, Richard of Middleton. Peter Lombard wrote, "There are some who, in order to glorify themselves, are driven to restrict God's power and to assign a measure to it. When they assert that God can do this much, but not more, what is it they are doing other than limiting God's power—which is infinite— and restricting it by some amount?"[11] Richard of Middleton came to ask whether God can make an infinity by commenting upon what Peter Lombard asserted about divine omnipotence. First, he denies that God can create a being infinite in every respect, infinite such that there is nothing finite in this being.[12] But his denial does not have the same sharpness when it is with respect to "whether God can produce something naturally infinite along some dimension"[13] or, in other words, infinite in some way without being infinite in all ways. To this question, he replies that "God can produce endlessly a dimension which is larger and larger yet, but on the condition that at each instant, the magnitude at that instant is finite. That is what one commonly calls infinite in actuality mixed with potentiality or infinite *in fieri*; but it is impossible for God to produce any dimension infinite *in facto esse* or, as one currently says, infinite *in actu simpliciter*."[14] Here is, according to our Franciscan, the metaphysical reason that makes the infinite *in actu simpliciter* contradictory for any creature:

> The words, *a creature's essence*, express something indifferent between existing or not existing in fact; that is evident, for the essences of creatures, which were known to God from all eternity, can easily not exist in fact, and many essences are even now known by God to which the Creator can give or not give existence. But this indifference is determined the moment when the essence is constrained to one of the two alternatives, to existence; a dimension existing in fact receives a determination by the effect of this existence itself. It is not, in any case, a determination by which it is placed in this genus and species; as soon as any surface exists in fact, the word surface designates no less than an essence pertaining to the genus quantity. It follows from this that an essence receives from its existence a determination of the same nature that it receives from division, that is, a determination by means of the limits imposed on its length, width,

and depth. Infinity is therefore repugnant to any dimension by the very thing by which it is given existence.[15]

Richard reasons somewhat like Thomas Aquinas and Henry of Ghent against the infinite in actuality. These philosophers agree in that actual existence is a determination. According to Thomas and Henry, it is a determination that form imposes on matter; according to Richard, it is a determination that essence imposes on existence. For each of them this determination carries with it the delimitation of extension and therefore excludes infinite magnitude. But because infinite magnitude in actuality is impossible, Thomas and Henry concluded with Aristotle that potential infinite magnitude is also impossible. But Richard, while denying the first possibility, allowed that God can produce the second. To those who would accuse him of misjudging Aristotle's axiom, he replied, "Any magnitude that is potential for an object is also actual for an agent which acts by means of preexisting things. But with respect to God who can produce from nothing, this assertion of the Philosopher is no longer true."[16] For Richard of Middleton the impossibility of an actual infinite magnitude entails the impossibility of the infinite multitude in actuality:

> God cannot produce something numerically actually infinite. In fact, any infinite multitude that God can realize by means of incorporeal things, He can realize by means of corporeal bodies; but God cannot produce an infinite multitude of bodies, for from these bodies whose multitude would be infinite, He can equally produce a continuum. Thus He can produce an actual infinite continuous volume, and in the preceding question we have shown that this cannot be.[17]

As support for the opinion that infinite multitudes can be actually realized one can cite the following argument: Any continuous magnitude is indefinitely divisible; there is therefore no impossibility in conceiving it as divided actually into an infinite multitude of parts.

> When one states that any continuum is divisible to infinity, I reply that it is true as long as one understands it thus: It can be divided without end, but in such a way that the number of parts already obtained is always finite. If one admits that it is thus divided, no impossibility results; the existence of an infinite *in facto esse* does not result, only the existence of an infinite *in fieri* which one commonly calls an infinite in actuality mixed with potentiality.[18]

Richard derives from the impossibility of the infinite in actuality the conclusion that the world cannot have existed from all eternity. His argument imitates that of the Mutakallimun which Maimonides reported and discussed, and which al-Gazali accepted. It is worthy of being reported here because it became the focal point of an ardent and important discussion between those who held the infinite *in fieri* and those who held the infinite *in facto esse*.

> If it is possible that the world was created from all eternity, God could have realized an actual infinity, either in number or in magnitude. He could have similarly created men from all eternity; from all eternity these men would have engendered other men, and their successors would have done the same up to today. Since their souls are incorruptible, there would actually exist an infinite multitude of rational souls.
>
> Similarly, God could have moved the heaven continually until today, and for each of these revolutions, He could have created a stone; He could have amassed the stones together. That done, there would be an infinite volume existing actually. But in the first book we proved that God cannot produce an actual infinite multitude or magnitude. God therefore cannot have created the world from all eternity.
>
> Again, if God could have created the world from all eternity, He could just as easily have moved heaven from all eternity, continually up to the present. God could have made it be, therefore, that an infinite multitude of days had passed. But it is impossible that God could have made a multitude of past days that was infinite *in accepto esse*; it is not possible, in fact, that He produced something that has been past today, and that has not been future. He could not have produced, then, a multitude of past days that was infinite *in accepto esse* if there was no infinity of future days *in accepto esse*. But God could not have made an infinity of days be future days *in accepto esse*, but only *in accipiendo esse* or *in fieri*. Seemingly then, God could not have produced a multitude of past days that was infinite *in accepto esse*, but only *in accipiendo esse*. It remains therefore that God cannot have created the world from all eternity.[19]

There is some antagonism between these two propositions of Peripatetic philosophy:

1. The universe has existed or could have existed from all eternity.
2. Something infinitely large in actuality is impossible.

The Arab philosophers recognized this antagonism; the Mutakallimun rejected the first proposition because of the second, and Avicenna and al-Gazali, if not denied, at least restricted, the second. After Averroes and Maimonides, Thomas Aquinas and his disciples, Giles of Rome, and Godfrey of Fontaines attempted to establish an agreement, or to mitigate the disagreement between the two propositions—a chimeric enterprise. The contradiction between the two propositions became more marked, was rendered more shocking, because of the creative power Christian dogma attributes to God. In virtue of this contradiction, Richard of Middleton, taking up the thesis of Mutakallimun, concluded from the impossibility of an actual infinite to the impossibility of the eternity of the world. Others, John of Bassols, for example, deduced the possibility of an actual infinity from the possibility of the eternity of the world. Each rejected half of the Peripatetic theory. Richard of Middleton's thesis, which affirmed that God can produce an infinite magnitude, at least potentially, was too new and too audacious to be accepted immediately; even among the Franciscans, it was not accepted at first. William Varon, who cites, on the subject of the infinite, the opinion of Saint Thomas *"in scripto,"* which is to say in the *Commentary on the Sentences*, appears indifferent with respect to the general disputations about the possibility of various infinites.[20] Duns Scotus seems not to have attached any importance to it. But Peter Aureol does not seem indifferent toward this matter; on the contrary, he attaches much importance to it. It is clear that he knows the doctrine of Richard of Middleton and intends to refute it. He attempts first to establish that there cannot exist in actuality either an infinite multitude or an infinite magnitude.[21] Of all the arguments that have tended toward the same end, none have reached the precision and clarity of the one he develops:

> The nature of the quantitative infinite is a mixture of actuality and potentiality. Whoever imagines that something is composed of parts and infinite, necessarily fixes his mind on something existing in actuality; then, to that which exists in actuality, he adds something equally in actuality, and continuing in this fashion, he asserts that there is no termination to this addition. But the mind cannot actually reach that, for if it did reach it, it would by that put an end to its operation. It is therefore evident that what the mind conceives as being something formed out of parts, or a quantity,

and something deprived of termination, is necessarily a mixture of actuality and nonactuality or potentiality; in fact, whatever is said to be deprived is not in actuality.

This reasoning is confirmed by the remark that end, termination, and actuality are one and the same. Therefore, conceiving a thing as composed of parts or a quantity and denying that it has an end, is conceiving it while denying that it has a termination and an actuality; it is conceiving it as incomplete and imperfect, and what is incomplete and imperfect is potential. The nature of the infinite is therefore a mixture of actuality and potentiality.[22]

We can draw the following consequences from these principles:

Our second proposition is the following: there is always a contradiction in terms when one unites together infinity and some permanent whole—when one says, for example, an infinite multitude, an infinity of souls, an infinite magnitude, and similar expressions; there is always a contradiction if one connects two formally opposed notions—if one says, for example, a rational stone, . . . or a successive permanent. For to say an infinite multitude or an infinity of souls, is to say that something permanent is successive or something successive remains. A multitude is in fact a quantity whose parts are permanent or conceived as permanent, as it was demonstrated in the *Categories*; on the other hand, we have shown that the infinite is that whose parts are necessarily conceived as succeeding one after the other. To say an infinite multitude is to speak of a whole whose parts succeed each other and do not succeed each other, remain and do not remain; it is an evident contradiction.

It is evident that whoever conceives an infinite multitude in his mind, by uniting the two terms, unites contradictories; consequently, he conceives nothing, but forms a fiction that cannot hold together in reality or in the understanding.[23]

Here is our third proposition: an actual infinite multitude cannot, by means of any power, be posited in actuality either outside the mind, in nature, or even in any intellect, in a purely objective manner.[24] This is a clear result of what has been asserted; in fact, that which implies a formal contradiction (*directe et in primo modo dicendi per se*) is absolutely impossible. Neither divine power nor any power at all can take hold of it. . . . Therefore in no way and by no means can an infinite multitude be posited in reality.[25]

What has just been asserted in order to demonstrate the impossibility of an actual infinite multitude is also valid against the possibility of an actual infinite magnitude:

> The existence of an actual infinite magnitude is absolutely impossible in itself, even with respect to any power whatsoever. . . . The demonstration clearly results from what precedes it. In fact, an actual magnitude has all its parts in actuality; the infinite, on the other hand, is never completed—it never has all its parts because it is something successive. It possesses an actuality mixed with potentiality, as we have already explained. A result of this [the existence of an actual infinite magnitude] would be that a single thing would possess all its parts and not possess them [at the same time], which is absolutely impossible.[26]

What Peter Aureol has declared so far, Richard of Middleton would have accepted willingly, but we are at the point where they part company. Against Richard, but with the whole Peripatetic school, Aureol teaches that the impossibility of actual infinite magnitude entails the impossibility of potential infinite magnitude. In order to justify this conclusion he has recourse to Aristotle's principle: what is capable of existing potentially is equally capable of existing in actuality; but he is not able to give Aristotle's demonstration of the principle since the belief in the creative power of God renders the principle inoperative, as Richard of Middleton showed. Therefore, Peter Aureol attempts to establish the principle on new foundations which Christian dogma cannot destroy. Here is the argument our Franciscan conceives:

> It is completely impossible, and no power can add a magnitude to another magnitude of equal amount, and so forth to infinity, such that any determined and given magnitude can be surpassed. It is in fact impossible, and an impossibility results from it; for if a magnitude can be indefinitely augmented, and progress to infinity by this augmentation, as a result, an infinite magnitude would be able to exist in actuality. And this has been declared impossible; therefore it is also impossible that a magnitude can grow to infinity by addition. . . .
> In fact, if a magnitude can be augmented to infinity, it would result that an [actual] infinite magnitude is possible and does not imply a contradiction.
> This is evident if one admits that any parts of a magnitude, parts by which the augmentation of the magnitude is

possible,[27] contribute to form a unique continuous magnitude; a numerically unique magnitude can, in fact, be constituted from these parts.

And an actuality is to another actuality as the potential [of the former] is to the potential [of the latter]. If the parts of the magnitude posited in actuality constitute a numerical unity so that the latter magnitude integrates all these parts, the possibilities and powers of all these parts are integrated into a single power which is the power of the whole; in this way, in the same fashion that all the actual parts concur to form an actual unique whole, the possibilities of all the parts constitute the unique possibility of the whole.

Let us pose the question, given a magnitude, can one add to it successively an infinity of parts of magnitude?

If one responds yes to the above question, then an infinity of magnitudes is possible; not, it is true, in actuality, but potentially. It is certain, on the other hand, that if these magnitudes were posited in actuality, they would constitute a unique actual magnitude. Therefore, if one posits the possibilities of these magnitudes in potentiality, these possibilities would constitute the possibility of a unique whole, and, consequently, this whole would be something possible. But a whole constituted by an infinity of magnitudes would be an infinite magnitude. Therefore the possibility constituted by the possibilities of these magnitudes would also be the possibility of an infinite magnitude so that the infinite magnitude would be something possible. And any possibility can be reduced to actuality, as was shown by our definition of potency—if it cannot be reduced to actuality it would not be a possibility, it would be something prohibited and impossible. Actual infinite magnitude would then be possible once one admits that a magnitude can be augmented into infinity.[28]

This argumentation might appear scholarly; it is actually only a perpetual play on words that is rendered possible by the abuse of the term potential by Peripatetic philosophy. If this term generally signifies possibility, ability to be actualized, then it no longer has any meaning in the expression, potentially infinite. If we followed the advice of Peter of Spain's *Summulae* with respect to this unfortunate expression and substituted for it the expression, syncategorematic infinite, which excludes any other meaning, we would see Peter Aureol's argument go up in smoke. Unfortunately, the designations the *Summulae* proposed to introduce in the discussions on the infinite took a long time before being received by the masters of Scholasticism; this delay allowed for numerous equivocal discussions to be conducted that a language defined with precision would have rendered impossible. The use of technical language, even a barbaric one, is often the best way to avoid the interminable disputes that are the result of mere misunderstandings.

Peter Aureol was, we believe, the last of the great Scholastics who rejected the possibility of potential infinite magnitude, or according to the terminology of the *Summulae*, syncategorematic infinite magnitude. William of Ockham was the immediate successor to Peter Aureol. Before examining whether something can be augmented to infinity, William of Ockham posits a few distinctions:

> There are two kinds of increases, increase *by extension* and increase *in intensity*.
> Increase by *extension* is itself of two kinds.
> The increase of the first kind is accomplished *by* the *addition* of one part to another; the second part, together with the first, constitutes a whole, but it remains distinct from the first part with respect to its place and location (thus it is when one adds some water to water, or better, when one paints a whole body white—one paints one part, then another).
> The second kind of increase by extension is *by dilation*; it happens when some substance or quality becomes rarefied. . . . This is not a case of addition of one quantity to another as in the previous case where one part is added to another.
> Further, there are increases *in intensity* which are accomplished by the addition of one part to another; the second part constitutes a single thing with the first, but it is not distinguished from the first by its place and location (thus it is when a body which is completely white becomes more white than it was).[29]

These distinctions posited, here is the first conclusion Ockham formulates:

> The increase in extension, given the first sense of the word, can progress to infinity; given any form whatever, capable of increase, God can make a larger one.
> The reason for this conclusion depends on three propositions, of which the first is as follows: God can indefinitely create individuals of the same nature, such that for any given individual whatever, God can make another individual of the same kind. . . .

The second proposition is as follows: any individual whatever having been posited, God can make an individual of the same kind without destroying the first. . . .

The third proposition is as follows: individuals of the same kind are capable of being united to form a homogeneous whole (*aequabiliter unibilia*). Thus any mass of water whatever can be united with others forming a homogeneous whole. . . .

As a result, God cannot make a form of this kind if He is not able to make a larger one; in fact, given any form whatever, God can make another individual of the same kind, and unite this individual with the first. The resulting whole would be larger than the first. I see nothing that would prevent God from creating a drop of water and joining it to some previously made finite mass of water. . . .[30]

One sees that the authority of the Philosopher should not be accepted for this question, for he imposed a very small termination to these objects capable of increase. He posits, in fact, that there would be a contradiction in the volume of water growing to the size of the sphere of the sublunar substances subject to generation and corruption. That is what must not be accepted, because God can create another world; even better, I believe that He cannot create so many finite worlds that He would not be able to create more. There would be no contradiction in His forming a single mass of water from the waters thus created.[31]

Let us leave aside what Ockham asserted about the two other modes of increase, and let us come to what our Inceptor asserted on the subject of the actual infinite. Does the power accorded to God to produce a potential infinite have as consequence the power to produce an actual infinite magnitude? Aristotle and Averroes maintained that this is a logical consequence; must one concede that to them? Ockham replies:

It is not true that in permanent things it is possible to realize by some operation a magnitude such that there is no smaller or such that there is no larger. I assert the following truth: In permanent things divisible to infinity, as are all continua . . . one cannot assign a minimum, for however small is the given part, divine power can produce a smaller; similarly, one cannot assign a maximum, for, however large a given quantity is, divine power can produce a larger.

Must we say that however large a quantity is, it can be produced by some operation? I agree. Similarly, if one gives any state of division of a continuum whatever, it can be actualized by a single operation.

Must we say that this possibility is not only a possibility of existence *in fieri*, but also a possibility of existence *in facto esse*? If by possibility of existence *in facto esse* one understands a possibility reduced to actuality such that there is no remaining further potential, I say that the possibility of existence *in facto esse* is not the case here.

Hence, one never attains in this manner an infinity, or a magnitude which is in actuality all that it is potentially; the potential can never be exhausted such that there is no possibility of a new creation. That is how I reply to the Commentator.[32]

Ockham affirms the possibility of infinite magnitude *in fieri*; he denies the possibility of infinite magnitude or infinite multitude *in facto esse*.[33] His doctrine is exactly that of Richard of Middleton. To deny that permanent things can form an actual infinite multitude can lead to a serious difficulty, one that had already preoccupied Avicenna and al-Gazali. And Ockham does not admit the proposition that the world existing from all eternity is contradictory. Without this assumption one can prove that infinities can exist in actuality, for God could have created a rational soul for each day passed. The multitude of these souls would then be actually infinite. The Venerable Inceptor replied to this objection using the distinction between the composite sense and the divided sense of a proposition; his reply takes on a form we have not encountered with his predecessors; therefore, let us reproduce his reasoning:

The assertion that God could have produced a soul each day is true, for each of the singular propositions [making up the universal proposition] is true; but it is not a result of this that God can produce an infinity of souls, for He would have begun producing them on a particular day.

But you assert that He could have produced a soul each day; let us admit that this can happen, and that the infinite multitude of these souls is the result. I reply that one must distinguish two senses in the proposition, each day God could have produced a soul; one has to distinguish a composite sense and a divided sense. I assert that the composite sense is false; this sense must be understood as follows: the proposition, every day God has produced a soul, is impossible. This sense is false because an infinite multitude results from it. Given its divided sense, the universal proposition is true, for each of the singular propositions corresponding to it is true. But a true proposition with a divided sense affirming a possibility must not be posited as realized. For example, here is a proposition which is false in its composite sense: each of the two parts of a contradiction can be true; however, it is true in its divided sense, for each of the two singular propositions is true. One cannot posit it as realized [and assert, each of the two parts of a contradiction is true]; in fact, one of the two parts considered as realized must be denied. It is the same with the case we are considering.[34]

Let us detach Ockham's consideration from the dialectical form in which it is presented; it is reduced to the following: God could have created the world from all eternity; since an infinite multitude of permanent objects cannot exist in actuality, He could not have created rational souls from all eternity. Necessarily the creation of such souls has had a beginning. Stripped bare, Ockham's thought resembles Richard of Middleton's thought. After Ockham no master of any renown dared to deny that infinite multitudes and infinite magnitudes are absolutely impossible; all affirmed that God can produce at least a syncategorematic infinity of distinct objects and a syncategorematic infinite continuous magnitude. Athough in agreement with respect to this proposition, they then separated into two factions. One of these factions, content with having allowed the possibility of the syncategorematic infinite, refused to allow the possibility of categorematic infinite multitudes and categorematic infinite magnitudes; it maintained the doctrine of Richard of Middleton and William of Ockham. The other faction, a more audacious one, affirmed that God can produce an infinite magnitude as well as an infinite multitude, whether one takes the word infinite syncategorematically or categorematically. It seems that one must rank Walter Burley into the first faction; it is certain that Joannes Canonicus is a member of this party. He taught that "actual finitude and potential infinity suit quantity and are not repugnant to it; but actual infinity is repugnant to quantity, and God cannot endow quantity with it."[35] It is also to this party that Durandus de Sancto Porciano rallied, not without some hesitation. To the question, "Can God produce an actual infinity either of number or magnitude?" Durandus replied:

The affirmative opinion seems probable; it was adopted by Avicenna, al-Gazali, and others. It sometimes seemed acceptable to me. But there is another response which seems more probable, that God cannot produce such an actual infinite, not because of some defect in His power, but because it is repugnant to reality.[36]

In order to justify the preference he has for the latter opinion, Durandus carefully examines the arguments that had been produced for and against actual infinity. In his discussion he seems to focus on the reasonings of John of Bassols, whose words he sometimes reproduces almost verbatim. If the Dominican Doctor was tempted to attribute to God the power to produce an actual infinite, was the tempter not (John of Bassols), the Franciscan Doctor? After Durandus de Sancto Porciano we ought to study the two very wise proponents of the thesis of Richard of Middleton and William of Ockham, namely, John Buridan and Albert of Saxony. But the doctrines of these two have already weathered the attacks of the doctrines of those who held the categorematic infinite, from the first attempts in the direction of the hypothesis up to the powerful systematic thought of Gregory of Rimini.

The Possibility of the Categorematic Infinite: First Attempts

One cannot confidently place John Duns Scotus either among those who hold the syncategorematic infinite only or among those who hold the categorematic infinite also. He said but a few words about these issues and he did so without taking sides. However, in these few words are indications that were developed by those who believed that categorematic infinite multitude and magnitude are possible. John Duns Scotus first issues a remark of some importance in favor of the actual infinite: the impossibility for our minds to conceive anything other than potential infinity does not necessarily entail the impossibility of actual infinity. In particular, the Subtile Doctor seems to admit that an hour contains an infinity of actual instants, even though our minds can conceive only a potential infinity of indefinitely decreasing parts. Duns Scotus also has something to say about the argument, so frequently used, that if the infinite existed, a part would be equal to the whole (and other similar arguments); he observes that several of these arguments are purely sophistical. He asserts the following principle, which John of Bassols treated with scorn, but to which Gregory of Rimini attached the greatest importance:

> The words, *equal, greater,* and *smaller,* are not suitable for large quantities unless finite. In fact, before one can apply

the words equal or unequal to a quantity, one has to divide it into a finite quantity and an infinite quantity. The reason by which a quantity is greater than another lies in the fact that it exceeds; the reason of equality, in the fact that it has the same measure (*commensurari*). Everything indicates that these concern finite magnitudes. One must therefore deny that an infinity can be equal to another infinity; *more* and *less* also designate differences between finite quantities, and not between infinite quantities.[37]

By these various remarks Duns Scotus blazes a path for those who wish to uphold that actual infinite multitudes and magnitudes are possible. Therefore one should not be surprised if among them one encountered two of the more eminent disciples of the Subtile Doctor, Francis of Mayronnes and John of Bassols. But before elaborating upon the doctrine of these two Scotists, we should summarize the doctrine of another proponent of actual infinity, the Carmelite Friar, John Bacon of Baconthorpe. The theories we are reviewing will then succeed each other in an order of increasing perfection. If one is to believe an ancient genealogy of the Bacon family, John Bacon of Baconthorpe was the third son of Sir Thomas Bacon of Baconthorpe, and grand nephew of Roger Bacon. He entered the order of the Carmelite Friars. In 1327 we see him take part in the General Chapter held by the order at Albi; in 1329 a Provincial Chapter held at London elected him prior of the province of England. It was with the title of provincial prior of England that he took his place in the General Chapter held at Valence in 1330. He remained provincial prior of England until 1333, the year he took part in the General Chapter of Nîmes. He died at London in 1346 and was buried in the Church of the Carmelites.[38]

We possess a voluminous commentary on the *Sentences* of Peter Lombard by John of Baconthorpe. This commentary was fashionable and had great authority among the Scholastics. The principle by which John of Baconthorpe intends to establish the possibility of the actual infinite is as follows: The world could have been created from all eternity, and God could have created a stone and maintained it in existence every day; therefore the multitude of these stones can be actually infinite. Some objections have been formulated against this argument, but our author finds none that carry weight; thus he concludes in this fashion:

For this reason I say that the philosopher and the theologian would speak differently on the subject of the eternity of the world.

The philosopher asserts that the world has existed from all eternity and that an infinity of individuals has preceded the present moment; but this production of past individuals was accompanied by the continuous destruction of these same individuals. These individuals were not conserved in existence, for that would imply an impossibility for the world. [Actual] infinity therefore does not result from this opinion.

For his side, the theologian would say that if the world had been from all eternity, an infinity of individuals would have preceded the ones existing today; but since he assumes that God could have preserved these individuals, the theologian would admit the possibility of the infinite in actuality, at least of past infinity and not of absolute infinity which is both past and future. . . .

Necessarily the theologian must agree that future infinity is not actual.

Let us assume that in the world as it is now there is an infinity of stones, produced continually from all eternity and preserved in existence until now. Let us also assume that the world is not limited by differences in space, that there is no void outside the world. Some of these stones could then have been created outside the world. Once this first empty space is filled with stones, if there is, outside it, another empty space, new stones could be created outside the first empty space now filled; and so on to infinity.

And theology supposes something like this; beyond the infinite multitude of these stones which have been preserved indefinitely, other stones can be created. Let us prove the minor premise: Let us assume that the world is already full of stones. Doubtless, outside the world there is no void; but it is true that outside the world there is nothingness and the negation of being. Similarly, when God alone existed, outside God there was no void, but it is true that there was, outside God, nothingness, a negation of all being. That is why it was possible for God to create the world from the beginning; it is because there was outside Him only nothingness and the negation of being. If we assume that the outside of this world can be filled by the stones that have been preserved, that form an infinite multitude in the past, there will still be nothingness and negation of being outside the world filled with stones; therefore God can create other stones, and so forth to infinity.[39]

We have just read the essential propositions of John of Baconthorpe's doctrine. What should concern us now is the defense against the objections that, from time immemorial, have been pitted against the possibility of the eternity of the world, or the categorematic infinite. However, in this defense, our Carmelite Friar shows himself to be an awful logician. Let us give an example. If the world existed from all eternity, said the Mutakallimun (and all the adversaries of this position after them), the number of revolutions now accomplished by the sun would be infinite; the number of revolutions accomplished by the moon would also be infinite. However, the latter number would be greater than the former number; there would then be two infinite numbers of which one is greater than the other. What does our author find to say about this classic objection? He responds that the revolutions of the sun "would not be infinite in number . . . for, being less numerous [than the revolutions of the moon] their number could increase by addition, which is an impossibility for an infinite number."[40] There is more logic in the considerations of Francis of Mayronnes on the possibility of the actual infinite. Let us indicate summarily what Francis of Mayronnes said about actual infinite multitude first, and then what he said about actual infinite magnitude.

Our author gave several reasons in support of God's being able to produce an actual infinite multitude of distinct and subsisting objects; among them is the following: "Given any finite multitude of individuals whose simultaneous production is not repugnant, God can produce these individuals all together; it is not possible to assign to the number of these individuals so great a value that God would not be able to produce as many. It seems then that God can simultaneously produce an infinity of such individuals.[41] The above reasoning would not be conclusive if one did not admit the following principle: The possibility to produce a syncategorematic infinite multitude entails the possibility to produce a categorematic infinite multitude. We shall see that this principle conforms with the thought of our author when we read what he wrote about infinite magnitude. That God can create an actual infinite magnitude results from the possibility of infinite multitude: "It has been demonstrated that God can create an actual infinite multitude; let us suppose that He has created an infinity of drops of water; He can unite these drops together, and consequently He can create an infinite mass of water."[42] But the possibility of an actual infinite magnitude can also be established directly, in the following manner:

It is not possible to progress to infinity in the domain of finite things; therefore if God cannot create an infinite magnitude in one blow, one could assign a magnitude such that God can create nothing greater in one blow, which is false. . . . For however large something created by God is, He can still create something larger,[43] which is not possible if He can only create a finite magnitude.[44]

Our author develops the same argument in another form:

A magnitude surpassing any finite magnitude is infinite; but the ultimate magnitude that God can create in one blow surpasses any finite magnitude. It is then infinite. Proof of the minor [premise]. Given any finite magnitude that God can create, He can create a greater one; that which surpasses all is therefore infinite.[45]

This reasoning is taken up again in the form of a dilemma:

The ultimate magnitude that God can produce is either finite or infinite; if it is infinite, the proposition is demonstrated. Let it be finite; that is impossible, for the termination [of the progression to infinity of magnitudes able to be created] can be located just as easily at another degree of magnitude, since these degrees are all of the same nature.[46]

The thought behind this reasoning is easy to understand and to exhibit; it is the naive thought of many a student beginning to think about the infinite. Francis of Mayronnes sees a tendency toward a limit in the series of finite magnitudes increasing indefinitely; this limit he names the ultimate magnitude that God can create (*illa magnitudo quam Deus ultimate simul potest facere*). Necessarily, it is a categorematic infinite magnitude. Ockham would have simply responded to Francis of Mayronnes that there is no ultimate magnitude among the magnitudes God can create; he would have thereby demonstrated the error contained in this new reasoning intended to restore the Peripatetic axiom, the possibility of potential infinite magnitude implies the possibility of actual infinite magnitude. Francis of Mayronnes was more inspired when he attempted to show the absurdity of the reasons one gave in order to convict actual infinite magnitude or multitude as guilty of contradictions. It is true that he derives his inspiration from the remarks formulated by Duns Scotus. "A multitude of things able to be created is a number if it is finite; but it is not if it is infinite—such a multitude is not a number—it does not belong to the genus of discontinuous quantity if it is not reduced."[47] This last expression means that one can define, under the genus of discontinuous quantity, an extended genus that can be immediately subdivided into two other subgenera, discontinuous infinite quantity and discontinuous finite quantity; only the latter is able to be called number. It is also under the subgenus of discontinuous finite quantity that the various notions of arithmetic are suitable; and it is so for the first such notions, the notions of equality and inequality: "equal and unequal are properties of finite multitude, and more and less are the two species of inequality. The two properties, equal and unequal, cannot therefore belong to infinite multitude."[48] Here is the real reason for the above assertion:

> Multitude is divisible into finite multitude and infinite multitude. The genus of finite multitude contains an infinity of species of numbers. The genus of infinite multitude contains a single member which is not a species, properly speaking. Similarly, if one divides being into finite and infinite being, only God will be contained under the title of infinite being. Besides, within the genus of finite multitude is an infinity of species of number, but the multitude of these species is not contained in this genus [since it is an infinite multitude]."[49]

Because Francis of Mayronnes thinks that there are no diverse species of infinite multitude, that all infinite multitudes are of the same species, it is clear that the words greater and smaller cannot be used in the comparison of these multitudes; it is also clear that no multitude can be augmented or diminished: "When one takes away any finite multitude whatever from an infinite multitude, there remains an infinite multitude which has not been diminished in any way.[50] Because of the principle he admits, Francis would not dream of creating an arithmetic of infinite multitudes based on the arithmetic of finite multitudes; the various species of numbers are in fact the grounds for this latter science. Francis of Mayronnes admits, although he does not state it, that continuous magnitude can also be subdivided into finite continuous magnitude and infinite continuous magnitude; he takes care to affirm that the many properties of finite magnitude must not be attributed to infinite magnitude:

> An infinite volume has no parts and consequently no aliquot parts. . . . A part is always something related to the whole. If one wishes to understand that in an infinite volume there is some portion of matter such that this infinite is a

whole with respect to this portion, I would reply that, in this sense, there is nothing in this continuum which is a part. If one wishes to speak about a part, not of the infinite but, on the contrary, of a certain [finite] volume included in the infinite, I would state that, in this sense, there is, in the infinite, a multitude of things that can play the role of part, and a multitude of things that can be wholes.[51]

It is not continuous magnitude in general, but only finite continuous magnitude that can have shape, place, and motion; one should not worry about objections such as these: An infinite body cannot have shape, position, or motion. In the same way that there are not several species of infinite multitudes, according to Francis of Mayronnes, is it not true that there are not several species of infinite volumes? He does not assert this, but doubtless he holds it; infinite volumes cannot be larger or smaller than each other. That is what explains the following:

> After having created an infinite quantity of water, can God create another quantity of water? . . . I reply that, beyond all this water, God can create another quantity of water, as we have asserted about infinite multitudes. But after having created an infinite quantity of water, can He still create another infinite quantity of water? I reply affirmatively, for He can still create another finite quantity of water,[52] another greater quantity, and so forth to infinity, as it has been already demonstrated for the first infinite water. . . . Is there a termination above which He cannot add more? I reply negatively, He would have encountered a termination in the production of the singly infinite magnitude; in fact, an infinitely infinite magnitude does not exceed a singly infinite magnitude.[53]

Francis of Mayronnes foresees an objection to his doctrine affirming the possibility of actual infinity for continuous magnitudes and for collections of distinct objects: "But how is it that infinity is something whose magnitude always leaves something outside it capable of being taken up?" He answers: "That is a definition relative to our intellect; when it takes up a multitude, no matter how large, there always remains something else to take up; in fact, it takes up a finite multitude, and never an infinite multitude."[54] That is a response in which the inspiration of Duns Scotus is capable of being detected. However, we no longer recognize the spirit of the Subtile Doctor in the final reflection of Francis of Mayronnes on the subject of actual infinite magnitude. This reflection, in any case, seems to go against the general tenor of the theory:

> Can God divide, in one blow (*simul*), a continuum into an infinity of parts? I reply negatively. In fact, the infinity of the parts of a continuum is not something immediate, as would be infinite multitudes and magnitudes in which God can produce an infinity of parts in one blow; on the contrary, the division of a continuum progresses indefinitely. It is by following an order that it passes from larger parts to smaller parts; therefore, this infinity is a successive infinity.[55]

John of Bassols's doctrine on the actual infinite is a close relation to Francis of Mayronnes;' it keeps the best parts of the latter, strengthens and renders it more precise, at the same time that it abandons some of its errors. John of Bassols admits plainly, as does Peter Aureol, Aristotle's axiom that potential infinitely large entails actual infinitely large. Although beginning with the same axiom, his reasoning is completely opposed to Aureol's. As a Christian, Bassols believes in the creative omnipotence of God, so that he cannot think of the production of the potential infinitely large as impossible; therefore, he cannot consent to actual infinity being contradictory, and declares that God can create it. The following passage enables us to see John of Bassols's train of thought:

> A quantity surpassing any determined magnitude is an actual infinite quantity; but given a quantity of determined measure, one can give a larger one—one can give an actually infinite quantity. Allow me, for example, any length you wish; two feet, for example, or three, or any other such particular measure; there is nothing it seems which is repugnant to my being able to give a greater, not only potentially and *in fieri*, but in actuality. Length, in fact, does not assign to itself such a determined measure. One can invoke Aristotle's assertion from the third book of the *Physics* to support this reasoning: if a magnitude can be increased indefinitely, it can be actually infinite. But a magnitude can be increased indefinitely, for if we are given any kind of creature, or some individual of any kind of determined species, God can produce a similar second creature, or a second individual of the same kind, and add it to the first creature or first individual. This assertion is confirmed by Aristotle himself, in his work, *De lineis indivisibilibus*, for there he teaches that any magnitude, as

long as it is finite, can be made to touch another magnitude and to prolong it; in the same fashion, one can progress indefinitely in the series of numbers. It is the same with forms, etc.[56]

John of Bassols claimed that actual infinity implies no contradiction, that God can give it existence. But first he posited a distinction:

> Actual infinity can be understood in two ways:
> First one can understand, by these words, absolute infinity (*simpliciter*), infinity according to all manners of being and all perfections.
> One can understand them as infinite not in all manners of being and all perfections, but only according to some manner of being or perfection of a special nature . . . for example, infinite length or some similar attribute.
> God cannot create an actual infinity in the first sense of the word, for there can be no other God [and that infinite would be God].[57]

But it is not the same with infinity taken in the second sense of the word; among the various kinds of infinity the second sense implies, there are four whose existence implies no contradiction and therefore can be realized by God. These are: infinite in geometric magnitude (volume, surface, length); infinite in number; infinite in intensity or magnitude of a nongeometric perfection or form (heat, for example); and finally, infinite in force (*virtus*).[58] The power to realize an actual infinity is reserved to God, in any case; no natural agent can produce such an effect:

> The increase of a magnitude progresses or can progress indefinitely; hence an infinite magnitude, potential as well as actual, can be produced by divine virtue, not by actual virtue. If the natural forces act alone, a limit is imposed on magnitude and its growth.[59]

The direct argument that John of Bassols uses in support of his thesis is always Aristotle's axiom, potential infinity would not be realizable if actual infinity were not. And potential infinity cannot be thought as contradictory in the case of continuous magnitude or number; any continuous magnitude, any number, can always be surpassed by another magnitude or by another number. But Bassols adds indirect arguments to this direct argument; he attempts to resolve the contradictions Aristotle and other philosophers thought to have discovered in the assumption of an actual existing infinite magnitude or infinite number. Moreover, he does resolve them, and with much wit, exposing the paralogism which is almost always at the bottom of these kinds of objections. For example, against actual infinite magnitude, a multitude of alleged impossibilities are derived from the shape one attributes to the body in which this magnitude is realized; the necessary existence of such a shape furnished Saint Thomas his greatest objection. But why must one attribute a shape to an infinite body?

> It is absolutely unnecessary for a body to be terminated and to have a shape; thus an infinite body has no shape, unless one prefers to say that its shape is infinite like its magnitude. But in this case, one must add that the definition of shape from which one draws these impossibilities is suitable only for finite shapes.[60]

Aristotle raised against actual infinity an objection derived from the impossibility for one to attribute finite parts to it. John of Bassols destroys this objection by means of a very simple remark: "The infinite has [finite] parts that are not aliquot parts. By taking any determined number of these parts whatever, it is always impossible to reproduce the whole."[61] Another common argument was as follows:

> From an actual infinite magnitude, it is possible, at least by means of God's power, to detach a first part of one foot, for example, or of two feet; I ask then, if the remaining part is finite or infinite. One cannot say that it is infinite, for, since the whole is greater than its parts, and since an actual infinity is thus given, another thing of the same kind could be greater—which is false and absurd. One cannot say that it is finite either, for with two finite magnitudes, one cannot form an infinite.[62]

Bassols replies:

> When you say, an infinite can therefore be greater than another infinite of the same kind, I reply that there is no difficulty with that unless it concerns infinity considered as absolute (*simpliciter*), which is infinite in all ways and respects; it is thus that a line having no eastward or westward termination would be greater than a line unbounded on its eastward side, but having a termination on its westward side.[63]

Educated on the most subtle dialectic by Duns Scotus, Bassols does not hesitate to point out inconsistencies even in the reasoning of the Stagirite. He goes further; he accuses the Philosopher of contradicting himself by denying the actual infinity of number:

> If Aristotle had put all his principles together, he would have admitted the actual existence of infinite number. In the eighth book of the *Physics* he admits that the world is eternal and that men engender themselves from all eternity. Secondly, he admits that the rational soul is the form and actuality of the body; the number of souls is therefore precisely the same as the number of human bodies. One can see that he has not admitted the absurd opinion upheld ever since the Commentator, the opinion according to which there is only one intelligence for all men. Had he admitted it, I contend that one could have shown him the contrary, once one assumed or demonstrated that the soul is the form of the human body. Thirdly, in the first book of the *De Anima* and in the second, third, and sixteenth book of the *Animals*, he admits that the soul is incorruptible, that it differs from that which is corruptible and extrinsic by its perpetuity. From these three propositions follows this inevitable consequence: The multitude of human souls is infinite. Therefore, if Aristotle attempts to deny the possibility of actual infinite number in the third book of the *Physics*, as maintains the Commentator, it follows that he contradicts himself, and that one can draw from his statements either of two assertions: There is an infinite in actuality. There is no infinite in actuality.[64]

Of the philosophers who wish to deny actual infinite magnitude and actual infinite number, almost all connect the impossibility of this infinite to the impossibility of an actual infinite division of a finite magnitude. If God can actually realize an infinite multitude, He can actually divide any finite magnitude into an infinity of indivisibles. Bassols, like these philosophers, denies that a finite magnitude can be actually divided to infinity; but he also denies that this impossibility carries with it the impossibility of actual infinite multitude:

> The division of any finite quantity into parts whose magnitudes follow a constant relationship can be pursued to infinity. It is the same with the increase of a quantity by the addition of similar divisible parts. Divine virtue itself cannot reduce this division or this increase to actuality *in facto esse*, but only *in fieri*, and this is because the reality or nature of things repulses this actualization. But this in no way constitutes an objection to our proposition.[65]

John Bassols's thought is, in this instance, very similar to

Francis of Mayrones'. Their thinking is analogous in many circumstances; there are few questions for which these two disciples of the Subtile Doctor are in disagreement. Mayronnes, following an indication given by his teacher, does not wish that the words, less, more, equal, and unequal, be applied to infinity; two infinites of the same nature, of which one is greater than the other cannot exist—there is only one infinite of each kind. That is not the opinion of John of Bassols; he sees no difficulty in admitting that two infinites of the same kind can be unequal. He knows, and even recalls, the principle formulated by Duns Scotus and adopted by Francis of Mayronnes: The comparison between greater and lesser quantities can only be accomplished between finite quantities. But he treats this principle with disdain. He states, "I do not heed this principle (*sed non curo*)."[66] Scotus's two disciples each perceived a facet of the truth. It is certain that if one wishes to admit the possibility of categorematic infinites, one must conceive relations between them as similar to those holding between finite quantities and expressed by the words, equal, greater, smaller, all, part, double, triple, etc. But, on the other hand, it is certain that one cannot always reason about these things as if they were relations that held between finite quantities. It is with some hesitation that we place Robert Holkot in the same category as John of Baconthorpe and Francis of Mayronnes, among the precursors to Gregory of Rimini. If, as it is asserted, the Dominican, Robert Holkot, died in 1349, it is probable that his *Questions on the Sentences* were redacted before 1344, when Gregory completed his. Further, what Robert Holkot asserts about infinity resembles passages written by Gregory of Rimini on the same topic;[67] in the writings of the two masters we often encounter similar thought expressed with the same language. The chronology suggests the following explanation: The shorter and less perfect account of the Dominican preceded that of the Augustinian and was its inspiration. We do not believe that this opinion, which seems so natural, can resist even casual reading of the texts. Compared with Gregory's theory, Robert's theory does not exhibit the kind of imperfection generally exhibited by the work of a precursor when one compares it with the completed work of the final innovator; its defects are of another kind—those of obscurity and disorder—the incomplete and indecisive thoughts of one who received them from another and did not understand them sufficiently. It seems as though Robert Holkot did not bother to penetrate completely the sense of the affirmations he adopts, nor did he fix firmly his convictions with respect to them; often, the doctrines he professes would be difficult to understand if one did not, in order to interpret them, have recourse to the clear and rigorous discourse of Gregory of Rimini. It would therefore not surprise us if the *Questions on the Sentences* attributed to Robert Holkot were more recent than one believes and the influence of Gregory had inspired them. Be that as it may, we provisionally leave this work in the place where the received chronology assigns it. The occasion Robert Holkot uses to develop his doctrine on the infinitely large occurs when he deals with the question: Can God have produced the world from all eternity? The Dominican Doctor holds for the creation of the world *ab aeterno*. Let us cite some of the objections he attacks and the responses by which he attempts to refute them. Here is the first objection: "It is repugnant to infinity to be surpassed; and if the world had existed from all eternity, an infinite multitude would have been surpassed. In fact, an infinite multitude of men would have already died, each of whom could have been a future man; the multitude itself could have been future while it is now past. An infinite multitude would therefore have been surpassed."[68] With a clarity worthy of Ockham, who is reputed to have been his teacher, Holkot bares the confusions such reasonings draw upon the verb "to surpass." At each instant of time the number of men already dead would be infinite, while the number of dead men between that instant and the actual instant would be finite. Therefore, if one understands by "to surpass" an operation having a beginning and an end, one cannot state that the proposition, the world has existed from all eternity, implies the proposition, an infinite multitude could have been surpassed. But our author adds: "One says that it is repugnant to infinity to be surpassed. . . . I assert, on the contrary, that there is no inconsistency in asserting the proposition, [an infinite multitude can be surpassed]; . . . every time that any amount of time whatever has flowed, an infinite multitude has been surpassed. In the same way, if a magnitude, however small, has been surpassed, one must concede that an infinite multitude has been surpassed, for any magnitude is an infinite multitude.[69] This overly brief reply becomes clearer when one relates it to the teachings of Gregory of Rimini; like him, and unlike Francis of Mayronnes and John of Bassols, Robert Holkot must allow that any limited duration, that any finite magnitude, can be considered as an actual infinite multitude of infinitely small parts. Whatever is the thought of the Dominican Doctor, we can understand it better by looking at his reply to another objection, which is as follows: If the world had existed from all eternity, "God could have created a soul each day, and preserved it; there would then exist a multitude of souls today which is infinite and in actuality"[70]—which is absurd. Holkot replies:

> This consequence relative to the actual existence of a multitude of souls can be conceded, as long as one distinguishes between actual existence and real and true existence in this world. In every continuum there is an infinity of parts distinct one from the other because of their locations, for example. . . . However, the set of these parts constitutes a unique whole. And Aristotle, in the third book of the *Physics*, calls this infinite multitude a potential multitude, because in his language, anything that is part of another is said to exist potentially.[71]

Holkot then ridicules Aristotle's theory on this topic; if one accepts it, then the sun would exist only potentially, since it is part of its orbit. "I believe, nevertheless, that in Aristotle's philosophy, there can exist no infinite multitude in actuality."[72] One could also formulate the following objection to the eternity of the world:

> It is repugnant for infinity to be surpassed; and if the world had existed from all eternity, there would be an infinite multitude surpassing another infinite multitude. In fact, there would be a greater number of fingers than men, and a greater number of revolutions of the moon than of the sun.[73]

To which Holkot replies:

> I deny that the infinite cannot be surpassed without contradiction . . . as for the proposition formulated in the proof, that there would be a greater number of fingers than men and a greater number of revolutions of the moon than of the sun, one can reply to it by denying it. With a thousand men, there is a greater number (*plures*) of fingers than men; but with an infinity of men, there is no greater number (*plures*) of fingers than men, because there is an infinity of men and an infinity of fingers.
>   Others express themselves otherwise; they state that an infinite multitude can be larger than another—they concede that there is a greater number of revolutions of the moon than of the sun . . . that an infinite multitude can be double or triple another . . . that one can add something to the infinite. That is the opinion Robert of Lincoln expresses in his writing on the *Physics*.[74]

The last assertion seems completely false; Robert Grosseteste says nothing not purely Aristotelian on the subject of the infinite in his *Summa* (which is so concise, so full of ideas), and nothing in particular that resembles what Holkot attributes to him. Of the two replies that our author has reported, the first was favored by Francis of Mayronnes, and the second was favored by John of Bassols. Holkot did not declare any preference between the two; he seems, however, to lean toward the first. Here is an occurrence where he expresses the same viewpoint that Francis of Mayronnes had expressed: "The sixth objection states that if there were an infinity of souls, God could not create a greater (*plures*) number of souls. I agree with this proposition, taking it literally (*de virtute vocis*). God cannot create a greater number of things (*plures res*) than He has created; but He can create other souls, even though there is already an infinity of them."[75] We also hear him contradict the language that Bassols used. One can formulate the following well-known objection against the possibility of the actual infinite: a result of the language would be that a part might not be smaller than the whole. Holkot does not hesitate to affirm the proposition; he thinks it evident when comparing a straight line extending to infinity in one direction with one extending to infinity in both directions. Bassols asserted that the former was smaller than the latter.

It is possible that the reasonings developed by Robert Holkot might have preceded those which Gregory of Rimini developed; but Gregory also could have discovered, within his own order, an influence that might have led him to admit the possibility of the categorematic infinite, that is, the opinion held by the general prior of the Hermits of St. Augustinian, Thomas of Strasburg. Thomas, who was of German descent, first wore the robe of the Hermits at the convent in Strasburg, and there began the studies he completed at Paris, according to the custom of the time. In 1345, the General Chapter of the order held at Paris elevated him to the generalship after Denys of Modena. He presided over three other General Chapters, that of Pavia in 1348, Bâle in 1351, and Pérouse in 1354. He died in 1357 at Vienne (Isère); during that same year, the General Chapter of Montpellier named Gregory of Rimini as his successor. We have a voluminous commentary on the *Sentences* of Peter Lombard from Thomas of Strasburg, a commentary often printed; and in this commentary Thomas teaches that God can create an infinite length and an infinitely large body. His philosophy, which is somewhat scanty, does not bother with such distinctions as infinite in actuality and potential infinite, categorematic infinite, and syncategorematic infinite; he does not even mention such distinctions, but it is evident that he intends to refer to the infinite in actuality and the categorematic infinite. In support of the opinion he maintains, our author does not have recourse to any sophisticated reasoning:

> If God made an infinite length having no termination, its length would not have the nature of line any less than a line terminated and limited by two points. It is not therefore the nature of line *qua* line which is limited in length, it is only the nature of line *qua* finite. Similarly, being limited in length, width, and depth is not of the nature of body *qua* body, although it is of the nature of body *qua* finite.[76]

Thomas does not shy away from illuminating the obscurities one encounters in the notion of categorematic infinite either. Somewhat earlier in the work, he touched upon one of these difficulties when he taught that the world could have been created from all eternity. He had then encountered the objection that if the world existed from all eternity, the number of revolutions accomplished by the moon and the number of revolutions accomplished by the sun up to today would be infinite, yet the former number would be greater than the latter number. Thomas of Strasburg gave to this classic objection a reply that seems to have been no less classic at the time: "Doubtless one can add nothing to the infinite *qua* infinite; but if an infinite is finite in some respect, it can be added to in that respect. . . . To such an infinite which is finite in some respect, one can introduce the considerations of greater and smaller, not by reason of its infinitude, but by reason of its finitude."[77] We will have dealings with a logician of another caliber with Gregory of Rimini.

It seems that during the time immediately preceding 1344, when Gregory of Rimini wrote his commentary of the first two books of the *Sentences*, the possibility of actual infinity was commonly conceded. We have seen this with respect to the work of the Carmelite Friar, John Baconthorpe, the Franciscans, Francis of Mayronnes and John of Bassols, the Dominican, Robert Holkot, and the Augustinian, Thomas of Strasburg. We shall also hear this almost unanimous agreement affirmed by the Franciscan, Nicholas Bonet. "The possibility of actual infinity does not seem to hold any contradiction for modern philosophers (*modernis philosophis non apparet aliqua impossibilitis quin sit possibilis infinitas actualis*),"[78] said Nicholas Bonet. Bonet gives to the affirmation of the possibility of actual infinity the greatest possible extension that can be given to it.

There are two ways of understanding actual infinity.

According to one, there are not so many objects that there cannot be more (*unus quod non sint tot quin plura possint esse*). A multitude of objects thus understood can nevertheless be infinite; thus it is that we assume that there is an actual infinity of stones or donkeys and that other stones or donkeys can however be produced. In such an infinity there are not so many objects that more cannot exist. The second way of understanding actual infinity is that there are so many objects in actuality that there cannot be more, because they are all posited in actuality (*quod tot sint in actu quod non possint esse plura, quia omnia sunt actu posita*). Thus it is if all possible stones are posited in actuality simultaneously, in such a way that it would be impossible to admit the existence of a new stone not included within the set of the stones already posited in actuality.

Infinity conceived in the first way appears possible, but so does the infinity conceived in the second way. In fact, if simultaneous actual existence is not repugnant for two or three objects, if there is no more repugnance for the coexistence of a greater number of objects than for a smaller, then there is no repugnance for the coexistence of an infinity of such objects nor for the universality of these objects. (*Si duo vel tria non habent repugnantiam existendi in actu, nec major pluralitas quam minor, concluditur quod nec tota universitas istorum nec infinitas.*)[79]

What is true for any finite number of objects is true for an actual infinity of these same objects. Such is the principle Nicholas Bonet likes to invoke. Basically this principle is only Aristotle's axiom, potential infinity requires actual infinity. But Aristotle used this axiom in order to deduce the impossibility of potential infinity from the impossibility of actual infinity; Nicholas Bonet, like John of Bassols, uses the axiom in the opposite way, to conclude for the possibility of actual infinity from the possibility of potential infinity. That is the reasoning allowing Bonet to formulate this daring proposition: There is no impossibility for the coexistence of an actual infinite set of causes related according to an essential order.

Let us first assert that causes related according to an essential order can exist together in actuality. Here is the reason for this assertion: They are permanent beings; if two or three of these beings can exist simultaneously in actuality, one can conclude from that that it is the same for all. In fact, if there is no contradiction when one posits in actuality simultaneously a lesser number of these beings, then there is no contradiction in a greater number. That is how the philosophers reason when they assert, if two bodies can coexist in the same place, then so can a hundred bodies and an infinity of bodies.[80]

Avicenna and al-Gazali affirmed that causes can form an actual infinite set of causes when there is only an accidental order between them; they denied that it is the same with causes whose order is essential. Nicholas Bonet attempted to show them that the proposition they rejected is implied by the one they accepted.

There can be an [actual] infinity of causes ordered essentially. This proposition is proven by the following reasoning:

An infinity of causes whose order is accidental can exist simultaneously. However, it is the same with causes whose order is essential.

Let us prove the premise: All schoolmen allow that causes whose order is accidental can exist successively in infinite numbers; they can therefore exist simultaneously, for they are permanent beings of which two or three can coexist. The following is an example derived from those who admit that the world has existed from all eternity.

According to these philosophers, a man, engendered in such a fashion, has been preceded by an infinity of successive men; this man has been engendered by that man, and that man by a third, and so forth to infinity. . . . Further, there is no repugnance in that these three, four, or ten men who have been engendered by each other be able to exist simultaneously in actuality. Hence, there is no repugnance in that an infinity of men, in that all men, be able to exist simultaneously in actuality either, since they are all of the same nature, and this nature is that of permanent beings.

Hence, if an infinity of causes whose order is accidental can exist successively, they can exist simultaneously. Our premise is proven.

Let us now justify our reasoning:

The causes whose order is essential are permanent beings just like the causes whose order is accidental; therefore, if coexistence is not repugnant to an infinity of causes ordered accidentally, it is not repugnant to an infinity of causes ordered essentially either.[81]

After having treated the infinitely large in his *Physics*, Bonet also took up the same question in his *Natural Theology* in order to define the power of the Prime Mover; his conclusions were no less absolute in the latter than they were in the former.

> The various senses of the word infinite are divided between infinite in intensity (*infinitum virtutis*) and infinite in magnitude (*infinitum quantitatis molis*).
>
> First, one calls potential infinity infinite; there what is taken is always finite. However, what is left to be taken is infinite. Aristotle speaks of this infinite in the third book of the *Physics*. . . .
>
> Second, one calls relative infinity in actuality (*secundum quid*) infinite. There what is already taken is not finite, but infinite; however, there remains something to be taken up, so that it is not an absolute infinite in actuality (*simpliciter*).
>
> Here is an example: Let us imagine a straight line infinite in two directions, and let us cut it at a point; it is evident that each of these two parts are infinite in actuality. . . .
>
> The third kind of infinite is as follows: One says that a thing is infinite in actuality according to the degree suitable for its own species; that thing is infinite in such a way that what is already taken is infinite in actuality and there is nothing left to be taken—there is no new degree to be added to that species. Such an infinite contains, in itself and actually, all degrees that it is possible to posit in actuality [for that species].
>
> Here is an example: Let us imagine a straight line actually infinite in two directions, constituted by all the straight lines it is possible to posit in actuality, in such a way that all the lines it is possible to posit are contained in this infinite straight line. Such a line would be infinite in such a way that no other line can be added to it, since all lines are already contained in it.[82]

The Prime Mover can produce either of the two kinds of actual infinities which in any case are concerned with infinite in number or infinite in continuous magnitude.

> We shall formulate the first proposition as follows: The productive force of the Prime Mover extends to the production of infinite in number (*infinita extensive secundum multitudinem*). Here is the proof.

Things that are not repugnant with respect to existing and being simultaneous, are not repulsed by being produced simultaneously, particularly since the production is the work of a power of infinite dimension.

But the singular individuals of the same nature whose multitude is infinite have no repugnance toward existing and being simultaneous, for there is no opposition between them.

The power of infinite dimension can therefore produce a multitude of individuals of the same kind.

One can prove the minor premise, that the singular individuals whose multitude is infinite do not repulse coexistence, in another manner: where there is no greater repugnance in a greater number than in a smaller, there is no repugnance toward an infinite multitude.[83]

One should not think that God's power is limited to the production of the infinite multitude in actuality which is only relative (*secundum quid*); it extends also to the production of the infinite multitude in actuality which is absolute (*simpliciter*). Bonet is careful to affirm this when recalling the distinction between the two infinities in actuality in the second book of his *Natural Theology*.[84]

Here is the second proposition: The Prime Mover can produce an infinite magnitude. Let us prove it.

The proof is apparent through our first proposition. In fact, if a numerically infinite multitude can be actual, it is the same with a magnitude. Let us prove the legitimacy of this deduction:

One can form a quantity, a magnitude from this infinite multitude [of objects], and it is evident that the magnitude resulting from this multitude of partial, numerically distinct quantities, would be an infinite magnitude.

Here is an example: From an infinite multitude of lines two feet long numerically distinct from one another, one can form a single continuous line by combining all these lines; one clearly sees that this line is not finite, but infinite.

This consequence is necessary: If an infinite multitude of finite quantities (*quanta*) is possible, an infinite magnitude is possible. Moreover, there can be a body of infinite magnitude; similarly, there can be an infinite line in the same fashion that there can be an infinite surface or infinite body.[85]

No one has ever had fewer reservations about the possibility of actual infinity than Nicholas Bonet. Nicholas Bonet is, as we have seen, a logician who never hesitates to pursue the consequences of a principle to their extremes. He brought forth no restrictions concerning the possibility of actual infinity. He even extended it to causes related by an essential order. He thereby destroyed the proof for the existence of a Necessary Being which Avicenna had given and Saint Thomas Aquinas had accepted.

The Possibility of the Categorematic Infinite:
The Doctrine of Gregory of Rimini

If the study of the categorematic infinite during the Middle Ages did not inspire anything other than the attempts of John of Baconthorpe, Francis of Mayronnes, John of Bassols, and Robert Holkot, it would not have merited the attention of the historian. These essays are not really doctrines; they are merely attempts at one. But the study did give rise to a theory in which the power of fourteenth century logic can be felt: Gregory of Rimini's theory. At the same time, the attempts of Gregory's predecessors seem more interesting because they appear as the beginnings of the doctrine which the Augustinian master completed. We have already detailed how Gregory defined categorematic infinite magnitude; it is not, for him, a magnitude such that no greater exists. He characterized the syncategorematic infinite by this formulation: *Quantocunque finito majus*; and the categorematic infinite by: *Majus quantocunque finito*, greater than any finite quantity, however large it is. One can state that, for Gregory of Rimini, categorematic infinite magnitude is *transfinite magnitude*. Objections against the possibility of the categorematic infinite were common in the schools. By various devices, one derived from its possibility conclusions of this kind: One can add something to infinity; there can be something greater than infinity; an infinity can be the multiple of another, etc. One calls these conclusions absurd and one deduces from them that the possibility of the categorematic infinite is contradictory.[86] Although they are applicable against infinity conceived as a magnitude such that there can be no greater, these objections are without force against the categorematic infinite as Gregory of Rimini has defined it. Already John of Bassols, foreseeing this definition in an obscure manner, did not hesitate to concede the above conclusions and to refuse to think them absurd; he openly allowed that an infinite can be greater than another infinite, that an infinite can be part of another infinite. He knew that one could give the following response to these objections: the comparison of greater and smaller quantities can only be accomplished between finite quantities. But he rejected this response by asserting that he does not heed the principle. Regardless of the indifference by which John of Bassols treats this question, it is worthy of being examined. Can the words *greater, smaller, whole*, and *part* be used legitimately with infinities, and do they then have the same meaning as when they are used with finite quantities? That is what Gregory of Rimini examined most thoroughly. Our philosopher applies himself first with respect to the words *whole* and *part*:

> These terms can be taken in two ways, according to their common meaning or according to their proper meaning.
>
> According to the first way, anything containing another (or a third thing distinct from the second and from what the second comprises) is said to be a whole with respect to the second; and anything contained in a whole is said to be a *part* of the whole in which it is contained.
>
> According to the second way, for a thing to be called a *whole* with respect to another, it is not sufficient that it contains the other, as is assumed in the first way, but it has to contain a determined number of things of determined magnitude (*tot tanta*) not contained within that which is included; similarly, something included is said to be *part* of the whole when it does not contain a number of determined things of determined magnitude which are contained within that in which it is contained.[87]

Thus according to its common meaning, the *whole* is a *part* and anything else not contained in the part; according to its proper meaning, the *whole* is a *part* and a number of things of determined magnitude. Gregory of Rimini pursues this further:

> Let us apply this distinction to multitudes. According to the first way, any multitude is a whole with respect to another multitude when the first multitude contains the second (and, consequently, all the objects making up the second) and when it contains, in addition, an object or objects distinct from all and each of these. In this way, an infinite multitude can be part of another infinite multitude.
>
> According to the second way, in order for a multitude to be a *whole* with respect to another multitude, it has first to contain the second multitude, as in the first way; in addition it has to contain a determinate number of things of determined magnitude (*tanta tot*)—that is, a determined number of groups of objects such that the quantity of each group is determined (for example, a determined number of groups of two or three

units)—which are not contained in the contained multitude. Inversely, the latter is said to be *part* of the former.

In the second way, an infinite multitude cannot be either *whole* or *part* of another infinite multitude; there is no determined number of groups of such units (*tot tanta*) contained in one of the multitudes and not in the other, for each of them contains an infinite number of groups of such units (*infinites tantum*) or an infinity of such groups in which one can count an infinity of units (*infinita tanta*).[88]

Gregory of Rimini introduces similar distinctions for the meaning of the words, *larger* and *smaller*:

These words can be taken in their proper meaning; thus a multitude is said to be larger than another when it contains not only a number of units as large as the other, but also a larger number (*tantumdem et plures*). A multitude is said to be less than another when it contains a lesser number of units (*pauciores*).

These words can also be taken in an improper sense; if a multitude contains all the units of another multitude and also some different units than the former, one says that it is larger than the former multiple, even though it does not contain a *larger number* of units (*plures unitates*) than the other multitude.

In this second sense, to say that a multitude is larger than another is simply to say that it contains another, that it is a *whole* with respect to this other, taking *whole* in its first sense.

If one adopts the first definition, the words *larger* and *smaller* must not be used in the comparison of infinites to one another; one must use them only to compare finite magnitudes. One can still say that an infinite is larger than a finite magnitude and that a finite magnitude is smaller than an infinite.

According to the second definition, on the other hand, an infinite can be larger than another infinite, in the same way that it can be a *whole* with respect to the second infinite, taking the word *whole* in the first sense.

The two ways of understanding *whole* and *part* are related in the following way: anything which is *whole* or *part* in the second way is also *whole* or *part* in the first way, but the inverse is not universally true.

It is not the same with the words *larger* and *smaller*, taken in the two manners just defined. In fact, a multitude containing a larger number of units (*plures unitates*) than another does not always contain the units that the other contains—a half score of men in Paris contains as many and more units as a half dozen horses in Rome, but it does not contain these horses—it is not true that everything larger in the first sense is larger in the second sense. Further, that which is larger in the second sense is not always larger in the first sense: that is evident when one compares an infinite multitude with another infinite multitude containing the first.[89]

These principles enabled Gregory of Rimini to dispel the objections that had accumulated against the possibility of actual infinity in a better fashion than had John of Bassols. After having analyzed the efforts by which our subtile logician attempted to detail the meaning of the words *whole, part, larger,* and *smaller* with respect to infinite magnitudes or multitudes, it is interesting to read the first few pages of Georg Cantor's *Theory of Transfinite Numbers.* There is a clear affinity between the thoughts of the two powerful logicians even though five-and-a-half centuries separate the times during which they were writing. Gregory of Rimini certainly glimpsed the possibility of the system Cantor constructed; he deemed that there was room for a mathematics of infinite magnitudes and multitudes next to the mathematics of finite numbers and magnitudes. He thought that the two doctrines were two divisions of a more general science:

> With respect to infinite multitude, we have used the two words *how much* and *so much (quot et tot)*; similarly, nothing restricts us from using them with respect to infinite magnitude *(quantum et tantum)*. If, for example, one follows the opinion of the Philosopher and one asked how much time has preceded the present instant, one could reply, an infinite time. The infinite can therefore reply to the question, *how much (quantum)?,* and it is a *quantity (quantum)*, if, as one asserts, everything that answers the question, *how much?,*is a quantity.
>
> But perhaps one wishes to use the words *how much (quantum)* and *so much (tantum)* only with respect to magnitudes that are finite in some measure. In that case, I would say that an infinite magnitude is not a quantity *(quantitas)*, but that it is a magnitude *(magnitudo)*; similarly, an infinite multitude would not be a quantity, but it would be a multitude. The word quantity *(quantitas)* would no longer designate the most general genus of the second predicament—we will have to construct a new name for this predicament. But this restricted meaning of the term *quantity* is neither useful nor timely.

> I therefore state that infinite magnitude is certainly contained in some species of quantity. Consequently, magnitude is divisible into infinite and finite magnitudes; finite magnitude is then subdivided into magnitudes of two cubits, three cubits, etc.[90]

Although Gregory of Rimini glimpsed and hoped for a mathematics of transfinite magnitudes, one must not, however, exaggerate his work to such an extent as to make him the forerunner of Cantor's theory. Gregory of Rimini was not able to conceive and define the notion that must play the role for the infinite multitude of concrete objects that the abstract number plays with respect to a set of finite concrete objects. Although wishing to enter into an arithmetic of infinite multitude, he did not know how to cross the threshold of this science. But due to the distinctions he was able to fashion because of his logic, Gregory was rid of the paradoxical consequences that one derived from the notion of actual infinity in order to show it contradictory; however, Gregory was not free from all the objections against the acceptance of the categorematic infinite. According to the disciples of Richard of Middleton and William of Ockham, to admit the possibility of the categorematic infinite is to go against the definition of infinity itself; this definition posits that infinity is a thing that can exist only *in fieri*, and not *in facto esse*:

> The definition of infinity is as follows: When one has already taken some part, there remains some other part to be taken; infinity is not, as thought the ancients, that which there is nothing beyond, but something beyond which there is always something, beyond which there always remains many similar objects. Consequently, to posit in the reality of nature the existence of a permanent thing having parts, and to admit that this thing is infinite, is, as one can plainly see, to posit a contradiction. Insofar as the thing is a permanent and actual thing, each part of the thing, and the thing itself, are complete and finished beings; on the other hand, insofar as the thing is infinite it is always incomplete and unfinished.[91]

On several occasions Gregory pits himself against this essential argument denying the possibility of the categorematic infinite; for example, he discusses it as an argument opposed to the hypothesis of a world created from all eternity:

> If the world existed from all eternity, there would be today an infinity of past time; this consequence is impossible.

Therefore, it has to be the same with the antecedent proposition. Moreover, the impossibility of the consequence is evident; in fact, it is the nature of the past to be taken definitively and completely, so that nothing of this past remains potential and to be taken in the future. On the other hand, it is the nature of the future to be always incomplete, not to be a whole taken once and for all and to be posited in actuality—it is of its nature that some of it is potential and remains to be taken.[92]

We rediscover Richard of Middleton's reasoning here, though more clearly formulated. But Gregory rejects this definition prohibiting infinity from ever being anything other than the syncategorematic infinite; he thinks that the definition is too narrow: "I say that it is not simply the nature of infinity (*simpliciter sumptum*) that something of this infinite exists only potentially."[93] The above definition is suitable for the syncategorematic infinite perpetually *in fieri*, but Gregory attempts to demonstrate the necessity to admit the categorematic infinite existing *in facto esse*. The possibility of an infinite magnitude in actuality would result in the assumption of an eternal world; the adversaries of this hypothesis, Richard of Middleton, for example, know this and against it they cite the following consequence: "God could have created a stone measuring a cubic foot each day and united it with a previously created stone; it is not doubtful that this infinite multitude of stones each measuring a cubic foot would form an infinite magnitude."[94] Our logician, like a number of his predecessors, does not see the possibility as an absurdity enabling one to conclude against the eternity of the world. But unlike his predecessors, he goes further; he attempts to prove that one would have to admit the possibility even when one held that the world had a beginning. In fact, one can divide an hour into parts whose durations decrease in a geometric progression, or, as the Scholastics said, into proportional parts. "If it is certain that God could have created a stone and acted as above, it is also certain that He could have created a stone in each of the proportional parts forming an hour and continued as above; since the multitude of these proportional parts is infinite, by the end of the hour there will result an infinite stone."[95]

This argument concluding for the possibility of the categorematic infinite without invoking the eternity of the world had great currency in the schools. Those who upheld the syncategorematic infinite only, John Buridan and Albert of Saxony, for example, considered it the strongest weapon of their adversaries. Did Gregory of Rimini originate it? We do not know; but at least we see him using it on several occasions and applying it to infinities whose natures vary considerably. Sometimes he uses it to show that God can actualize a rectangle whose base is invariable and whose height is categorematically infinite.[96] Sometimes he uses it to prove that God can create an infinite charity *in facto esse*;[97] for he assumes with Richard of Middleton and William of Ockham that each form susceptible of different intensities, that charity as well as heat, attains its various degrees by the addition of parts of the same nature to one another. These examples, where we see God give actual existence to the infinite, do not serve to show that the opinion that infinity is in essence something unfinished, a mixture of actuality and potentiality, is erroneous, but they do serve to lay bare the cause of this error.

> When one says, infinity is something never completed, I reply that it is so if its infinitely numerous parts are acquired in equal durations; if, for example, each part of this infinity were acquired after an hour or a moment, or some other determined quantity of time. In that case, it would have to be that the time would have an infinity of equal parts and, consequently, that it would be infinite. Since, in any case, it is impossible that an infinite time whose first part is given becomes a past time, an infinity could not be totally completed or surpassed in this manner.[98]

"But that assumes that there is in this infinity a first part acquired or surpassed. . . . If one notices this remark"—which we have already heard from Robert Holkot—"one sees that the impossibility would cease once one could not give either a first part of duration or first part of the infinity to be surpassed."[99] And that is what Aristotle himself is obliged to concede, as John of Bassols observed; if the world existed from all eternity, an infinity of men would have lived so far, and heaven would have accomplished an infinity of revolutions.

> One says, infinity is something such that when one takes any part of it whatever, there always remains another part to be taken. I reply that this proposition must be understood as the previous one, by admitting that the parts taken successively are all of the same magnitude and that they are all taken in equal times. If one takes, in some time, a portion of infinity, then in a time equal to the one in which the first

part was taken, one takes an equal portion, and one continues in this fashion, there will always remain something to be taken of this infinity, and it will never be taken in its totality. . . . But once equal parts of the infinity are not taken in equal times, but in times whose durations decrease in geometric progression . . . there is no longer any inconsistency in the infinity being taken in its totality, as long as there is no obstacle of some other nature to this. Similarly, there is no inconsistency in that the infinite multitude of parts of time, in which the successive parts of the infinite are taken, come to be completely past, as we have already stated. Not only is there no inconsistency in this, but it is necessary that it be.[100]

For Gregory of Rimini, then, the possibility of the categorematic infinite is no more difficult logically than the following proposition: If one considers an infinite series of durations, for example, a half hour, a fourth of an hour, an eighth of an hour, etc., at the end of an hour, the infinite multitude of these durations would have been surpassed. Further, similar assertions can be formulated not only for duration but also for a multitude of variable magnitudes; if, for example, a path has been taken by an agent in an hour, one can divide this hour into proportional parts and consider the paths traveled during each of these proportional parts of duration; at the end of an hour, the infinite multitude of these paths would have been completely traveled. One can detail similar considerations with respect to forms variable in intensity, the form of heat passing in an hour from one degree to another, for example. The objections that one can raise against the process by which God can create an infinitely large body, a surface of infinite area, or a form of infinite intensity, in an hour, can also be raised against the propositions just formulated; in either case they can be dispelled in a similar manner. The objections that Buridan and Albert of Saxony attempted all rest upon the same principle: If one assumes a continuum divided into proportional parts, one is not allowed to state that one takes *all* the proportional parts of this continuum, parts whose multitude is infinite, for there will have to be a part taken last, "and there is no proportional part which is the smallest or the last."[101] Gregory of Rimini admits this principle, but on the condition that it be understood syncategorematically (*distributive*). And, as he has shown, this condition implies another—that the successive parts of a continuum are assumed to be taken *in equal times*. If one does not restrict oneself in this way, then the principle cannot be invoked. There can be corresponding propositions, that are true *collectively* or categorematically, to the propositions that are false *distributively* or syncategorematically. If one considers an infinity and the parts of the infinity *distributively*,

> it is impossible that *all the parts* of this infinity can be taken all together; whatever number of parts already taken and in whatever manner they have been taken, they are still parts of the whole containing them and, consequently, they have a part, or other parts, outside them. The parts thus taken are not *all the parts* of this infinite. The proposition is then false in its proper sense *(distributive)*. It is the same with respect to these other propositions: *All the parts*, taken simultaneously, form the whole; the whole is identical with *all its parts* taken simultaneously. However, these propositions are true: *The set of things* [*omnia* instead of *omnes partes*] of which each is a part of this whole constitutes the whole; inversely, the whole is *the set of things* of which each is one of its parts. In these propositions, the words *set of things (omnia)* are taken collectively.[102]

The logicians insisted on the truth of: a proposition which is true syncategorematically (or in its divided sense) can be false categorematically (or in its composite sense). Inversely, Gregory of Rimini demonstrates by numerous examples that a true proposition understood collectively can correspond to a false proposition understood distributively. Concerning an hour divided into proportional parts, and the instant terminating it, the following distributive proposition would be false: before the instant, every part of the hour was past. But the following collective proposition is true: every part of the hour was past before the instant. Is it the same with respect to a form going with constant speed from one degree to another, growing by proportional parts corresponding to the proportional parts of the hour? "These two propositions are both true: every proportional part of the form that exists at the final instant of the hour, existed before this instant; at no instant and at no time before this final instant, did there exist an infinity of proportional parts of this form."[103] By a similar distinction Gregory resolved the paradox of Achilles and the tortoise,[104] and by a similar distinction he was able to agree with Burley's principle, there is no final portion of an hour divided into proportional parts, and with the proposition, in such an hour God can create a rectangle of infinite height:

> At the end of the hour there will not be a rectangle or a total figure; there will be an infinite magnitude containing

an infinity of rectangles, none of which is the last. Similarly, when a form grows in a continuous fashion, for each of the instants terminating the successive proportional parts of the hour, there is always a greater number of parts of the form; however, at the end of the hour, there is no number that is the number of these parts, but there is an infinite multitude containing an infinite number of parts, none of which is the last part.[105]

If God can create an infinite magnitude by the addition of equal parts during a finite time divided into proportional parts, He can just as easily divide a continuum into proportional parts during the same time; the possibility of the categorematic infinite therefore implies that a continuum can be actually divided to infinity. Except for Francis of Mayronnes and John of Bassols, most of Scholasticism conceded this correlation between the two propositions; denying that a continuum can be actually divided to infinity, they concluded for the impossibility of the categorematic infinite. Gregory of Rimini also conceded this correlation, but in an opposite sense; since he admits the existence of the actual infinite, he also admits the actual divisibility of any continuous magnitude. Whether the word infinite is taken categorematically or syncategorematically, our logician teaches that "any magnitude is composed of an infinite multitude of partial magnitudes equal among themselves."[106] He explicitly formulates the following two propositions: any magnitude has an infinity of equal parts, the word infinity being taken syncategorematically . . . any magnitude has an infinity of equal parts, the word infinity being taken categorematically. . . . The latter proposition even furnishes Gregory another argument by which he can prove that the actual existence of an infinity is not contradictory: "I say that it is not simply the nature of infinity that something of this infinite remains potential always; that is able to be seen clearly in the infinite multitude of parts of a continuum. Each of these parts is in actuality as each of the others; it is not true that some of these parts are in actuality while others are only potential. (*Non est de ratione infiniti simpliciter sumpti quod aliquid ejus sit tantum in potentia; quod patet in multitudine infinita partitum continui, quarum quaelibet est actu sicut aliqua earum; nec ejus est aliqua pars in actu, aliqua vero in potentia tantum.*)"[107] We end our report on Gregory of Rimini's system with the above quote. The thought it contains— which Robert Holkot already expressed—is truly the main thought of the system: one can conceive a continuous magnitude and a method of subdivision capable of discerning an infinity of parts in this magnitude. One must not say that these parts exist only potentially in this magnitude; they are there in actuality, although they are not separated. One can take the words in their collective sense, categorematically, so that one can say that the magnitude in question is a set of its parts—that giving this magnitude is giving the set of its parts—that surpassing this magnitude is surpassing a set of its parts. Once one concedes this principle, all of Gregory of Rimini's doctrine follows. That is the principle which the adversaries of this doctrine must attack if they want their own arguments to have any credibility.

Gregory of Rimini's Adversaries: John Buridan and Albert of Saxony

John Buridan devotes his six final questions on the third book of Aristotle's *Physics* to the problem of infinity.[108] In these six questions he discusses Gregory of Rimini's doctrine step by step, in order to prove that, with every kind of magnitude, the syncategorematic infinite is possible, while the categorematic infinite is impossible. Among these six questions is one with the following title: "Is there a spiral line which is infinite, taking the word infinite categorematically?"[109] The question concerns the example of an infinite line not occurring in Gregory of Rimini's commentary on the *Sentences*, but clearly proceeding from the spirit of this commentary. The example is as follows: Let us take a cylinder of given height, divide it into proportional parts, and place planes parallel to the points of division; we thus split up the whole cylinder into an infinite series of partial cylinders whose heights decrease in a geometric progression by one half. That accomplished, let us trace a spiral, on the surface of the first partial cylinder, whose path is to the height of the first cylinder. Let us continue this line by drawing another spiral, on the surface of the second cylinder, from the point where the first spiral stopped, whose path is to the height of the second cylinder. Let us continue indefinitely according to the same rule. Such is the manner of construction for the *linea gyrativa* which will have a categorematic infinite length once it is drawn. This manner of constructing a categorematic infinite magnitude is closely connected with what Gregory of Rimini was conceiving when he divided an hour into proportional parts and assumed that God can create a stone of one cubic foot during each of these parts. But it has the advantage of not having to call upon God's creative power, and therefore it does not have to entangle two questions which do not seem linked to one another. Is it true that one can obtain an actual infinitely long line in this fashion? Buridan responds that this question is extremely difficult (*ista quaestio est mihi valde difficilis*). In order to respond to it, he formulates the following conclusions:

> First conclusion: If one begins from one of the ends of the cylinder and proceeds toward the other end by proportional parts, there is no final proportional part. Each proportional part, in fact, leaves behind it another part equal to it; the latter, in turn, is divisible into two halves whose first part is a proportional part to the preceding parts. . . .
>
> The second conclusion follows from the above: Let A be one of the ends of cylinder B, and C be the other end. Let us begin to form proportional parts from A to C. Our conclusion is that there is no proportional part that attains C following this method. There is no proportional part nearer to C than all the others either. Such a part would, in fact, be the final part, and there is no part which is the final part. . . .[110]
>
> Twelfth conclusion: To any spiral line drawn along the proportional parts, there corresponds a straight line crossing the proportional parts.[111]

In fact, it would be sufficient to take the projection of the spiral line on the axis of the cylinder.

> The thirteenth conclusion follows from that: If a spiral line is traced in the above manner along all the proportional parts of the cylinder and does not extend beyond these proportional parts, there corresponds to it a straight line carried across these proportional parts and not extending beyond them. . . .
>
> Fourteenth conclusion: No straight line can be carried across all these proportional parts unless it extends beyond all these proportional parts. If it extends across the whole cylinder so that it touches C and touches the external body with which the cylinder is in contact, then it extends beyond all the proportional parts, since none of them attain C, as it is stated in the second conclusion. If, on the contrary, the straight line does not extend up to C, . . . it results that it must stop at some termination before C so that between this termination and the external body which the cylinder contacts there remains something of the cylinder. But beyond this termination and before C, there remain some proportional

parts. . . . This straight line is therefore not drawn across all the proportional parts. Thus there is no straight line drawn across all the proportional parts unless it extends beyond all these parts. . . .[112]

From this conclusion and the preceding one results the sixteenth conclusion: There is no spiral that can be drawn, in the above manner, across all the proportional parts of the cylinder; no such line, in fact, is traced above all these proportional parts, as can be seen by the definition of the case we are considering (*ut apparet per casum*); and it cannot be drawn across the length of all the parts unless it is extended beyond all these parts, as one can see by the two preceding conclusions. . . .

From these propositions I draw the following principal conclusion: no spiral line drawn across the length of the proportional parts is infinite in length; one would not assume that such a line were infinite unless it were drawn along all the proportional parts of the cylinder, and there is no such line that can be drawn along all the parts.[113]

As we can see, Buridan's argument rests on the truth for which Walter Burley gave so clear an exposition, the one reputed not to be known by many. Buridan continues in this fashion:

There is therefore no spiral line drawn along all these proportional parts. When one objects that there is, however, such a line circling, not only three or four proportional parts, but a hundred, a thousand, I agree; and whatever number you call forth, there is a spiral line circling this number of proportional parts. But when you assert that since there is a spiral line drawn along so many proportional parts, there is no reason for such a line not to circle all of them, I reply: on the contrary, there is a very serious reason; I would, in fact, agree to this collective proposition (*copulativa*): there is a line circling three proportional parts, there is another circling ten parts, another circling a hundred, another circling a thousand, and so forth to infinity. But I do not agree to this categorematic proposition in which the ultimate term is collective (*de copulato extremo*): there is a spiral line circling three parts, ten, a hundred, a thousand, and so forth without end. Similarly, I would agree with this [syncategorematic] proposition: along all the parts, a spiral line is drawn; and I would not agree with this [categorematic] proposition: a spiral line is drawn along all the parts. Moreover, even though there is a spiral line circling a hundred proportional parts, or a thousand, or any number of parts whatever, there is none

drawn along an infinity of parts for there are no parts which
are an infinity of parts, and there are no parts which are all
the parts, whether we take the word *all* in its collective sense
or in its distributive sense (*quia nullae sunt infinitae et nullae
sunt omnes, sive sumamus: omnes collective sive distributive*).
That is what we shall soon see.[114]

Here then is a clearly formulated principle which is the contrary
of the one Gregory of Rimini required. Buridan then refuses to
see the spiral line as defined as an example of a categorematic infinite
length. All that he agrees with is that, syncategorematically, "there
is a spiral line of infinite length, for given any of these lines, there
is a longer line."[115] But one must understand the exact meaning
of the words utilized. The proposition, a spiral line is
syncategorematically infinite in length, would be false if one
understands the term *line* with a completely determinative
signification which is incompatible with any notion of collectivity.

> If the proposition is understood of a line determined or
> capable of being determined, a result of the proposition would
> be that the line would be longer than itself, which is impossible.
> Doubtless, a longer line corresponds to any given spiral line,
> but there is no spiral line [which is always the same] which
> is longer than any other given line.[116]

The above passage shows one the precision that Buridan
introduces to these discussions. Whoever would reproach him of
being too careful would be ill-prepared for the examination of such
problems. What John Buridan asserts about the spiral line prepares
us for what he asserts about another question to which Gregory
of Rimini answered affirmatively: Can God create a categorematic
infinite magnitude by creating a stone of one cubic foot during
each of the proportional parts of an hour? It seems that we must
answer affirmatively to this question:

> All the singular propositions are possible and compossible;
> the universal proposition can therefore be realized. I reply
> that the conclusion does not follow from the premises. What
> is true is as follows: it is possible that at the same time all
> the singular propositions are true, and it is impossible that
> all the singular propositions are true at the same time. It is
> never legitimate to conclude, with respect to possibility, from
> a universal proposition taken in its divided sense to the same
> universal proposition taken in its composite sense.[117] It is not
> possible for God to create a stone of one cubic foot in each

of the proportional parts of an hour, for He would have to do it by a process that would distinguish all the proportional parts from each other, and would enumerate them, the word *all* being taken here not only in its distributive sense, but also in its collective sense, in such a way that after these parts there are no others, that there are no others of these parts; and that is impossible, for in this sense, there are no parts that are all the parts (*quia sic nec omnes sunt aliquae nec aliquae sunt omnes*).[118]

When one takes a whole magnitude, one is not allowed to say that one takes *all* the proportional parts into which this magnitude is divisible; to say that is to speak nonsense. Such is the principle which, with unshakeable firmness, Buridan pits against the arguments constructed by Gregory of Rimini or by his disciples in favor of the categorematic infinite. Albert of Saxony shows himself to be a faithful disciple of Buridan in this discussion. Albert expounds upon the process by which, according to Gregory of Rimini, God could realize an actual infinity: In each of the proportional parts of an hour, God can create a stone of one cubic foot; the hour elapsed, He can combine these stones. Our author does not conceal his admiration for the ingenuity of this process: "If an infinite magnitude can be realized in actuality, it would be by this process."[119] But the process implies a contradiction; in fact, among the stones that God created, there is one created after the others, during the final proportional part of the hour. But time is a continuum, and in the division of a continuum into proportional parts, there is no ultimate part. Albert does not fail to note with respect to this matter that the same proposition can be true or false depending upon whether one takes it syncategorematically or categorematically; such is the proposition, God can create a stone of one cubic foot in each of the proportional parts of an hour. Albert relates the proposition to another: If Socrates is capable of lifting a weight measuring eight units, then Socrates is capable of lifting any portion of a weight whose measure is eight units. In each of the two cases one must be careful not to conclude the composite sense, which is false, from the divided sense, which is true. Each of the singular propositions is true and compatible with each other, but they are not all compossible; they cannot all be verified at the same time, so that the universal categorematic proposition is false. A similar argument can be used to demonstrate the error of those who wish to realize a categorematic infinite length by means of a spiral line. Albert of Saxony concedes that if this curve is drawn, its length would be infinite; but it cannot be drawn in its entirety. In order for it to be completed, the spirals would have to encircle all the proportional parts of the cylinder, and "no parts are all the proportional parts of the cylinder (*nullae partes sunt omnes partes proportionales columnae*)."[120] The *Quaestiones super libris de Caelo*, written in accordance with Buridan's teaching (and attributed to Buridan), defend the same opinion as Buridan and Albert of Saxony with respect to the problem concerning us; but while upholding the opinion of these philosophers, the author of these *reportata* has kept neither their clarity nor their precision. He reports that some thinkers attempt to prove the following proposition: God can create a categorematic infinite body in an hour.

> In fact, in each of the proportional halves of this hour, God can create a body of one cubic foot, and conserve these bodies together; hence, since there is an infinity of proportional parts in the hour, there will be an infinity of bodies of one cubic foot constituting an infinite body at the end of the hour.[121]

The author of the *Quaestiones* attempts to demonstrate that this thought implies a contradiction.

> I assert that one has to concede that a stone is created last, for these proportional parts follow one another according to some order, one preceding the other, no two of them existing at the same time, beginning at the same time, and ending at the same time; moreover, no two of these stones are created at the same time—taking any two, one was created before the other.[122]

Hence, the above process implies a contradiction, "for one cannot assign the final proportional part of any hour (*non est dare ultimam medietatem proportionalem alicujus horae*). One would have to assign the stone that was created last, and this stone cannot be the one created last unless it was created during the ultimate proportional half."[123] The whole argument rests on the axiom formulated by Burley which Buridan made use of in his *Physics*: when one divides a given magnitude into proportional parts, there is no ultimate proportional part. Buridan and Albert gave this axiom a more penetrating and more general formulation: when one divides to infinity, by any process whatever, there are no parts of which one can state that they are all the parts of this magnitude. Buridan said: *"Nullae sunt infinitae et nullae sunt omnes, sive summamus:*

omnes collective sive distributive"; and *"Nec omnes sunt aliquae, nec aliquae sunt omnes,"* which Albert of Saxony repeated as: *"Nullae partes sunt omnes partes proportionales columnae."* That is the principle Buridan and Albert of Saxony invoked in order to refute the arguments in favor of the categorematic infinite. Thus one can grasp the essential point of the controversy between them and Gregory of Rimini. Let us imagine a rule dividing a finite magnitude into an infinity of parts—the division into proportional parts being an example of such a rule. Gregory of Rimini asserts the following: To give the said magnitude is to give *all* the parts into which the said rule can divide it; thus, to last an hour is to last *all* the proportional parts of this hour. John Buridan and Albert of Saxony reply: when one gives the whole magnitude, one does not give *all* the parts into which the said rule can subdivide this magnitude, for the expression has no sense; there are no parts for which one can assert that they are *all* the parts of the given magnitude. The whole debate revolves around these two opposed theses; according to whether one holds one or the other, one is forced to accept its consequences, to assert with Gregory of Rimini the possibility of the categorematic infinite or to deny this possibility, to accept only the syncategorematic infinite with Buridan and Albert of Saxony. Moreover, we should not think that the debate is over. If we focus our attention on what is essential in this discussion, we might be surprised to find that it is debated among our geometers today. Within the University of Paris, during the fourteenth century, two schools opposed each other with respect to the infinite. We can designate these two schools by the epithets of *finitist* and *infinitist* that Couturat uses when he wishes to classify contemporary mathematicians.[124] The finitists of the fourteenth century, those who upheld the syncategorematic infinite only, William of Ockham, John Buridan, and Albert of Saxony, for example, might easily summarize their doctrine by the following: "The notion of the infinite—which one should not render mysterious in mathematics—is merely that, for each whole number, there is another."[125] The infinitists, on the other hand, those who, along with Gregory of Rimini, attempted to construct a mathematical science of categorematic infinite quantities, would welcome the theory of transfinite sets as the finished form of the doctrine they barely began. If we also sought for the point of departure between finitists and infinitists today, we would note, not without some surprise perhaps, that this point remains where the logicians of the fourteenth century had placed it. Let us listen to the fundamental objection of the finitists against the infinitists as phrased by Baire:

As soon as one speaks of the infinite, even the denumerable infinite, the conscious or unconscious comparison with a bag of marbles one gives from hand to hand just disappears completely . . . especially when a set is given. . . . *For me, to think that the parts of this set are given is false.* I therefore refuse to attach a sense to a choice made in each part of the set. . . . At the bottom line, despite appearances, everything must be brought back to the finite.[126]

Is that not the language of some distant disciple of John Buridan? And when, in some room of the new Sorbonne, the Mathematical Society of France disputes the Cantorian antinomies, do not the opposing arguments echo those heard, not too far away in the schools of the Rue du Fouarre, four and-a-half centuries ago.

Gregory of Rimini's Followers: Nicole Oresme and Marsilius of Inghen

Buridan's vigorous argument against Gregory of Rimini's doctrine plainly convinced Albert of Saxony; it did not completely hold the allegiance of the other Parisian masters who followed, sometimes steadfastly, sometimes with reservation, the opinion of the general prior of the Hermits of Saint Augustine. Nicole Oresme, in his *Traité du Ciel et du Monde*, written in French at the command of Charles V, refers to some questions he had examined on Aristotle's *Physics* and in his own commentary on the *Sentences*. Doubtless, either of these two works would have taught us what the author thought about the categorematic infinite; unfortunately, neither of these two works survived. The only thing we find in the works of Oresme that has survived is the solution of some crucial problems about infinity. This solution, which feels like the exercise of a mathematician, does not enable us to conclude about the philosophical opinions of Oresme; it seems however that Oresme speaks the language of a disciple of Gregory of Rimini, of a defender of the categorematic infinite. Oresme first conceived and resolved these problems dealing with infinity in his important work entitled *De difformitate qualitatum*; he took up these matters again in his *Traité du Ciel et du Monde*. We cite the latter work. In the first book of the *De Caelo*, Aristotle gives various reasons against the possibility of an infinite body. An infinite body would be infinitely heavy, which seems impossible to him.

Oresme replies:

> But it seems to me that the reason given above is not evident without adding another assumption. For, in accordance with the second reply, I assume a body to be infinite, and I take or assign in this body a finite portion, spherical in shape, called A. Next I take another sphere B from the same section, and of the same shape, and then another sphere C, exactly like A and B, proceeding in this manner without stopping. In this way, it appears that there are, in this infinite body, infinite equal parts A, B, C, D, and so on without limit.
>
> Now I posit that in the portion called A there should be distributed the weight of one half-pound, and in B there should likewise be distributed one-half of another half pound, and in C one-half of the residue of a pound, and in D half of this remainder, which would be one-sixteenth part of a pound, and so on without end.
>
> It appears then that the entire infinite body will weigh only one pound, while A will weigh as much as all the other portions, however many, taken together.[127]

Pursuing Aristotle's argument, Oresme writes the following:

> But it seems to me that, if a body is infinite, it does not follow that it must be infinite in all its parts, and likewise in the case of a line or an area.
>
> For one can imagine a body of absolutely infinite size and greater beyond any ratio than any finite body. However, such a body will be infinite in length in only one direction, and still it will be no less than a body infinite in every direction.
>
> Such a body would be one, of which the first part A would be a foot in every direction, and the next part B equal and like A, and the next, C [equal and like B] and so on to infinity.
>
> With this, I indulge my imagination and posit that A shall be a body measuring one half-foot in all directions and B another body exactly the same, that one-half of B be taken and made flat and round or circular, that the semidiameter of B be made one foot, and that one-half of the remainder of B be made still more tenuous until, when it is added to the circular half of B, it will be as wide as the semidiameter of the first half. Now, let half of the remaining portion of

B be treated in the same way, and so on to infinity. The resultant imagined body, let us call C.

Then let there be an infinite body neither greater nor smaller than this body we have imagined, which we called ABCD, etc., and let us call this infinite body D.

I say, therefore, that C is infinite in length and breadth in every direction, but not in depth, and that D is infinite only in one direction, that is, in length. However, C is finite absolutely and equal to an *a* which is only one foot in all directions, while D is infinite absolutely in all directions and not less than any infinite body which would occupy all space.

One could push this concept further and extend it and arrive at conclusions still more marvelous, but this will suffice for the present.[128]

Let us cite one last response by Oresme to Aristotle. The Stagirite asserted that "it is not possible for an infinite body of uniform parts to move circularly, for such a body has neither middle nor center, and all bodies that move circularly have a center."[129] The bishop of Lisieux responded to that as follows: "Perhaps this reason is not absolutely clear, for one could say that in such a body the middle is the center of movement, but not the middle of its mass, unless we add that in such a body the center is everywhere and the circumference is nowhere."[130]

Whatever side Oresme took in the dispute about infinity, whether he allowed the possibility of the categorematic infinite or not, the problems he treated for entertainment (*par esbatement*) can show the holders of the two opposed opinions the minute precautions one should use when reasoning about the equality or inequality of infinite magnitudes. There was no doctor during the fourteenth century at the School of Paris, after Nicole Oresme, who could have proposed a clear and original solution to this serious debate about the infinite. Marsilius of Inghen is but a faint echo of his predecessors. Marsilius merely repeats the propositions of John Buridan in his *Abbreviationes libri physicorum*:

> It is impossible that a piece of wood be divided into all its proportional parts following some determined procedure (*consequenter se habentes*). . . .
> It is impossible that a power can produce a stone of one cubic foot in every proportional part of a forthcoming hour; that is evident because that is no more possible than dividing a continuum into two proportional parts during each proportional part of the hour. . . . The proposition is therefore

contrary to the one which has been demonstrated before . . .
therefore, an actual infinite magnitude cannot exist. . . .
If it were possible, it would be produced in the following
manner: God would create a stone of one cubic foot during
each proportional part of an hour; and that cannot be
accomplished, as was already stated.[131]

One should not construct the following objection either: God can
make a stone of one cubic foot in any proportional part of the
hour; therefore, He can create such stones in every proportional
part of the hour. It would not be correct to maintain here that
the truth of each of the singular propositions entails the truth of
the universal proposition; "that would be correct in the case where
the singular propositions are *all* the singular propositions
corresponding to the universal proposition; and that is not the case
here. . . . Therefore, an infinite body cannot be produced by divine
power unless one understands the word infinite
syncategorematically.[132] But strangely, this conclusion is followed
immediately by another conclusion that not only contradicts
Buridan's teaching, but also all the reasons just given; it is as follows:
"An infinite length exists in fact; that is rendered evident by the
spiral line. (*Infinita longitudo de facto est; patet de linea
gyrativa.*)"[133] This last sentence already indicates that Marsilius of
Inghen's opinion leans toward the doctrine upheld by Gregory of
Rimini. This leaning is more marked in the *Questions on the Physics*
of Marsilius. Marsilius often shows more independence from his
Parisian teachers, John Buridan and Albert of Saxony, in these
Questions which were probably composed much later than the
Abbreviationes—this is quite evident with respect to the problem
of the infinite. Marsilius devotes two questions on this problem:
"If an infinite magnitude can be actually realized"; and "If, in
fact, an infinite body actually exists in nature."[134] The author
enumerates various objections against actual infinity (some better
than others) in these two badly ordered questions. He also presents
Gregory of Rimini's argument in favor of this infinity: In every
proportional part of an hour God could create a stone of one cubic
foot.[135] But he does not refer to Buridan and Albert of Saxony's
objections to this argument. In this discussion, the logical rigor
of Buridan and Albert's arguments completely disappears; the
distinction between syncategorematic and categorematic
propositions is not even made, although it seems that the author
intends to refer to the categorematic infinite. If the discussion is
vague and hesitant, the conclusions are precise and bold:

First conclusion: In fact, there exists an infinite curve [the spiral considered by Buridan and his successors].

Second conclusion: In fact, there exists an infinite surface. One proves this by imagining a surface surrounding the various proportional parts of a continuum in a manner analogous to what one has imagined for the line.

Third conclusion: It is possible that an infinite body exists actually. Aristotle did not admit this conclusion, because he did not conceive an infinite active power, while we believe in the existence of such a power because of faith. This conclusion cannot be proven except that one is not led to any contradiction by assuming that it is true. That is what we will see clearly by resolving the objections posited in the following way: For each, we will show that the reasoning is not conclusive or that the conclusion is not impossible.

But does there exist, in fact, an actually infinite body? Can such a body be produced by a natural power? That is what we shall see in the next question.[136]

Marsilius's response is given at the end of his tenth question and phrased as follows:

There does not exist in fact and actually a body of infinite volume. Nevertheless, this proposition cannot be demonstrated; one can only say that it agrees better with our experience than any other. All bodies we perceive are finite, and in fact, no reason constrains us from positing an infinite body. (*De facto nullum est corpus actu infinitum; et licet ipsa non possit demonstrari, tamen magis concordat sensui, quia quodlibet est finitum quod sentimus, nec aliqua ratio cogit ad ponendum infinitum.*)[137]

Aristotle objected to the existence of such an infinite body with what he considered to be impossibilities: Such a body would be infinitely heavy; it could not move. Marsilius replied,

One can allow that it would have an infinite gravity and that it would be absolutely incapable of locomotion in its totality, although its parts would be mobile, in conformity with what we think today about the total mass of the earth. It is therefore evident that the reasons by which one proves the non-existence of the infinite are not demonstrable; they are only probable and better than those that can be given in favor of the contrary opinion. (*Patet igitur quod rationes praedictae, quibus probatur non esse infinitum, non sunt demonstrativae, sed probabiles, et meliores quam possint fieri ad oppositum.*)[138]

Immanuel Kant will not conclude otherwise with respect to the antinomy: the world is infinite; the world is limited. When Marsilius of Inghen was writing his *Quaestiones secundum nominalium viam*, perhaps he was no longer in Paris. The University of Paris was breaking up; its most brilliant masters, Marsilius of Inghen and Henry of Langewstein, for example, were leaving it to establish new universities, such as Heidelberg and Vienna. It rushed to distribute to all corners the life which the religious schism, foreign and civil wars, and epidemics were exhausting and almost drying up. After the departure of Marsilius of Inghen, the classrooms of the Sorbonne, the schools on the Rue de Fouarre, did not hear any new opinion worth noting on the infinitely small and the infinitely large; the teachings of the old masters—of William of Ockham, Gregory of Rimini, John Buridan, and Albert of Saxony—were forgotten or served as fodder for unintelligent, rote repetitions. The fate befalling the problem of infinity also befell all the cosmological problems that were the subjects of impassioned debate in Paris during the fourteenth century. The hour marking the start of the Western Schism also marks the end of the mission to initiate modern science that the University of Paris had received.

3
Infinity in Fifteenth-Century Cosmology

Paul of Venice

Paul of Venice's *Summa totius philosophiae* is clearly a later work than his *Expositio super libros physicorum*; thus, one cannot go from one work to the other without remarking that Paul of Venice becomes less faithful to Peripatetic Averroism and more favorable toward the ideas of the Parisians. One can note this with respect to what our author asserted about infinity. The doctrine of the *Expositio* about the infinitely large is exactly Aristotle and Averroes' doctrine. It begins by demonstrating with much detail that the actual existence of an infinitely large body is impossible; then, along with the Peripatetics, it concludes from this impossibility to the impossibility of a potential infinitely large body. Following the example of Aristotle and his Commentator, it carefully distinguishes the question as it is posed by the geometer and as it is posed by the physicist (*naturalis*): "The geometer considers magnitude in an abstract fashion and by what is in reason itself; the physicist considers it as realized in qualified matter, as termination or passion of a natural body."[1] The geometer and the physicist can therefore give different responses to the question: Given any magnitude, can one give a greater? The geometer will respond affirmatively: "but by that he understands only the following: Given any magnitude one can give a greater one in the imagination."[2] The geometer's proposition is denied by the physicist; the physicist denies that, given any magnitude, one can always give a greater one which is realized in matter by a natural body. "The words *to give* are equivocal; sometimes they mean to imagine, sometimes to exist. In the geometer's proposition, they are taken as the words to imagine and in the philosopher's proposition as the words to exist. . . . Thus the geometer accepts the proposition that the philosopher rejects."[3] The Peripatetic doctrine is most clearly formulated here. The postulate upon which the whole doctrine rests is as follows: there is a finite universe bound by an immutable surface; no power would be able to add even the least body to this universe. But Christian belief in a God whose omnipotent creative power knows no limit other than contradiction, forces a reexamination of the problem of infinity. Paul of Venice is well aware of this. Aristotle's theory affirms that the existence of a magnitude which is potentially infinite requires the existence of a magnitude which is infinite in actuality. Against this proposition, our author foresees the following objection:

> One can imagine that A is a body, that one adds to it a body of one cubic foot to it the next day, that one adds another body of one cubic foot to it the day after, and so forth, to infinity. It is evident that A will grow indefinitely without ever being infinite in actuality.[4]

The reply to this objection is as follows:

> One can imagine that each of these additions is accomplished by the creation (*generatio*) of a new quantity; hence, the reasoning that such a magnitude is infinite potentially therefore the magnitude is or will be infinite in actuality is invalid.
> One can also imagine that the addition is accomplished only by subtracting from some other body the volume that one adds to A; in this way, doubtless, the reasoning that such a magnitude is infinite potentially therefore there exists an infinite magnitude in actuality is valid. In fact, such a subtraction cannot be accomplished in an infinite time at the expense of some body without that body being infinite.
> The theologians allow that God can create an infinite multitude of absolutely new quantities and material masses; they therefore would deny that the reasoning in question is valid. Aristotle, on the other hand, assumes that prime matter is excepted from generation and destruction, that it is always of the same quantity, of the same magnitude. Hence, one must allow that his reasoning is valid; for him, such an addition cannot be accomplished by the creation of a new quantity, but only by the subtraction of this quantity from another body.[5]

Catholic theology thus opened, for the theory of infinity, entirely novel perspectives, and the Parisian physicists then attempted to embrace its whole extent and to penetrate its depth; thus they were led to construct some singularly clairvoyant predictions about the syncategorematic infinite and the categorematic infinite. Our Parisians even came to realize that they could dispense with the appeal to the creative omnipotence of God as long as they limited themselves to lengths and surfaces and did not extend their doctrines to three-dimensional bodies; a spiral line served as a useful example for their penetrating analyses. The voluminous *Expositio super libros physicorum* says nothing about these analyses; it stops at the threshold that would have to be crossed in order to perceive them. This threshold was crossed by the *Summa totius philosophiae*, without looking back. The *Summa* develops a theory extremely similar to John Buridan and Albert of Saxony's theory with respect to infinity; it no longer says anything about Aristotle's theory. It continues to teach that a body of actual infinite magnitude cannot exist in nature, but it allows that "in each of the proportional parts of an hour to come, God can create a length one foot long."[6] Nevertheless, it does not wish that one derives the conclusion that "the resultant line would be infinite." The reasoning would not be conclusive because one would have to go from a divided sense to a composite sense. Paul of Venice marks the passage from one sense to another with the help of the language conceived by the Parisians:

> In every proportional part of the hour to come, God can create a line one foot long; it does not result that God can create a line one foot long in every proportional part of the hour to come.
>
> Let A be a cylindrical body; this body contains an infinity of proportional parts. To each part there corresponds a spiral which is longer than the circumference of the base; the line composed in this manner is therefore infinite.[7]

To which Paul of Venice replies with Albert of Saxony:

> Neither body A nor any part of this body has a spiral, except potentially. Let us admit that in the first proportional part of an hour, one traces a spiral encircling the first proportional part of cylinder A, that in the second proportional part of the hour one continues the preceding spiral so as to encircle the second proportional part of A, and so forth indefinitely. I do not admit that in this case one obtains an infinite line, no more than I conceded it in the previous case. What I admit is this: In each proportional part of the hour one can draw a line encircling a proportional part of A.[8]

And the order of the words of this sentence announces to us that its author takes it, according to the Parisian rule, in its divided sense, syncategorematically. It is also with Albert of Saxony that Paul of Venice admits that "the existence of an infinite magnitude does not imply a contradiction."[9] And following his example, he enumerates some of the surprising properties that such a magnitude would possess. We can also recognize the Parisian doctrine that Albert of Saxony taught in the chapter where Paul of Venice examines whether an active power has as termination a maximum it can accomplish, and whether a passive power has as termination a minimum it can no longer suffer.

First let us note how one would present the question.

One formulates this proposition: A is the maximum Socrates can carry; Socrates can therefore not carry either weight A or any weight equal to it, but if one gives any weight less than A by any amount, one will find a weight that Socrates will be able to carry, and another greater than it. . . .

Once these premises are posited, here is our first conclusion: One asks whether there exists a maximum weight that Socrates can carry or a minimum weight that he cannot carry; that is the weight which is Socrates' power.[10]

The definition of what one should mean by limit is given here with a rigor our modern mathematicians will not surpass. The Averroist who wrote the *Expositio super libros physicorum* entered the camp of the Parisian nominalists in order to compose the *Summa totius philosophiae*.

II
Place

4
Theory of Place before the Condemnations of 1277

Arabic Theory of Place

Peripatetic theory of place rested upon two essential propositions: According to the first, the place of a body must contain the body.

According to the second, the place of a body must be a motionless thing, for it is the fixed term to which all local movement is referred.

Moreover, these two propositions are condemned to be unreconciled in the framework of Peripatetic physics.

In virtue of the first proposition, the ultimate celestial sphere cannot have a place since nothing contains it. Having no place, it should not be capable of local movement in virtue of the second proposition, since all local movement requires a place, meaning a fixed term to which movement is referred.

Moreover, not only is the ultimate sphere of the Peripatetic system capable of local movement, it actually moves, since diurnal movement is its movement.

These are the contradictions that Peripatetic physics attempted to resolve. The history of these resolutions can be divided into two periods having distinct character.

Until 1277, one proceeded by repairing Aristotle's system; these repairs were partial solutions and, as such, insufficient, since the contradictions one attempted to resolve had their roots in the essential principles of the Peripatetic theory of place.

In 1277, the decrees brought forth by Etienne Tempier, bishop of Paris, formulated a proposition contradicting Aristotelian philosophy with respect to the mobility of the ultimate sphere and of the whole universe. Afterward there appeared theories of place that broke with Aristotelian tradition; these theories put forth and developed ideas to which modern philosophers have often returned.

We have already detailed [in the *Système du monde,* vol. I, chap. 5] what the Greek commentators taught Christian Scholasticism about place; it remains for us to detail what they taught the Arabic philosophers.

One should not expect to find the depth and originality of thought of Damascius and Simplicius in the work of the Arabic philosophers; with respect to the nature of place and its immobility, they mostly limited themselves to commenting upon Aristotle's doctrines, making use of the reflections of Alexander of Aphrodisias and Themistius, with varying degrees of success. They hardly ever mentioned Joannes Philoponus's theory except in order to reject it summarily, and they appear not to have bothered with Damascius's theory which Simplicius developed.

In his opusculum *On the Five Essences*—on matter, form, place, movement, and time—Jacob Ibn Ishak al-Kindi repeats some of the aphorisms borrowed from Aristotle with respect to the nature of place. Al-Kindi thinks with the Stagirite that place is separable from body and that it remains immobile. Place is not destroyed when one removes a body from it; air rushes into the place one has emptied, and water fills the place air has left.[1]

Avicenna defined place in the same manner as al-Kindi, that is, in the same manner as Aristotle. Avicenna expressed himself as follows in his treatise, *The Fountains of Wisdom*: "The place of a body is the surface surrounded by what is next to the body, in which the body is contained."[2] And in the *Nadjat* he wrote: "Place is the limit of the container touching the limit of the contained; that is real place. The virtual place of a body is the body surrounding the one we are considering."[3]

The above definition of place necessarily brings Avicenna into the same conflict that confronted Aristotle. Because of the definition, the ultimate heaven has no place. How can it move then? Averroes has preserved for us Ibn Sina's [Avicenna] reply to this embarrassing question.

According to Avicenna, the revolution of a sphere on its axis is not a movement from one place to another; it is a movement in place. In order that a body be animated by such a movement in place, it is not necessary for it to have a place. The eighth sphere then is not in a place either by itself or by accident; but it can, however, rotate on itself.[4]

Averroes did not have to work hard to exhibit Avicenna's error; the sphere rotating on itself can be divided into parts and each part passes from one place to another in the course of its movement.

The problem of the eighth sphere, which was so embarrassing for the Peripatetics, was the occasion for Avempace (Ibn Bajja) to develop an interesting theory. Averroes, who describes this theory, thinks that Avempace received it from al-Farabi, who conceived it in order to refute Joannes Philoponus.[5]

In any case, this theory clearly has the imprint of Themistius's influence. Themistius, we know, conceived the place of the eighth sphere in another manner from the place of all other bodies in the universe. Each body has as place the body surrounding it; the eighth sphere, however, has as place the body contained inside it, meaning Saturn's orb.

This opposition, slightly changed, is the point of departure of Avempace's theory.

With respect to place, one must, according to Avempace, distinguish two categories of bodies.

In the first category are the elements that suffer generation and corruption, whose natural movement is a rectilinear, centripetal or centrifugal movement. In the other category are the celestial spheres, eternal bodies whose natural movement is a uniform rotation.

The straight line is not like the circle, a line complete in itself to which nothing can be added; it can be shortened and lengthened. Therefore, in order to limit the movement of an element capable of generation or corruption, one must enclose it in a container. That is why bodies whose natural movement is rectilinear, meaning the elements and their mixtures, must be contained *from without*; the place of one of these bodies is the part of the container immediately contiguous to that body.

The celestial spheres do not need to be lodged in this fashion; thus they are not lodged *from without,* but *from within.* Each of them has as its place the convex surface of the body it contains and around which it rotates. With respect to this, there is no distinction between the ultimate heaven and the other orbs. All the celestial orbs have a place essentially, and not accidentally; and place is defined in the same manner for all.

As for the whole universe, its manner of lodging consists in that each of its parts has a place.

Such is the theory of Ibn Bajja [Avempace]. Averroes has no problem showing that it is not in conformity with Aristotle's thought. But would the Stagirite have accepted as his own the commentaries of Averroes on this subject?

In some passages of the fourth book of the *Physics*, Aristotle appears to identify place with the immobile body which is the term with respect to which one can recognize and determine the movements of other bodies. This identification, still confused and almost latent in the writings of the Philosopher, is distinctly affirmed in the writings of the Commentator.

When, for example, Aristotle affirms the immobility of place, Averroes also adds: "Place is immobile essentially; in fact, place is that toward which something moves or in which something rests. If something were to move toward a term which is itself in movement, the thing would be moving in vain."[6]

The principle serving as a point of departure for such a theory is, according to the evidence, the following proposition: the local movement of any body supposes the existence of some concrete immobile body about which or around which the first body moves. Every time that Averroes formulates this principle, he invokes the authority of the book, *De motibus animalium*, attributed to Aristotle;[7] in this respect he is imitating the behavior of Alexander of Aphrodisias, Themistius, and Simplicius. We have seen, however, how much the meaning of the proposition formulated in the *Treatise on the Movements of Animals* differs from the meaning given to it by these commentators.

Alexander, Themistius, and Simplicius had recourse to the above principle, though not given its true meaning, in order to establish Aristotle's conclusion that the rotation of heaven requires the existence of an immobile central body. Averroes makes similar use of the proposition.[8]

Aristotle's conclusion seems to gain importance for Averroes since he can use it to argue against Ptolemy's system of eccentrics. He repeats the proposition in order to reject Ptolemy's system: "A body moving circularly must move around a fixed center."[9] The *Almagest* also seems to be the target of the following passage:

> It is absolutely impossible that there are epicycles. A body moving circularly must necessarily move so that the center of the universe is the center of its movement. If the center of its revolution were not the center of the universe, there would then have to be another center outside this center; there would then have to be another earth outside this one, and that is impossible according to the principles of physics. One can say the same for the eccentrics that Ptolemy assumes. If celestial movements required several centers, there would have to be several heavy bodies external to this earth.[10]

The impossibility of Ptolemy's system is thus connected by Averroes to the principle he claims to have drawn from the *De motibus animalium*: All moving bodies suppose the existence of a body at rest. But this does not exhaust the consequences of this principle. Ibn Rushd also uses it to deduce a solution to the problems relative to the eighth sphere.

After having recalled what Alexander, Themistius, Joannes Philoponus, Avicenna, and Avempace assert about this "important question," the Commentator adds:

> Here is what one must say about this subject: Any body moving of its own movement, *per se* [and not *per accidens* in virtue of the movement of another body to which it is attached], requires a motionless body with respect to which it moves; that is affirmed by Aristotle in the treatise, *On the Movements of Animals*. Doubtless, this immobile term constitutes the place *per se* of the mobile body when it contains the body within it; on the other hand, when it does not contain all the parts of the mobile body, this immobile term is the place *per accidens* of the mobile body. That is what is produced for the celestial bodies. One sees then that in order for a body to move *per se*, it is not necessary that it is in a place *per se*.[11]

In this way the ultimate orb possesses a place, but a place *per accidens*, meaning the central immobile body required by its rotation.[12]

The Questions of Master Roger Bacon

We have previously expounded upon the difficulties with respect to place that began to preoccupy the masters of Christian Scholasticism [in the *Système du monde*, vol. VI]. We have reported on the suggestions contained in the writings of William of Conches and Gilbertus Porretanus [in the *Système du monde*, vol. III, chaps. 3 and 4]; the statements held by the latter in his *Treatise on the Six Principles* are often cited. One has to come to Roger Bacon to discover an exposition of the theories of place which Aristotle and his commentators developed.

The *Questions* of Master Roger Bacon on Aristotle's *Physics* indicate to us that the discussion about the theory of place got a very ample development at the Faculty of Arts of Paris during the middle of the thirteenth century.

Of the two series of *Questions on the Physics* conserved in the manuscript of Amiens, it is the second series that furnishes us the important details about the doctrine with which we are concerned.

Bacon's study of the fourth book leads him to some uninteresting considerations about place in general. He then examines a series of problems on the place of various bodies; there he seeks to discover what are the natural places of fire, air, water, and earth. Finally, passing to the celestial bodies, he examines the following questions successively:

I. *Queritur de loco celi et primo queritur an celum habeat locum.*

II. *Queritur ergo utrum celum habeat locum per se vel per accidens.*

III. *Habito quod celum habeat locum per accidens, queritur quomodo istud per accidens habeat reduci ad per se.*

IV. *Queritur utrum celum habeat aliquo modo locum in quo.*

V. *Queritur ergo de toto universo utrum habeat locum.*

VI. *Queritur de orbibus planetarum et de loco istorum orbium, an habeant locum per se vel per accidens, scilicet planete.*[13]

"Any body having local movement necessarily possesses a place, and heaven moves in this fashion; therefore heaven has a place."[14]

Except that the place of heaven is not a place *per se*; it is only a place *per accidens*.[15] A body which of itself (*per se*) moves by a rectilinear local movement requires a place *per se*, but it is not the same for a body moving only by a circular local movement; such a body needs only a place *per accidens*. There is a great difference between these mobiles; the mobile body having a rectilinear movement moves from one place to another. On the other hand, the mobile animated by a rotation does not change place. That is why, as we shall see, it does not need a place *per se*.

Assuming that heaven does not have a place *per se*, but only a place *per accidens*, it remains for one to specify this assertion and to explain in what way heaven is lodged *per accidens*: that is the issue to which Bacon devotes his third question.

He examines three theories successively, starting with the one he likes least in order to finish with the one he thinks best.

The theory he presents first—and which he thinks least

satisfactory—is Themistius's theory: heaven is in a place *per accidens* in that its various parts are in a place *per se*.

> But the most elevated celestial sphere has no parts other 'than those juxtaposed. . . . It has no body outside it holding it and containing it. . . . Its juxtaposed parts have no more place than heaven itself has a place.[16]

Doubtless heaven is formed of eight spheres containing each other so that each inner sphere has a place *per se*, but the ultimate sphere does not have one; hence when one says that heaven has a place *per accidens* because each of its parts has a place *per se*, one merely plays on words: "what we prove about heaven, we are intending to prove about the ultimate sphere."[17]

Having rejected Themistius's thesis, Bacon touches upon another thesis he says is better: "Heaven has a place, but since it does not need one, one says that it has it *per accidens*. . . . The place of heaven is the ultimate convex surface."[18] That is the thesis that Gilbertus Porretanus maintained in his *Treatise on the Six Principles*.

One can object to this thesis as follows:

> Place is separable from the lodged body; but the ultimate convex surface of heaven is not separable from heaven, which is the lodged body, since it is its termination.[19]

Bacon does not think this objection holds completely; one can reply as follows:

> The ultimate convexity can be considered in two different ways. One can consider it as the termination of heaven; in this way it is not separable from it and is not its place. One can also consider it *per se* and define it as an abstraction from heaven, according to its own essence and insofar as it is a surface; in this way it is the place of heaven.[20]

In spite of this, Bacon does not think that Gilbertus Porretanus's position can be held as the true position:

> Any body in the world has a surface distinct from the body, essentially and by its definition; however, one cannot say that this surface is the place of the body. It is therefore equally false that heaven has a place for this same reason.[21]

Our author finally arrives at the solution that receives his approval; it is the truly Peripatetic solution, the one defended by Averroes:

"Heaven has a place *per accidens* because its center has a place *per se*.[22]

Here are the clearly Peripatetic considerations with which Bacon corroborates his doctrine:

> It is said in the eighth book of the *Physics* that heaven derives its fixity from the fixity and rest of its center; heaven, in fact, remains fixed as a whole while its parts move. That is why Aristotle and the Commentator say that heaven holds the same place by its substance, although its form changes. It is evident that a circular body has by itself no relation to anything [external] while it has an essential relation with its center. Its movement is a constant movement around its center, so that the only place it needs to have is the place around which it is placed (*locum circa quem*); this place is first (*primo*) and *per se* the place of the center of heaven. It does not contain heaven. Since this place does not hold and contain heaven and since it is only a place around which heaven rotates, one says that it is the place of heaven *per accidens*. But it is first and *per se* the place of the earth, because it holds, contains, and conserves the earth.[23]

One can object to this solution that

> the local movement of heaven precedes the movements of the lower things, as it is stated in the eighth book of the *Physics*. In any case, rest is only the privation of movement. The movement of heaven therefore precedes the rest of any sublunar body; hence it precedes the [rest of] the center. But once heaven has a local movement it must have a place; the place of heaven would then have to precede any rest and any movement of the earth. And that which precedes something cannot be brought about by that which succeeds it.[24]

It is therefore inadmissable for the existence of the place of heaven to be deduced from the immobility of the earth.

Bacon replies as follows to this argument:

> There are two kinds of earthly rests.
> The earth can be considered as a whole, insofar as it is the center of the world; considered as such, it is at rest, but this rest, which is of the whole earth in its sphere, is not a privation of movement. This rest precedes the movement of heaven and it is necessarily required for the movement of heaven.
> There is another kind of earthly rest. One can consider

the earth in one of its parts which first moves and then remains at rest; this rest is a privation of movement—it succeeds movement. The movement of heaven precedes this rest and similarly precedes the rest of the other lower things.

But the first rest is the earthly rest *per se*; and the movement and place of heaven must be related to this place [of the earth] as being what constitutes them *per se*.

(*Ad aliud dico quod quies terre duplex est, quia terra secundum se totam in quantum est centram mundi potest considerari, et sic quiescit, sed non quiete que sit privatio motus, quia talis quies est totius terre in sua spera. Ideo ista quies precedit motum celi et de necessitate exigitur ad motum celi.*

Alia est quies terre secundum quod terra consideratur pro aliqua ejus parte, que prius movetur et deinde quiescit. Talis quies est privatio motus et sequitur motum; et talem quietem precedit motus celi et quietem similiter aliorum inferiorum.

Sed prima quies in loco est per se. Ideo motus celi et locus ejus ad istum locum habent reduci tanquam ad suum per se.)[25]

We do not think that any Christian Scholastic more clearly perceived Aristotle's thought concerning the place of heaven; we do not think that any expressed it with as great a precision. Wrestling with one of the more essential, though subtle, theories of Peripatetic physics, Bacon was able to master it as early as when he taught at the Faculty of Arts at Paris, and gave thereby a striking proof of his perspicacity.

The whole universe has a place.

The universe contains the eight celestial spheres and the four elements; and each of these parts has a place, at least *per accidens*. "Heaven, the eighth sphere, has a place *per accidens* because its center has a place *per se*."[26] Hence one can say that the universe whose parts have at least a place *per accidens* also has a place *per accidens*.

A final question relative to the place of the lower spheres requires the attention of our author.

It seems that one can say that each of these spheres is in a place because it is contained and enveloped by the orb located immediately above it.

But it is not natural that a nobler body is contained by a less noble body, and the orb of Mars is less noble than the sun's orb; it cannot therefore contain the sun's orb and be its place.[27]

Here is Bacon's reply:

> I suppose that the orbs are contiguous to one another and that they differ from one another by a numeric difference and a specific difference.
>
> That posited, I notice there are two kinds of places.
>
> There is the place around which the body is disposed (*locus circa quem*). Each orb, in the same manner as the set of spheres, has the center as its place, for each circular body has a place around which it is disposed.
>
> There is the place in which a body is contained (*locus in quo*).
>
> There can then be a question about the place one says lodges, contains, perfects, conserves, and limits the body. The celestial orbs do not have a place of this kind, while that kind of place is the natural place of lower bodies.
>
> There can also be a question about a place that only contains. In this sense, each lower orb has a place, for the orbs contain each other; but since they have no need of it, one says only that they have such a place *per accidens.*
>
> The celestial spheres therefore have the least possible lodged nature: they are perfect in themselves, they are terminated, and their conservation is assured without needing any place.[28]

In many ways Albertus Magnus's opinion was similar to Roger Bacon's, but in general, it was expressed less concisely.

Albertus Magnus

Albertus Magnus said nothing truly original with respect to the nature of movement and place; he limited himself to commenting upon Aristotle and Averroes.[29]

Averroes did not write a commentary on the treatise *De motibus animalium*; Albertus Magnus, on the other hand, wrote two paraphrases of this work.

In one of the paraphrases he shows himself faithful to the meaning of the propositions as formulated by its Greek author. He admits that there must exist a fixed body outside a mobile body; but the fixed body is not required as the term to which one relates the movement. It is necessary as support which the motor can use while it produces its effort. This truth is illuminated by means of examples invoked from the *De motibus animalium*. Following the doctrines of this treatise, Albertus proves that the immobility of the earth is not intended to offer a point of support for the motor producing the celestial revolutions.[30]

The second paraphrase, freer than the first, expresses itself ambiguously about the fixed body all progressive movement supposes; it would not require much effort to read these words as the principle Themistius, Simplicius, and Averroes believed to have derived from the *De motibus animalium*.

Albertus first remarks that any mobile part of an animal's body in movement supports itself on some other part of the body; if the latter part is not fixed, it has a support, and so forth. Somewhere we reach a fixed part of the body.

We can reason similarly about any movement; since the series of mobile bodies cannot go to infinity, one necessarily reaches the conclusion that any progressive movement supposes an immobile body.

> Such a movement is surely analogous to the movement of the compass. . . . When a compass moves, it moves in virtue of its form, in virtue of its configuration as a compass, which gives it its existence and which specifies it. But at the same time, during its movement, one of its parts remains attached to an immobile center; the mobile compass describes a circle around this immobile center.[31]

This example derived from the compass does not bring to mind the necessity of providing a fixed support for the motor that moves a body; it seems more fit for bringing to mind the immobility of the central body needed for a rotation.

It is the latter idea that presented itself to Albertus Magnus's readers; Peter of Auvergne is evidence for this. A prominent master of the University of Paris and rector of the university at the end of the thirteenth century, Peter of Auvergne was one of the closest and most illustrious disciples of Albertus Magnus and Thomas Aquinas. He presents the following considerations in his commentary on the treatise *De motibus animalium*, in which the influence of the bishop of Ratisbon can be clearly discerned:

> In the same fashion that heaven would not be able to move if there did not exist something fixed and immobile, the movement of an animal requires that there is, outside the animal, an immobile support which it can use in order to move. . . . But let us understand that the reason for which an immobile foreign body is necessary is not completely the same for heaven and for the animal. There is, however, a

common theme in the two cases. In fact, for a body to be in movement, there must exist another body with respect to which the moving body is disposed in another fashion than it had been before; this second body is immobile or, if it moves, at least it differs from the first by the form and speed of its movement. Therefore if it moves, it would have to be that either the series of mobiles goes to infinity, or that one can finally reach a completely immobile termination. The animal and heaven share that reason in common, but there is another reason specific to the animal. In order to move, the animal must push and pull.[32]

We can prove that Peter of Auvergne grasped Albertus Magnus's thought when he cited the example of the movement of the compass since we can find the same thought expressed in various other writings of the bishop of Ratisbon.

We can display it as written in his *De Caelo*.

As did Aristotle, Albertus seeks the reason why the whole celestial sphere does not move by a unique movement; during this inquiry, guided as always by the example of the Philosopher and the Commentator, he wrote:

> Let us take as point of departure the conditions required by circular movement. According to what has been demonstrated in the *De motibus animalium*, let us say that no body can move circularly unless it does so on another body that is fixed and immobile; if the latter body moved, the former would not be able to inscribe a circle remaining always in the same place. The virtue of the immobile body adds at least the fixity of circular trajectory to the virtue of the mobile; for none of the parts of the orbit accomplishing a revolution remains stable, fixed, and deprived of movement. The poles appear to be immobile, but they move in place, without going from place to place or from location to location. Similarly, it is evident that a body rotating is not of the same nature as a body fixed at its center [if it were of this nature its center would be its natural place and its various parts would descend to it, which they do not do]. . . . Therefore if it must be that every body moving circularly moves on something fixed and stable, then it must be that a fixed and immobile body is located at the center of the universe; and that body can only be the earth.[33]

The existence and fixity of the earth are thus required by the rotation of the celestial orbs.

As for the multiple movements of the heavens, Albertus, as a faithful Peripatetic, assigns as their object the generation and corruption of things corruptible which the regions of the elements contain.

Three things contribute to this generation and corruption: first, the perpetuity of existence; second, the continual opposition between birth and death; and third, the variety of the forms of engendered species. Of these three things, the first is dependent on the diurnal movement, the second stems from the revolutions accomplished according to the ecliptic, revolutions causing the sun and other generative stars to ascend or descend, and the third is caused by the particularities of the paths of the planets which sometimes cause these planets to be near each other and sometimes cause them to be far from each other.

In his *Physics*, Albertus also invokes the necessary existence of an immobile body at the center of an orb animated by a rotational movement.

After having paraphrased what Aristotle asserted about the nature and immobility of place, the bishop of Ratisbon then broaches the debated question about the place and movement of the eighth sphere.[34]

With respect to the "important question" about the place of the ultimate orb, Albertus's avowed goal is to expose and clarify Averroes' solution, which he adopts.

> Averroes states that the first mobile is in a place *per accidens* while its movement is *per se*, and not *per accidens*. One says that this orb has a place because its center is in a place *per se*; the orb then is in a place *per accidens*. In fact, Aristotle declared in his treatise *De motibus animalium* that any movement proceeds from an immobile body. The movement of the eighth sphere must therefore proceed from something immobile. This something would be a place *per se* so that the orb would be in a place *per accidens*.[35]

When invoking the principle which, following the example of Alexander, Themistius, Simplicius, and Averroes, he attempted to derive from the *De motibus animalium*, Albertus Magnus also submitted it to a discussion that his predecessors did not submit it to. According to Albertus, the principle does not come into play with respect to the natural movements of heavy or light bodies. It must be restricted to movements produced by a soul, like the movements of animals, or by an intelligence, like the movements of the stellar bodies. This discussion illuminates the profound deformation of the thoughts expressed in the *De motibus animalium* by the commentators. The author of the treatise in question taught, in fact, that the motor of the heavens has no need of fixed support since it is a pure intelligence.

The eighth sphere does not move *secundum subjectum,* meaning it does not move as a whole; taken as a whole, this mass keeps an invariable position. It moves with respect to its form, meaning according to the relative position which its various parts effect with respect to the immobile body occupying the center; this form, this relative position, changes from instant to instant.

This, then, is Averroes' solution: the first mobile is in a place insofar as it is around its place, and this place is the convex surface of the immobile body located at its center. According to Albertus Magnus, "this solution appears subtle because it is presented obscurely; it must be understood according to the explanation just given."[36]

But an objection may be formulated to this explanation. In order for heaven to rotate, the earth must remain immobile, containing the center of the movement; that is the proposition upon which the whole previous deduction depends. Does not this proposition affirm that the immobility of the earth is the cause of the movement of the eighth sphere? Moreover, this affirmation seems unacceptable for a Peripatetic. The ultimate orbit receives its movement from the first motor; the immobility of the earth is the effect, not the cause, of the celestial movement.

Here is Albertus Magnus's reply to this argument:

> The movement of the ultimate orb can be considered from two distinct points of view. One can see in it the movement of the first mobile. One can also study it as a revolution accomplished in place. If one considers this movement from the first point of view, the movement derives from the first motor which presides on the eighth heaven, and not from the central body. If, on the other hand, one considers it from the second point of view, the rotation of the final orbit stems from the immobility of the central body.[37]

Although Albertus Magnus showed himself to be the faithful disciple of Averroes with respect to these questions, he did not follow the path traced by the Commentator completely. Since all rotations suppose a body whose immobility fixes the center, Ibn Rushd had declared that the epicycles and eccentrics imagined by Ptolemy were physical impossibilities. While holding on to the principle formulated by Averroes, Albertus Magnus refused to accept its consequences; he safeguarded Ptolemy's system, but to do so required him to accept an inconsistency.

Saint Thomas Aquinas

Albertus Magnus's doctrine seems to have been influenced only by Aristotle and Averroes. The theory of place and local movement that Thomas Aquinas develops bears the evidence of other influences. One can recognize in it some of Themistius's thoughts. One can also glimpse in it, although more vaguely, some similarities with the theory proposed by Damascius and completed by Simplicius. These similarities were strengthened and clarified by some of Saint Thomas's successors to such a degree that the doctrine of Damascius and Simplicius triumphed over Aristotle's doctrine in the Scotist school. Saint Thomas shows himself to be a more scrupulous commentator than his predecessors when exhibiting the doctrine of the Stagirite that the rotation of the heavens requires the immobility of the earth; he does not invoke the propositions formulated in *De motibus animalium*:

> There has to be something remaining immobile at the center of a body moving circularly. It is evident that any circular movement occurs around a fixed center. And it needs be that this center is located in a fixed body, for what we call center is not something subsisting in itself. It is an accident belonging to something corporeal; this center can only be the center of a body.
> This fixed body must be part of the world . . . but it cannot be part of the mobile orb, meaning the celestial body. . . . That which is at the center is eternally immobile, as heaven moves eternally. . . . And that which is naturally immobile at the center is the earth. . . . Therefore, if heaven revolves eternally, the earth has to exist.[38]

Simplicius's influence, which is recognized at the end of this *lectio*, is evident in the consideration by which it is proved that an immobile center cannot exist anywhere other than in an immobile body.

The rotation of the ultimate celestial sphere supposes the existence of an immobile central body; must one say with the Commentator that this immobile pivot constitutes the place of the ultimate orb and that this orb is in a place *per accidens* because the central pivot is in a place *per se*? Saint Thomas refuses to accept this interpretation of the words *per accidens* by which the Stagirite qualifies the place of the final celestial sphere.

Aristotle does not say that a body is in a place *per accidens* when another body which is completely foreign to it is in a place. It seems ridiculous to me to maintain that the final sphere is accidentally in a place by the mere fact that its center is in a place. Thus I prefer to give my approval to Themistius's opinion according to which the ultimate sphere is in a place because of its parts.[39]

In support of Themistius's opinion, Saint Thomas Aquinas develops considerations that merit being reproduced completely; we do not recognize in them the influence of the Greek commentator, but that of Aristotle himself. We can also recognize in them the influence of Avempace's theory which the Angelic Doctor related before reporting Averroes's theory.

Place would not be investigated if it were not for movement; movement calls attention to place because bodies succeed each other in one place. Hence although a body does not of necessity have a place, nevertheless, a body moved with respect to place does have a place of necessity. Therefore, it is necessary to assign a place to a body moved in place insofar as one considers in that movement a succession of various bodies in the same place. Thus in things moved in a straight line, it is clear that two bodies succeed each other in place with respect to the whole. For the whole of one body leaves the whole place and into that whole place another body enters. Hence it is necessary that a body moved in a straight line is in place with respect to its whole self.

But in circular movement, although the whole comes to be in various places by reason (*ratio*), nevertheless, the whole does not change place in the *subject*. For place always remains the same in the *subject*, and it is diversified only by reason (*ratio*). . . . But the parts of the mobile change place not only according to reason (*ratio*), but also in the *subject*. Therefore in circular movement attention is directed to the succession in the same place, not of whole bodies, but of parts of the same body. Hence for a body moved in a circle, a place with respect to the whole is not due of necessity, but only in respect to the parts. . . .

Moreover it is much more suitable to say that the ultimate sphere is in place because of its own intrinsic parts than because of the center which is altogether outside of its substance; and this is more consonant with the opinion of Aristotle.[40]

Let us now describe Aquinas's general doctrine about the nature and immobility of place.

We have seen that Aristotle, when treating this question, successively adopted two incompatible definitions of place. First he called the place of a body the portion of matter immediately contiguous to the body; but place defined in this way is not immobile. In order to assure immobility, Aristotle declared the place of a body to be the first immobile surface surrounding the body.

To avoid this change of definition, which is a serious breach of logic, was the principal objective of several Scholastic commentators. Toward this end they distinguished two senses of place: according to the first sense, place is mobile; according to the second sense, place is immobile.

Such a distinction is already implied, so briefly that its clarity suffers, in the extremely concise *Summa* that Robert Grosseteste, bishop of Lincoln, wrote on Aristotle's *Physics*.

Robert Grosseteste remarks that the place of a body is an accident of the body, so that it has to move with the body. Having taken up the difficulty, he devoted only one phrase to it: "Materially, place is mobile; formally it is immobile."[41] The bishop of Lincoln does not tell us what constitutes *formal* and *material* place.

Saint Thomas Aquinas does fill in the details, however.

First, one can call the portion of matter immediately contiguous to a body, the *place of the body*. This place, insofar as it is formed from some matter, is mobile. The body in question is surrounded by some air or water, and a little later the surrounding air or water can have changed.

Next to the place thus understood, which is mobile, we ought to consider another place; this latter place is bounded by the extreme parts of the ambient bodies that serve to delimit the former place, but it is constituted by a relation holding between the extreme parts of ambient bodies and the set of celestial spheres. It determines the order or the situation of the body contained by these parts to the whole immobile universe; this place is the rational place (*ratio loci*).

> Although the container is moved insofar as it is a body, nevertheless, considered according to the order it has to the whole body of heaven, it is not moved. For the other body that succeeds it has the same order and site in comparison to all of heaven that the body which previously left had.[42]

The rational, immobile place is a fixed relation with respect to the whole heaven; heaven itself is determined by its central body and its poles, so that one can define the rational place as the situation with respect to the central body and the poles. "It is clear that the whole *ratio* of place in everything that contains anything is from the first container, that is, heaven."[43]

Here is an example, suggested by Aristotle's text itself, showing how any *ratio loci* is drawn, in the final analysis, from the ultimate sphere: In the domain of heavy or light elements, the difference in place between above and below is determined by a comparison between the center of the world and the concave surface of the lunar orb; moreover, we have seen that the fixity of the central body is required for the rotation of the ultimate orb.

> Although the concave surface, which from our perspective terminates the celestial spheres moving in a circle, moves in a circle, nevertheless, it remains permanent insofar as it is similarly related, that is, it is the same distance from us [meaning the immobile center].[44]

Saint Thomas also expressed the same thought in the opusculum, *De natura loci:*

> That is the way in which we ought to understand that the extreme parts of natural bodies form the place of other bodies; they form it in virtue of the relative position, the order, and location that they present with respect to the set of celestial bodies. The latter is the natural container, the principle of all conservation and all location (*primum continens et conservans et locans*).[45]

The above final sentence can already be found in Thomas's *Commentary on Aristotle's Physics*; the only word missing being *conservans*. The presence of the word in the passage we have just cited is not something fortuitous and of little importance; it is an indication of the difference between the theories of the opusculum *On the Nature of Place* and those of the *Commentary on Aristotle's Physics*. Moreover, the theories of the former merit our attention; they carry the seed of several of the doctrines Scotus and his disciples later professed.

In any case, it has been sufficiently demonstrated that the opusculum, *De natura loci,* is not Thomas Aquinas's,[46] that it has to be ranked among the apocryphal treatises which over the ages have been attributed to the *Doctor Communis*. Nevertheless, we should analyze it here, for it appears to us to be the work of an immediate or very close disciple, so that it seems proper to use it, in some circumstances, as an elucidation of Thomas's thought.

Here is the doctrine developed in the opusculum:

The place of a body is the extreme part of the container (*ultimum continentis*); what, then, is the difference between the place and the surface of the container? The surface is the limit of the container considered intrinsically; it becomes the place when one considers it extrinsically, not as the limit of the body to which it belongs, but as the boundary of the contained body it surrounds. The surface of the container and the place are the same thing materially, that is, they are the extremity of the container. The characteristics differentiating them are purely formal.

This formal characteristic, extrinsic to the container, upon which the extreme part of the container becomes the place of the contained body does not simply consist in enveloping the body; it also implies an aptitude to conserve the body. Place does not merely contain, it also conserves.

The conservational virtue of place explains why its *ratio*, from which the permanence of place is derived, consists in the location it occupies with respect to heaven; in fact, among the bodies capable of generation and corruption, matter cannot have the property of conserving another portion of matter if it did not derive it from heaven. And this virtue or this influence which it receives from heaven depends upon its distance from it and its location with respect to it. That is why, in all containers, the *ratio loci* is obtained by comparison with the supreme orb, the *primum locans*, the body lodging all other bodies.

The opusculum *De natura loci* ends with an article entitled: In What Manner is the Final Sphere in a Place?

With respect to this problem, the author first reproduces the solution given by Thomas Aquinas in his *Commentary on Aristotle's Physics*. He addresses an objection to this solution, also to be found in the *Commentary*: actual existence and movement is proper for the whole but not for its parts; the way a body is in a place depends upon the way it is in movement. Therefore a body ought to be in a place because of its entirety, not because of its parts.

This objection receives the following response in the *Commentary*: the parts of the supreme orb do not exist in actuality, but potentially. In the same manner they are not actually in a place, they are there potentially; if one were to distinguish a part from the rest of the orb, it would be in the totality of the orb in the same fashion as it would be in a place. Thus the ultimate sphere is in a place accidentally because of its parts, which are themselves lodged potentially; this manner of being in a place suffices for a movement of revolution.

Not only does Thomas Aquinas find this response conclusive, but it seems to him to evince a harmonious gradation among the beings.

Above the supreme orb there are only, according to Aristotle's teaching, substances barren of place and essentially immobile. Within the eighth sphere are bodies, each of which is in a place in its totality and actually; these bodies move or can move by the total transportation of their substances from one place to another. Between these two kinds of beings is the ultimate sphere; it is not in a place because of its totality, but because of its parts; these parts themselves do not have an actual place, but a potential place. Thus this orb cannot displace itself wholly; the movement of revolution is the only movement proper to it.

The reply suggesting this view is repeated by the pseudo-Thomas in his opusculum, but doubtless it does not seem to him capable of resolving the objection that provoked it, for he follows this immediately with the sentence, "If we wish to hold on to the portion of truth contained in this opinion, we must say that the ultimate heaven is not in a place purely and simply, but it is in a place accidentally, in that it encircles its place."[47]

Here, then, is an odd turn of events: Saint Thomas Aquinas's theory being linked with Averroes's theory, the theory Thomas ridiculed! That alone would be sufficient to indicate the apocryphal character of the opusculum, *De natura loci*.

In support of his theory, the author develops considerations in which the influence of Ibn Bajja, an influence avowed by him, can be noted more clearly than it can be noted in the authentic writings of Thomas Aquinas.

Any body naturally at rest is in a place; in fact, in order for a body to be naturally at rest, it must be surrounded by bodies suitable for its nature; having a container, it has a place.

On the other hand, it can come to be that a body moving naturally does not have a place. A distinction is necessary with respect to this.

There are bodies whose natural movement has as its end the maintenance of existence and the increase of perfection. These bodies move toward the bodies suitable to their nature and able to offer them a natural place because they are actually surrounded by bodies repugnant to their nature; therefore, when these bodies move, they are in a container, in a place.

Other bodies do not move for their existence or their perfection; they are moved by an intelligence, and their movement has as its end the development of the causality of the First Cause; these bodies are the celestial orbs. It is not necessary, in order for them to move, that they be surrounded by bodies contrary to their nature, nor that they aspire for a container in conformity with their nature. They have no need of a place.

In other words, place does not merely contain the lodged body, as we have seen; it exercises an action of conservation with respect to it. The elements and the perishable composites need to be conserved; they require a place. The celestial bodies are imperishable; they do not need to be conserved and have no need of a place.

A difficulty can be noted here. What has just been asserted is true not only for the ultimate sphere, but for all celestial orbs; none of them need a place. However, with the exception of the last orb, each has a container and, hence, each is in a place.

In fact, one can say that the lower orbs are in a place, as long as one does not give these words the sense one gives them when they are used for the corruptible elements and their composites. The place of these bodies does not merely contain, but it also conserves; the place of the lower orbs contains them without conserving them.[48]

Giles of Rome

Without bothering too much about chronological considerations, we shall now turn to what Giles of Rome has written with respect to place; the thoughts of this author demand to be compared with those of Thomas Aquinas.

Thomas does not state that the mobile and proper place of a body can be called *material place*, while the term *formal place* suits the immobile *ratio loci*, but one can easily conclude this from a comparison he uses: "Similarly, one says that fire remains identical with respect to its form even though the combustion of part of the wood and the addition of some new wood varies it with respect to its matter."[49]

Although Thomas Aquinas did not use the terms *material place* and *formal place*, which had already been used by Robert Grosseteste, Giles Colonna—called Giles of Rome—did not hesitate to introduce them into his philosophical vocabulary. With respect to the difficult problem of the immobility of place, he took up textually the aphorism of the bishop of Lincoln: *Locus est immobilis formaliter, mobilis vero materialiter.*[50]

The example he uses in order to explicate his thought is the same example that Aristotle already considered, the example of an anchored ship in flowing water. Materially, the water bathing the ship and constituting its place is renewed constantly. Formally, one says that the ship remains in the same place, because the flowing water bathing it retains its location with respect to the immobile shores of the river.

The immobile place is therefore not, as Aristotle wished, the fixed container within which the ship is contained—it is not the banks and bed of the river—it is a certain fixed disposition of the ship with respect to a reference which is itself immobile.

What is this immobile reference to which the location constituting the formal place has to be reported? This immobile reference is the universe.

The various parts of the universe are mobile, but the universe itself as a whole (*secundum substantiam*) is immobile. The location that constitutes formal place, whose permanence stems from its immobility, is its position with respect to the whole universe. "Let us suppose that a man is immobile on the surface of the earth, and that the wind carries off all the air surrounding him; we would not say that he has changed place, even though he is now in some other air than he was once. We would say that he stayed at the same place, because he kept the same location with respect to the universe."[51]

The immobility of the universe *secundum substantiam* entails the immobility of the center of the world; the whole world would have to be displaced in order for the center to change. When speaking about an immobile center, Giles of Rome, like all his predecessors from Aristotle to Saint Thomas Aquinas, refers to a central fixed body, not actually a point. A little further on, when designating the invariable pivot for the celestial revolutions, he uses the words *center* and *earth* indifferently.[52] The fixity of the central body entails the fixity of the spherical surface limiting the universe and the fixity of the surfaces delimiting each of the celestial orbs, for each of these spheres has an invariable radius; the fixity of the poles then results from the immobility of the central body and the ultimate surface of the world.

Therefore, instead of defining the formal place of a body as its location with respect to the whole universe, we can say that it is the position it occupies with respect to the center and the poles of the world. But the first definition is preferable to the second, since the fixity of the central body and the poles derives from the fixity of the universe.

Let us return to the doctrine:

> The material place of a body is the surface of the body containing the first body; what is formal in place is its location with respect to the universe, for the position of the universe itself is absolutely immobile. . . . From the formal point of view, place is not mobile by itself or accidentally; from the material point of view, the place of a body is not mobile by itself but it is so accidentally [since the ambient bodies that form the place can be displaced].[53]

In all of what we have reported, Giles of Rome did no more than follow Saint Thomas Aquinas's thought. He modified it at only one place: he affirmed that the *ratio loci* was formal place. The Angelic Doctor only insinuated it. But Giles deviates from his master's doctrine with respect to his response to the noted question, What is the place of the final celestial sphere? In his solution of the problem, the Angelic Doctor, at least in his *Commentary on the Physics*, sided with Aristotle against Averroes; Giles sides with Averroes against Aristotle.

He first recalls the Thomist objection against Averroes's system:

> Is heaven in a place because of its center? It seems not. In fact, the center seems completely extrinsic to heaven; it appears not to have anything of the essence of heaven. Hence it would be ridiculous to maintain that heaven is in a place because its center is in a place.[54]

Here now is the response to this objection; the Commentator's thought is formulated in it with unusual clarity:

> Any movement proceeds with respect to an immobile object. We can never imagine a movement if we do not imagine a fixed term with respect to which we can affirm that the body moves. Moreover the *ratio loci* is conceived as something immobile; we can therefore not imagine a local movement if we cannot conceive of an immobile object to which the *ratio loci* is related. Further, in order to fix a sphere, one has to fix its center, so that the immobility of the sphere is derived above all from the immobility of its center. Similarly one judges the movement of a sphere by comparison to its

center. Therefore, one should fix the place of a sphere by comparison to its center. . . .

The ultimate heaven is, at the same time, wholly at rest and wholly in movement.

It is wholly at rest because, taken as a whole (*secundum substantiam*), it never changes place; that results from the continual immobility of its center.

Further, the ultimate orb moves as a whole in that its disposition changes endlessly. The earth that remains at rest in the center of heaven is not always seen in the same way from a region of heaven.

Therefore, one judges the immobility of heaven as well as its movement with respect to the central body. And one only asks about place in order to judge about rest and movement. One could only seek the place of heaven by considering its center.[55]

Thomas's thoughts and language are called upon by Giles in support of the Averroist solution that the Angelic Doctor rejected; Giles also uses these thoughts and language in order to refute Aristotle's solution, to which his glorious predecessor rallied.

Let us compare Averroes's solution, which has just been exposed, to Aristotle's solution; the advantages of the former will contrast favorably with the disadvantages of the latter.

The movement of heaven endlessly modifies the location of the parts of heaven with respect to the parts of the central body; the part of heaven that not long ago was related to some part of the earth is now related to some other part of the earth because of the movement of heaven. The whole heaven is related to the whole earth, but it is not always related in the same way; at the same time the various parts of heaven do not remain constantly in the same relation with the same parts of the earth. Therefore, if we compared heaven to the central body and the parts of heaven to the parts of the central body, we would find that the whole heaven moves by changing its proper disposition within its place, and that each of its parts is displaced *secundum substantiam*.[56]

Let us now assume that, in accordance with Aristotle's theory, we compare the parts of heaven to one another.

Heaven is continuous. Its movement does not affect the disposition of its parts within the whole. Then if we derived our definition of the place of heaven from this disposition, the result would be that heaven does not change place in its movement.

The various parts of heaven do not change in the position that each of them occupies with respect to one another either; two celestial parts united at one instant remain always united. Hence if one attempts to assign the place of heaven to this connection of parts, then not only will heaven, taken as a whole, not change place, but the movement of heaven will not change the place of the various celestial parts.[57]

Let us conclude: If one admits the hypothesis Saint Thomas Aquinas admitted, "heaven would not move in its totality, nor in its parts, nor by the transportation of substance, nor by change of disposition."[58]

Graziadei of Ascoli

In this exposition of theories of place, we are continuing to maintain an account that follows the development of doctrines rather than a temporal sequence; that is why we now discuss the middle-fourteenth century opinions of Graziadei of Ascoli. In fact, the aim of Graziadei appears to have been to keep and develop the Thomistic aspects of Giles of Rome's doctrine, and to correct Giles's deviation from Thomas's doctrine with respect to the place of the eighth sphere.

Graziadei first teaches us that

the perfect constitution of the nature (*ratio*) of place requires two things. First, the lodged body has to be contained by its place; since that which immediately contains the body is the surface of the enveloping body, it seems that the surface of the ambient body contributes to the constitution of place. Second, a well-determined order contributes to the constitution of place within the corporeal universe.

When something is in a place, which in itself implies these two characteristics, the thing is in a place perfectly and properly; the bodies moving by a rectilinear movement are necessarily lodged in this way.

But a body moving by rotation is not necessarily in a place that implies these two characteristics; one can find, as we shall show later on, a body of this kind not contained by the surface of another circular body.[59]

Graziadei called, as did Giles of Rome before him, the two elements whose contributions produce the perfect and complete nature of place, the matter and form of place.

There are two things that contribute above all others to making up the nature of place; one of these is the formal element of place, and the other is the material.

Formally, place does not appear to be anything other than a determined order in the corporeal universe (*ordo determinatus in universo corporeo*). Here is the demonstration of this: as long as a mobile keeps the same order with respect to the universe, one says purely and simply that it remains in the same place; once this order changes, one says then that the body changes place.

For example, a ship anchored in a river remains in the same order with respect to the universe; even though the surface of the water containing the ship changes, as long as the order remains the same, one says that the ship remains, purely and simply, in one and the same place. If on the other hand the ship were to change from one order to another with respect to the universe at the same rate as the surface of the water containing it, even though the ship would remain in the same surface of the ambient water, one would say, nevertheless, that it has changed place because of this change of order. That suffices to show that this order formally constitutes place. . . .

From this the following conclusion can be derived evidently: everything considered in itself and wholly has a determined order within the corporeal universe; therefore everything has a formal place. Everything has its own *ubi*, meaning it exists in a place taken formally.[60]

While he develops with great precision the teaching of Saint Thomas Aquinas and Giles of Rome, our author attempts to demonstrate that formal place is immobile while material place can move.

There are two things to consider about place, the surface of the ambient body which comes in as the material in the nature of place, and the order within the universe, from which the formal nature of place is derived.

Place does not possess immobility through the first constituent, for the surface of a natural body moves by the effect of the movement of the body in which it exists. Place derives its immobility from the second constituent, the order of the universe.

In fact, this order is absolutely immobile. Here is the demonstration of this:

The order that the locations—not the natural things— effect in the universe has as foundation the distance between the center and the circumference of the universe. And this distance always remains the same, for the center remains always at the same distance from the circumference, and reciprocally. Therefore, it is the case that the order of locations of the whole universe (*ordo situalis totius universi*) retains an absolute immobility, and it is the same in each part of this order.

Let us see how this order can possess immobility, even though it is only an accident.

For that, one has to consider that a relational accident never changes unless through the effect of a change in its foundation. If its foundation remains the same, it is necessary that the accident also remains the same. . . .

And the order of locations in the universe (*ordo situalis universi*) designates a relation that has as immediate foundation its distance from the center or the circumference; this order has no reference to the body's movement or to the surface of the moving body, or else, if it has a reference to it, it is only because it is [presently] conjoined to the extremity of the said distance, that it is [presently] united with a marked sign in the space included between the center and the circumference, a sign that does not always coincide with the surface of the natural body.

The bodies and their surfaces can therefore move in the neighborhood of this sign in the said space; since the sign always keeps the same distance from the center and the circumference, and since this distance is properly the foundation of the order about which we are speaking, it is necessary that the order remains immobile. Doubtless the order befalls the surface of a natural body accidentally (*accidit*), not as an absolute accident, but as a relative accident. It does not derive its immediate foundation from this surface; that befalls it only insofar as this surface coincides with the particular sign of the space which always keeps the same distance from the center [and the circumference].[61]

Graziadei's thought is extremely clear: Within the universe, the location of each point is marked by the distances from the point to the center of the universe and to some references with respect to the spherical surface enclosing the world; the set of geometric locations thus determined is what one calls the order of the universe. To know the geometric locations that coincide from moment to moment with the surface of a natural body, is to know the formal aspect of the place of the body. There is no doubt that Graziadei's thought reproduces faithfully the thought of Thomas Aquinas and Giles of Rome.

But in order that the order of the universe be immobile and that one can affirm the immobility of a location marked by its distance to certain references, these references must be fixed. What has just been asserted therefore assumes the immobility of the center of the universe and of the surface enclosing it. That is something Graziadei realizes and asserts with his usual clarity:

> You may ask whether the distance [from the center to the circumference] is immobile; one responds that it derives its immobility from the immobility of the center and the circumference. In fact, this distance is an accident of the center and the circumference or else it is part of the distance between the center and the circumference. And the distance between the center and the circumference remains immobile, as does each of its parts, for the center is always stopped in the same location and the circumference is always stopped in the same location (*quia centrum stat semper in eodem situ, et semper etiam in eodem situ stat circumferentia*).[62]

However, the final assertion from Graziadei's passage above is false. The surface enclosing the universe is the convexity of the ultimate sphere, and, as our Dominican carefully reminded us, the surface of a natural body moves by the effect of the movement of the body in which it exists. The surface embracing the universe, then, is not immobile, but mobile, and that is what leads to the destruction of the Thomist theory of the immobility of formal place.

In order for the above theory to hold, one must admit an immobile celestial sphere at the ends of the world, as did Proclus and the theologians who believed in the existence of the empyrean, or else one would have to think, with Damascius and Simplicius, that the locations constituting the order of the universe are related to purely ideal immobile points of reference.

What Graziadei asserts about the formal element of place furnishes an answer for the difficult problem about the place of the ultimate sphere.

> The ultimate sphere is by itself in a determined order with respect to the corporeal universe; therefore it has a formal constituent of place, and a *ubi* taken formally. That is why we assert that the ultimate sphere, taken in its totality, always remains really in one and the same *ubi*.
> The various parts of the ultimate sphere do not remain really in the same *ubi*; while the movement lasts, each is from instant to instant in a different *ubi* because each of them is in a different order [with respect to the corporeal universe].

Place considered materially is the surface of the ambient body; however, since no body surrounds the ultimate sphere, it is not itself in a place materially. It is there only accidentally.[63]

Moreover, the ultimate sphere has no need to be in a place considered materially:

If a body goes from one order with respect to the universe to another, because of its local movement, it must be that the body is by itself in a place considered materially. But a body that does not, by its local movement, change its order with respect to the universe requires only a place formally; that is the case of the ultimate sphere.[64]

The definition of formal place given by our author, although repeating Giles of Rome's propositions, suffices to demonstrate, contrary to Giles's conclusion, that the ultimate sphere is not in a formal place because its center is the immobile earth:

It does not result from the fact that the ultimate sphere moves around its center and other bodies that its place considered formally is this center or these other bodies, but only that the place considered formally is the order that the sphere effects with respect to the center and the other bodies; and this order consists in the location occupied by the circumference. . . .

Doubtless the sphere derives its immobility from the fixity of its center, because, once the center is fixed, the ultimate sphere keeps an invariable order [with respect to the corporeal universe] and consequently, it remains in the same place; but this order [which it conserves] is not the order of the center, it is the order of the circumference.[65]

One must therefore reject the Averroist formulation taken up by Giles of Rome: the ultimate sphere is formally in a place because of its center, the immobile earth.

The ultimate sphere does not have a material place by itself, but it does have one accidentally. How is this possible? It has a material place accidentally because each of its parts has a material place.

If in fact we were to cut up this sphere into sections, each of the sections would be at one time in the east and at another time in the west; each section therefore actually passes from one order to another with respect to the corporeal universe. In other words, each section passes from one formal place to another formal place:

And we have already repeated many times that if a thing moves locally so that it can pass in actuality from one place taken formally to another taken formally, it must be that the thing has also a place taken materially. . . . Hence, each section of the ultimate sphere must itself be in a place materially; the sphere itself is in a place by reason of its parts.[66]

It is evident, in any case, that each section is in a place materially by itself since each section is contained by the two sections contiguous to it.

In any case, by saying that the ultimate sphere is in a place [materially] by reason of its parts, we are not assigning this place to it simply in virtue of the order that the parts keep within the whole sphere, rather—if one can say this— in virtue of the order that the parts keep with respect to the order regulating the whole universe; it is this latter order that formally constitutes place.[67]

Moreover, the surfaces separating the various sections of the ultimate sphere from one another are not actually marked; the sections are not actually separated from one another, and it is even impossible for them to be separated, since the celestial substance cannot have any breaks. The divisions obtained on the ultimate celestial sphere's body exist only potentially; hence the sections of the celestial sphere have material places only potentially, and if the celestial sphere has a material place, not only does it have it accidentally, by reason of its parts, but it also has it purely potentially.

If the ultimate sphere is in a place materially, it is not so actually, but potentially. That is a consequence of the nobility of its nature; Saint Thomas mentions this in his lesson.

The ultimate sphere, in fact, is the highest of all the bodies in the order of nature, as in the order of location; it is therefore, among all the bodies, the one which must most approach the uniformity of spiritual substances.

And no spiritual substance is contained by the surface of an ambient body, either in actuality or potentially, for nothing that is not a body can be embraced by a body.

Bodies below the ultimate sphere are more distant from spiritual substances and thus they are contained within the surface of an ambient body actually as well as potentially. . . .

If the ultimate sphere is in a place, it would be accidentally, and not actually, in the manner of spiritual substances. However, since it is inferior to them, it would be in a place potentially. This potency would not be reducible to actuality; it could not be actualized except by imagination.[68]

Aided by Giles of Rome's terminology, Graziadei presented, clearly and completely, an exposition of the Thomist theory of place; he unified the scattered thoughts of the Angelic Doctor and tied the theory of the particular place of the ultimate sphere in the most logical and most natural way to the general principles that dominate the whole doctrine of place.

Graziadei rendered more precisely than his predecessors how one must understand the order of the corporeal universe in which Thomas Aquinas perceived the *ratio loci* and Giles of Rome perceived the formal place. We have seen more clearly what Thomas and Giles merely glimpsed, that the order is a set of measurements capable of marking geometrically the location each point in space occupies with respect to the spherical surface delimiting the universe.

In consequence, the truth that Graziadei only implicitly enunciated became manifest to us, that is, in order for the *ratio loci* or the formal place to remain immobile, the universe must be bounded by an immobile spherical surface.

Thus the Thomist theory of place came to express the requirement that Peripatetic theory would also formulate if it were pushed to its ultimate consequences: the universe must be bounded by an immobile sphere. But since Thomist physics agrees with Peripatetic physics in not recognizing the existence of such a sphere, both would be equally incapable of producing a satisfactory theory of place.

Roger Bacon

One cannot find anything concerning the theory of place in the published writings of Roger Bacon. However, one can find a lengthy study about place in the major work that until recently remained in manuscript form, the *Communia naturalium*.[69]

Bacon's study differs from the theories of place the Scholastic masters gave before him and the ones they gave after him. These theories attempted to understand the properties of place under a single definition from which the various properties can logically follow. Bacon does not attempt to attain such unity; on the contrary, he asserts that the word *place* is capable of bearing several meanings. Bacon enumerates five such meanings.

Among the five meanings that are attributed to the word *place* there is only one proper meaning (*secundum esse potissimum*); all others derive from it by equivocation. One can classify them in the order of their greater equivocation and distance from the proper meaning.

The study about the concept of place that Roger Bacon develops following the plan just sketched is not, then, a metaphysical theory; rather it more closely resembles the analysis of a grammarian when he wishes to classify methodically the various significations of the same word. The spirit of nominalism guides the noted Franciscan in this circumstance.

In order to define the proper meaning of place, Bacon latches onto the formulation: the extremity of the lodging body (*ultimum locantis*).[70]

If one considers the extremity of the lodging body in itself as the termination of the container, it is a surface; surface is a name truly and properly suitable for it.

This surface is capable of containing a body within it; when one pays attention to its potential containing, one sees fit to call it "cavity" (*concavum*).

But what makes something a cavity is not in itself sufficient to make it a place; in order for a cavity to become a place, it must actually contain a body.

This actual containing is not, in any case, sufficient to characterize proper place (*secundum esse potissimum*) which requires two relations in order to be defined.

The first relation is the relation between the surface of the container to the volume it contains, which the contained body occupies.

The second relation is the location of the surface of the container relative to the terms of the world (*termini mundi*). Bacon does not say what he means by this expression, but from the various considerations he develops about place, one can infer that the terms of the world is for him the center and ultimate surface of the universe; moreover, what he says about the center of the universe makes no sense unless one understands by these words a central body of finite dimensions, not a geometric point.

The relationship with the terms of the world is one of the essential elements defining place *secundum esse potissimum*; "in fact, as long as the lodged body keeps the same relation with the terms of the world, it keeps the same place; when the relation changes, the body changes place. This relation therefore belongs to the essence of place."[71]

The concept of place *secundum esse potissimum* as defined by Bacon presents clear analogies with the concept of place that Saint Thomas conceived and Giles of Rome adopted.

The proper sense is not the only meaning of the word *place*; if we suppressed or altered one of the defining elements, we would obtain a sense for which the word *place* is suitable only by equivocation.[72]

The preceding definition requires a unique containing body that remains unchanged.

A body may be contained by several differing matters that do not change from instant to instant; it can be partly in air and partly in water. By a first equivocation, we would say that the extremities of air and water are the place of this body.

A body can be enveloped by one and the same matter at each instant, but this matter can change from one instant to another; thus we say equivocally that an immobile tower remains in the same place, even though the air around it is constantly carried off by the wind.

We can join the two preceding equivocations together; a body can be contained by several different media, one or more of these media flowing from one instant to another. That is the case with one's foot bathed by the flowing waters of a river.

Place is a name suitable for these three derivative meanings only equivocally; the proper sense of place concerns a unique and temporally invariable surface. Here we have considered successively several invariable surfaces, then a variable surface, and finally, several variable surfaces. But the equivocation is of another kind when we are referring to the ultimate heaven.[73]

The ultimate heaven has a place, since we say that its parts move by local movement, that they change place, and that one of its portions is east at one time and west at another time. Even if heaven were immobile, it would be in a place, because its various parts would be at rest locally.

But no body surrounds the ultimate heaven; no body lodges it. Hence, when we speak of its place we are not referring this place to any surface, simple or multiple, invariable or changeable. We merely intend to designate by this a relation between the ultimate heaven and the center and terms of the world.

    I assert that this place is nothing other than a relation
with the center and terms of the world. When a star is at
the extremity of a line drawn from the east to the center of
the world, one says that the place of this star is east; if the
star is at the extremity of a line drawn from the west to the
center of the world, one says that it is lodged in the west.
When it is at the extremity of another line drawn from the
center of the world, one says that it is in another place, since
it has another relation with the terms of the world; the
proposition has therefore been demonstrated.[74]

Here the word *place* does not imply any relation between the
containing body and the contained body, but a unique relationship
with the well-determined terms of the world.

    Bacon does not hesitate to affirm that Aristotle took the word
place in the derivative and equivocal sense when he asserted that
place is immobile: "For a unique place corresponds to a unique
relation with the terms of the world, while different places
correspond with different relations. On the contrary, when he said
that place is *ultimum corporis continentis immobile*, Aristotle
understood the word *place* as *secundum esse potissimum.*"[75]

    Alone among the masters of Scholasticism, Bacon clearly noted
that, in order to understand Aristotle, one needed to distinguish
two meanings of the word *place*, the Philosopher having used one
or the other according to the circumstances.

    Bacon appends to these considerations about the place of the
ultimate orb a critique of the opinions put forth by various authors,
and differing from his.

    The first opinion he refutes is the opinion Albertus Magnus
wrongly attributed to Gilbertus Porretanus:

    One must not assert, as many have, that the continuous
    surface that terminates the ultimate heaven can be considered
    as the place of heaven. This surface, in fact, is not separate
    from the lodged body, it is an accident of it, while place is
    an accident of the containing body, since it is defined as the
    extremity of the containing body.[76]

Moreover, this convex surface moves exactly as the heaven which
it terminates; it must therefore have a place in the same way that
this heaven has a place.

    Hence, if one cannot obtain a place without supposing
    the existence of a containing body, it must be that this convex
    surface has a container; however, either it contains itself or

it would be contained by some other surface. But these two alternatives are both impossible.[77]

Some wish to impose Averroes's opinion, according to which the center of the world is the place of heaven; but this opinion does not satisfy me.[78]

Doubtless, in fact, the parts of heaven are in a place when they have a relation with the center of the world; when this relation changes they are said to change place. This relation with the center of the world therefore constitutes the place of these parts. But the relation is not the center of the world. It is therefore true to say that the place of heaven results from some relations between the parts of heaven and the center of the world, but it is false to maintain that the place is the center of the world.

In spite of this divergence, of language more than thought, between Averroes and Bacon, it seems that the two philosophers agree with the following proposition: in order that the ultimate orb be in a place, in order that it be possible for it to move by local movement or to be in a state of rest depriving it from local movement, there must exist at the center of the universe an immobile, concrete body. Assuredly, this fundamental axiom of Averroist philosophy is not enunciated anywhere in the theory of place developed by Bacon, but it seems to be understood everywhere; if one denied that the celebrated Franciscan wished to designate as *centrum mundi*, a finite, immobile, and concrete body such as the earth, one would render unintelligible a good number of his propositions.

Let us not forget that Bacon elsewhere formulated the proposition that "heaven itself will stop one day, or at least it is possible that it will stop."[79] We cannot say whether this assertion predated or postdated the similar assertion brought forth by the theologians of Paris in 1277; we cannot say this because we do not know the date of composition of the *Communia naturalium*.

The Place of the World in the Firmament: Campanus of Novara and Piere d'Ailly

Most of the discussions we have summarized are due to the difficulty of reconciling the following Peripatetic propositions for all the bodies making up the universe:

The place of a body surrounds it.
The place of a body is something immobile.

Christian theology appeared ready to defuse this issue by placing around the universe an extra sphere exempt from any local movement. It was thought that the Scriptures affirmed the existence of this ultimate fixed orb which Proclus already wished to take as the place of the universe, the reference for all movements.[80]

Guided by some passages from the Scriptures, a good number of theologians wished to posit an ultimate immobile heaven above the various mobile heavens that the astronomers imagined; Isidore of Seville, the Venerable Bede, Raban Maur, the pseudo-Bede, Saint Anselm, and Peter Lombard admitted this assumption. And theologians sought physical reasons to support this theological opinion. Michael Scot, William of Auvergne, Saint Bonaventure, and Vincent of Beauvais blazed a path for this. Some physicists, bothered by the question of the place of the ninth sphere, thought to find its solution in the hypothesis of a tenth immobile sphere. This empyrean sphere or aqueous heaven enveloped the ninth sphere and provided a place for it. It was the fixed reference for the movements of heaven; it assured the fixity of the two poles about which the other orbs rotated.

It seems that this theory already had currency at the time of Saint Bonaventure; some of Bonaventure's comments allude to the role of the universal place attributed to the empyrean sphere. In fact, he speaks of it as an immobile orb "which contains and is not contained."[81]

Some of Saint Thomas Aquinas's expressions would also lend themselves to a similar interpretation.

In any case, the theory at stake is clearly formulated in the *Theorica Planetarium*, which Campanus of Novara wrote at the request of Pope Urban IV. Here is how the astronomer that Urban IV had taken as his chaplain expressed his theory:

> Whether there is anything, such as another sphere, beyond the convex surface of this [ninth] sphere, we cannot know by the compulsion of rational argument [alone]. However we are informed by faith, and in agreement with the holy teachers of the church we reverently confess that beyond the ninth sphere is the empyrean heaven which is the dwelling place of good spirits.[82]

Is the empyrean heaven the tenth heaven, the heaven directly contiguous with the ninth sphere? Or must one place between this sphere and the empyrean heaven an aqueous heaven which would make the ultimate heaven the eleventh heaven? Campanus seems to hesitate between these two options, but he formulates the following conclusion with assurance:

> The empyrean heaven's convex surface has nothing beyond it. For it is the highest of all bodily things, and the farthest removed from the common center of the spheres, namely, the center of the earth; hence it is the common and most general place for all things having position, in that it contains everything and is itself contained by nothing.[83]

The above final phrase *"omnia continens et a nullo alio contenta"* reproduces almost verbatim the formulation which Bonaventure used.

We can find a singularly clear exposition of Campanus's theory in the *Summa Philosophiae* which some manuscripts attribute to Robert Grosseteste, but which, as we have stated, is the work of some disciple of Roger Bacon. Here is what one can read about the empyrean heaven in the work:

> It is necessary that the first mobile move on something completely immobile; that has been demonstrated in physics as well as in mathematics. And this immobile thing is not originally the center of the world as Aristotle and other Peripatetics have thought; it is the empyrean heaven which is naturally immobile in all its parts. It is with respect to this orb that the various parts of the first mobile and the other mobile spheres can be mobile; it is also with respect to it that they actually move. And, as we have said, it is also the reason that the world has a center and that this center is fixed; it is not the existence and fixity of the center which is the cause of the fixity of this heaven, which does not move and cannot be moved—it is the other way around. If not, then what is lowest and most vile would be a cause of what is in nature most noble and highest; that is impossible. It is therefore impossible that the circular movement of the sphere cannot occur unless one conceived an absolutely immobile center—and not only a mathematical center, but a natural center about which the sphere would move—once one admits the existence of the empyrean heaven containing all other corporeal things, preceding them all, either in time or in nature. If one still has to admit a center or something playing the role of center (*ratio centralis*), a center upon which the mobile heavens would necessarily move, it is clear that the rest of the empyrean heaven would be the universal cause of any change suffered by the beings capable of generation and corruption, rather than the first mobile and the lower spheres

playing this role. It is in the same way that the first cause is more of a cause than the secondary causes.[84]

This hypothesis that located at the ends of the world the necessarily immobile place Averroes sought in the center, appears to have been poorly received during the fourteenth century. Let us first say a few words about its reception so that we will not have to return to it.

Already Duns Scotus in his *Quaestiones Quodlibetales* exposed the absurdity of such a theory: "To say that the ultimate sphere does not move would be to affirm that it does not move by the local movement of which it is capable; but of what local movement would it be capable if it were not in any place?"[85] The hypothesis of an immobile empyrean heaven merely forestalls, without resolving, the difficulty about the place of the ultimate orb; that is the natural corollary of the Subtile Doctor's remark.

Joannes Canonicus, like John Duns Scotus, his teacher, alludes formally, but briefly, to this theory: The question of the place of the first mobile gives rise to philosophical difficulties, but not to theological difficulties; according to the philosophers the first mobile is not surrounded by any body, though it contains all of them; but according to faith, on the contrary, it is surrounded by the empyrean heaven. He adds, judiciously:

> But the difficulty that philosophers encounter in order to give a place for the first mobile is encountered by faith when it has to attribute a place to the empyrean heaven; in fact, although this heaven does not move, God can move it. However, it would not be contained by any body during the course of its movement.[86]

Albert of Saxony, who like Joannes Canonicus rejects the hypothesis of an immobile tenth heaven, also exposes the reasons invoked by the proponents of the assumption.

> Any body moving of local movement must be by itself (*per se*) in a place. Since the ultimate sphere is in movement by itself, it must be in a place by itself; and that cannot be if there did not exist an immobile sphere containing it from above. Place is the ultimate part of the containing body, and place must be immobile; therefore, there must exist a fixed sphere above all the mobile spheres.
> It is true that some physicists attempt to resolve this difficulty in another manner; they say that what assures the ultimate orb a place is its position with respect to the earth.

But this solution is worthless; the earth does not possess the properties it would need to be the place of the ultimate sphere—it does not contain the lodged body, it is not equal to it, etc. Moreover, natural movement must be referred to a place and its nature, and the natural movement of heaven is in no way determined by the earth.

No body mobile by itself has its fixed support in itself. An immobile body outside it is required in order to furnish it this fixed support, as it can be seen in the book, *De motibus animalium*. And the celestial orbs cannot find the principle which fixes them in the earth; the inverse is more likely true. One must therefore posit among the celestial orbs an immobile body which is fixed *per se* and from which they receive their fixity.[87]

The above are the reasons the proponents of the new hypothesis invoked in order to substitute it for Aristotle and Averroes's hypothesis; but the arguments provided by Albert of Saxony against the latter were just as forceful against the former. The first mobile moves in place, by a rotation, so that its fixity needs no extrinsic support, whether the support is the earth or the empyrean sphere; if it has no movement of translation, that is due to "its nature and the will of God."

However, toward the end of the fourteenth century, the doctrine acquired a resolute defender in the person of the noted Pierre d'Ailly.

In one of his *Fourteen Questions on the Sphere of Sacrobosco*, which had great currency and a powerful influence on the teaching of astronomy, Pierre d'Ailly wonders about the number of celestial orbs.

Probably one can posit an immobile sphere above the mobile spheres. Several reasons can persuade us of this. Here is the first: one supposes first, that a body which can move by local movement changes place either as a whole or in its parts. . . . A result of this is that any body moving by local movement is in a place, without which it cannot move. These principles posited, one can reason thus: By hypothesis, any mobile sphere moves by local movement; therefore, according to the first principle, it changes place either as a whole or in its parts. Therefore also, according to the second principle, it is in a place. However, each of the mobile spheres must be in a place. None can be in a place by the sphere lower than it, for place must surround the lodged body; each of the mobile spheres must therefore be lodged by a sphere above it, so that, above the mobile spheres there must be another sphere remaining at rest.[88]

The above argument clearly demonstrates the natural conclusion of the Peripatetic theory of place, the hypothesis of a necessarily immobile empyrean sphere.

The opinion of Campanus of Novara and Pierre d'Ailly about the place of the universe does merit being noted; in fact, it is identical to the position Copernicus accepted. Copernicus took an immobile sphere circumscribing the universe as the place of all bodies in the universe, as the immutable reference for all the local movements of these bodies—the only difference being that this immobile sphere was not the empyrean sphere, but the sphere of the fixed stars.

5
Theory of Place from the Condemnations of 1277 to the End of the Fourteenth Century

A Proposition Condemned by Etienne Tempier:
Richard of Middleton

With respect to place and movement, Averroes, Albertus Magnus, Saint Thomas Aquinas, and Giles of Rome proposed theories differing greatly from one another in several ways. These theories, however, all agreed about the truth of a single assertion: the ultimate sphere has no movement other than a movement of rotation, and its fixed center belongs to an absolutely immobile body, the earth.

The presence of this assertion in all these theories, and the preponderance of the role it plays, appears more neatly yet if we attempt to rid from each theory what distinguishes it from the others in order to leave only what they have in common. Here is what these theories reduce to:

1. It is impossible to conceive any local movement if one does not imagine a reference, fixed by definition, with respect to which the bodies are said to move, or to remain at rest, according to whether their position changes in time compared to the fixed term.
2. This invariable term is a concrete body, actually existing.
3. In particular, the revolution of a celestial orb requires that its fixed center be incorporated by an entirely immobile mass.
4. This body is the earth which remains perpetually immobile in the center of the world.

These propositions are the support and framework of the doctrines that Arabic and Christian Scholastics put forth on the subject of place and movement; if one were to deny these propositions, the doctrines would be destroyed, carrying the whole of Scholastic physics with them.

Among the consequences of these propositions, there are some that the Scholastics—astronomers as well as theologians, particularly those of the University of Paris—were forced to deny.

One of these consequences was formulated by Averroes: if all celestial circulation is produced necessarily around a central immobile body, the astronomical system of Ptolemy is inadmissible; one would have to imagine an earth at the center of the eccentric of each planet and one would have to place another at the center of every epicycle.

Moreover, at the beginning of the fourteenth century, the astronomical system of Ptolemy reigned uncontested among the Franciscans who followed Duns Scotus and among the masters of the Faculty of Arts at the University of Paris. Doubtless, the ingenious arrangement of orbs conceived by Ibn al-Haytham and extolled by Bernard of Verdun answered most of Averroes' objections against Ptolemy's system. But one remained unanswered, and it is precisely the one we have just recalled. Among the three orbs Bernard of Verdun attributes to each planet, there is one, the intermediate orb, that describes a revolution around a simple geometric point which is incorporated by no mass. If one wishes to put the astronomical theory of the *Almagest* beyond reproach, one would have to renounce the following axiom: the rotation of a celestial orb requires an immobile earth occupying its center.

According to the doctrines we have just related, the immobility of the earth at the center of the world is necessary, not only by a physical necessity, but also by a logical necessity; to deny this would be to deprive the concepts of place and movement of any sense—it would be to proclaim an absurdity.

To affirm the immobility of the earth at the center of the world is to affirm the immobility of the universe *secundum substantiam*. The various parts of the universe can exchange the places they occupy in such a way that the world is mobile *secundum dispositionem*, but the universe cannot submit to any displacement as a whole; it remains enclosed in a sphere which is invariable, for its center is absolutely fixed. To speak of a displacement of the universe as a whole would be to speak of a logical impossibility. God's omnipotence itself cannot produce this displacement, which implies a contradiction.

But Christian orthodoxy grew angry with the numerous fetters Peripatetic philosophy and Averroism imposed in the name of logic upon divine omnipotence; it decided to break the fetters. In 1277, at the request of Pope John I, Etienne Tempier, bishop of Paris, convened an assembly of doctors of the Sorbonne "and other wise men." Without exception, these theologians condemned every proposition that refused God the power to accomplish an act, under the pretext that the act is in contradiction with the *Physics* of Aristotle and Averroes.

Among the condemned errors, one is formulated in these words: "*Quod Deus non possit movere Caelum motu recto. Et ratio est quia tunc relinqueret vacuum.*"[1]

In order to deny God the power to impose on the universe a displacement as a whole, the condemned author invoked a reason no Peripatetic would have invoked; according to the Philosopher, outside the world there is no place, there is no void. But what the doctors of the Sorbonne censored was the proposition itself, not the reason invoked in its favor; if the proposition was upheld by arguments that were more closely Peripatetic, it would, no doubt, have met with the same treatment in the hands of the doctors of the Sorbonne.

Although the dogmatic validity of Etienne Tempier's decisions was contested from the start, the condemnations brought forth by the doctors of the Sorbonne carried a great influence at the University of Paris and in the English and German universities that followed the example of the University of Paris. In any case, even those who contested the validity of the condemnations we have just reported did not dare uphold that the Assembly of 1277 formulated something nonsensical; they were constrained to admit, in contradiction with Aristotle's opinion, that one can attribute a movement to the universe as a whole without speaking words that signify nothing.

Thus astronomy and theology united their efforts in order to compel the philosophers to take up again the theory of place and of local movement.

The new doctrine, erected on the wreckage of the Peripatetic theory, recalled, with respect to most of its features, the doctrine of Damascius and of Simplicius; the Franciscan Scholastics were the principal workers on the edifice that needed building.

One of the first theologians in whom we can note the influence of the condemnations brought forth in 1277 by Etienne Tempier against the *Articuli Parisienses* is Richard of Middleton. He seems particularly eager to examine the question: "Can God have given the ultimate heaven a movement of translation?"[2] He takes care to place the following reason in support of the arguments justifying an affirmative reply: "That God cannot move heaven by a rectilinear movement has been excommunicated by My Lord Etienne, Bishop of Paris and Doctor in Sacred Theology."[3]

According to Richard of Middleton, God can impart a movement of translation to the whole heaven. Doubtless, there is no place, no space, outside the ultimate heaven, and nothing can be moved by a movement of translation by any power whatever, even divine power, unless there is some space outside it; but God can create a space outside the world.

Further, God can move a portion of heaven by a rectilinear movement, without having to create any space; God can, for example, make a portion of the empyrean heaven descend down to the earth.

The thought that a rectilinear displacement of the world would bring with it the production of the void does not frighten our Franciscan. He asserts that God can produce the void; He can annihilate all the bodies existing between heaven and the earth, without moving either heaven or the earth. That done, there would no longer be any distance between heaven and the earth, for the distance between two bodies is constituted by the creatures in between. But heaven and earth would not be conjoined to each other either, for without modifying either in any way, God can create some bodies between them—therefore a distance. For any two bodies, not to be distant is therefore not the same as to be conjoined; there is no contradiction in affirming that they are neither distant nor conjoined, or, in other words, that there is a void between them.

In any case, Richard of Middleton remarks that it would be wrong to pit the possibility of a rectilinear displacement of the world against the impossibility of the void. In fact, heaven is not in a place; a translation of heaven would not produce a void.

Richard of Middleton does not present to us, with respect to the question we have just examined, anything that would capture the attention of a philosopher. But the passages we have just analyzed merit the notice of a historian of philosophy. We see here that the decrees brought forth by Catholic theology constrained physicists to take up anew the examination of propositions which Peripatetic philosophy had bequeathed them. The possibility of the void, so firmly denied by Aristotle, was, as we shall see in part IV, one of the principal questions submitted to this discussion; a new theory of place and movement also emerged from this critique.

John Duns Scotus

The new theory was inaugurated by John Duns Scotus.

He did not give a single exposition to his ideas on place and movement; he issued them here and there, incidentally, with respect to theological discussions. This fact suffices to render difficult the task of understanding them fully; their extreme subtlety is not such as to make this task less arduous. Still, let us attempt it.

The study of place is, for Duns Scotus, the study of a relation between two terms, the contained body and the containing body.

The idea of place requires first the idea of surface;[4] but surface is not sufficient to constitute place. One must join to it some consideration about the matter forming the container. Surface alone, having abstracted away this consideration, cannot be thought of as delimiting a place; the necessity of having to refer, not only to the limiting surface, but also to the ambient matter, when defining place, is designated by the Peripatetics with the expression *ultimum continentis.*

But a body can only be a container with respect to the contained body; therefore place has a counterpart. Place corresponds with the action of lodging, *locare*; the counterpart corresponds with the passion opposed to this action, to be lodged, *locari*. Duns Scotus designates this counterpart of place by the word *ubi*. He borrows the definition of this word from the author of *Treatise on the Six Principles: "Ubi est circumscriptio corporis a circumscriptione loci procedens."*[5]

In the treatise of Gilbertus Porretanus, the definition of the *ubi* that Duns Scotus has just cited is followed by the essential remark that place is an attribute of the containing body and that the *ubi* is an attribute of the contained body.

The relation we have just studied is therefore a relation between two terms; one of these terms, *place*, is intrinsic to the containing and extrinsic to the contained body; the other, the *ubi*, is intrinsic to the contained and extrinsic to the containing body.[6]

In addition to place and the *ubi*, Duns Scotus considers still a third element he names *positio*;[7] this word can be translated by *disposition*. The parts of a body are arranged in a certain order within the whole body; when the body is in some place, when it possesses its *ubi*, its various parts occupy the various parts of the place. The *disposition* indicates the order in which the parts of the body are located with respect to the various parts of the place or of the ambient body. The disposition is a set of quantitative

givens, of geometric elements that specify the *ubi* of the body. That
is how Damascius and Simplicius thought of it.

Having set aside these preliminaries, Duns Scotus can approach
the difficult question of the immobility of place. Let us examine
various cases in succession.

Let us first imagine that the containing bodies remain the same
while the bodies contained by them change. Can we say that place
remains and that different bodies come to occupy the same place
successively?[8]

Such an assertion would seem to be in contradiction with the
above. Place is a relation between the containing and the contained
body; if one of the two terms changes, the relation changes. Even
when the containing body remains invariable, one cannot assert
that place remains the same, if what is contained does not remain
the same.

Duns Scotus replies that place is not *the whole* relation existing
between the containing and the contained body; it is [the whole
relation] with respect to the containing body. As for the contained
body, it does not figure in a specific way, but in a general way;
in order to define the place formed by such containing bodies, one
must consider a contained body, but it is not necessary to designate
it specifically, and to state whether it is this or that body. Hence
if we were to change the contained body without changing the
containing body, we would be modifying the relation between the
containing and the contained body, but we would not be changing
this relation with respect to what constitutes place. When the
contained body moves alone, without a change in the containing
bodies, place remains immutable.

Let us take a second case: the contained body does not move,
but the containing bodies are constantly renewed.[9] According to
Aristotle's example, that is the case of a ship anchored in a flowing
river. Would we say that the place of the ship does not change?

Here the reply cannot be doubted. For Peripatetics the place
of a body is an absolute attribute of ambient bodies; for Duns Scotus
it is a relative attribute of these bodies—it consists of a relation
of these bodies to the contained body. For the one as for the others,
it is an accident of the containing bodies. And no accident can
remain if the subject of this accident comes to be replaced by another
subject. Therefore it is not possible for the place of a body to remain
the same when the matter surrounding it renews itself, even though
the body in question remains immobile.

In order for a body or a set of bodies to be in an immutable place, the enclosure containing must be composed of bodies incapable of any movement; Aristotle saw clearly that an immobile place cannot be obtained in any other fashion. But where in the universe can one find invariable bodies making up such an enclosure? There are none.

In desperation, some philosophers retreat to the limits of the universe in order to find this immutable enclosure; they believe to have discovered it in the spherical surface delimiting the universe. No doubt, they assert, the celestial orb for which it is the extremity moves, and in that way the surface is variable; but as limit of the universe it is invariable, for the universe taken altogether is immobile. We can recognize in this the opinion Albertus Magnus falsely attributed to Gilbertus Porretanus.

This reasoning is not valid. The spherical surface cannot limit the universe unless it first limits some of its parts; if the part changes from instant to instant, the surface limiting it also changes from instant to instant. Therefore, it would not remain identical to itself as the limit of the universe.

Hence, one must renounce the search for the enclosure which is incapable of movement and which alone would constitute an immutable place; the matter surrounding a body is always capable of some local movement.

Therefore, when this surrounding matter, the subject of the accident we call place, becomes animated by a local movement, the place of the fixed body which is contained by the matter changes constantly. Not that the place is animated by local movement; it is not susceptible to that movement. But at each instant, the place of the body perishes, is corrupted, and a new place is engendered. Incapable of local movement, place is susceptible to generation and corruption.

However, one commonly says that the body in question remains in the same place. What does one mean by that? According to what we have just stated, the body is truly in some place at some instant, and in another place at another instant. To each of these truly distinct places corresponds a rational place (*ratio loci*), and, in truth, these two rational places are also distinct; but they are *equivalent from the point of view of local movement*. It is this equivalence that one calls upon when one says that the place of an immobile body remains invariable even when the surrounding bodies are moving.

What is this rational place, this *ratio loci*? It is a relation with respect to the whole universe. When two such relations are numerically distinct, but specifically identical, they correspond to two equivalent, but distinct places; a body that occupies successively these two places does not move locally. When two rational places have not only a numerical but also a specific difference, the places corresponding to them are no longer equivalent; the body occupying these two places successively moves locally.

When a body moves, one commonly says that another body occupies the place the first leaves; that is not true if the surrounding bodies also move. The place of the second body is not identical with the place of the first; the former place perished while the latter place was engendered. The second rational place lost by the first body, numerically distinct from the rational place acquired by the second body, is specifically identical to it, in such a way that the place engendered is equivalent to the place that perished; from the point of view of equivalence, one can say that place is incorruptible.

According to this theory, when a body moves locally by driving away the body whose place it takes, one can distinguish four changes in the two bodies;[10] two of these changes are produced in the body driven away, and two in the body replacing it. Since each of these changes operates between two terms, eight different terms can be enumerated.

Let us consider, for example, the body that drives away the other. A first change has as initial term (*a quo*) the old *ubi* of the body, and as final term (*ad quem*) the privation of this *ubi*; this change is the loss of the old *ubi*. The second change has as initial term the privation of the new *ubi*, and as final term the new *ubi*; this second change is the acquisition of the new *ubi*.

Two entirely similar changes have their seat in the body that was driven out.

Duns Scotus's theory on the immobility of place does no more than develop what Saint Thomas indicated, particularly in his opusculum, *De natura loci*. However, one must notice a divergence between the doctrine of the Angelic Doctor and the doctrine of the Subtile Doctor, a divergence to which the Scotists attached great importance. When an immobile body is within a variable medium, Thomas Aquinas attributes to it a unique rational place, and Giles of Rome likewise considers the formal place of this body as invariable. That is a doctrine Duns Scotus strongly denies; for him, this body finds itself in two different rational places from instant to instant. Numerically distinct, the successive *rationes loci* are only equivalent among themselves. It is the influence of Damascius and of Simplicius that we clearly perceive here in the doctrine of the Subtile Doctor.

The distinction between the fact of lodging and the fact of being lodged, between place and the *ubi*, is the foundation of the explanation of the movement of the final celestial sphere.

The final celestial sphere is not contained by any body.[11] It is not in a place; it does not have a *ubi*. How then can it move locally? Perhaps one can maintain that the final celestial sphere is immobile. That would not help much. To say that the final sphere is immobile would be to affirm that it does not move by the local movement of which it is capable of moving. But of what local movement would it be capable of moving if it is not in any place?

According to Duns Scotus, the solution of this difficulty lies in a distinction.

The local movement of bodies other than the ultimate orb consists in the continual destruction of a certain *ubi* which is replaced by another *ubi*; the body ceases *to be lodged* in a certain way in order to become lodged in another way. It is not the same for the final orb; its manner of *being lodged* does not change. It is never lodged; what changes from instant to instant is the manner in which it *lodges* the contained body. The other bodies move *secundum locari*; it moves *secundum locare*.

According to Duns Scotus, that is the meaning one should attribute to the noted proposition of Averroes: the final heaven is in a place because of its center.

The Subtile Doctor formulates the following conclusion with respect to the above considerations:

> In the same manner that heaven can rotate even though no body contains it, it can rotate even though it contained no body; it can even rotate, for example, if it is formed of a single sphere homogeneous throughout its whole extent. The movement of rotation, taken in itself, is therefore a certain form flowing endlessly (*forma fluens*), and this form can exist by itself, without needing to be considered with respect to any other body, whether it be container, or contained. It is a purely absolute form.[12]

This conclusion which posits the absolute character of movement formally contradicts everything the Scholastics taught until then; it assuredly deserves some explanation. But Duns Scotus refuses us this explanation; he presents his surprising assertion as a kind of enigma: "Search for the answer," he says, *"quaere responsionem."*

John of Jandun

After Duns Scotus, many attempted to find a reply; however, some did not, because, in spite of Etienne Tempier's condemnations, they continued to declare, with Aristotle and Averroes, that the central body of the world remains necessarily immobile, and that the ultimate sphere cannot receive any movement other than its uniform rotation. It is among these philosophers faithful to the Peripatetic tradition that we find John of Jandun. Although his theory of place is extremely respectful of the past, it does include some new thoughts, which are sometimes included next to Scotus's thoughts in the teachings of the masters of Paris.

John of Jandun expounded his theory of place in several of the questions he wrote on Aristotle's *Physics, De Caelo et Mundo,* and the treatise *On the Movement of Animals.*[13]

The Averroist master defined place as Aristotle did: the place of a body is the ultimate part of the matter containing the body.[14] But by "ultimate part" he did not understand, as Ockham affirmed later, a certain volume of the containing body confining the contained body. Place has length and width, but it has no depth; from the material and quantitative point of view it is a simple surface.

It resides in the containing body, and not in the contained body; in this regard, it should be distinguished from the *ubi.* The *ubi,* whose definition John of Jandun borrows from the author of the *Six Principles*—as did Duns Scotus—is the essential and intrinsic term for local movement; place is not the term, or else it is the term in an extrinsic and mediate way, by the intermediary of the *ubi* of which it is the cause.

Is place simply a surface? John of Jandun replies to this question by borrowing the opinion of the pseudo-Thomas from the *De natura loci.*

However, he discusses and rejects the first part of the above opinion; he is not satisfied with the distinction between the place and the surface of the containing body, that "the surface is the limit of the containing body considered intrinsically to the body, and place is this same limit of the containing body considered extrinsically, as the limit of the contained body."[15] Although the distinction seems inadequate to him, he fully admits the considerations that the pseudo-Thomas joined to it. Place is not only the ultimate surface of the container, it is also a virtue of the surface—which it acquires from heaven—able to conserve the contained body. There are therefore two elements to consider: the surface, which is in some way the material element and which takes its place in the category of quantity; and the virtue capable of conserving the contained body, which plays the role of the formal element and which must be ranked within the category of quality.

After having analyzed the nature of place, Jandun studies its immobility. In what way can one say that place is immobile?[16]

Two theories that have attempted to safeguard the immobility of place hold Jandun's attention; one is Saint Thomas's, which attributes mobility to the material place and immobility to the *ratio loci*; the other is Giles of Rome's, which attributes to place an immobile matter and a mobile form. The canon of Senlis rejects equally these two theories; he opposes them by the arguments by which the Scotists and William of Ockham later objected to them.

Jandun concludes that place is not mobile by itself because it is not a body; but it is mobile by accident. As attribute of the ambient matter, it is mobile with the matter. This conclusion is also the one Walter Burley developed, inspired no doubt by the Averroist master.

What then is the meaning one should attribute to the proposition that place is immobile? John of Jandun indicates two such meanings.

First, one can state that the place of a body is immobile because the movement of the body does not necessarily carry with it the movement of the place. Thus rivers and their beds are the immobile place of the ship floating on the waters of the stream, because the ship can move without the river and its bed changing place. Similarly, one can say that the concavity of the lunar orb is the place of fire; if a portion of fire comes to move downward, it is not necessary that the portion of the lunar orb which contained this fire follow it in its descent, even though the lunar orb itself moves of another movement.

The above example leads to the second meaning that Jandun attributes to the immobility of place: when a body moves toward a certain place and the place is where it would be at rest naturally, the place is not animated by the same movement as the mobile body.

Jandun insists that this second meaning of the word immobile is characteristic of natural place only, whereas the first meaning can be understood with respect to place in general. If we are to believe Joannes Canonicus, Franciscus de Marchia put forth similar opinions, but understood them to apply to place in general. Walter Burley, on the other hand, who seems to be inspired by John of Jandun on this point, restricted his considerations to natural place only.

The noted problem of the place of the ultimate orb takes up much of Jandun's discussion; he reviews the various opinions on the subject and discusses them in detail. He takes up the arguments of Giles of Rome against the theory proposed by Saint Thomas in his commentary on Aristotle's *Physics*. At the same time that he rejects this theory, he refutes the objection that the Angelic Doctor used against Averroes's solution; it is true that the central body is alien to the supreme sphere by its substance, but it is not entirely extrinsic to it since it is contained by it.

Among the replies given to the difficult question Jandun examines, there are two which appear defensible to him: one is Avempace's formulation which, according to Averroes, takes up al-Farabi's formulation; the other is the Commentator's. The canon of Senlis refuses to choose between these two replies; it seems, however, that he leans in the direction of Averroes's solution— he applies himself toward dispelling any doubt that can be suggested by it.

Among the difficulties capable of engendering such doubts is the one the pseudo-Thomas had examined in his opusculum, *De natura loci*: "A beginner," says John of Jandun, "can be stopped by the following doubt: if the ultimate orb is in a place because of its center, . . . it is the same for other orbs, for the same reason; . . . each orb is thus lodged *per accidens*. But if one excludes the ultimate orb, each orb is thus lodged *per se*, since another orb surrounds and contains it. One and the same body would therefore be in a place *per se* and *per accidens*."[17] In his opusculum, Thomas Aquinas did not hesitate to regard this conclusion as deduced logically and acceptable. John of Jandun seems more hesitant: "Perhaps," he says, "there is nothing wrong in this, as long as it is so [the lodged body is in a place *per se* and in a place *per accidens*] with respect to various bodies; it would be impossible if it is related in the same way with respect to the same body."[18] That is not the only difficulty John of Jandun examines; the others are related to problems he examines in his *De Caelo* and *De motibus animalium.* Let us then refer to this latter treatise, since there Jandun examines with great detail the question of the relation which, according to the Peripatetics, unites the fixity of the earth and the movement of heaven.

For an animal to progress, must there exist a fixed body outside it?

The reason the movement of heaven requires a fixed body outside heaven proves also that the movement of an animal requires an immobile term; and, according to the Philosopher, the reason is more powerful in the latter case.

Here is the reason common to the movement of heaven and the movement of animals: to move is to behave now in a manner other than before; there must therefore be a reference by which the manner of being of the mobile can be referred from one instant to the next. But what moves, moves in geometric space (*super magnitudinem*). Therefore there must exist in geometric space an object with respect to which the situation of the mobile changes with time. And if one can say that the mobile behaves differently, with respect to some object, during various times, it is because the object is immobile. In fact, this object can only be either mobile or immobile. If it is immobile, the proposition holds. If it is mobile, one would have an infinite series of mobiles—which is impossible.

If the volume that must serve as reference moved completely with the same movement as the mobile, in the same fashion, in the same direction, with the same speed, the manner of being of the mobile would not change from instant to instant with respect to this reference. Thus, for a body to move, there must exist outside it an immobile body, or at least a body that does not move with the same movement and with the same speed.[19]

John of Jandun develops these considerations on three separate occasions;[20] in any case, they reproduce almost textually what Peter of Auvergne wrote when commenting upon the same work.

The objective of the above considerations is to establish the Peripatetic axiom: any movement assumes the existence of a fixed reference. And John of Jandun invokes this axiom in several other writings.[21]

Albertus Magnus did not admit the axiom without restriction; he wished to restrict it to the movements caused by an intelligence (such as the movement of the heavens) or caused by a soul (such as the movements of animals). Natural movements—the fall of heavy bodies; the rise of light bodies—did not seem to him to require the existence of a fixed reference.

John of Jandun, on the other hand, upholds the universality of the principle formulated by Alexander, Themistius, Simplicius, and Averroes; the natural movements of heavy or light bodies are not an exception.

> I reply affirmatively to the question, does a weight require the existence of a fixed body toward which it moves? Heavy and light bodies move in order to obtain rest; all natural movement has as end that the mobile rest in its proper place. If there did not exist an end capable of serving as termination of movement, the movement that could not attain its end would be in vain, or else the movement of heavy or light bodies would go to infinity. Both these assumptions are impossible naturally. Further, it is clear that if the place toward which a body moves were in movement, and not at rest, the body would move toward the place in vain. . . . It is therefore manifest that the place serving as the term of natural movement must be immobile. Therefore every animated body that moves requires the existence of an immobile term toward which it moves.
>
> But perhaps you still doubt the proposition that the place serving as term for natural movement must remain immobile. It seems, in fact, that the proposition must be false; the first heaven is the natural place of the lower elements, yet it moves. One can say the same for fire, air, and water. One must understand that place must either be mobile in an absolute way, or at least be exempt from the movement by which the body moves toward it, movement with respect to which it plays the role of natural place. Although the first heaven moves constantly in a circular movement, it is exempt from any centripetal or centrifugal movement, which is what allows it to be the place for heavy and light bodies and to serve as term for their movements.[22]

Joannes Canonicus attributes these same considerations, in almost the same words, to Franciscus de Marchia. We have read them in John of Jandun's *Questions on the Physics*, which he must have written after his *Quaestiones de motibus animalium*. We will read them also in the commentaries of Walter Burley, who probably borrowed them from the canon of Senlis. Here they are presented in conjunction with the reflections that appear to have been their source; we mean to refer to the reflections of Peter of Auvergne in his commentary on the *De motibus animalium*.

The axiom, whose necessity for any movement is proclaimed by John of Jandun, is to be applied specifically to the movement of heaven. Heaven therefore needs a fixed reference to which its movement can be compared.[23]

This reference cannot be an indivisible. It must be immobile. To say that it is immobile is to say that by nature it cannot move. And nothing is capable of moving unless it is a body.

This body cannot be formed of celestial matter; no part of celestial matter can be immobile. It cannot be outside heaven, because there are no bodies outside heaven. It is therefore surrounded by heaven.

> This fixed body is the earth, with respect to which heaven behaves differently at different times, when in movement. Considered in its totality, heaven changes in relation to the earth with respect to its disposition, but not in its totality; as for the parts of heaven, each experiences, with respect to the earth, both a change in disposition and a total displacement. Such is the opinion upheld by the Commentator in the fourth book of the *Physics*.[24]

John of Jandun analyzes this opinion of the Commentator more fully than did any of his predecessors. He takes up again, point by point, the whole preceding argument, summarizing with rare precision the Peripatetic tradition that went from Aristotle to Averroes:

> [First] heaven moves in a uniform and perpetual movement. . . .
> Second, I assert that this movement requires the fixity of a corporeal object. In fact, to move is to behave now in a manner different from before. But if there is no corporeal object fixed with respect to heaven, one cannot say that heaven behaves now in a manner other than it behaved before. . . . To behave differently, in fact, can only be accomplished by comparison with something fixed, for it is by comparison with uniformity that all diversity can be recognized. Consequently, there must exist a fixed object with respect to which one can say that heaven behaves differently now than it behaved before. And that thing is necessarily a body; with respect to an indivisible, heaven would always behave in the same manner, and not in a manner that is variable from instant to instant. It is therefore required that the object be a body.
> Third, I assert that this fixed reference does not belong

to heaven. . . . It has to be alien to heaven and what we call the center of the world.

But, you would say, what is the center of the world? One can understand by this a point such that all the lines drawn from this point to the circumference of heaven were equal; that is not what one intends to designate when one speaks of the object that remains fixed with respect to heaven. One can, using another interpretation, understand that the word *center* designates the whole earth. . . . It is the whole earth that plays the role of center with respect to heaven. It is not, however, a mathematical point; it is endowed with some volume. And that is necessary, as we have stated above; if the earth were not a body of some extension, one could not say that heaven behaves in various ways at various times with respect to it, because, with respect to an indivisible, its situation would always be the same.[25]

The above final remark to which Jandun returns with insistence was worth making; by an oversight, no doubt, Burley thought that one can speak of the change in situation of heaven with respect to an indivisible center.[26]

The canon of Senlis describes with much precision this change of disposition of heaven with respect to the earth:

Heaven can be at the same time the first fixed body and the first mobile. But a body may be mobile in two ways: it can be mobile according to its substance (*secundum subjectum*) or only according to its form (*secundum formam*). One says that a body moves according to its substance when it suffers a total displacement from one place to another. . . . It moves according to its form when it suffers only a change of disposition. Let us consider that heaven does not change place with respect to the earth; . . . given two different instants, it is clear that heaven is not disposed in the same fashion with respect to the earth. . . . Let us divide heaven by means of an infinity of meridians, and let us also divide the earth by means of an infinity of meridians; let us correspond the first meridian of heaven to the first meridian of the earth, the second to the second, and so forth. A moment later, each heavenly meridian would correspond with another earthly meridian. Heaven is therefore immobile with respect to its substance, for its total mass is never transported from one place to another; but it is mobile according to its form, meaning according to its disposition, for its situation with respect to the earth, around which it moves, changes from instant to

instant. It is therefore in different senses that heaven is said
to be the first mobile and the first fixed body.[27]

John of Jandun thus resolved an apparent antinomy residing in
the theory of the movement of heaven; similar antinomies offer
themselves to whomever meditates upon this theory.

Among these antinomies, the most serious is the following,
which already attracted the attention of Albertus Magnus: according
to the preceding doctrines, the earth constitutes the place of heaven,
and the movement of heaven cannot be produced if the earth is
not immobile; it seems therefore that the existence and immobility
of the earth are the causes of the fixed position heaven occupies
and of the movement animating it.[28] Is it not impossible that the
cause be less noble than its effect?

But it is not the position of the earth that fixes the position
of heaven nor the immobility of the earth that produces the
movement of heaven.

It is the position occupied by heaven that determines the
situation of the center of the world; it is heaven that confers upon
the various parts of the earth the gravity by which they move toward
the center of the universe. It is therefore the position of heaven
that determines the position of the earth—"if we were to displace
heaven, we would, by the same fact, be displacing the earth."[29]

The immobility of the earth is the effect, not the cause, of
the movement of heaven. "According to Aristotle, it is because of
the movement of heaven that all the parts of the earth tend toward
the center. . . . One can reason thus: the earth is immobile by the
effect of gravity; but heaven is the cause of gravity; heaven is therefore
the cause of terrestrial immobility."[30]

This doctrine agrees with the principle that Aristotle formulated
in the first book of the *Meteorology*, and which dominates all
medieval astronomy and astrology: the world of the elements is
governed by the movements of celestial bodies; any virtue of this
world is derived from these movements.[31]

From the above principle follows a corollary that was
universally accepted by Peripatetic philosophy: in the world of the
elements, all generation and corruption of a new being or of a
new quality is dependent upon changes in the aspects of heaven.

The above proposition serves John of Jandun as the point of
departure for a new argument by which his *Quaestiones in libros
de Caelo* attempts to link the earth's immobility with the mobility
of heaven.[32]

The generations and corruptions produced in the region of the elements require that an object exists with respect to which the disposition of heaven changes from one instant to the other, meaning that there is a central immobile body in the concavity of heaven. The movement of heaven therefore requires the immobility of the earth in order to impart to this movement the diversity required by the generation of lower beings, animals in particular.

This argument allows the rebirth of an objection which appeared to have been dispelled; it seems, in fact, that the generation of lower beings—and therefore the immobility of the earth—is the final cause of the movement of heaven; the least noble thing in the universe is being proposed as the cause of the movement of the noblest body.

The above conclusion does not absolutely repel John of Jandun. No doubt the generation and conservation of the beings in the region of the elements is not the final direct and principal cause of celestial movements, but one can allow that it is the final cause in an indirect and secondary way.[33]

The earth's immobility is not the cause of the movement of heaven, but it is no less a necessary condition of it; the heaven's motor requires an immobile earth in order to exercise its action.[34]

From that results the fact that it is absolutely impossible for the earth to move or to stray from the center of the world.[35]

In order for heaven to accomplish its uniform revolution, the earth must remain immobile at its center. If the earth were to move, heaven would have to stop. If it were chased from its place, heaven would also have to be displaced; either heaven would be displaced, or else its movement would end.

But the above two hypotheses are impossible. Heaven, properly speaking, is not in a place, cannot submit to any displacement as a whole. It cannot stop its rotation either; if it were to stop turning, it would stop existing, and its motor would also stop existing. These propositions are an essential part of the Averroist doctrine; here is how John of Jandun justifies them:

> If objects are directed to some aim, they would cease to exist the moment this aim were missing.
> And the motor of heaven and heaven itself are directed to the movement of heaven; here is why: the aim of the celestial motor is to extend its goodness among the beings. But it cannot extend its goodness without the intermediary of movement; by itself, in fact, the first motor would only be able to exercise

a uniform influence. In order that it may exercise a variable influence, it must be assisted by some object whose way of being changes from one instant to another; heaven, because of its movement, furnishes it this object. Thus the celestial motor would not be able to extend its goodness among the beings without the intermediary of heaven, whose way of being must change from one instant to another for this end; and the way of being of heaven changes from instant to instant only for the movement of this body. It is therefore correct to say that the celestial motor and heaven itself are directed to the movement which is their end.

Hence, if the movement were missing, heaven and its motor would cease to exist, . . . which is impossible.[36]

God, who is this first motor of heaven, would therefore not be able to move the earth; the consequences that follow from this movement, which we have just detailed, are contradictory.

In this whole argument, there is almost no proposition which is not among those the doctors of the Sorbonne, under the direction of Etienne Tempier, did justice to. By the condemnations that they brought forth in 1277, the theologians of the Sorbonne traced out a path to the system of Copernicus. How, in fact, could this system have been proposed if the philosophers, taking the side of John of Jandun, regarded the movement of the earth as a logical absurdity, defying even God's omnipotence?

The Scotist School

God can impose upon the earth, as upon heaven, any movement that He wishes to impart upon them; far less from restraining God's freedom in the name of a theory of place, one has to try to construct a theory of place that safeguards this freedom. Such is the research program Duns Scotus drew up for his disciples. Further, he traced a path to follow in order to accomplish it. In attributing to place an immobility *by equivalence*, a notion clearly borrowed from Damascius and Simplicius, he indicated that it would be better to follow these philosophers in this discussion rather than Aristotle and Averroes.

Peter Aureol also appears to have been inspired by Damascius and Simplicius in the theory of place he develops in his commentary on the second book of the *Sentences*:

"The place of a body," stated Aureol, "is nothing more than the determinate position that the body occupies here or there. (*Locus per se et primo non est aliud quam positio, puta hic et ibi.*) Place is accidentally the surface of the containing body.''[37] Our Franciscan summarized his whole thesis in these two propositions.

Let us suppose that it suffices to posit something in order that the body to which it is referred occupies a determinate place in the universe, that it suffices to change it in order that the place of this body is changed; this thing, assuredly, will be formally identical to the place of the body. Now let us put a body in the same position at several occasions; it would be in the same place. If on the other hand we were to change the position of the body without modifying the matter surrounding it, if, for example, we were to carry with it the vase that contains it, it would change place. The place of a body is therefore nothing more than the position or the situation of the body in the universe.

This definition dispels the difficulties with respect to the movement of the ultimate sphere. The ultimate sphere that no body surrounds is not in a place in the sense that Aristotle gives the word; it does not have a *ubi*, according to the language of Gilbertus Porretanus and Duns Scotus. But it does have a position, a situation. And local movement does not consist in a change of *ubi*, but in a change of situation; therefore nothing prevents the ultimate sphere from moving locally.

This theory, as one can easily recognize, is a return to the ideas of Damascius and Simplicius; the *positio* or *situs* that Peter Aureol considers the essence of place is identical to the position [*thesis* instead of *topos*] of the two Greek philosophers.

This *positio*, on the other hand, differs from the one by which Saint Thomas defined rational place (*ratio loci*) and which Giles of Rome identified with formal place. The position these two authors consider is that of the parts of the container touching immediately those of the contained body; the position Peter Aureol speaks of is, on the contrary, that of the contained body. Even though the two positions are fixed by means of the same geometrical magnitudes, so that the mathematician would not distinguish one from the other, they are nevertheless very different for physicists. In the reasoning of Saint Thomas Aquinas and Giles of Rome, position is an attribute of the container;[38] in Peter Aureol's theory it is an attribute of the contents.

It is this point that Joannes Canonicus seizes upon to condemn the theory.[39] Like Aristotle and like all his faithful disciples, Canonicus wishes that place inform the container, not the content; the place of a body cannot therefore be the position of the body.

Joannes Canonicus judged Peter Aureol's attempts harshly; he was not any more indulgent with respect to the doctrine of Giles of Rome, whom he refers to, with some disdain, by the words "a certain doctor."

Formal place, as defined by Giles of Rome, is an attribute of the parts of the container touching the content; an accident cannot remain when one changes the subjects in which it exists. Formal place cannot, despite Giles of Rome's assertion, remain immutable while the matter containing the body is renewed.

The argument Joannes Canonicus opposes to Giles Colonna's theory was the argument by which Duns Scotus objected to any theory maintaining the absolute immobility of place.

If the Scotists are in agreement when condemning the theories of Saint Thomas Aquinas, Peter Aureol, or Giles of Rome, they are less in agreement when interpreting the subtle doctrines of their teacher.

They all recognize that the surface is the matter and the support of place, but that place is not simply identical to the surface; all wish that place be an actual entity having its foundation in the surface separating the container from the content. "But what is the nature of this entity? Today that is something in doubt for many philosophers," asserts Joannes Canonicus.[40]

Some hold Duns Scotus's opinion almost verbatim: the entity added to the surface in order to constitute place is the action by which the container circumscribes the content, or a relation deriving from this action. To this action constituting place is opposed the passive operation which, according to the definition of the author of the *Six Principles*, constitutes the *ubi*. In order to indicate this opposition better, Joannes Canonicus goes so far as to call the *ubi* considered by Gilbertus Porretanus and Duns Scotus the *ubi passivum*, while he proposes to give place the name *ubi activum*.[41] The local movement of most bodies is then a movement whose two terms belong to the species of passive *ubi*, while the terms of the movement of the ultimate sphere are ranked in the category of active *ubi*.

Others do not believe that the operation by which the container circumscribes its content is the entity constituting place; they believe the entity to be only an attribute. As for the essence of this entity itself, it remains unknown.[42]

The distinction between the *ubi activum* and the *ubi passivum* was borrowed by Joannes Canonicus from a Franciscan whom he cites in each of his two questions about place,[43] and from whom, here as elsewhere, he gains frequent inspiration—we are referring to Franciscus de Marchia.

In his commentary on the *Sentences*, Franciscus de Marchia has a lengthy discussion about the question, "Is the first mobile or the ultimate sphere in a place?"[44] This discussion is worthy of attention; it is one of the most important discussions of place among those raised within the Scotist school.

From the beginning, our author declares the following, in which we clearly recognize the inspiration of the Subtile Doctor:

> I assert that this question presents a preliminary difficulty. According to our faith, since the first mobile is contained by the empyrean heaven, it is in a place. According to the philosophers, on the contrary, it is contained by nothing but contains all things; however it moves locally (*movetur localiter*). That posited, the difficulty arises: How can one put it in a place? This difficulty remains, in any case, if one follows the teaching of faith concerning the empyrean heaven; doubtless it [the empyrean] is not moved by local movement, but God can move it by local movement, and it is not in a place because it is contained by nothing.[45]

After having set forth, discussed, and rejected the three solutions proposed by Avicenna, Themistius, and Averroes, our author gives us his own solution, which he formulates as follows:

> I assert that heaven is not in a place and is not moved in a place; however, it is moved by local movement because its movement has a *ubi* for term. (*Dico quod celum non est in loco nec movetur in loco; movetur tamen localiter, quia motus ejus terminatur ad ubi.*)"[46]

It is in order to justify this somewhat surprising response that Franciscus de Marchia develops his theory of the *ubi* and of local movement.

> As evidence for this doctrine, one must know that there are two kinds of *ubi* as there are two kinds of circumscriptions. There is an active circumscription, of the body circumscribing, and there is a passive circumscription, of the body being circumscribed and contained. Similarly, there are two kinds of *ubi*: the *ubi passivum*, which is of the lodged body, and the *ubi activum*, which is of the lodging body.
> Moreover, these two *ubi* are of the same kind; therefore, if one of them is capable of serving as the term for local movement, the other is also. In fact, when two things are of the same kind, if one of them can play the role of term

with respect to a movement, the other can do so equally. But [local] movement can terminate with a passive *ubi*; it can therefore also terminate with an active *ubi*, which is of the same kind as the passive *ubi*.

Secondly, I assert that there are two kinds of local movements. There is a local movement having a passive *ubi*; a body lodged and contained in a place moves with the movement in the place. Another local movement has an active *ubi* for term; that is the movement by which a lodging body moves around its place, not in its place.

Replying then to the proposed question, I assert that the movement of the first mobile does not move by a passive *ubi*, since it is not contained in a place; it moves by an active *ubi*. The first mobile does not move in a place but around a place.[47]

Within this theory that captured the approval of Joannes Canonicus, one can distinguish a curious mixture of thoughts suggested by Duns Scotus and thoughts suggested by William of Ockham.

We will soon find William of Ockham teaching that a body can move locally in two different ways, according to whether it moves with a view toward acquiring a new place within which it is contained, or whether it is to become the place of a new lodged body. And the Venerable Inceptor will tell us that heaven moves in the latter, not the former way. He will refrain from asserting that the first movement tends toward the passive *ubi* and the second toward the active *ubi*, for he truly dislikes the notion of the *ubi*; he banishes it without pity from his physics. But a Scotist, having confidence in the reality of the *ubi*, proposing to define Ockham's distinction between the two kinds of local movements, would express himself exactly as Franciscus de Marchia did.

The influence of Ockham on Franciscus de Marchia should not surprise us; we know that Ockham also influenced him in the theological domain, and that the Franciscan of Ascoli became liable to the errors of the English Franciscan.

After having formulated his response, Franciscus de Marchia attempted to dispel the doubts that might remain in one's mind. There is a doubt, he rightly remarks, that can be addressed as effectively against other theories as against his own: the place of a body must be the ultimate part of an immobile container. However, is not the concavity of the lunar orb thought unanimously to be the place of fire, and does not the lunar orb move? Here is what our Franciscan replies to this difficulty:

Place must be immobile, of an immobility opposed to the local movement with which the body lodged by the place moves toward it; the reason for this is that place is the term of dependence with respect to which the body is lodged. If a place moved or were susceptible to moving by the same movement that the lodged body moves toward it, then place would not be capable of serving as term of dependence for the lodged body; the lodged body would not be capable of being fixed by place, limited by place. But although place must be immobile in this way, it does not have to be immobile universally and in all ways. Although it can move in some way, as long as it does not move of the same movement by which the lodged body moves toward it, it can still serve as the term of dependence with respect to which the body is lodged.

For example: the lunar orb is the place of fire; however it moves by a rotation. But it is immobile with respect to the rectilinear movement by which the body it lodges (meaning fire) moves; if the lunar orb moved by a rectilinear movement, it could not be the place of fire.[48]

Joannes Canonicus does not find this reply satisfactory;[49] although valid against the objection derived from the movement of the lunar orb, it is not valid against similar objections. Thus the supreme orb is thought of as the place of the lower orbs, although like the other orbs it moves by revolution.

With respect to this problem, Franciscus de Marchia and Joannes Canonicus propose a distinction which is not a solution; moreover this distinction is borrowed from the opusculum, *De natura loci*, attributed to Thomas Aquinas: There are perfect places that not only surround the lodged body, but also support it by the pressures they exert on it. These places are absolutely immobile, or at least immobile with respect to the local movement of the body they circumscribe. There are also imperfect places circumscribing contained bodies without supporting them. The places of the celestial orbs are among these, since these orbs have no need of support in order to remain in their places. These places can dispense with satisfying the condition formulated by Franciscus de Marchia.

With Simplicius and with his teacher, Duns Scotus, Joannes Canonicus admits fully that place can be engendered and can perish.[50]

A stake is driven into the bed of a river; the water bathing the stake flows constantly. At some instant the volume filled by the wood of the stake is surrounded by some parts of water; these parts form, at that instant, the proper place of the stake. A little later these same parts are downstream from the stake; they no longer circumscribe the foreign body; they have become contiguous with each other; they are no longer the place of anything. The place they formed has perished. In the meantime, some other parts of the flowing water have come to surround the immobile stake; a place not existing in them at first has been engendered.

These two places are really distinct, even though they have the same disposition with respect to the center and the poles of the world, which renders them equivalent places. Further one should not pretend that these two places are the same formal place. As has already been stated, where the subject varies, the attribute cannot remain identical to itself.

But against a similar doctrine "the masses will scream; for in the end no one will dare pretend that a house changes place because the wind is blowing. . . . Let us not worry about the masses when reason is against them; in this matter, the masses are not very competent. We do not stop at the opinion of those who pretend that a body remains in the same place even when the container changes; that is the idea of very aged people (*imaginatio vetularum*)."[51]

If we exclude Francis of Mayronnes, whose theory will occupy us later, all the Scotists seem to have embraced the opinion that the place of an immobile body is not immutable, unless it is by equivalence. Some of them seem to have admitted this doctrine even while they ignored many other elements of Duns Scotus's theory of place—such was Antonio d'Andres.

Among the many writings of Antonio d'Andres, there is a commentary on the *Treatise on the Six Principles* of Gilbertus Porretanus.[52]

Writing on the *Six Principles*, Antonio d'Andres deals almost exclusively with the study of categories; however, one of the questions devoted to the study of the predicament *ubi* treats the noted problem of the place of the ultimate orb.

Let us reproduce here what the faithful disciple of the Subtile Doctor asserts in this short question:

Various philosophers and commentators have held various doctrines for they wished to uphold the proposition, the ultimate heaven does not have a proper place, but it is in a place in some manner. Some authors, such as Averroes, have stated that the ultimate heaven is in a place because of its center; others, like Themistius, that it is in a place because of its parts; others still that it is lodged by its terminal surface. This question concerns the topics of the fourth book of the *Physics*.

Whatever the opinion of these philosophers is, I hold for certain that, properly speaking, the ultimate heaven is in no place, and that for the reasons given by the author [of the *Six Principles*]. In fact, everything in a place is surrounded by some body outside the lodged body, which is distinct and separated from it, as demonstrates the fourth book of the *Physics*; but there is no body outside the ultimate heaven; if there were one, it would not be the ultimate heaven.

One must note here that the bodies of the universe are ordered by each other in such a way that they are locally containers and contents; earth is contained by water, water by air, air by fire, fire by the lunar orb, the lunar orb by another orb, and so forth until the ultimate orb. Therefore, in the same way that one can easily find a body within the universe, the earth, which is contained but which is not the place of any other body and which contains nothing, one can also find a body that plays the role of place, containing another body but which is in no place and is not contained by any body; such is the supreme orb or ultimate heaven, whether it is the first mobile, as think the philosophers, or the immobile empyrean heaven, as think the theologians, and as is the truth. In the empyrean heaven is the place of the blessed; beyond it there is no place, no movement, and no time, as said Aristotle in the second book of the *De Caelo et Mundo*.[53]

Joannes Canonicus also denied place to the supreme orb, but as the faithful interpreter of Duns Scotus's thought, he attributed a *ubi* to it; Antonio d'Andres said nothing about this *ubi*. Further, in the course of the three questions suggested to him by what Gilbertus Porretanus wrote on the predicament *ubi*, d'Andres frequently repeats the word *place*, but not once does he pronounce the word *ubi*.[54] It seems that in opposition to Duns Scotus, his teacher, he attributes no reality to the *ubi*.

When Antonio d'Andres, along with Joannes Canonicus, denies that the supreme sphere has any place in the proper sense of the word, he seems to have been influenced by Roger Bacon, an influence that was extremely powerful in the Franciscan school during the fourteenth century; when he sets aside the notion of the *ubi* in order to concentrate on place alone, he is preparing the way for the Parisian philosophy of Gregory of Rimini, John Buridan, and Albert of Saxony.

In a later chapter we shall have the occasion to analyze another writing of Antonio d'Andres; in this writing he alludes to the *ubi*, but there, more than in the above, we shall see him distance himself from the teachings of Duns Scotus and Joannes Canonicus.

On the other hand, in a third work Antonio d'Andres expresses himself in almost the same words as Joannes Canonicus with respect to the immobility of place.

> According to the Philosopher in book IV of the *Physics*, place is the ultimate part of the container; it is immobile and incorruptible. Some explain the immobility of place by saying that material place is mobile, whereas formal place, which conveys order to the various parts of the universe, meaning to the center and circumference of the world, is immobile and incorruptible. . . .[55]
>
> I assert that such a [formal] place is corruptible.[56]

In support of the proposition, place is incapable of local movement, but it can be engendered or destroyed, our author develops an argument similar to that of Joannes Canonicus. He pursues it as follows:

> Further I assert that place insofar as it expresses a relation is corruptible, but insofar as it designates the ultimate surface of the containing body, it possibly is incorruptible. That is evident concerning the concave surfaces of various heavens, for these surfaces are incorruptible. However, as they are mobile, the relation that each has to a lodged body is corrupted by the effect of the movement of the surface itself. Here I am not referring to the immobile empyrean heaven, for Aristotle did not know of it.
>
> I therefore assert that place is immobile, as thought the Philosopher, in the sense that it possesses immobility opposed to local movement; besides, it is incorruptible by equivalence. . . . It is clear that it is incorruptible by equivalence; in fact, if the lodged body moves, there is an immediate acquisition of a relation between the place and the lodged body which has moved, similar in every way to the relation it had with the abandoned place.[57]

Under a confused and overly precise form, we recognize the notion of place persisting *by equivalence*, engendered by the teachings of Damascius and Simplicius, to which the Scotist school and the Nominalist school attached an equally great importance.

John of Bassols

How can God move the whole universe by local movement if there is no immobile term to which this movement can be related? That is a difficult question which one of Etienne Tempier's decisions imposed on the consciousness of the philosophers. "Search for the answer (*quaere responsionem*)" said Duns Scotus; but none of his disciples whose work we have just analyzed were able to unravel this enigma in any way.

It seems that the first to have prepared the way for the solution of this thorny problem was John of Bassols.[58]

John of Bassols's whole argument is directed against Giles of Rome's theory. He denies that the form of a body's place is the distance of the place from the center and the poles of the world; he also denies that this place remains immobile when the lodged body does not move. Like Joannes Canonicus, John of Bassols admits that this distance is an attribute of the intermediary bodies between the lodged body and the center or poles of the world. Like Joannes Canonicus, he admits that this distance, and therefore the place whose form it is, can be corrupted by the corruption of the intermediary bodies. Further he admits, in opposition to Joannes Canonicus's opinion, that the local movement of these bodies has as consequence the local movement of the place, except that around an immobile body; the places succeeding one another have a certain relation of equivalence (*aequipollentia*) to one another.

> The place that follows is equivalent to the previous place from the point of view of local movement; one can combine each of them with a third place and furnish the same term as the other for the local movement directed toward a third place; . . . with respect to the same straight line issued from one or the other place directed toward a third place, the movement is the same.[59]

With respect to what can this equivalence be understood? Joannes Canonicus makes it consist in a similar disposition with respect to the center and the poles of the world, but in his argument against Giles of Rome he denies the immobility of this center and these poles, so that his theory seems to turn around a vicious circle.

John of Bassols breaks the circle. The real poles of heaven and the real center of the world are bodies capable of movement; one cannot understand the real equivalence of two places—if one prefers, the immobility of a place—with respect to these mobile references; but the immobility and equivalence of which we are speaking is a purely fictive immobility and equivalence with respect to a center and poles existing only in the imagination of the geometer.

> In effect, the mathematician, with a view to the exposition of science, and without pretending that it is so in reality, imagines a line drawn from one part of heaven to another, passing through the center of the world, which is itself an imagined point. This line terminating in one part of heaven and the other, receives the name axis of the world; its extremities, or in other words, the points terminating it, are called the poles of the world. They are merely points that one imagines in heaven. It is with respect to such poles and such a center that place is said to be immobile, of an imaginary immobility and not of a real immobility; in reality this place is mobile and corruptible, but the places succeeding one another maintain a certain equivalence among themselves.[60]

Therefore, when a body remains at rest, the place of the body is located, with respect to some references, at some distances whose value always remains the same; these references have no reality, and do not exist outside the imagination of the geometer. Such is John of Bassols's opinion concerning the immobility of place; such is also, on the same topic, the essential proposition of the Ockhamist doctrine.

William of Ockham

William of Ockham composed an opusculum entitled *Tractatus de successivis* whose first few lines tell us its subject: "The common opinion is that movement, time, and place are things distinct from the mobile or lodged body; one has to see what the intention of Aristotle and his Commentator was on this subject."[61]

Presented modestly as a simple explanation of the thoughts of Aristotle and Averroes, this opusculum put forth, with respect to movement, time, and place, the most profound and original views.

The *Tractatus de successivis* is a masterpiece which we believe has never been printed. But the second part of this treatise, expounding the theory of place, was taken up again by Ockham with insignificant variations in his *Summulae Physicorum*; it was printed twice with these *Summulae*.[62]

The Ockhamist theory of place has great similarities with the Scotist theory; it conserves its essential doctrines. However, there are some important differences worth noting.

First of all, Ockham distances himself from Duns Scotus with respect to the nature of place.

For the Subtile Doctor, place is an entity whose foundation is to be found in the surface of the container, in contact with the contents; this surface of contact is the matter of the entity, whose form is an active relation of the container with the contents. Duns Scotus defined place by means of similar considerations several times, and Joannes Canonicus related to us the efforts by which the Scotist school attempted to elaborate upon this definition.

However, every element of this definition is repulsive to William of Ockham's philosophy.

John Duns Scotus could, without being illogical, declare that the surface of contact between the container and the content was the support, the subject of the entity which, according to him, constituted place; in effect, he did not hesitate to attribute a reality to the surface, to consider it as the seat of some physical properties— of color, for example.

William of Ockham, on the other hand, affirmed with some persistence that there is nothing real, nothing positive, in the notions of point, line, and surface. Only volume, magnitude given three dimensions, extended in length, depth, and width, can be realized. Surface is a pure negation, the negation of the volume extending above a certain term; similarly line is the negation of the surface extending beyond a certain border, and point the negation of the line extending beyond a certain limit.

The limiting surface of the container, itself having no reality, cannot be the matter of an entity constituting place.

Moreover, the Venerable Inceptor could not admit such an entity without going counter to his most powerful tendencies; to suppress as much as possible the entities that Scotism multiplied profusely was a primary principle of his method.

In conformity with this principle, Ockham takes up Aristotle's definition of place and renders it back to its initial simplicity by ridding it of any parasitic additions. Place is the part of the containing body touching the contained body.

But one must understand that this part is a body extended in length, width, and depth. One can trace, within the containing body, an enclosed surface that entirely surrounds the cavity filled by the contained body; this surface divides the containing body into two other bodies, of which one, encased within the other, encloses the contained body. The part of the containing [body] enclosed within the other part constitutes the place of the contained body.[63]

By a completely analogous operation, one can again separate the two places into two enclosures encased within one another; the enclosure within the other enclosure is now the place of the contained body.

One can proceed indefinitely in this way; we would be giving the contained body a thinner and thinner layer borrowed from the containing body. Each of these places would be a part of the preceding place; each would be a body, and not a simple surface.

One can see that modern mathematical language would have allowed Ockham to express his opinion about place with great precision: we would say that place is an infinitely thin layer from the containing body always contiguous with the contained body.

William of Ockham took up this proposition in one of his quodlibetal discussions and, in order to formulate it and explain it, he resorted to the rigorous precision which, because of him, the School of Paris brought into any discussion about the infinite:

> Place is what is ultimate in the container, meaning the ultimate part of the containing body. Not that there is an ultimate part which is in its totality distinct from other parts. I call ultimate part any part that extends up to the lodged body—touches the lodged body in its place; according to this way of speaking, the ultimate part has, itself, a multitude of parts that do not touch the lodged body.
>
> But, you would say, I take the ultimate part, the one called place; it does not have some parts that touch the lodged body and others that do not; if this were not so, it would not be this part, but a part of this part that would be called place.
>
> I reply that one must make some distinctions with respect to the ultimate part.
>
> First, one calls ultimate part any part that extends up to the lodged body and that touches immediately the contained body in its place; in this sense there is an infinity of ultimate parts which are all its place. If an ultimate part touches the lodged body by its right side, the right half of the tangent part is also an ultimate part, and the half of this half is also an ultimate part, and so forth to infinity.
>
> Second, one calls an ultimate part contiguous to the lodged body the part located after every part contiguous to the lodged body. In this sense there is no part that is an ultimate part.[64]

The above is the principle by which, in the course of their discussions on the infinite, John Buridan and Albert of Saxony opposed Gregory of Rimini.

In disagreement with the Scotist system on the subject of the essence of place, the Ockhamist system is also in disagreement on the subject of the immobility of place.

For Duns Scotus, place is an entity; this entity can be engendered or destroyed; place is therefore declared capable of generation and corruption:

> Insofar as it considers the place of a body as replaced by another body when the ambient matter moves locally, this opinion is true, but insofar as it admits the corruption of place because of this local movement, it is false—it proceeds from the false thought that place is a relation really distinct from the containing body.[65]

The same false thought leads Duns Scotus to another erroneous proposition, namely, that place is incapable of local movement. Place is a body; therefore it is itself in a place and it can move.

It is even capable of moving in two ways:

The place of a body can move in order to become the place of another body; if, for example, a stake driven into a flowing river is followed immediately by a stone, the water which at one instant touched the wood and formed its place, a moment later touches the stone and provides its place.

The place of a body also moves so that instead of becoming the place of another body, it simply finds itself in another place, without lodging a foreign body; after having bathed the stone, the parts of water that the stone separated from one another are drawn nearer and conjoined—they are no longer a place, but they are in a place.

Upholding firmly the first definition of place given by Aristotle, Ockham is led to the logical consequence that place is mobile.

Consequently, he rejects Giles of Rome's theory, whose essential passages he reproduces verbatim.[66]

The order and situation of the universe that Giles of Rome calls the *formal place*, is the order and situation of the container, not the content; if it is otherwise, he would be contradicting Aristotle, for whom place must be attributed to the containing body and not the contained body. This principle posited—and Peter Aureol's theory is thereby also rejected—there remains the question, how is it possible that formal place remains immobile, while its content, which is material place, is displaced?

When the contained body does not move, its distance to the fixed parts of the universe does not change; but this distance is not what constitutes formal place. In order to constitute it, one must consider the distance of the parts of the container which surround the content to the fixed references. And these parts can move even though the contained body does not move.

Ockham also attacks, like Joannes Canonicus before him, and in the same fashion, the immobility of the center and the poles of the world to which Saint Thomas Aquinas wished to relate the immobility of place.

> What one says about the immobility of the poles and the center proceeds from false thinking, namely, that there exist immobile poles in heaven and an immobile center in the earth. That is impossible. When a subject is animated by local movement, if its attribute remains one numerically, it is moved locally. But the subject of the accident which are the poles, meaning the substance of heaven, moves locally; or else the poles would be constantly replaced by other poles, numerically distinct from the first poles, or else they would be in movement.
>
> Perhaps one can say that a pole, which is an indivisible point, is not part of heaven, for heaven is a continuum, and continua are not composed of indivisibles.
>
> But if the pole exists, and is not part of heaven, it is then a corporeal or incorporeal substance. If it is corporeal, it is divisible, and not indivisible. If it is incorporeal, it is of intellectual nature, and we have come to the ridiculous conclusion that the pole of heaven is an intelligence.[67]

Therefore, neither material place nor formal place are immobile; the only immobility possessed by place is immobility *by equivalence*, such as Duns Scotus and Joannes Canonicus defined it. William of Ockham attributes the greatest importance to the notion of immobility by equivalence; he believes that it expresses in an explicit manner what Aristotle and his Commentator expressed implicitly. The notion seems to him proper to interpret all they asserted about the immobility of place.

Ockham even seems to derive from the notion of equivalent places the solution of difficulties which the notion manifestly would not be able to dispel.

Some Scholastics wished to find in the ultimate surface of the universe the immutable reference required by the immobility of place; Duns Scotus condemned their error. According to the Venerable Inceptor, their manner of thinking can be saved as long as the identity they attribute to this surface is interpreted as a simple equivalence.

> In this way one can understand what is meant when it is said that the totality of heaven is the place of some body; in fact, when this body remains at rest, each of its parts is at an equal distance from heaven. At each instant the distance of a part of the body to the ultimate parts of heaven is always measured by the same magnitude. . . . That is why a body is said to be at rest on earth, in spite of the movement of the air or heaven. . . . It does not matter if heaven moves, as long as the movement is not rectilinear, but circular. In this way, one can also explain the rest of this body and the constancy of its distance to heaven, whether heaven moves or not; one can explain it as easily as if there were immobile poles, as some have imagined. The immobility of poles, therefore, does not affect the question.[68]

Ockham's oversight is too obvious to be insisted upon; it is clear that what he asserts about an immobile body can be repeated about a body revolving around the center of the world. As erroneous as it is, this argument is no less interesting at one juncture; the argument rests upon the assumption that heaven is animated by a movement of rotation only, and not a movement of translation. The Venerable Inceptor repeats this hypothesis several times, and with insistence. "One cannot make use of the center of the world," he repeats with Giles of Rome, "to recognize the immobility and the identity of place by equivalence, except for one condition, namely that this immobility can first of all be concluded with respect to the absence of any movement of translation in heaven; it is because heaven has no movement as a whole on one side, or the other, that the center of the world is said to be immobile by equivalence."[69]

The oversight that we have just pointed out in Ockham's exposition is, in any case, corrected in his *Questions on the Physics*. The great precision of these *Questions* on this point and on others leads us to believe that they were written after the *Summulae*.

The seventy-seventh question of the *Questions on the Physics* is entitled: "When the body surrounding the body at rest moves continuously, does the place of the [latter] body remain at rest?" Here is his reply:

When the body surrounding the body at rest moves continuously, the immobile body is in a different place constantly; in fact, as we have already shown, it is the surrounding body which is its place. And this surrounding body changes constantly; therefore, its place also changes constantly.

However, in order to conform with the Philosopher's intention (*pro intentione Philosophi*), here is what I assert: no doubt it is true that the place constantly changes numerically, but the place is unique by equivalence. In fact, in order *to save* the rest of this body, and everything said of place and the lodged body, these two distinct places are as if they were a single place by equivalence. (*Tantum valent ista loca, distincta esse unum secundum equivalentiam.*)[70]

Ockham therefore formally considers the Scotist notion of equivalent place as Aristotle's.

With the help of this concept, he answers the seventy-eighth question: "Is place immobile?"

I posit two conclusions for this question.

The first is that literally (*de virtute sermonis*) one must agree to the truth of the proposition, place is mobile. In fact, every substance encountered among the creatures here below, whether it is some matter, or form, or a composite of matter and form, is mobile in some way or another; as for an accident, since the subject which affects it is mobile, it is also mobile by its own movement or by accident (*per se vel per accidens*). I therefore assert purely and simply, and truly, that place is mobile.

Second conclusion: according to the intention of the Philosopher, place is immobile by equivalence. Here is what I understand by this: multiple places, numerically distinct and mobile, have the same value for saving the properties of place, the end for which place is posited, as if there were a numerically single and absolutely immobile place.

What one should understand by the words *being immobile by equivalence* is the following: one considers place in view of the rest and movement of natural bodies. And their movement can be saved whether place is mobile or absolutely immobile. . . .

As for rest, even though the air surrounding the earth or heaven is in movement, one says of a body placed on the earth that it remains immobile when it keeps always the same distance to the ultimate parts of heaven. . . . Once these propositions [that affirm the constancy of such distances] remain true, whether the ambient body moves or not, whether each part of heaven moves or not, as long as heaven moves in a rotation and not a translation, one says that the body in consideration remains immobile. It is not possible, in fact, for a body put on the earth to maintain invariable distances to all the parts of heaven and to move by a movement of translation; although it can rotate and conserve these constant distances. (*Licet posset movere motu circulari et aequiliter sic distare.*)[71]

Ockham here perceives the objection that the *Summulae* did not signal, but he says nothing to dispel it. How one must respond to it, we will see clearly when we detail his doctrine on the movement of the ultimate heaven.

The passages of the *Summulae* and *Questions* we have just cited already prepare us for the examination of the problem of the movement of the ultimate heaven.

It would be useless for a disciple of Aristotle and Averroes to formulate the hypothesis that heaven does not have a movement of translation, since the contrary assumption would be an absurdity. Obviously, it is no longer so for Giles of Rome or for William of Ockham; to attribute a movement of translation to the celestial spheres and to their center no longer appears to be nonsense to them. To refuse movement to them is a postulate which it is necessary to formulate explicitly. Here as elsewhere, the philosophy of the Venerable Inceptor comes to the aid of doctrines which the theologians of the Sorbonne, under Etienne Tempier's guidance, wished to defend: *philosophia ancilla theologiae.*

We have just recognized the first crack, the first hint of the coming ruin, in the edifice raised by Aristotle and his Commentator; we shall now discover a second, larger and deeper.

The passage of the *Summulae* we cited earlier continues as follows:

The center of the world is said to be immobile by equivalence, but it is mobile in reality, even though the earth never moves as a whole. Notice that the places designated by the words above and below are marked by comparison with the center. For the distinction between places above and places below, the immobility of an indivisible center imagined by some physicists does not matter . . . it only matters that the center not be animated by a movement of translation.[72]

Thus the center of the world is the geometric point located at the same distance from all other parts of the celestial sphere; as long as heaven has no other movement than a rotation, we are assured that the center is always identical with itself *by equivalence*, even if the body within which it is found at every instant were mobile. The earth does not move as a whole, but it could. Some contemporaries of William of Ockham even held that it rotates on itself every twenty-four hours; others attributed to it some constant small movements which Albert of Saxony considered to be of great importance. According to these physicists, the part of the earth containing the center of the world changes from instant to instant; in reality, this center moves, but the new center is in an equivalent position with respect to the celestial sphere as the one it had occupied. The center of the world remains the same by equivalence.

Neither Aristotle nor Averroes would have been satisfied with this immobility by equivalence for the center of the world. According to them, the revolutions of celestial bodies suppose a center that is truly immobile, and in order for this center to be truly immobile, it must be located in a body deprived of any movement; thus the rotation of heaven would require the existence of a truly immobile earth.

This argument, Ockham is careful to recall, becomes void once immobility by equivalence is sufficient for the center of revolution of celestial orbs. The leader of the Terminist school recognized that his theory carried with it a consequence, which he formulated as follows:

> The celestial body moves around the earth, which remains at rest in the center of the world; let us note however that one can suppose that the earth is in movement, and that the center of the world would still remain immobile, even though in fact heaven would no longer be moving around an immobile body; however, it would continue to move. It would behave itself in the same way as if there were an immobile body at its center; its parts would constantly get nearer and farther from the parts of this immobile body.[73]

The *Summulae* end on this reflection; they could not have ended on a more important one.

Ockham takes up Scotus's proposition as his own, but under a somewhat enigmatic form. It is not necessary to be able to compare the changing positions of the parts of heaven to an immobile body actually existing in order for heaven to accomplish its revolution. The Venerable Inceptor does not merely formulate the proposition, he also indicates what the principle explaining it is, erasing its paradoxical character: in order for the movement of heaven to be possible, it is sufficient that one can conceive a fixed reference with respect to which the position of heaven changes from instant to instant. The immobile term, without which we cannot conceive local movement, does not need to be a concrete and actual body, as Aristotle and Averroes wished; it suffices that it is an ideal body, as Damascius and Simplicius announced.

This doctrine can be found again neatly formulated in one of Ockham's quodlibetal discussions.

The debated question is, "Does the term *ubi* designate a thing distinct from absolute things?"[74] (meaning distinct from the body to which one attributes the *ubi*).

> It would seem so, for something is truly acquired by local movement; but nothing absolute is acquired by local movement; therefore, it must be that the thing acquired is a relation.[75]

But he continues as follows:

> I reply wth a simple no to this question, and here is how I prove my reply:
>
> One would not admit such a relation, except to explain local movement, so that something is acquired or lost with every local movement. But still one should not admit it for this, for the ultimate sphere moves locally, and yet it does not acquire a new *ubi*, since there is no body circumscribing the eighth sphere that can serve as term for this relation.
>
> Perhaps you would say that the ultimate sphere has a variable relation with the center, for the earth remains immobile at the center in such a way that the ultimate sphere moves around it.
>
> I reply that, on the contrary, one deduces from this our proposition, that local movement can take place without the acquisition of such a *ubi*, for it is manifest that heaven is not in the earth as in a place, so that there is no such *ubi* there.
>
> Besides, if heaven as a whole were continuous with its content, so as to make one and the same body with it, God could still rotate this body, and yet, there would be nothing remaining at rest.
>
> Finally, if God made a body barren of any place, He could

still move it; however, nothing would remain at rest and no *ubi* would be acquired. . . .

You would say that in every moment something is acquired by the mobile. I reply by denying the proposition; it suffices that a place be acquired or lost, and the mobile is not the subject of this place. That is the special character of local movement.

You would say that a place cannot be acquired by something unless it informs the thing. I also deny that proposition. To say that something acquires a place is simply to say that nothing is interposed between the lodged body and the place by the effect of local movement. Moreover, sometimes the local movement can occur without there being the acquisition of anything that informs or does not inform the mobile; the following suffices: a place would be acquired if there existed a surrounding body. We have as example the ultimate sphere which moves locally but acquires nothing; however, if there were an immobile place surrounding this sphere, it would acquire a place, but in fact it does not acquire any place. Yet one says that it moves by local movement.[76]

We can detect the influence of Etienne Tempier's decree and Duns Scotus's teaching in the arguments by which William of Ockham opposes the Scotist theory of the *ubi*. But in its conclusion, in the reply that the Venerable Inceptor gives the enigma proposed by the Subtile Doctor, what we should recognize is the influence of Damascius and Simplicius; anyone who would doubt this should read the question that follows almost immediately.[77]

Someone having heard Ockham, and having read the *Commentary on the Categories* composed by Simplicius, objected, no doubt, that the order of the universe is the term by which all local movement must be appreciated. Ockham is quick to establish that the term, unity (or order) of the universe, implies no relation distinct from the absolute things that the world contains.

The order or unity of the universe is not a relation similar to a bond uniting the various bodies ordered in the universe to one another; in this way, if this relation did not exist, these bodies would not be ordered and the universe would not be truly one, as thought Simplicius in his writing on the *Predicaments*. This order implies only absolute things that do not make up something numerically one; among these things, one is more distant and the other less distant from the same thing—one is closer to some thing and the other is more or less distant to that same thing—without there being any relation inherent in them. It happens that between some of them there is an intermediary and between some others there is not. Thus the connection of the universe is better saved without this relation than with this relation.[78]

It is therefore Simplicius's lecture—and therefore the ideas of Damascius—that suggested to Ockham his essential thesis about place: The fixed term to which one relates all movement does not have to be an immobile body, concrete and actually existing in the universe. And Ockham intends to affirm that this fixed term is nothing but a conception in our mind. He does not wish to follow Damascius and Simplicius and take for the fixed term an order of the universe to which one would attribute a kind of ideal existence, distinct from the existence of the bodies forming the universe.

The doctrine that the immobile place to which one relates the movement of heaven is a pure conception of the mind, and has no need to be realized in a concrete body, was started in the *Summulae* and completed in the *Quodlibeta*; it was then taken up by the *Quaestiones super libros Physicorum*, and formulated with great precision.

In order to find this precise formulation, one must read the reply to the eightieth question: "Does the eighth sphere move, properly speaking?"[79]

What one doubts is as follows: How can this sphere rotate, properly speaking (*per se*), when it is not in a place, properly speaking?

One can reply that this heaven moves, properly speaking, because each of its parts does not now have the same distance that it had before with respect to the various parts of the immobile earth; and that is moving, properly speaking, for this sphere.

But what if one objects that the earth moves? (*Si dicatur quod terra movetur?*) I reply that if one admits this assumption, then, in fact, heaven no longer moves around something immobile; but one would not say that, in this case, it does not move, if it moves. It behaves in such a way that if there were something immobile in its center, its various parts would have continually variable distances to some determinate part of the earth.

Moreover, one can say that heaven moves, properly speaking, because if there existed some immobile body that surrounded it constantly, each of the various parts of heaven would have a variable distance to each of the parts of this immobile body, and that is to move, properly speaking.

Perhaps you would say: heaven moves, properly speaking, therefore it acquires a place. I reply: your reasoning is not valid. In fact, the movement of heaven and the movement of a vase full of water is the same, up to a point, as we have said before. In fact, the vase moves locally, properly speaking, by the various parts of its concavity as well as by the various parts of its convexity; it does not move with a view toward acquiring a new place for these parts, but it moves locally by these parts in order to become the place of another content. In the same fashion, heaven does not move locally in order to acquire a new place to contain it, or to contain its parts; it moves with a view toward becoming the place of some different immobile thing, when there is something immobile, as is now the case. Or else, when there is nothing immobile, it suffices that it moves to become the place of some different, immobile thing, if there were some immobile thing.[80]

This passage summarizes, with much clarity and firmness, Ockham's novel doctrine; its extreme importance as well as the difficulty in attaining it, leads us to cite the Latin text:

Utrum octava spera movetur per se. . . .

Sed dubium et quomodo movetur per se motu circulari cum non sit per se in loco.

Potest dici uno modo quod ista movetur per se quia partes celi aliter adproximarentur parti terre quiescenti quam prius, et hoc est ipsam moveri per se.

Si dicatur quod terra movetur. Respondeo: Isto casu posito, tunc de facto non movetur celum circa aliquod quiescens; nec propter hoc diceretur non[81] moveri si tunc moveretur, quia taliter[82] se habet quod, si esset aliquod quiescens in medio, partes[83] sue diverse aliter continue adproximarentur isti determinate parti terre.

Aliter potest dici quod movetur per se quia, si esset aliquod corpus quiescens circumdans continue, partes celi aliter adproximarentur diversis partibus illius corporis quiescentis, et hoc est ipsum moveri per se.

Si dicas: Celum movetur per se localiter; ergo acquirit aliquem locum. Respondeo: Conclusio non valet. Quia simile est quodammodo de motu celi et vasis repleti aqua, sicud prius dictum est; nam sicud vas per se movetur localiter tam secundum partes concavas[84] quem convexas,[85] non tamen movetur ut secundum istas partes acquirat[86] novum locum, sed movetur localiter per istas[87] partes ut sit locus alterius;

ita celum non movetur localiter at acquirat novum locum continentum ipsum secundum se vel per suas partes, sed movetur localiter ut sit locus alterius quiescentis, quando est aliquod quiescens, sicud est modo de facto; et[88] si ullum esset quiescens, sufficit quod sic[89] moveretur ut esset locus alterius quiescentis, si esset aliquod quiescens.[90]

Let us now relate this language to the doctrine which Ockham held, particularly in his *Questions on the Book of the Physics*, with respect to *equivalent* place, and we will clearly see what he intended by this place. If the bodies surrounding the ones we are observing are changing and mobile, we can replace them, he asserts, by immobile and immutable places and these would be equivalent to them for the study of the rest or movement of the body in consideration. An *equivalent place* is a purely conceived, fixed reference that no concrete body realizes, a term similar to the one Ockham has just imagined with which he judged the movement of heaven if there were no immobile earth at the center of the world.

As we have noted, the whole Peripatetic theory of place rested, in the final analysis, on this proposition: there necessarily exists an immobile earth at the center of the world. Ockham does not deny the immobility of the earth, even though this immobility was already contested by some of his contemporaries, but he no longer considers the affirmation of the earth's rest as a necessary proposition; the proposition is a truth of fact for him (*sicut est modo de facto*). Were one to formulate a theory requiring the movement of the earth, the theory of place is already constituted in such a way that it would offer no resistance.

Walter Burley

It would be difficult to link Walter Burley to any particular school; he inaugurates the eclecticism so distinctive of the Terminists of the University of Paris during the fourteenth century. His theory of place derives inspiration from Duns Scotus, as well as Saint Thomas Aquinas and Giles of Rome, and William of Ockham as well as Joannes Canonicus and Peter Aureol;[91] he borrows some thoughts from each, and he addresses a critique to each. His faults, as his virtues, derive from his eclecticism; he sometimes lacks the dogmatic neatness and logical rigor that might have been possessed by a less open and less receptive mind.

How is one to understand the Aristotelian definition of place as *ultimum continentis*? Should one admit with William of Ockham that place is the containing body itself or a volume included in the containing body? Burley rejects this interpretation.[92] As Ockham himself asserts, if one admits this, one can attribute an infinity of different places to any body: one can cut out a layer from the thickness of the body enveloping the lodged body that also envelops the lodged body; from the thickness of the second layer one can cut out a third layer, and so forth indefinitely. Given Ockham's interpretation, each of these layers is the place of the body, in the same way as the layer from which it has been cut out, and in the same way as the layer which will be cut out from it.

Moreover, Walter Burley does not have the same repugnance to the reality of the surface as does the Venerable Inceptor; admitting that a body is extended in all dimensions, he also thinks, with Duns Scotus, that an accident of the body can be attributed to its surface only, without affecting its depth in any way. He therefore does not hesitate to attribute to the Peripatetic formulation of place as *ultimum continentis*, the interpretation that it is the surface of the container.

As the two words indicate, place is not simply the surface; it results from the union of two elements, the surface and the action of containing (*continentia*). Burley's opinion is in conformity with Duns Scotus's opinion.

Also in conformity with Duns Scotus's opinion is Burley's distinction between place and the *ubi*;[93] the *ubi* is the effect produced in the lodged body by the action of containment which, united with the surface of the ambient body, constitutes place. In an immediate and intrinsic fashion, the *ubi* is not place, but the *ubi* is the term for local movement.

After having cleared up his definition of place, Burley tackles the question, is place immobile?

The first reply to this question that he examines is the one proposed by Giles of Rome.[94] One must distinguish the matter and the form of place: the matter, the surface of the containing body, moves at the same time as the body; the form, on the other hand, remains immobile when the contained body does not move, for it is the distance of the surface from the supreme orb, or, better yet, from the poles and center of the world.

"Others"—meaning Saint Thomas Aquinas—"say, and it comes almost to the same thing, that the ultimate part of the container does not possess a *ratio loci*, except in virtue of the order and position it occupies with respect to the celestial sphere."[95]

Walter Burley objects to these theories using the same arguments that Joannes Canonicus and William of Ockham have already used to oppose them; he also raises a new argument against them:

Let us imagine that a body remains immobile in midair, for example, and let us suppose that divine power imposes on heaven as a whole and to the set of elements a rectilinear movement eastward. The portion of the universe that was west of the body gets nearer to it, and the portion that was east of it gets farther away; one of the poles of the world comes nearer and the other gets farther from the body.[96] The distance of the body from the center of the world gets smaller or larger than it was. And, since the body remained immobile, it must have remained in the same place, and consequently, its place must have remained invariable. However, the situation of the place in the universe, its distance from the poles and the center, did not remain identical; the situation, the distance from the poles and the center, are therefore not the formal element of place.[97]

To maintain that God can give the universe a movement as a whole was, for Peripatetic philosophy, to affirm an absurdity. The condemnation brought forth in 1277 by the theologians of the Sorbonne accustomed minds to think of the proposition as a truth. Thus we have seen Giles of Rome insinuate and William of Ockham affirm that any theory of place where the center of the universe is regarded as immobile must indicate this postulate explicitly; Walter Burley shows us how, in fact, the denial of this postulate rendered absurd the doctrines by which Saint Thomas Aquinas and Giles of Rome attempted to save the immobility of place.

These doctrines are flawed at their very core. According to Walter Burley, one cannot distinguish a matter and a form in place.[98] Place is a simple form, similar to any accidental form such as whiteness, cold, and heat.

Duns Scotus, Joannes Canonicus, and William of Ockham reduced the immobility of place to an immobility by equivalence; Walter Burley knows this theory and exposes it, as follows:

Let us suppose that I reside here, in this house of the Sorbonne, and that a great wind blows around me in such a way that it constantly renews the air surrounding me; if, however, I remain at rest, it is certain that I remain at an invariable distance from heaven as a whole, from the center of the world, or from any immobile body whatever. For example, at every instant there are as many leagues between

England and myself as there was before. Consequently, the place I find myself in does not remain the same numerically; but this place remains the same by equivalence with respect to the distance to the immobile things. It is equivalent to a single place with respect to the production or the reference of a movement. Thus, when the lodged body remains immobile, either its place remains numerically the same, or it is replaced by a place equivalent by its distance to other immobile objects, and equivalent for any local movement beginning or continuing.[99]

Walter Burley then states that he will examine the sense one ought to attribute to this theory at another time, and he returns to the question of the immobility of place, properly speaking.[100]

Nothing is mobile by itself other than bodies. Burley, who rejected William of Ockham's interpretation of place as a body, also rejects Ockham's proposition that place is mobile by itself (*per se*). On the other hand, he agrees that place is mobile by accident, that is, because of the movement of certain bodies. The place of a body is the surface of the matter surrounding the body; it therefore moves when the matter moves.

The above proposition leads to undesirable consequences. A body can change place without moving; it can move without changing place. The undesirable consequences arise from a confusion.[101] One thinks of place as the term for local movement, but that is not true. Local movement is not a change of place, but a change of *ubi*. Thus it is true that a body cannot change *ubi* without moving, and that it cannot move while keeping the same *ubi*. But the same *ubi* can correspond to different places, and the same place to different *ubi*.

The above theory, which is in conformity with the thought of Duns Scotus and his most faithful disciples, such as Joannes Canonicus, serves well the eclecticism of Walter Burley; the substitution of the *ubi* for place happily puts back into favor the systems he ought to have rejected.[102]

Such is the case for the systems of Saint Thomas Aquinas and Giles of Rome which assume immutable the *ratio loci* or formal place.

The rational place of Thomas Aquinas and the formal place of Giles of Rome change by the fact that the container moves, even when the content remains immobile. That is so because the situation relative to the universe which constitutes the *ratio loci*, or formal place, is an attribute, not of the contained body, but of the ambient matter, and an attribute cannot remain immutable when its subject varies.

But Walter Burley proposes a modification to this theory that seems to render it acceptable to him; it consists in asserting that "the order the lodged body presents with respect to the ultimate sphere, and to the poles and the center of the world—its distance to these references—is the formal element of the *ubi*, not of place, or, better still, this order and distance are the *ubi* itself. . . . It is true that one says that the *ubi* is caused by place, but it is not necessary for the *ubi* to vary every time place varies; a new place does not cause a new *ubi* unless the new place corresponds to a new order and a new situation with respect to heaven as a whole, and the immobile poles."[103]

The definition identifying the *ubi* of a body with the distance of the body to the other immobile bodies is in agreement with the one assigning the *ubi*, instead of place, as the term for local movement:

> One must not say: a body moves locally when it behaves differently with respect to place, from one instant to another. One says: a body moves locally when it behaves differently with respect to a second body deprived of local movement, from one instant to another. Therefore any body whose distance to a body deprived of local movement changes from instant to instant, becoming larger or smaller, is a body that moves locally.[104]

Walter Burley's transformation of the theories of Saint Thomas and Giles Colonna is far from new; already Peter Aureol proposed to attribute to the lodged body the characteristics which his predecessors attributed to the ambient matter under the name *ratio loci* or formal place. But he conserved the word *place* as an attribute of the lodged body, while the Peripatetics agreed to consider place as the container. Burley adopts Peter Aureol's reform, but he takes care to leave the word *place* its Peripatetic meaning; what Aureol defined under the word *place* he identified with the *ubi* of the author of the *Six Principles* and of the Scotists.

After having examined the theory of the permanence of place by equivalence, Burley deferred the discussion and interpretation of the theory. If we compared what he stated before with what he just stated about the *ubi*, we would easily recognize that we now possess the right meaning, the *bonum intellectum*, of the words: *two equivalent places*. Two equivalent places are evidently two places specifically distinct, but causing the same *ubi* in the lodged body.

Thus the eclecticism of Burley, which already reconciled the various formulations of Saint Thomas Aquinas, Giles of Rome, and Peter Aureol into a single doctrine, also succeeded in uniting this doctrine with the theory of equivalent places formulated by Duns Scotus, Joannes Canonicus, and William of Ockham.

We shall see Burley walk farther along the path traced by the Subtile Doctor and the Venerable Inceptor.

According to the definition given by Burley, the *ubi* of a body is the position of the body with respect to other immobile bodies; local movement, which is a change of *ubi*, is a change of the situation the mobile body occupies with respect to the fixed bodies.

> All movement supposes an immobile body, as it is stated in the book *On the Movements of Animals*. In fact, for a body to move, it must be, at each instant, other than it was before; for that, there must be a fixed reference by which it behaves, at each instant, other than it behaved before. This reference must be absolutely immobile, or else it must possess rest opposed to the movement of the mobile; either it does not participate at all in this movement, or it participates in it, but with the least speed. If a man directed himself toward Saint-Denis, and if another followed him on the same road with exactly the same speed, the relative position of the two men would not be changing at all.[105]

From a similar observation, Franciscus de Marchia concluded that a body can serve as the place of another even when it is not immobile; it suffices, according to him, that it possesses the immobility opposed to the movement that a body lodged by it can take, movement with respect to which it must serve as a reference. Following John of Jandun, whom he sometimes followed but more often opposed, Burley restricts the domain of this proposition; he applies it only to natural place, not to place in general.

> The concavity of the lunar orb is the natural place of fire, and yet this orb moves; but its movement is not the natural movement by which fire directs itself toward its concavity. It is true that one says that a movement whose end is itself in movement is futile. I reply that if a body were to move toward an end which moved in the same direction that the body moves, and with the same speed, the movement would be futile; in fact, the mobile could never attain its end. And that is how one should interpret the Commentator. But this

movement would not be futile if the end did not move with the same speed; thus it is when fire moves toward the concavity of the lunar orb. Therefore when the Philosopher wishes that natural place be immobile, one can understand by this that natural place ought not move by the same natural movement by which the body it must lodge moves toward it.[106]

One should not conclude that the reference that serves to define the *ubi* of a body and to determine its local movement must not be held to be absolutely immobile. What is true about natural place and natural movement is not necessarily true about the *ubi* and local movement: "One attributes to bodies a natural place because of their natural rest, rather than their local movement."[107] Without exception, when Burley defines the *ubi* of a body, when he determines local movement, he supposes that this definition and this determination are made by comparison with an absolutely immobile reference.

Must this reference be a concrete body, actually existing, or does it suffice that it is conceived without being realized? That the latter opinion is Walter Burley's cannot be denied; in fact, he admits the opinion of William of Ockham with respect to the place and the movement of the final celestial sphere.[108]

The ultimate heaven is in a place *per accidens* because of its center which is within the immobile earth.

> If someone were to say to me, heaven would still be in a place, as it is now, if the earth moved, I would agree. If one formulates the objection that heaven cannot be in a place because of its center unless the central body was immobile, I respond that heaven is in a place by its center, whether the central body remained at rest, and whether it moved. In fact, heaven behaves in such a way that the situation of each of its parts with respect to the parts of the central body would change from instant to instant if there existed an immobile central body. In fact, the central body by which heaven is in a place [meaning the earth] is an immobile body; but if we supposed that this central body moved, heaven would still be in a place because of its center. And that because, in this case as well, the manner of being of heaven would be such that if the central body were immobile, the disposition of the parts of heaven with respect to the parts of this latter central body would be variable from instant to instant.[109]

Walter Burley adds to the above a reflection that lacks logic:

If the earth moved with the same speed as heaven, one could still say that heaven would be in a place by the indivisible center of the whole universe, for, with respect to this center, the various parts of heaven would behave differently from instant to instant.[110]

A more careful reading of John of Jandun's comments would have enabled him to avoid this error.

Moreover, Walter Burley does not seem to have always followed, in a rigorous fashion and completely, the theory whose principles he posited.

Duns Scotus said that the universe can rotate even if it contained no body; it can still rotate, for example, if it is formed of a single sphere, homogeneous throughout its whole extent; the movement of rotation, taken in itself, is therefore a *forma fluens*, and this form can exist by itself, without needing to be considered with respect to another body, whether it is container or content. It is a purely absolute form. Search for the answer.[111]

The question posed by the Subtile Doctor seems to have lost its enigmatic character because of William of Ockham's theory, which Burley adopted. But, far from seeing the solution to the enigma in this theory, Burley appears to be singularly intrigued by it. He asserts:

God has created a discontinuous world, formed of distinct parts. It is in virtue of this discontinuity that each of the parts of the world is in itself in a place. But God could have also created a world continuous in all its parts; He could therefore have created only an absolutely homogeneous sphere. Let us then imagine that, at the moment of creation, God instead of creating this universe, created an absolutely homogeneous sphere. Every body is in a place; therefore this spherical body would be in a place. It would not be in a place because of its parts; none of its parts would be lodged, for place is a container separate from its content, and there is no separation in this continuous body. This body must therefore be in the void. Those who believe in the creation of the world must therefore admit the void.

One can reply to this in the following way: those who speak according to faith uphold that God could have created such a spherical body perfectly continuous, occupying the whole space that our universe occupies. Then speaking as physicists, they are driven to recognize that such a body cannot be lodged by its parts, nor by the terminal region of the containing body, since no body can exist outside it; they simply conclude that it is not necessary for a body to be in a place.

But can one say that God can move such a world, either by a rotation or by a translation, transporting it to another place? Every local movement requires a place. . . . Therefore if one imagined that there existed such a continuous body, and nothing outside this body, God could not impart to it a movement of translation unless He created at the same time a place toward which it could move; God could not move this body by a rotation, or else one would have to admit that the movement of rotation that He would impart to it is not a local movement, but a movement relative to the situation.[112]

Albert of Saxony will soon show us that all these difficulties can be dispelled using the same principles that Walter Burley upholds. Let us leave these doubts aside, then, and return to our author's theory in order to emphasize its essential characteristics.

Walter Burley's doctrine about place and local movement is a synthesis of the various attempts of the more eminent doctors of Scholasticism. Let us stop for an instant in order to contemplate this synthesis and let us note its dominant characteristics.

Aristotle defined the place of a body by the formula, the ultimate part of the container. He wished to impose immobility to this place, in order that it serve as term of comparison in the determination of local movement. But, the immobility of place was clearly incompatible with the definition—from which arose the necessity to modify the definition. The Stagirite accomplished the modification of this definition only implicitly, and one can say, surreptitiously, by equivocal uses of the word *place*. Inconsistencies arose in the theory of place and local movement because of this.

This doubling of the notion of place, which the Philosopher unfortunately practiced almost in secret, is clearly revealed in the theory that Walter Burley exhibits in its almost completed form.

In this theory, place keeps the definition Aristotle attributed to it at first, but place so defined loses its immobility. One refuses to utilize place in the description of local movement.

The fixed element serving as reference to movement is not the place, but the *ubi* of the mobile. The *ubi* of a body is the position of the body with respect to other fixed bodies. Moreover, these fixed bodies, serving as terms of comparison for the definition of the *ubi* and the determination of local movement, do not need to be real and concrete bodies; it suffices that they be conceived by reason.

If place, as Walter Burley considers it, is place as Aristotle first presented it, the *ubi* he conceives is completely identical to the *thesis* of Damascius and Simplicius. In the synthetic doctrine traced by Walter Burley, the doctrine of these two philosophers finds itself harmoniously united to the doctrine of the Stagirite.

Nicholas Bonet

Walter Burley is a conciliator; he suppresses everything that seems too cutting from the various doctrines he is comparing. Deprived of their prominent characteristics—rounded off and blunted, so to speak—these doctrines become less easily recognizable, but in return, they resemble each other more, and consequently, they allow themselves to be more easily assembled into a single system.

The intellectual tendencies of Nicholas Bonet are exactly opposite those of Walter Burley. Bonet excels in discovering within each doctrine principles that can produce extreme consequences; he presses these principles and forces them to bring to light the corollaries hidden in their shadows. In his hands the characteristics of a theory become so prominent that the author of the theory would not always recognize the legitimate offspring of his thought. Thus, the contrasts between the various theories are brought forth; one can easily understand why a synthesis would be impossible for Bonet.

Burley attempted to reconcile the opinions of Duns Scotus and William of Ockham with those of Aristotle with respect to place. Bonet, on the contrary, shows us that the ancients and the moderns pursued, in this theory, the solution to two essentially different problems, without having perceived this. There is nothing to reconcile; their discussion is an irreconcilable misunderstanding.

Our Franciscan writes:

Let us say in what way place is immobile. Let us begin by exhibiting the opinion of the modern philosophers; let us then return to the ancients.

Among the assertions about the immobility of place which are the objects of discussion among the modern philosophers, the first doctrine is as follows: place is immobile with respect to the poles of the world, which are immobile. The relation of place to these poles is a relation immobile with respect to the subject (*subjective*) and inversely.

But this doctrine lends itself to a criticism. In fact, it is not repugnant for the poles of the world to change and to move. The firmament moves from east to west, but there is no repugnancy or contradiction in its turning from the north to the south; and the poles of the world would be changed by this. . . .

The second assertion discussed about the immobility of place is the following: place is immobile with respect to the center of the world, which is absolutely immobile. . . . But this assumption also is not true, that the center is immobile. The earth is not incapable of movement, either in its totality or in each of its parts. These parts are, in fact, all of the same nature; if one of them is capable of movement, all of them are also.

Perhaps one might object that the center of the world is simply a point imagined at the center of the earth, at equal distance to all the parts of the circumference of heaven; this point is absolutely immobile—even if the earth were to move, this point, which is purely conceived, would always remain immobile.

I reply that if heaven displaced itself in a straight line, this center also would be necessarily displaced by the same length, as if it were drawn by the traction of heaven. Heaven displacing itself along a straight line does not imply a contradiction, and it is possible for it to be.[113]

Bonet's whole argument (exhibiting and refuting the opinions of Saint Thomas Aquinas and Giles of Rome) rests, as one sees, on the axiom commonly admitted since the condemnations of 1277: if God wishes to impose upon heaven a movement of translation or rotation other than the movement it possesses in fact, He can do it, for this assumption does not entail a contradiction.

After having wrecked the proposition of Saint Thomas Aquinas and Giles of Rome on the immobility of place, Bonet takes up the theory of Francis of Mayronnes. We will repeat what he asserts later, when we exhibit this theory.

He finally comes to the doctrine about the immobility of place by equivalence, a doctrine he seems to take under the form presented by Duns Scotus:

The fourth opinion about the immobility of place is the following: One says that place is immobile by equivalence, from an immobility opposed to local movement. . . .

It is evident that this opinion is presented with great subtlety; nevertheless, it is not completely satisfactory. First, in fact, one could object to it with what one has objected to in the other opinions: this place, which for the moment remains single by equivalence, would experience a change of place if the whole celestial sphere were displaced in a straight line. . . .

Let us conclude that the philosophers of our faith (*in nostra lege loquentes*) did not explain sufficiently what is the immobility of place.[114]

We must now examine the immobility of place according to the tradition of the ancients.[115]

Instead of summarizing the texts of Aristotle and his Commentator, our Franciscan attempts to penetrate the depth of their thought and to bring out their essential principle, the directive idea hidden in it, and which they themselves had barely discerned. The attempt is novel and daring; perhaps we should recognize that it largely succeeded.

One should note how Aristotle proceeds in his book of the *Physics*. No doubt he discusses movement in general there, but he also discusses specifically the movement of the first mobile; there he demonstrates the passions of this movement— that it is uniform, regular, and eternal. He also establishes the other properties of this movement. Similarly he proves that time is a passion of this movement, for time [according to him] is a property of [the movement of] the first mobile, and not of other movements; he does not discuss time in a determinate fashion, as a passion of other movements.

Thus, in conformity with this method, he defines the place of the first mobile, and not place in general, when he states that place is the surface of the containing body, which is immobile.

Place, I assert, is the surface of the other containing heaven, for the first mobile is contained by another heaven, which is elsewhere, and whose surface is absolutely immobile, as this heaven—meaning the one immediately touching the first mobile—is itself the first immobile. The surface of this body is therefore absolutely immobile, and that is because of the subject in which it resides, since it is absolutely [immobile], incorruptible, and eternal.

That is the first way of assigning an eternal place; and it seems that that is what the words of Aristotle and the philosophers signify.[116]

That Aristotle proposed to find, first of all, a place for the first mobile, that, in order to resolve this problem, he conceived of an absolutely immobile ultimate sphere, and that one can derive these conclusions from his own writings, are propositions that are manifestly false. But that the theory of place Aristotle constructed requires the existence of such a heaven, and that the theory remains somewhat incomplete because it is prohibited by the other doctrines of the Stagirite, is what Nicholas Bonet can proclaim legitimately.

"Our ancestors (*progenitores*) have another way of assigning immobility to place."[117] Bonet is alluding to this second way, at the start of the chapter we are analyzing, when he writes: "The ancients affirmed that natural place is absolutely immobile, natural movement being, in any case, either centripetal, centrifugal, or a rotation around a center; they do not affirm the immobility of a place occupied by violence, for they did not suppose that a place acquired by violence is immobile."[118]

This passage seems inspired by Damascius and Simplicius. However, it is the theory of natural place, as it is given by Aristotle and Averroes, that Bonet attempts to exhibit:

> They admit that the places of the elements are by nature distinct from one another. The concavity of an element is the immobile place of another element; never can the former move naturally above the concave surface of the latter. The containing element also cannot move naturally beyond the convex surface of the contained element. The limits of the various elements are therefore immobile by nature.
>
> Let us give in succession each element as example.
>
> The center of the world is the lower limit; the concavity of the lunar sphere is the upper limit. The four elements within these two limits have determined places that are absolutely immobile, both above and below.
>
> Let us take, as first example, elementary fire. The concavity of the lunar orb is absolutely immobile; in fact, it is manifest that this concavity moves by a rotation. It is therefore immobile in the sense that its movement always leaves it in the same position, that it cannot be displaced upward or downward. Consequently, the place of the ultimate surface of fire is immobile. Moreover, if the impossible were to happen and the lunar orb were completely annihilated, fire would still not move higher than its present position; there is, in fact, an upper limit for its place above which it cannot climb. Thus the purely conceived (*ymaginata*) surface surrounding the final spherical layer of the sphere of fire is immobile, and the place of fire cannot be displaced, either upward or downward.

Air also has an immobile place—which is the concavity of the sphere of fire; it cannot climb naturally above this surface. Moreover, even if the sphere of fire were completely annihilated, air would not climb above this purely conceived surface.

One can say the same about the ultimate surface of water, which is its immobile place, and which is the concavity of the sphere of fire.[119] One can say the same for the earth with respect to water.

As it is manifest, the elements also have immobile places downward.

We have as example earth, whose ultimate and immobile place is the center of the world, above which it cannot move in any way. Whether it is possible or impossible, let us imagine that the whole earth is pierced in part following a diameter passing through the center of the earth; if a stone, placed in midair, fell through this hole, it would descend no farther than the center. There it would stop and remain at rest; it would never move beyond, unless by violence. The center, in fact, is its proper and immobile place, toward which it tends. Similarly, whether possible or impossible, if the whole earth were annihilated and one left something heavy suspended in midair, the heavy thing would fall and move precisely until it reached the center, purely conceived (*ymaginatum*); there the heavy thing would remain naturally at rest even though nothing upheld it.

It is evident that the element, water, also has a termination, a limit, a place absolutely immobile, which its nature assigns it in the universe; it does not move naturally beyond this place. And that place is the ultimate surface of the earth. No doubt water is heavy, and consequently it moves downward; nevertheless, it does not descend purely and simply down to the center, but only down to the place nature has assigned it, which is the ultimate surface of the natural place of the earth. The following conclusion is entailed from that: although impossible, if the terrestrial element were annihilated in its totality, water would not move toward the center of the world in order to occupy its natural place; it would move only to the surface limiting the place of the earth. There it would remain suspended, naturally at rest, and the space constituting the natural place of the earth would remain void. . . .

One must express oneself similarly with respect to air and fire; they also have their proper places, limited above and below, beyond which they do not move naturally.

It is clear that that is the intention of Aristotle and his Commentator, in comments 41 and 42 of the fourth book of the *Physics*.[120]

Such a rigorous formulation of the theory of natural place would normally startle a Scholastic; it is too clearly inadmissible. When one digs a hole in the earth and one pours some water in it, one knows that the water will fall to the bottom, even before we dig the hole. The distinction posited by Aristotle between absolutely heavy bodies and relatively heavy bodies, and the necessity to avoid a vacuum, are invoked in turn to mitigate what, in the doctrine of natural place, would clash too violently with our daily experience. Nicholas Bonet makes no use of such palliatives; what he wishes to do is to bare Aristotle's directive thought, Aristotle's "intention." Within this framework, should we not recognize that he has succeeded?

According to our Franciscan, the natural place of the elements (the purely conceived, concentric spherical surfaces of the world that delimit place) is the only kind of place for which Aristotle intended to affirm immobility. He never considered the other place, the ultimate part of the containing body as immobile; in order to distinguish it from the abstract and immobile natural place, he named it the vase:

> In fact, that is how we distinguish place from the vase— by its immobility; according to Aristotle and his Commentator, the vase is a mobile place. That is why the water of the stream surrounding the ship is not the place of the ship; it plays the role of vase. The whole stream is called the place of the ship.
> From what has just been stated, it appears clearly that the philosophers of our faith (*loquentes in nostra lege*) have said nothing in conformity with the intention of the ancients; they have worried about the immobility of the vase, in fact, while the ancients have denied this immobility when they asserted that the vase is a mobile place. They did not worry about the immobility of place; thus what they asserted was insufficient. The air surrounding a body plays the role of vase; thus it is easily susceptible to movement. But the totality of air is absolutely immobile because of its place, because the [natural] places of the elements are absolutely immobile.[121]

Hence, according to our author, the problem of the immobility of place was subject to a serious misunderstanding. What Christian Scholastics called place is the ambient body, the vase; this vase is not immobile in any way. Aristotle attributes mobility to it, and the Christian philosophers attempted vainly to discover an illusory immobility in it.

Bonet attempts to dispel the misunderstanding completely by positing a distinction.[122]

According to him there are two kinds of places.

The first is the place of the physicist. It is a concrete place, realized by the containing and lodging body; it is properly a receptacle, a vase, and this vase is mobile.

The other is the place of the mathematician. It is an abstract surface that has no reality outside our mind; this place, which is purely conceived, is the only one that is immobile.

Moreover, Bonet puts forth this doctrine as one in conformity with the true opinion of the ancients.

Here is how he formulates it:

> The final affirmation of our ancestors on the subject of the immobility of place is the following one:
> There are two ways of considering place; one is mathematical, the other is physical (*naturalis*).
> Aristotle considered place from the mathematical point of view when he asserted that place is the surface of the containing body, which first of all is immobile.
> He who considers place from the physical point of view, on the contrary, would define it in this way: place is the surface of the containing body, which first of all is mobile. The physicist, in fact, does not consider, as physicist, the nature of place, properly speaking, but the nature of the vase; and any place, insofar as it plays the role of the vase, is mobile.
> Therefore let us remark that place, considered from the mathematical point of view, is the surface of the containing body, which first of all is immobile, taken without considering the physical body to which this surface belongs. When the mathematician considers the airy surface surrounding and containing us, he does not consider in what body it exists at all; he ignores whether it is air or another thing. He considers this surface separately, as if it were detached from all physical bodies. Such a surface is then immobile; all mathematical beings are immobile, in fact, for they are separated from sensible matter and movement by abstraction. . . . Thus place, as the mathematician considers it, is absolutely immobile. The airy surface surrounding us, when we consider it as separated from the air and any other body, does not change when the body surrounding us changes. . . . That is why place is immobile when one considers it as the mathematicians do.

On the other hand, when one considers it as the physicists do, it is mobile; then it plays the role of receptacle (*vas*). Physical place, in fact, is considered as a surface belonging to this or that body; such a surface can move, subjectively and objectively, in the same fashion as the natural body to which it is part. . . .

And, according to the Commentator, in the thirteenth comment on the fourth book of the *Physics*, the study of place is more mathematical than physical.

Let us therefore conclude that that was the intention of our ancestors on the subject of the immobility of place. As for him who is not satisfied with the three kinds of immobility of which we have just spoken, let him seek something else.[123]

In order to appreciate better the opinion that Bonet put forth lastly, the opinion that reserves immobility for the abstract place of the mathematicians, it is important not to forget what kind of reality our author attributes to the concepts of mathematics. He explained himself clearly with respect to this in one of the chapters of his *Metaphysics*:

The separation of magnitudes can be understood in three ways. The first is when one separates a singular magnitude from the subject that bears it. The second is when one separates the universal magnitude from singular magnitudes. The third is when one separates the universal magnitude from any subject. In each of these three ways one can understand that the separation is accomplished in the intellect, that is, in conceptual existence (*in esse cognito*), or else outside the intellect, in real existence.

Let us first speak of the separation practiced in conceptual existence, and let us assert that the separation is possible in each of the three ways. Objectively,[124] in fact, the intellect can abstract a particular magnitude from any subject, for magnitude can take conceptual existence without the subject in which it inheres having also to exist conceptually; thus to abstract is nothing other than to consider this without considering that. He who accomplishes such an abstraction is not lying.

Such an abstraction that separates a magnitude from any subject and sensible matter is properly mathematical; mathematicians, in fact, consider the magnitudes of bodies without having to know in what way these magnitudes exist. . . .

Let us now speak about the separation of magnitude in real existence.

Let us first assert that it is impossible to separate absolutely magnitude from its subject, by natural means.[125] It is, in fact, proper for an accident to subsist in an actual way only in something else, to depend on a subject, and consequently, to exist subjectively in this subject.[126]

These passages leave no doubt about the thinking of our author. The surface that the mathematician detaches from the ambient body in order to construct the place of the contained body has no real existence outside the intellect; the only existence it possesses is conceptual existence—*esse cognitum*—within the intellect. The immobile place the mathematicians consider when they speak of local movement has no existence other than that; it is a pure concept. Outside our understanding, in real existence, there is no immobile place, there are only mobile bodies.

Bonet wished to have the ancients, especially Aristotle and Averroes, endorse this doctrine he so neatly formulated; it is possible to think that he lent to them what they did not teach, and what they doubtless would have fought against. The real initiator of his theory was someone who immediately preceded him, William of Ockham. It is William of Ockham's endorsement that he should have sought, had Ockham's name not been, at the time, the opposite of an authority.

Ockham clearly affirmed that a body can move of local movement even when there was not, in reality, any immobile body to serve as term of comparison for this movement; lacking a concrete reference, an abstract and purely conceived reference would suffice. Bonet pushes this affirmation to its conclusion. It is not only when lacking an immobile, really existent term that one can compare local movement to a purely conceived term, it is always thus. In every circumstance, the immobile place to which one refers local movement is a simple mathematical abstraction; it has no existence other than a conceptual existence—*esse cognitum*. To seek it among the concrete and really existent bodies is to err.

John Buridan

The questions in which John Buridan discusses place perhaps form in their totality the most extended and most detailed theory that any master of Scholasticism composed concerning the notion of place. Many influences can be discerned when reading the discussions which form the theory; the ones deserving to be noted above all are the influences of Roger Bacon, John Duns Scotus, William of Ockham, and Walter Burley, whether these influences carry the assent of John Buridan, or whether, on the contrary, he opposes them.

John Buridan adopts the classical formula in order to define place, properly speaking: *superficies ultima corporis continentis.* He comments on this formula as a faithful disciple of Ockham. By *superficies* he means, as do all Nominalists, not a surface having only two dimensions, but a layer having some thickness. As a result, the containing body has an infinity of ultimate surfaces:

> Let us imagine, in fact, that the lunar orb is divided into two halves by means of concentric surfaces, or into three thirds, or into a hundred hundredths, and so forth. Among these parts, there will always be one that will be the final one on our side, that will touch our lower world while touching the sphere of fire; it will be the last of the two halves, or the last of the ten tenths, or the last of the hundred hundredths, and so forth indefinitely. Each of these parts is the ultimate surface of the lunar orb, from our side, and there is no reason for one rather than the other to be called that, so that each of them is the proper place [of fire].[127]
>
> But a difficulty remains: if any surface is a body, why do we say that place is the surface of the containing body, and not that place is the containing body?[128]

Clearly, place, properly speaking, is a body; but it is not called place and body under the same relation, while it is called place and surface under the same relation.

A line is a body, but one calls it a line when one considers it is divisible along a single dimension, length, without accounting for its divisibility along the other two dimensions, namely, width and depth. Similarly, a body is called surface when one conceives it as divisible along two dimensions, length and width, without considering its divisibility along the third dimension. One calls it body only when it is conceived as divisible along three dimensions, its length, width, and depth.

Moreover, the contact between the lodging body and the lodged body is only along two dimensions; because of the mutual impenetrability of these bodies, depth is not brought into play, so that it is legitimate to say that place is with respect to the terminal surface of the contained body and the terminal surface of the containing body; it is proper to say that in this sense, place, properly speaking, is constituted by the latter surface.

From the preceding, it follows that the term *place* is to the term *surface* as a passion is to the subject it affects.[129] Place is defined, as any passion should be, by the definition of the subject and by the terms that explain the particular *connotation* of this subject affected by such a passion.

These principles posited, John Buridan attempts the difficult question of the immobility of place.[130] What must one understand by "place is immobile"? A first reply, Giles of Rome's, has been given: there are two elements in place, a formal element and a material element. The matter of place is the surface of the containing body. The form of place "is the distance of this surface to heaven, the earth, and various parts of the world, which are at rest; heaven, in fact, exempt from all rectilinear movements, can be regarded as being at rest in a certain way, for it can serve as comparison for judging the rectilinear movements of other bodies."[131] Material place is mobile; but formal place is immobile in the sense that a body at rest always keeps the same formal place even when the ambient substances change.

Like all the Scotists and all the Nominalists, Buridan absolutely rejects this theory; the arguments with which he opposes the theory are those of William of Ockham and Walter Burley.

The distance between two bodies is nothing more, for Nominalists, than that various bodies are interposed between the two: "the distance of this stone to the earth and heaven is nothing more than this stone itself or the intermediary bodies that separate it from heaven."[132] The distance between two bodies changes, therefore, when the interposed substances change. If one defined formal place as Giles of Rome defined it, it cannot be held to be immobile.

Further, such a formal place can be, in certain cases, more mobile than the material place considered by Giles of Rome. This material place, the ultimate surface of the containing body, is never mobile *by itself*; it is only mobile *accidentally* and by the effect of the movement of the containing body. On the other hand, the distance between a body and the earth, which is the formal place of the body, can be realized in a single and whole interposed body; since this latter body is mobile by itself, formal place is then mobile by itself.

It seems, moreover, that the language that Giles of Rome uses is not well justified; there may be greater justification in giving the name formal place to the surface of the containing body, and material place to the distance between this surface and heaven or the earth. This distance, in fact, can be a body taken in its totality; the extremity of the container, on the other hand, is necessarily a part of the body. Does it not seem more reasonable to think this part of the body as the form of place rather than to attribute this role to a body which is taken as a whole and has its own existence?

It is therefore not possible to accept Giles of Rome's interpretation for rendering true the proposition, place is immobile.

Moreover, what was Aristotle's intention when he introduced "immobile" in the definition of place? According to Buridan, the Stagirite had no object other than to distinguish between place and the vase. These two are, in fact, the same body—the containing body—which at the same time plays the role of place and the vase with respect to the contained body; one calls it vase or place depending upon the point of view from which one considers it. One calls it vase when the content is capable of flowing or of expanding. The vase then puts a barrier to this diffusion. The movement of the vase alone permits the transportation of the content from one place to another. The name *vase* is therefore attributed to the containing body by reason of a certain mobility that one considers in the body. The name *place*, on the other hand, is given to the container by reason of a certain immobility the body demonstrates in comparison with the contained body. The contained body can move, in certain cases, while the containing body remains immobile.

Has John Buridan grasped the essence of the Stagirite's thought by means of this analysis? We do not think so; but instead of commenting at length on this question, it is more fruitful to ask our Parisian master to exhibit his own theory of the immobility of place.

Place, properly speaking—the place about which Aristotle's definition applies—is a body; as such, it is mobile. It as mobile as the lodged body; place can move even when the lodged body remains at rest: "the air surrounding the towers of Notre Dame can move and change while the towers remain in place."[133] In some cases, also, the lodged body can move without the place being displaced in any way.

One cannot, without error, maintain that place, *properly speaking*, is immobile; this assertion can only be produced with respect to place, *improperly speaking*.

One can, in fact, use the word *place* in a variety of different senses, as with most words; with the word *place*, as with other words, there is a first sense to which the other senses are related by attribution.

Buridan's idea, to distinguish in his theory of place the proper and derived senses of the word *place*, appears to have been borrowed from Roger Bacon; here is how Buridan makes use of this idea:[134]

It is impossible for us to perceive, *at least by means of the senses*, that a body moves by local movement if we do not perceive that this body behaves differently, from one instant to another, with respect to some other body; that this change consists in a variation of distance or in a variation of situation; that the two bodies change in totality with respect to each other; or that the parts of one are disposed in some other way with respect to the parts of the other.

The above assertion is not a philosophical conclusion; it is a simple judgment of common sense, which the whole world holds. Further, it seems impossible for us to judge with certainty that one of the two bodies, which from one instant to another behave in a different manner with respect to each other, moves if we do not know, by some other means, that the other is immobile, or at least, that it does not move by such a movement or such a speed.

That posited, let us imagine a lodged body and its place, properly speaking, meaning, according to Aristotle's definition, the ultimate part of the lodging body; let us suppose that this latter body remains immobile and that we know this. If, from instant to instant, we perceive that the lodged body behaves differently with respect to its place, we say that it moves locally; if, on the contrary, we notice that the lodged body always keeps the same relation with the lodging body, we say that the first body does not move locally, that it is at rest.

Extending the above, we say that an object is the place of a body, or else that it plays the role of place with respect to this body, when the object serves as term of comparison for the movement or rest of this body; we say that the body is immobile, or that it is moving, depending upon whether, from one instant to another, it behaves in a similar way or dissimilar way relative to this object. But immobile place so defined is *place, improperly speaking*.

These observations dispel the previously presented objections.

It is a common thought, to which all agree, that the towers of Notre Dame are today in the same place they were when they were built, even though the air surrounding them is renewed constantly, even though the intermediary bodies that make up the distance between these towers and heaven have changed frequently. This appears to be a difficult matter, but in reality it is a very simple matter: in fact, the terms, *the same*, which we apply to the place of these towers must not be taken in their proper and essential sense; one must admit that the words, *the same*, here designate the equality of distance either to the earth, or to heaven, or to some body, with respect to which we judge the rest or movement of other bodies.[135]

John Buridan does not assert, as did Duns Scotus, Joannes Canonicus, William of Ockham, and Walter Burley, that the words *the same place* designate two equivalent places between which there may not be a numerical identity; but if he does not use the language of his predecessors, the thought he expresses is no less identical to theirs.

It is by understanding the word *place* in its improper sense, not its proper sense, that one can formulate the proposition, the earth is the place of heaven. We shall encounter this proposition during the examination of the question: Is the supreme sphere in a place?[136]

> I believe that this question has been thought to be very difficult because we have not recognized the equivocation in the word *place*. As we have said previously, the word *place* can be understood in a proper sense, as signifying what contains the lodged body and touches it immediately, while being distinguished from it. It can also be understood in a less proper manner or a completely improper manner; it then designates the object by means of which one judges that a certain body moves . . . if one grants this distinction, the question becomes very simple.[137]

In the proper sense, the ultimate sphere has no place, since no body contains it; in this proper sense it does not move locally, since it has no place.

But taking the word *place* in its improper sense, if one designates by that the reference that allows one to appreciate the movement or rest of a body, then the ultimate sphere has a place, and this place can be the earth, or some wall, or some stone.

John Buridan, therefore, accepts Averroes's aphorism: the supreme sphere is not in a place *per se*, but it is in a place *per accidens*; however, he also accepts the following condition, that Averroes, no doubt, would not have accepted: place *per se* is place, properly speaking; place *per accidens* is place, improperly speaking.

John Buridan also accepts Avicenna's opinion: the ultimate sphere does not move of local movement, but of movement relative to the situation, for if it has no place, properly speaking, it has a situation that changes from one instant to another; its various parts, in fact, are located at variable distances from the various parts of the earth. Averroes and Saint Thomas Aquinas rejected this doctrine of Avicenna; Buridan declares the objections they formulated against it as ill-founded.

Even though the supreme sphere has no place, properly speaking, it moves; but it has a place, improperly speaking, the immobile earth, a term of comparison that allows us to appreciate the movement of the ultimate orb. Is this place, improperly speaking, indispensable to the movement of the ultimate heaven? Can the movement of heaven continue even when this place, improperly speaking, does not exist? Averroes would deny this; for him, the existence of an immobile earth is the necessary condition for the movement of heaven.

But that is not Buridan's opinion.[138]

Let us imagine that divine power transforms the world into a homogeneous and continuous whole; for such a world, there would be no place, properly speaking, nor any place, improperly speaking. Similarly, there would be no place for a single stone if God were to annihilate the rest of the world.

Can God communicate to this homogeneous sphere, deprived of every kind of place, the movement by which the supreme orb is actually animated? Averroes denies this, and John Duns Scotus affirms it; John Buridan agrees with Duns Scotus.

> By making use of one of the articles condemned at Paris, I prove that God can impress a rotation upon this world. This article states that it is an error to maintain that God cannot move the world by a rectilinear movement. There is no reason to think that He can move it by a rectilinear movement rather than by a circular movement. In the same way that He impresses a diurnal movement on all the celestial spheres at the same time as on the supreme orb, He can give to the whole world, including the sublunar bodies, a rotation as a whole, while the various spheres remained distinct from one another. But He can, just as easily, move this world after having transformed it into a homogeneous and continuous whole. God can therefore move the whole world, even when the world no longer has any place.[139]

We see that Buridan credits fully Etienne Tempier's decision with respect to the problem occupying us. He also invokes this decision in another work, in his *Questiones super libris de celo et mundo.*

> One must know that these reasons are sufficiently convincing with respect to natural powers and natural and violent movements; but they are not sufficiently convincing with respect to supernatural powers or the movements stemming from supernatural power, that is, divine power.
>
> It is true, in fact, that everything by nature mobile by rectilinear movement must be in a place, and is, necessarily, in a place, unless it is absolved of this requirement by divine power. . . .
>
> But, with respect to divine power, it has been decided by the Bishop and University of Paris (*per episcopum et studium parisienses*) that it is an error to assert: God cannot move the world as a whole and all together by a rectilinear movement. The whole world taken all together is not in a place, however, for there exists no body containing it outside it. Thus, no place is required for something to be moved by a rectilinear movement by divine power, which is what is supposed by the preceding reasons—that there exists no place and that God can nevertheless move this stone by a rectilinear movement as easily as He could move the whole world. I have spoken of this at greater length in the book of the *Physics.*[140]

Let us therefore return to John Buridan's *Questions on the Physics.*

The debate that our author has started in the two passages we have just cited exceeds the scope of the problem of place; another question is joined to it, which we will treat at length later on. This question concerns the nature of local movement; with respect to this topic, we will see Buridan uphold the doctrines of Duns Scotus against the doctrines of Gregory of Rimini and William of Ockham. In particular, the formal intention of the philosopher of Béthune in the passage we have just cited is to refute the theory of local movement proposed by Gregory of Rimini; local movement cannot be formally identical to the place which the mobile acquires at each instant.

Can local movement be nothing more than the mobile itself which, from instant to instant, behaves differently with respect to a fixed reference, as thinks William of Ockham? Buridan understands, and he has told us, that no local movement can be perceptible to the senses if the mobile body does not constantly change position with respect to a fixed body or if the parts of this mobile do not dispose themselves variously with respect to the parts of the fixed reference. But he does not agree that the local movement is reduced, in reality, to what allows our senses to notice its existence and to study its particularities.

> If the ultimate sphere moves, it is not simply because it always behaves differently with respect to the earth or some other body. I prove it thus: it would move no less even when all the other bodies turned with it without suffering any movement different from it; in this case, however, there would be no object with respect to which it could have behaved differently from instant to instant. Similarly, for a body to move by a rectilinear movement, it must behave differently from one instant to the other with respect to some object, in the same way as it is necessary for a rotational movement. However, for there to be a rectilinear movement, it is not necessary for the mobile to behave differently from instant to instant with respect to some other body; in fact, if God were to move the whole world by a rectilinear movement, the world would not suffer a continuous change of disposition with respect to the earth.[141]

It is true that William of Ockham foresaw and predicted this objection, and sought to avoid it; a moving body is not for him simply a body that behaves differently from instant to instant with respect to an immobile body actually existent, for it is possible that one might not find any immobile body. According to him, a moving body is essentially a body that, from instant to instant behaves differently with respect to an immobile object, if there existed such an object.

This conditional form of the definition of movement does not satisfy Buridan:

> Some might respond that to move is to behave differently from one instant to the next with respect to an immobile body, either absolutely (*simpliciter*), if a body remains immobile, or under a condition: if something remained immobile, the moved body would behave differently, from one instant to another, with respect to this thing.
> This subterfuge is worthless. It does not prevent the ultimate orb from moving in fact even when there does not exist in fact any immobile body; in this case, therefore, this sphere cannot, in fact, behave differently one instant to another with respect to some immobile body or some extrinsic object. Hence, if it does not behave differently from one instant to another in an intrinsic manner, it can in no way behave differently from one instant to another, in fact; and

consequently, in fact, it would not change in any way, for in order to change, it would have to behave differently from one instant to another, in fact and absolutely, and not merely under some condition.[142]

How Buridan conceives this intrinsic change of a body which, in the absence of any immobile place, can still move of local movement, we shall see later on.

The simple conception of a place, of a fixed reference not existing in fact, would not suffice to constitute the reality of local movement, according to Buridan; but it can suffice for the description of such a movement, and it must suffice, if there does not exist any body absolutely fixed to which this movement can be related. Nowhere does Buridan announce that his doctrine requires this restriction; but that he admits it, is shown to us by his manner of reasoning in some circumstances.

Buridan does not believe that one should place an immobile empyrean heaven above the mobile celestial spheres considered by the astronomers; in his *Questiones super libris de celo et mundo* he devotes a whole question to exposing and refuting the reasons given in favor of the existence of this heaven.[143] Among the arguments he presents and rejects is the following:

> Everything moving properly (*per se*) of local movement must be in a place, properly speaking (*per se*). And the ultimate celestial sphere moves properly of local movement. It is certain, in fact, that it moves properly, and that the movement is nothing other than local movement. This ultimate moving sphere must therefore have a place, properly speaking. But it cannot have this place if there does not exist above it an immobile sphere containing it; in fact, place is defined as the termination of the containing body.
>
> That place must be immobile is a truth that comes to the aid of the previous assumption. In fact, the following solution that some people maintain is not valid: the ultimate sphere has a place in virtue of the order that it presents with respect to its center, which is the earth. The properties of place in no way fit the earth considered with respect to the final sphere; it is not its property to contain the lodged body, nor to be equal to the lodged body, and so forth for the others.[144]

Buridan therefore does not admit the opinion that Pierre d'Ailly admitted. He does not assume an immobile empyrean heaven as the proper place capable of serving as reference for all the movements of the universe.[145]

In the above passage Buridan speaks in the same way that William of Ockham would have spoken; in order to reason about the local movements with which he is dealing, he is satisfied with a purely conceived, fixed reference. But if he believes with Ockham that this fixed reference suffices for us to discover these movements, he does not think, as does the Venerable Inceptor, that such a purely hypothetical place suffices to confer reality to these movements.

Albert of Saxony

Albert of Saxony almost always follows faithfully the opinions of John Buridan with respect to place.

Like John of Jandun and Walter Burley, Albert of Saxony defines the place of a body as the surface of the container touching the body; but he does not give to this formulation the same sense that Jandun and Burley gave it: "Those who consider the surface as an indivisible reality added to the body take this proposition literally."[146] Albert of Saxony is not among those who follow the opinion of Duns Scotus in this; with respect to this subject he ranks among the followers of William of Ockham. He refuses to consider the various magnitudes that the geometer considers—lines surfaces, and points—as realities distinct from the body:

> It is a sin to account for things by invoking a greater and greater number of realities, when one can account for them by means of a lesser number; moreover, if we suppose that magnitude is not a reality distinct from the extended body, we invoke a smaller number of entities than if we made of magnitude and body two distinct realities, and yet we explain all things as well.[147]

Therefore when Albert of Saxony defined place as the surface of the container, he did not take the formulation literally; in actuality, place is a body.[148] If he substitutes the word *surface* for the word *body*, it is in order to denote that the container is the place by virtue of the fact that it touches the content, and that this contact is established only along the two dimensions of a surface, without depth playing any role.

Place is a body; therefore place is mobile, in spite of the assertions of the Commentator and his followers.[149]

This movement of place does not necessarily result in the movement of the contained body. The contained body can rotate without its place changing: "the wine can turn in its bottle even though the bottle remains at rest."[150] But that is true only for a movement of rotation; if the contained body suffers a movement of translation, the place of this body moves necessarily: "if a stone falls in the water, the inner surfaces of water that formed its place would conjoin behind it at each instant."[151]

Place moves when the contained body moves; it does not follow from this that the contained body moves at the same time: "otherwise the towers of Notre Dame would move constantly, for the air surrounding them changes at each instant."[152]

But there we are concerned with the movement of the material place; can we not, with Giles of Rome, assert that the formal place of Notre Dame's towers does not change because the formal place is constituted by the distance of these towers from the celestial orb or some other fixed body, and that this distance always remains the same?

It is not true that the distance of an immobile body to the celestial orb or to some other immobile body always remains the same. The Terminists do not admit that a mathematical magnitude, considered in isolation, has any reality; the distance between the two bodies is nothing more than the bodies located between the two bodies. When these intermediary bodies change, the distance does not remain the same; it becomes another distance:

> The towers of Notre Dame have been immobile for a long time; yet during this time their distance to the lunar orb has not remained the same; the intermediary bodies have changed. The air and the water located between these towers and the lunar orb have moved constantly. And distance is nothing more than the intermediary bodies between the two distant bodies.[153]

The distance between two immobile bodies does not always remain the same, but it remains the same *by equivalence*. At two different instants the distances of the two bodies are numerically distinct, but they are equivalent among themselves; the geometer attributes to them the same measure.

This is how one should modify Giles of Rome's definition of formal place:

> One calls the distance of a lodged body to the lunar orb, or to the objects of this world remaining immobile, *formal place*: . . . when one speaks of the distance to the orb or to immobile bodies, one means to say that the same place always

corresponds to an equal distance, and a distance of differing magnitude, to another place; one considers a distance as remaining the same *by equivalence,* and not the same numerically. . . . One can then say that a body remains immobile when it remains in the same place, understanding the word *place,* in its formal sense, and taking the words, *the same,* not literally but as signifying *equivalent.* . . . In this sense I can say that I am now in the same place as when I started the lesson, because the distance between the lunar orb and me has a length which equals the length it had then, and that it is the same for the distance between one of you and me.[154]

Albert of Saxony now tackles the problem that Averroes called a large question: Is the final sphere in a place?

Inspired by the definition of place given by William of Ockham, Albert's response is formulated still more neatly than the Venerable Inceptor's; the wish to dispel the doubts that bothered Walter Burley surely contributes to the precision of his response.

The ultimate sphere, the ninth sphere according to the opinion of the astronomers, has no place since it has no container.[155] It has no place either by itself, taken in its totality, or by its parts, in contradiction to what so many others since Aristotle and Themistius and up to Saint Thomas Aquinas have upheld.[156] Can we at least assert with the Commentator that the supreme orb is in a place by accident, meaning by its center? While the opinion of the Commentator can be given a correct meaning, as we shall soon see, the expressions he makes use of are improper;[157] properly speaking, the ninth sphere has no place, even accidentally.

The Scotists, such as Joannes Canonicus, did not attribute any kind of place to the final orb; but they attributed a *ubi* to it, a special kind of *ubi* that they named active *ubi.* Does Albert also attribute a *ubi* to the ninth sphere?

As a disciple of Ockham, Albert of Saxony does not admit the existence of the entity that the Scotists designate by the name *ubi.* According to the disciples of Scotus,

the predicament *ubi* designates a certain real relation, distinct from substance and quality; this relation stems from the circumscription of the contained body by place. According to them, in order that one might say that a body has a *ubi,* there must exist a real relation distinct from place and the body it contains; the lodged body would be the subject of this relation, which would be in the place only under the relation. . . . But this opinion is not correct. . . . It superimposes a new reality on place and the contained body uselessly. . . . The terms of the predicament *ubi* must not be thought of as things distinct from substance and quality.[158]

Hence, if one says that a body has a *ubi*, that it is somewhere (*alicubi*), one is simply saying that it is above, below, to a side, or around some other body; in this sense, one can say that the ninth celestial sphere has a *ubi*, for it is true that it surrounds the other spheres and that it is above the other spheres.[159]

One can also say that the body is in a place when there is a term of comparison such that we can recognize that the body moves; it is in this sense that the Commentator was able to say that the earth is the place of heaven. In fact, it is the position of heaven with respect to the earth that allows us to know the movement of heaven. "But this way of speaking is improper."[160]

How can the final sphere, which has no place, move by local movement? That cannot be known; moreover, "the final sphere moves of a movement of the same kind as local movement, but which is, nevertheless, not a local movement."[161]

This movement, which is not a local movement, but is of the same kind as a local movement, is the movement the universe would be animated with if the First Cause impressed a translation on it;[162] the universe, in fact, has no place, so that it is incapable of local movement. It is true that Aristotle and his Commentator would deny that the universe can suffer a translation,[163] but one of the articles decreed by the theologians of Paris upheld that God can displace it thus.[164]

Moreover, do we not have the proposition, formulated in the *De motibus animalium*, that any moving body requires a fixed body eternal to it in order to demonstrate the impossibility of such a displacement? With infinitely good judgment, Albert of Saxony rejects the authority of this text that so many commentators have invoked: "In the *De motibus animalium*, Aristotle is only talking about the progressive movement of animals; every animal has need of fixed support for its movement. . . . But heaven has no need of such a support."[165]

But can one not otherwise demonstrate the impossibility of a translation of the universe? "To move is to behave at each instant differently with respect to some fixed object. If there did not exist any fixed object, it would appear that heaven would not be able to move."[166]

The above argument that Walter Burley accepted does not convince Albert of Saxony:

> For a body to move, it is not necessary that it behave differently with respect to some extrinsic object from instant to instant; it suffices that it behaves differently in an intrinsic manner. If God imposed a movement of translation to the entire universe, which one of the articles formulated at Paris declares possible, the universe would not change from instant to instant with respect to some extrinsic object; but it would suffer an intrinsic change—at each instant, in fact, there would be a new portion of movement in it.[167]

We see the arguments, by which the Peripatetics and the Averroists concluded the necessity of an immobile earth at the center of the world from the movement of heaven, falling one by one.

In any case, the bond these arguments attempted to establish between the uniform rotation of a celestial orb and the presence of an immobile body at the center of this orb manifestly does not exist: "According to the astronomers, the epicycle rotates around its own center, and yet there is no immobile body in this center. The spherical mass of the epicycle moves in its totality."[168] The Peripatetics and the Averroists attempted to oppose the system of Ptolemy with the proposition they proudly derived; the system of Ptolemy is now invoked in order to condemn this proposition.

It is therefore false to maintain that the rotation of heaven requires the presence of an immobile earth at the center of the world, with respect to which the position of heaven can change from one instant to another. "Heaven and earth can both move, and yet as long as the earth is not at rest, the position of heaven with respect to the earth changes from instant to instant. It is only in the case where earth and heaven rotate in the same direction and with the same angular speed of rotation that the position of heaven with respect to the earth remains invariable."[169]

There remains an argument, among the arguments concluding for the immobility of the earth from the movement of heaven, that receives the approval of Albert of Saxony; it is the argument proposed by John of Jandun: the generation and corruption of sublunar beings require the situation of heaven with respect to the earth to change from instant to instant; since heaven moves, the earth must remain immobile. But Albert adds, "it is not necessary for that, that it remains immobile in an absolute manner; it suffices that it not rotate in the same direction and with the same angular speed of rotation as heaven."[170]

Finally, it is true that astronomers, when they consider the rotation of the supreme heaven, relate this rotation to a fixed term. But, like William of Ockham, Albert of Saxony thinks that it suffices for them to conceive this fixed term without needing any body to realize it concretely; that is evidently what he intends to assert in the following passage: "[In circular movement] speed is to be measured by means of linear space, either real or imagined (*verum vel ymaginatum*), over such and such time, by the point of the moving body moving most rapidly. . . . Notice that I say real or imagined in this conclusion because of the ultimate sphere, which does not move over a real space, but merely over an imagined space."[171]

In no way then does the movement of heaven require the immobility of the earth; if the earth is immobile, its rest must be proven by other means.

We are not surprised to see Albert of Saxony demolish all the obstacles derived from the theory of place that the Peripatetics had accumulated against the hypothesis of terrestrial movement; soon we will learn from him that one of his teachers held the diurnal rotation of the earth. He himself attributed some very slow but constant movements to our globe. The disciples of Scotus and William of Ockham are in agreement here: they can believe that the earth does not move, but none of them considers the movement of the earth as an impossibility.

Marsilius of Inghen and John Buridan II

Toward the end of the fourteenth century, the doctors of the Sorbonne and the masters of arts of the Rue de Fouarre were divided between the Scotist doctrine and the Ockhamist doctrine.

These two doctrines, in any case, had very important theses in common.

On the nature of place, however, the thought of Scotus's disciples was opposed to that of the Terminists.

For the former, the surface of the containing body is a reality distinct from the body itself; this reality serves as support, as subject for an entity constituting place. For the latter, the surface has no reality independent from the body; place is not an entity added upon the surface, but a supplementary indication—in reality, body, surface, and place are but one and the same thing.

Although profoundly divided about the nature of place, the Scotists and the Ockhamists unite into one doctrine with respect to the role that place plays in local movement; they express the same thoughts in the same words with respect to the immobility of place, the localization of the supreme orb, and the relation between the immobility of the earth and the movement of heaven. Starting from two different and opposing metaphysics, they end up with the same consequences in the domain of physics and astronomy.

Some of the masters of the University of Paris adopted the Ockhamist doctrine with respect to the theory of place, and others the Scotist doctrine; some of them even—and not the least of them—might have hesitated between these two doctrines and might sometimes have given their assent to one and then the other. One of the more illustrious such masters at the end of the fourteenth century, Marsilius of Inghen, was in turn, with respect to this question, the disciple of Ockham and then the disciple of Scotus.

What Marsilius of Inghen asserts about place in his *Abbreviationes libri Physicorum* is no more than a faithful summary of Albert of Saxony's doctrine:

> The word *place* can be taken in two ways: properly and improperly. Properly speaking, place is the internal surface of the containing body immediately contiguous to the contained body. Improperly speaking, place designates the immobile body or the body moved by another movement which is used as the term of comparison in order to perceive that a body is in movement. . . .
> Place, properly speaking, is not a surface without depth. . . . All surfaces have depth. As a result, any body whatever has an infinity of places, properly speaking; in fact, each surface layer cut out of the containing body and contiguous to the contained body constitutes such a place, properly speaking. Moreover, there is an infinity of such surface layers; one can take the last third of the containing body that touches immediately the contained body, or the last fourth, or the last thousandth, and so forth without end.[172]

This doctrine is clearly William of Ockham's, John Buridan's, and Albert of Saxony's. Marsilius of Inghen, who adopts it in his *Abbreviationes*, rejects it in his *Quaestiones*:

> There are two opinions with respect to this problem.
> The first opinion admits that surface is not something real, indivisible in depth, differing from body, but that surface is body itself considered and measured in two dimensions only. Those who admit this opinion assert that place is the containing body considered as the part touching the contained

body; when they define place as the ultimate part of the container, they understand by that the final part of the container on the side of the contained body. From this principle they conclude that a single body has an infinity of places; for a single container, the last third of the container is a place, and so is the last fourth, and the last hundredth, and so forth to infinity. . . .

The second opinion admits that surface is something real, indivisible in depth, having only length and width; it admits that line and surface are things distinct from body.

I believe that the second opinion is truer than the first, for it agrees better with the sayings of mathematicians, and with what the Philosopher wrote in the sixth book of the *Physics*. Therefore one must not assume that place is a body, but that it is the surface of a body.[173]

This conclusion is in conformity with the doctrines of Duns Scotus and Walter Burley.

However, Marsilius of Inghen does not follow further the path traced by the Scotists; he does not make of place an entity above and beyond the surface of the containing body. Faithful to the teaching of Albert of Saxony, he admits that place has with surface the same relation as a passion has with its subject; but he understands by that simply that the expression *place* designates something more than the expression *surface*, in that it implies the idea of containment with respect to the lodged body.

We have just noted a divergence between the theory that the *Quaestiones* exhibits and that which the *Abbreviationes* summarizes; it is the only divergence that can be noted between the passages that these two works devote to the theory of place. It is also the only divergence that separates the teachings of Marsilius of Inghen and Albert of Saxony on this topic; besides this point, the agreement between these two teachings is perfect, so perfect that it would be useless to analyze here what the disciple repeats in questions whose order and titles he borrows from the order and titles Albert gave his own questions.

We shall be content with indicating a precision that Marsilius added to the questions formulated by his predecessor.

Albert declared on several occasions that the movement of a body in no way supposed the concrete existence of an extrinsic immobile body; in order for the body to be moving, it suffices that its manner of being suffers an intrinsic change.

Also, it is certain that we could not imagine this change, if it were not a change of position with respect to a term of comparison considered as immobile. Albert of Saxony's opinion therefore consists in the assertion that the term of comparison does not need to exist in an actual and concrete manner, that an abstract existence suffices for it. But this opinion does not deny that every movement supposes the possibility to conceive an ideal term of comparison to which our reason can refer the positions of the mobile. Albert of Saxony neglected to point out what William of Ockham and Walter Burley already provided with respect to this theory.

This point is taken up with greater emphasis by Marsilius of Inghen: "One says that a body moves locally when it changes its whole position or the position of its parts with respect to another immobile body, from instant to instant, or at least when it behaves in such a way that it would change its position with respect to an immobile body, if there were one."[174]

Moreover Marsilius understands the importance of this restriction, for he formulates it a second time, in almost the same words:

In order for a body to move locally, it is not necessary that it is in a place; it suffices that it has a position different from the one it had, at each instant, this position being referred to as an immobile object—or at least, this body would have to behave differently, from instant to instant, with respect to an immobile object, if there existed such an object. I say this for the case where one supposes that the whole universe is moved by a movement of translation or a movement of rotation.[175]

One cannot therefore conceive the local movement of a body without conceiving a fixed reference to which one relates the position of this body at each instant; but, in order for the movement in question to be realized, it is not necessary for the term of comparison, the immobile reference, to exist actually and concretely. This fundamental principle, posited in antiquity by Simplicius, was taken up during the fourteenth century by the most noted Parisian masters, by William of Ockham, Walter Burley, Albert of Saxony, and Marsilius of Inghen.

The University of Paris knew in succession two masters named John Buridan.

The first was born at Béthune, and, according to tradition, in the diocese of Arras; his life did not last much later than the middle of the fourteenth century. He is the great philosopher to whom we owe the *Questions on the Physics* and the *Questions on the Metaphysics* so frequently cited in this work.

The second taught at Paris toward the end of the fourteenth century and the first few years of the fifteenth century. We have from him a large number of works that have been mistaken as the work of the first John Buridan. Several of the works of John Buridan II have been printed, such as the *Quaestiones in libros de Anima* and the *Quaestiones in parva naturalia* (published in Paris in 1516 and 1518 in a collection that held also the *Questions* of Albert of Saxony on the *Physics* and the *De Caelo*, and the *Questions* of Themonis on the *Meteorology*), and such as the *Questions on the Ethics* and the *Questions on the Politics*, which have had several editions.

But the most original work of this John Buridan II, his *Questions* on Aristotle's *Meteorology*, has remained in manuscript form.[176]

In this writing, the author, following Albert of Saxony, studies on several occasions the small movements that the earth can suffer, and the slow movements of the oceans and continents resulting from this. In order to speak logically about these movements, one must relate them to a fixed reference, and this reference cannot be the earth, whose movements are to be analyzed. He therefore takes as fixed term a *real*, or *possible* heaven, which can be the empyrean or some other heaven; it is to this *caelum quiescens* that he relates the position of the earth and the seas: "in order to avoid any subterfuge, since a number of the parts of the earth can move or be engendered, I posit this hypothesis, which is true or merely possible (*pono ymaginationem possibilem vel veram*), that there is a heaven which is always immobile, whether it is the empyrean or some other heaven."[177]

It is the use of the empyrean heaven, whether real or fictional, that allows John Buridan II to formulate assertions such as these:

> If one admits that the ocean constantly withdraws on one side while it advances on the other, one must change constantly the position of the mean meridian of the habitable earth with respect to the heaven one has imagined immobile (*in ordine ad caelum ymaginatum quiescens*).[178]

The use of an immobile heaven, whether real or merely conceived, to which the movements of the earth can be related, is in conformity with the principles posited by the most eminent doctors of the Nominalist school, by William of Ockham, Albert of Saxony, and Marsilius of Inghen. It cannot be confused with the opinion professed by Campanus of Novara, fought against by Duns Scotus and Joannes Canonicus, treated with disdain by Albert of Saxony, and taken up again by Pierre d'Ailly, during the time when John Buridan II taught.

The Immobility of Place and God's Immutability: Thomas Wilton, Francis of Mayronnes, Nicholas Bonet, and Nicole Oresme

Except for the philosopher of Béthune, John Buridan, all the Parisian masters whose writings we have just analyzed subscribe to the axiom of John of Bassols and William of Ockham: in order for a body to move locally, it must, and it suffices that it does, behave from instant to instant in some other manner with respect to a fixed term, whether real or merely conceived.

We do not think that there was any exception to this other than John Buridan. Some masters did not want local movement to be judged by comparing the various positions of a mobile body with a reference which is merely conceived. Like Aristotle and Averroes, they wanted the fixed term to exist actually; but instructed by the failures of Aristotle's and Averroes's attempts at finding it within the body of the earth, they sought elsewhere for the being whose immobility would allow one to judge local movement.

We find a preliminary indication of such thought in the theory of place of Thomas Wilton, whom Joannes Canonicus calls Thomas the Englishman.

According to Thomas Wilton, the place of a body in air is, as thinks Aristotle, the set of the parts of air immediately contiguous to the body. Insofar as they are parts of air, they are mobile in the same fashion as the air to which they belong; but that does not mean that the place of the body is mobile. The portion of air contiguous to the body does not constitute its place because it is air, it constitutes its place because it is in a certain order with respect to the center and the poles of the world, or with respect to the intelligence moving the first mobile, an intelligence which is immutable. According to this theory, the place of an immobile body does not change when its ambient matter is displaced.

Except for the strange idea to require the intelligence moving the supreme heaven to serve as the fixed term for determining the immobility of place, this theory is purely Thomist. As a Scotist, Joannes Canonicus rejects it. I do not understand, he asserts, the role that it attributes to the poles.[179] There is nothing immobile in heaven; the poles therefore cannot be immobile. If they are mobile, how can they serve to fix the immobility of place? Further, "one can say as much about the center and the intelligence."[180]

Thomas Wilton mentions the immutable intelligence that moves the first mobile, an intelligence he evidently distinguishes from God as only one of the immobile places one can adopt. Francis of Mayronnes was more assertive.

Francis of Mayronnes exhibits his theory of place by means of his usual concise formulations when answering the question: Is heaven in a place?

One must first see what is place. After having summarily recalled and rejected the various definitions of place, our author concludes as follows:

> I therefore assert that formally place is the relation that any lodged body has with the first immobile Motor; we attribute two predicates to this Motor, namely, that it is everywhere and eternal. It is not everywhere and eternal because of the effect of a relation with place or with time, but because every creature has a relation of presence (*respectus praesentialitatis*) and a relation of presence established successively (*secundum periodum*) with respect to the First Being. . . .
>
> If God made other worlds, He would be everywhere; He therefore possesses potentially infinite locality; each of these worlds would, in any case, be present to God in another way as each of the others, and that is how He would be in actuality.[181]

Francis then tackles this second problem: In what way is place immobile? He criticizes harshly the doctrine of immobility by equivalence almost every master adopted.

> Scotus said that place is immobile by equivalence with respect to local movement.
>
> But against this opinion, when one pushes a relation to its final resolution, one stops at something absolute, and not at a relation. And equivalence is a relation; one must therefore resolve this relation into something else from which immobility stems.
>
> Second, any negation must be resolved into an affirmation that precedes it. Why is place equivalent? Because it is immobile, and not the inverse.

Third, what does the response, [place is immobile] because it is equivalent for local movement, teach us? It teaches us that it is called immobile because it comes to the same as if it were immobile. That is the same as if one replied to the question, "Why was this made?" with, "Because it comes to the same as if it had been made."

Fourth, all difficulties remain; from what does this equivalence stem?

Therefore I can only see the immobility of the first Motor, for every creature is mobile.

[It is therefore] this First Motor which is the term for local movement.[182]

Before pursuing Francis of Mayronnes' lecture and reporting what he replies to the question he posed ("Is the ultimate heaven in a place?"), let us stop and consider what we have just heard him assert. These words, in fact, captured the attention of Nicholas Bonet.

Nicholas Bonet, as we shall see in the next chapter, fully admits Francis of Mayronnes's thesis about time; he sees a relation of presence that changes successively between things and the First Intelligence. On the other hand, he seems tempted not to admit the correlative theory with respect to place that the Provencal Franciscan proposed. From this stems some fluctuations in his teaching that we cannot allow to pass in silence.

Here is a first occasion in which Bonet discusses the doctrine of Francis of Mayronnes.[183] He asks himself, "To what does the property of furnishing a place, of lodging another thing, essentially belong?" This question is announced by Scholastic terminology thus: "What is the formal reason of active locality?"[184] Our author replies to this question as follows: What is essentially capable of furnishing a place,

what is the formal reason of active locality, is the common nature of the ten categories. Here is the reason for this assertion: Anything that can be the term for a presence acquired by local movement can play the role of place actively; and nature, insofar as it is common to the ten supreme kinds, can hold a presence acquired by local movement. Nature is therefore the formal reason of all active localization.[185]

What Nicholas Bonet here calls the common nature of the ten categories is clearly what Avicebron calls the substance that bears the ten predicaments; it is all substance outside the Supreme Being, who is in no category, and about whom one is not allowed to attribute any category. Therefore, according to the preceding answer, what can lodge, what can serve as place, is everything that exists outside God.

But against this reply one can formulate this doubt: as the formal reason of active locality, nature is not, it seems, the first subject [that has this property]; by first subject I understand first in appropriate priority. One proves this thus: the First Intelligence can, it seems, be the term of a presence acquired by local movement and, consequently, the reason for [active] place; and this Intelligence is not part of nature, for nature is a limited being and the First Intelligence is not. . . .

One can easily resolve this objection: you say, in fact, that the First Intelligence can serve as term for a presence acquired by local movement; but perhaps one might deny this. The First Intelligence is present everywhere; no distance separates it from any position. We cannot therefore see how one can move toward it by local movement.

However one can admit the following: something can acquire some presence or another with respect to the First Intelligence by local movement; one would have to say, then, that the aptitude to furnish a place—active locality—is not a property that one can exchange with nature, so that nature is the same thing as [active] locality, locality taken as the most universal concept. One would have to say that nature is not experienced in all its active locality.[186]

It is with some hesitation that Bonet rejects Mayronnes' theory, a theory that makes of God Himself a presence that can differ from a thing, one world to the other, and consequently, that can be the place of these things. In fact, what Bonet asserts against this presence differing from one thing to another, one can repeat with respect to the other relation that Mayronnes admits, the relation that from one instant to the other renders the same thing present differently to God, and constitutes time for that thing. One can assert that God is eternal and immutable at every instant, and is present to everything in the same way, that there can be no change, no succession in this presence, nothing that can be the formal reason for time. It is difficult to admit only one of the two theories proposed by Francis of Mayronnes (a theory of time and a theory of place) without admitting the other, to reject one without rejecting the other. And, as we have said (and as we shall see in detail in the following chapter), Bonet agrees with his fellow Franciscan to conceive of time as a relation of presence with the First Intelligence, changing from instant to instant. It is therefore not surprising that he does not reject without some hesitation the supposition that grants to place a similar nature.

Further, in another passage, in order not to compromise the doctrine about time that he borrows from Mayronnes, he expresses himself exactly as Mayronnes does with respect to place.

Our author replied to the question, "Is the First Intelligence a place?" that "It is somewhere by *negative indistance*, meaning, by the negation of all distance."[187]

> But, [he adds] what has no manner of being with respect to place cannot be said to be here or there.
>
> And, as we have said, the First Intelligence has none of these manners of being with respect to place that constitute proximity or distance. It cannot therefore say of anything whatever: it is here, or: it is there.
>
> I reply that in order for someone to be able to say of something that it is here or that it is there, it is not necessary for the one who speaks thus to have by right of subject a manner of being with respect to place, or that it be the foundation of proximity or distance of this manner of being with respect to place; it suffices that it has, by right of term, a manner of being with respect to place, that it is the term of this manner of being constituting the proximity or the distance from place and position (*ubi*).
>
> And the First Intelligence can have, by right of term, a manner of being with respect to place. It can be the term of a proximity or a distance. . . . It can therefore designate things by saying: here, and: there, understanding that it has a manner of being by right of term, not by right of subject, with respect to place.
>
> From what has been said, we conclude against those who refuse to the First Intelligence the power to know the past as past and the future as future. In fact, it is not subject to the succession of time. They therefore say that it would not be able to designate events by saying: this happened at this time, or: that will happen at that time.
>
> But from the preceding, we can see that this is false.
>
> No doubt the First Intelligence is not included, by right of subject, in the temporal line of succession; it is not the foundation of the distance from time past or future time. But it falls within this line by right of term, because it can serve as term with respect to the past and the future; it can serve as term for place and for position in the same way. Thus it can designate various objects in place by saying: here, and:

there, and it can designate various events in time by saying:
at some past instant, and: at some future instant.[188]

Bonet does not underestimate the difficulties with this kind of
doctrine that speaks of the distance of a lodged body, a corporeal
object, from something which is not a body, from an intelligence,
from the First Intelligence; he knows what can be opposed against
this:

> In order that something may be used as subject for this
> presence by which one says that a thing is here, or that it
> is in such place, one must be able to say, there is some distance
> between this being and this thing; some linear distance must
> be able to be interposed between the being and the thing in
> which this relation of presence is founded. Here is the proof,
> I can say that I am here and that you are there because we
> are the two extremities of the same linear distance. But such
> a distance cannot be interposed between two intelligences or
> between an intelligence and a body.[189]

Bonet does not think this argument valid; he says,

> One calls it the Achilles of a certain author; it is not
> conclusive. One imagines something here which is a pure
> fiction, namely, that there is always some distance between
> place and the lodged body.[190]

It is true that the proximity or distance of bodies whose manner
of being in a place consists of being surrounded by other bodies
is measured by the interposition of a straight line; but that is not
true generally. Distance or proximity between an intelligence and
a body is a certain relation to which the intelligence serves as term
without serving as subject or as foundation, and without in any
way supposing the existence of a length between this body and
the intelligence.

This amounts to saying that distance, such as we conceive it
between bodies, is a particular case of a more general relation that
keeps the name of distance and that can be established between
a body and an intelligence, or between two intelligences. But what
does this distance, taken in this general sense, consist of? What
Bonet asserts does not explain this; however, it would be the one
thing that would render conceivable Francis of Mayronnes' theory.

Let us return to Mayronnes' text. We arrive finally at the object
of the question, is heaven in a place?

> I reply that heaven is not in a place materially, but that it is in a place formally, for the whole universe has a relation of presence with God. . . . God can move the universe indefinitely, and the world would not gain a new relation to God by that; one can, in fact, admit nothing other than this, neither the void, nor an actually existent space (*spatium positivum*).[191]

Taking as place a space, which is a reality although empty of all bodies, which is infinite and immobile, was what the Stoics did (Joannes Philoponus, in particular). Francis of Mayronnes rejects this theory in the question we have just cited: He states, "if there existed such a separate space, it possibly would be place; but according to our faith, we cannot admit such an actually existent infinite."[192]

We shall see this Stoic theory of place, which Francis of Mayronnes does not see fit to accept, taken up and merged with the theory that he upheld; this synthesis was accomplished by Nicole Oresme.

Here is what Oresme teaches in his *Traité du ciel et du monde*:

> Thus, outside heaven is an empty incorporeal space quite different from any other plenum or corporeal space, just as the extent of time called eternity is of a different sort than temporal duration, even if the latter were perpetual. . . .
>
> Now this space of which we are talking is infinite and indivisible, and is the immensity of God and God Himself, just as the duration of God called eternity is infinite, indivisible, and God Himself. . . .
>
> Also we have already declared in this chapter that, [since] our thinking cannot exist without the concept of transmutation, we cannot properly comprehend what eternity is; but nevertheless, natural reason teaches us that it exists. In this way the scriptural passage, Job 16, which speaks about God can be understood: *Qui extendit aquilam super vacuum.*
>
> Likewise, since knowledge of our understanding depends upon our corporeal senses, we cannot comprehend nor conceive this incorporeal space that exists beyond heaven; however, reason and truth inform us that it exists.[193]

Here then is the empty and infinite space of the Stoics posited outside the world; but it is also identified at the same time with God's immensity, which is God Himself.

The bishop of Lisieux alludes several times to this doctrine in his *Traité du ciel et du monde*:

But God's duration is eternity, indivisible, and without succession, of which we spoke in chapter 24 of book I; His position is immensity, indivisible and without extension, of which we spoke about in chapter 2 of book II.[194]

A little later, referring to the assumptions one can construct about the future of glorified bodies, he writes:

The space where they will be is now absolutely empty, and when they are there, no body will contain or surround them, for it is a place imagined void and infinite, the immensity of God and God Himself . . . as Job might have had in mind when he said of God: *Qui extendit aquilonem super celum.*[195]

This infinite space, or in other words, God's immensity, was, according to Oresme, the immobile place that serves as reference for all local movement; this hypothesis seemed to him capable of avoiding all the objections to which the other theories fell prey.

The second consequence was that, if heaven moved perpetually, the earth must be at rest in the middle of heaven.

I say no. First, because we observe that a wheel—like a mill wheel—moves completely without its center resting or remaining immobile in any part, save in some indivisible point which is nothing more than imagination, although there is something outside the wheel upon which it rests and upon which it is moved.

Therefore it does not follow that because heaven moves in a circle that the earth or something else rests at its center, for supposing that this is so and the consequent is true, still the consequent is not valid, because circular movement as such does not require that any body rest at the center of a body so moved.

It is not absolutely impossible, nor does it imply a contradiction, rather it is possible to imagine that the earth moves with heaven in its daily movement, just as fire in its sphere and a great part of air participate in this daily movement, according to Aristotle in his first book of the *Meteorology*. Although nature could not move the earth thus, it is however possible according to the second meaning of *possible* and *impossible* given in chapter 30 of book I.

Therefore, assuming that the earth moves with or contrariwise to heaven, it does not follow from this that celestial movement would stop; so, in and of itself, this circular movement of heaven does not require that the earth should remain motionless at the center of the world.

Indeed it is not impossible that the whole earth moves with a different movement or in another way: Job 9, *Qui commovet terram de loco suo.*

For otherwise the parts near the center would never reach the place where they are destroyed and would be perpetual, which Aristotle holds to be impossible naturally, although enough has been said of this matter at the end of book I. It is necessary, according to Aristotelian philosophy, that the earth should move occasionally and impossible that it always remain immobile. And celestial movement is eternal according to Aristotle's opinion; therefore, it does not follow that, if heaven moves, the earth remains at rest. . . .

Against this objection and against the principal argument is the manifest evidence of heaven itself, for to save appearances and from our observations of celestial movements, we have to confess that there are spherical bodies called epicycles in heaven, and that each epicycle has its own proper circular movement around its center—a movement different from the movement of the heavenly sphere in which the epicycle is found. Clearly it is impossible, according to philosophy, that any body should be at rest in the center of this epicycle. So again, it is not necessary that a body be at rest at the center of this epicycle. . . .

Someone might say that the definition of local movement is: to have a different place with respect to some other body at rest; then if no body were at rest, no body could move.

I say that the above hypothesis does not hold, for rest is privation of movement, as Aristotle says in this chapter. Therefore rest is not the essence of movement and ought not be included in its definition. Perhaps someone will say that local movement means to be otherwise with respect to another body, whether moving or not; again I say that this does not hold because, in the first place, beyond this world is an imaginary space infinite and motionless, as stated at the end of chapter 24 of book I, and it is possible to say without contradiction that the whole world moves in this space with a right movement. To say the contrary is to maintain an article condemned at Paris. With this assumption, no other body exists with which the world could exchange places; so the above description is not valid.

Now, let us imagine and assume it to be possible that God in His omnipotence created two bodies separated from each other, which we will designate as *a* and *b*, and that there are no other bodies save these two, and both are moved exactly alike so that the two bodies are always in the same position with respect to one another, neither *a* nor *b* having any connection with any other bodies. Therefore, to move is not to be otherwise with respect to another body.

Let us suppose that *a* moves and *b* rests; then *a* and *b* would be otherwise to each other just as though *b* moved and *a* rested. In such a case it would be impossible to explain why *a* should move rather than *b*, or vice versa, if moving implied being otherwise with respect to another body. . . .

Now, let us imagine the earth moving for a day by diurnal movement while heaven remained at rest, and afterward that they resumed their normal movements—the earth immobile and heaven revolving. I say that during that day heaven and earth would not be otherwise with respect to each other before this time or afterward, but they would be exactly as they were before with no difference whatever. Therefore, if to move is to be otherwise with respect to something else, we could not say that heaven rested at one time rather than at another. . . .

To become warmed or altered in some way is not to be otherwise with respect to another body, but if the body that does the heating becomes otherwise with respect to another, this change is accidental and beyond the essence of this alteration or movement; it is an internal change within the body itself. Likewise, to change place is an internal change with respect to the imagined motionless space, for it is with regard to this space that the speed of the movement and of its parts are measured. As a result, it appears that no movement, celestial or otherwise, requires in and of itself either the immobility or the movement of another body, and that Aristotle's consequence stating if heaven moves, the earth rests, is not valid.

Still it is clear from what has been said that local movement is something other than the body that is locally moved, because it involves the body becoming otherwise with respect to the space it occupies, imagined immobile. Such a movement is an accident, and cannot be separated from other things and cannot stand alone; it would be impossible because it implies a contradiction. It would be like the curvity or straightness of a line or rod, for such a thing cannot be imagined without any subject.[196]

Such is the theory of Nicole Oresme; it is written clearly enough to render superfluous any commentary.

When he imagines an indefinite immobile space whose existence is real and independent from any body, when he keeps this space as the term of comparison to which one must relate all local movement in the final analysis, he formulates an opinion that was defended by the Stoics—Joannes Philoponus, in particular—but also one that became Newton's and Euler's. When he identifies this space with God's immensity, he may be submitting to the influence of Francis of Mayronnes, but surely he is preceding Clarke who later upheld the same doctrine against Leibniz, and he is preceding Spinoza who later formulated as an axiom that extension is an attribute of God.

Thus we see in 1377 the bishop of Lisieux develop, in perfectly lucid French, some thoughts that emerged again in modern times and excited among the princes of philosophy some debates destined for fame. And perhaps, when we see these thoughts reappear, we should not think of them as truly novel conceptions; perhaps we should think of them as echoes of ancient assertions. If Spinoza repeats what Nicole Oresme asserted, doubtless it is because Hasdai Creskas transmitted it to him.[197]

In any case, all the theories of place which are championed during the modern centuries have already been encountered during the fourteenth century within the school of Paris. When Pierre d'Ailly, imitating Campanus of Novara, places around the universe an immobile sphere to use as fixed reference for all movements, he does what Copernicus will do. When John of Bassols and William of Ockham proclaim that local movement does not require the real and concrete existence of an immobile body, that it suffices, for such a movement to be possible, that one could conceive a fixed term to which the changes may be related, they are composing the system to which most everybody adheres to today;[198] they are preceding the luminous definition that Carl Neumann gave to the *Alpha body* toward the end of the nineteenth century.

Thus we see, in this problem of space, Scholastic philosophy proposing, instead of the Aristotelian solution which is forever ruined, solutions with respect to which, from the Renaissance on, various thinkers will be divided.

Two forces were united in order to reject the doctrine that Peripatetic philosophy had proposed with respect to this problem; these two forces were positive science and Christian theology. Positive science denied, by means of the system of Ptolemy, which was its most perfect expression, that a rotation requires the existence of an immobile body at its center. And theology through the bishop of Paris and the doctors of the Sorbonne, condemned the proposition that refused God the power to displace the whole universe. Neither positive science nor theology agreed that the mobility of the earth is contradictory; Peripatetic physics therefore had to be declared erroneous.

The history of the problem of place during the fourteenth century presents a kind of summary of the whole history of Parisian science during this period.

6
Place in Fifteenth-Century Cosmology

Nicholas of Orbellis

The manual of Nicholas of Orbellis is written *secundum viam Doctoris Subtilis Scoti*. Hence one should not be surprised if the theory of place developed there is but a summary of ideas scattered throughout the works of Duns Scotus. In particular, the professor of Poitiers insists on the following proposition: An immobile body within a mobile medium changes places continuously, but all these successive places are equivalent.[1]

Although he cites only Aristotle's opinion, he espouses Averroes's opinion with respect to the place of the eighth sphere because he formulates his conclusion thus: "One should assign a place to the sphere insofar as it is around something, its middle or its center. One says rightly that heaven is in a place because its center is in a place."[2]

This conclusion does not contradict Duns Scotus's opinions; however, it does not reflect them very well.

Nicholas of Orbellis adds the following proposition to the above: "One should note, nevertheless, that according to faith, the first mobile is in a place *per se*, for above it is the empyrean heaven, which the philosophers have not known; as for the empyrean heaven, it is not in a place, because there is nothing beyond it."[3]

The above passage, although too brief to be clear, appears to adhere to the theory of Campanus of Novara and Pierre d'Ailly; its author seems to admit that the empyrean is not in any place, and that it does not need to be lodged because it is immobile. If this is the author's thought, it should be submitted to the perspicacious criticism of Duns Scotus, of whom Nicholas of Orbellis in this instance seems to be an unfaithful disciple.

269

George of Brussels and Thomas Bricot

When he develops his theory of place, George of Brussels seems, at first, to attempt to remain faithful to the Parisian tradition. Examining how one must define the immobile place to which one relates local movement, he rejects the two theories of Saint Thomas Aquinas and Giles of Rome in order to introduce, like Burley, the notion of place that remains the same by equivalence;[4] his exposition seems to imitate that of Albert of Saxony.

But he distances himself from Albert of Saxony and he breaks with the Parisian tradition, inaugurated by John Duns Scotus and William of Ockham, when he discusses the question, is the supreme sphere in a place?[5] Taking up Campanus of Novara's solution, which Pierre d'Ailly endorsed, he seeks to find the place of the universe in the empyrean heaven surrounding the world.

Nicholas of Orbellis had almost invited his readers to receive this solution when he wrote "one should note, nevertheless, that according to faith, the first mobile is in a place *per se*, for above it is the empyrean heaven, which the philosophers have not known; as for the empyrean heaven, it is not in a place, because there is nothing beyond it."[6] But following Joannes Canonicus and Albert of Saxony, John Hennon formally rejected this immobile heaven intended to contain the universe.

However, George of Brussels and Thomas Bricot formally admit the existence of a supreme immobile sphere; they admit it because "the theologians place the empyrean heaven around the mobile heavens."[7] They admit it also for an astrological reason that John Hennon noted, but to which he did not attribute any validity:

> The heavens in movement, [they stated], cannot save all the appearances and diversities occurring in the various regions of the earth; these diversities must therefore be related to an immobile heaven located beyond the heavens in movement. . . . It happens, in fact, that the parts of a mobile heaven are now in the east and that they will be in the west later; there is no reason that they do not have the same effects in the east as they did in the west, and vice versa. One must therefore admit an immobile heaven, whose parts are varied, in order to save this diversity of effects.[8]

Having been admitted by means of the above wretched reasoning, the immobile heaven then resolves the most serious difficulty for the Peripatetic theory of place: is the ultimate heaven in a place?

One should note [asserts George] that the Philosopher means the first mobile by his expression, ultimate sphere. . . . But in reality the ultimate sphere is a celestial body absolutely incapable of movement (*simpliciter immobile*) by nature, so that it is a body that cannot be placed at rest (*non est quiescibile*). We would therefore assert that the first mobile, the ultimate sphere according to the Philosopher, is actually in a place, and that it is the same with all the spheres contained by it. . . . As for the celestial body, which in reality is the ultimate sphere, it is not in a place by itself (*per se*) or in a place by accident. That is evident, because a body that by nature, in its totality as well as in its parts, is actually incapable of movement (*secundum rei veritatem immobile*) is not in a place; and that is how this celestial body is.[9]

We should understand the intent of these reasons.

From Duns Scotus to Albert of Saxony, one did not fail to raise the following objection to those who wished, with Campanus of Novara or Pierre d'Ailly, to take as the place of the world an immobile ultimate sphere and to relate all local movements to this fixed term: this celestial sphere is immobile in fact, according to them, but it is capable of movement; one must therefore judge its rest, as one must judge the rest or movement of any mobile body, by comparison with a fixed term; but where is this fixed term? And we are back to the beginning.

George replies to this argument that the state of the ultimate sphere is not merely rest, a rest that is the privation of a possible movement. The ultimate sphere is something absolutely and by its nature incapable of movement, something for which movement cannot be conceived. One cannot therefore examine or judge whether it is in movement or at rest. One does not need to refer it to a fixed term in order to declare it void of movement.

That is clearly what the Peripatetic theory of place required; Aristotle did not dare posit it. George of Brussels extended Aristotle's thought to its logical conclusion.[10]

Fifteenth-Century Albertists and Thomists

When the *Sententiae uberiores* treat the place of heaven, they summarize briefly, but exactly, the opinions of Themistius, Avempace, and Averroes;[11] with respect to Themistius's opinion, they add that "Saint Thomas agrees with him."[12] But the manual of the students of Cologne says nothing about the problem related to the preceding problem, the determination of the immobile place, the reference for all movement.

Lambertus de Monte gives a brief summary of the various suppositions put forward with respect to the place of the supreme orb,[13] but he adds to it a reflection that Albertus Magnus, and perhaps Saint Thomas Aquinas, would have rejected. This reflection might have the authority of Campanus of Novara and of Pierre d'Ailly, but not the authority of the former doctors; it is the following:

> One must consider, nevertheless, that according to the truth taught by theologians, this question poses no difficulty. Theologians admit, in fact, that the first mobile heaven about which Aristotle speaks is in a place properly speaking (*per se*); the first heaven, they say, is absolutely in a place (*simpliciter*), because this first mobile heaven is contained by an immobile heaven, the empyrean; thus this first mobile heaven is in a place, properly speaking (*per se*), since it is contained by another body which is external to it. As for the absolutely first heaven (*simpliciter*), the empyrean, it is not in a place in any way, but it remains at rest, because it is subordinated with respect to the rest of the blessed.
>
> The physicists (*philosophi naturales*) do not know of this bliss; they assume that every heaven is mobile. Thus Aristotle stated that the first celestial sphere itself is mobile; that is what gave rise to the preceding question.[14]

That is very true; by introducing an immobile empyrean heaven, the theologians completed the Peripatetic theory of place in a fashion that was called for and that Aristotle's physics refused to give to it.

Parisian Doctrines in Germany

The theory of place, which was the subject of so numerous and important controversies among the Parisians, seems not to have excited the interest of the German masters. They were content with summarizing the theories developed previously; they often neglected the most essential and most fertile propositions of these theories.

Gabriel Biel avoids as much as possible taking up questions of theology as pretexts for discussions of physics; thus when he examines in what way an angel can reside in a place or move by local movement, he says nothing about the theory of the place of bodies.[15]

When Summenhart puts forth his theory of place, he shows himself to be a Scotist; he even exaggerates his master's tendency to multiply entities.[16]

When analyzing place, he finds four absolute realities and four relations. The four absolute realities are the contained body and its terminal surface, and the containing body and its terminal surface. The terminal surface of the container is the subject of two relations: the first is the aptitude of this surface to lodge the contained body, the *locativitas*; the second, the *locatio*, consists in what it lodges actually. The terminal surface of the contained body is the seat of two similar relations. Like Pierre Tataret, Summenhart distinguishes place *pro per se denominato*, which is one of the four absolute realities enumerated here, namely, the terminal surface container, from place *pro per se significato*, one of the four relations, the one by which the surface of the container actually lodges the contained body. He gives this relation the name "active *ubi*" and reserves the name "passive *ubi*" for the analogous relation having the surface of the contained body as subject. During this analysis, he takes care to give as support the authority of Gilbertus Porretanus and Duns Scotus.

Moreover, he does not treat the difficult question of the immobility of place; he limits himself to referring his reader to what the Subtile Doctor had stated.[17]

Although expressed extremely concisely, the opinions of Gregory Reisch with respect to the subject of place are closely related to those of Summenhart; more exactly, they are a short summary of what one can read in the *Questions* of Joannes Canonicus.

To the *ubi*, which he defines as Gilbertus Porretanus did and which he qualifies as a passive *ubi*, he adds an active *ubi*, such as Joannes Canonicus has characterized.[18]

He distinguishes material place, the extreme surface of the containing body, from formal place, a relation founded on the containing body with the contained body as a term.[19]

He teaches that place cannot either by itself nor by accident move by local movement, although it is susceptible to generation and corruption because of the movement of its subject. "When a place is corrupted, the place succeeding it is identical to it, not in reality, but by equivalence."[20]

Such was essentially the opinion of Joannes Canonicus.

Like Albert of Saxony, and doubtless like the majority of the masters who taught at the Rue du Fouarre during the fourteenth century, Sunczel is more interested in physics than metaphysics; the innumerable entities that the overly subtile Scotism multiplied seem to him somewhat chimerical.

> There are philosophers who increase relations and forms at will; they posit six entities, distinct from one another, of which three are in place and three are in the lodged body. In place there is first the surface or entity of the surface, then there is the *locativitas* by which it can receive and contain the body, and finally there is the active *locatio* by which the place actually contains the lodged body. In the lodged body, there is first the entity of the lodged body, that is the contained body, then there is the *locabilitas*, the aptitude of the body to be lodged or contained, and finally the passive *location*, by which the body is actually contained and lodged. . . . But these relations do not all exist in reality; they exist merely in the mind of those who imagine them. Is place a relation or not? Is a relation something or is it nothing? Has Aristotle mentioned relations or not? These questions are the object of disputes among the metaphysicians, but they are not among naturalists or physicists.[21]

Sunczel is clearly not a metaphysician; perhaps he is not enough of a metaphysician. Most of the Scotists distinguished before him two elements combined to constitute place, a formal and a material element; the professor of Ingolstadt also wishes to consider a material and a formal place, but he establishes too crude a contrast between them.[22] His material place is the containing body itself; his formal place is the surface by which the containing body confines the contained body. "The concave surface of the lunar orb is the formal place of fire; the material place of fire is the lunar orb taken in its entirety."[23]

Moreover, he is not the innovator of this unsubtile distinction; it was provided to him by Paul of Venice, as we shall see.

After this, we should not be surprised that Sunczel has not provided any original solution to the Scholastics's difficult problems with respect to the theory of place. Is place mobile or immobile?[24] Does the supreme orb have a place, and what is it?[25] These questions are merely the occasion for the Nominalist of Ingolstadt to summarize in a dry and empty manner the theories of Albert of Saxony.

Jodocus of Eisenach conserves better than his predecessors the memory of the discussions about the theory of place which had been heard in Paris.

As a faithful disciple of Ockham, he rejects useless entities and relations; he numbers the distinction between the material and formal place among the rank of these superfluous complications:

> Some understand in the signification of the word *place* a material being or *denominatum*, which is the surface of the containing body, and a formal being or *per se significatum*; this latter being is not something absolute, but a relation with the lodged body. This relation is distinct from any absolute thing; it is the order with respect to the universe, meaning the first containing or lodging body of heaven, or else it is the distance from the poles and center of the world. But we are not in agreement with those who uphold this opinion; we reject as superfluous all relations of this kind. . . . We therefore state that place is the surface of the containing body, without any other relation added.[26]

In this way our author rejects the theory of Saint Thomas and Giles of Rome.

He then broaches the problem of the immobility of place:

> Nobody defines the unity, invariability, and immobility of place in the same manner. However, here are two points with which everyone agrees:
>
> If the lodged body keeps the same order and situation with respect to the center and poles of the world, or with respect to the concavity of heaven, or with respect to some other fixed object, one says that the lodged body remains immobile in the same place, even when the surface of the containing body changes; thus it is with a stake driven within a flowing river or a tower surrounded by air blown by the wind.
>
> Even when the surface of the containing body remains the same, if the situation and order we have just spoken of changes, one says that the lodged body moves, that it does not remain in the same place; thus is it when one transports some wine contained in a keg from one town to another.[27]

After having discussed the opinion of those—like Saint Thomas— who wish to account for the general opinions by the distinction between material and formal place, one of which remains immutable while the other changes, our author adds:

> Others like Scotus and those who follow him state that if the surface of the containing body gets destroyed or changed, the relation of order or distance of which the surface was the subject also gets destroyed; in itself and actually it does not

remain the same numerically, whether one considers it materially or whether one considers it formally. It is continually other than it was. But because the lodged body always remains the same only by equivalence, this saves everything one says about place as well as if it remained one and identical to itself. In fact, to save the rest of the lodged body, and everything one says about place and the lodged body, all the numerically distinct places succeeding one another have the same value as a numerically single place, permanent and immobile; and that is what one understands when one says that these various and distinct places are one and the same place by equivalence. . . .

One does not say of every body constantly in a new place that it moves locally; one says this only when the body would behave in this fashion even if its place were immobile or, in other words, if the body is continuously in places that differ by equivalence, if it does not remain in a place that is the same equivalence.[28]

This theory's interpretation by the Scotists is that the relation they give to a formal place is destroyed and replaced by another formal place when the lodging surface changes, but "that it remains incorruptible by equivalence."

We approve and praise this opinion of the identity and immobility of place, which states that one must consider them from the point of view of equivalence and not from the point of view of real existence; but we deny what it asserts about a distinct relation of the surface, and what it asserts about the surface as an accident of the body, distinct from the body, as we have just asserted.[29]

We would have liked it if Jodocus of Eisenach's faithfulness to Ockham's doctrine would have led him to assert that local movement does not suppose the real existence of a fixed term, that this term can be merely conceived. One can say that he insinuates this truth in several passages of his discourse; in the following, for example: "In this way it is no longer necessary to assume a center and immobile poles."[30] But he never formulates it explicitly.

The problem of the place of the ultimate sphere would have given him occasion to formulate it. But in what he states about this problem we can recognize the belief in the existence of an immobile empyrean heaven; this belief in the empyrean, which was neglected from Albertus Magnus to Albert of Saxony, took on new strength during the fifteenth century. Jodocus of Eisenach holds the same doctrines as the other masters of the fifteenth century with respect to this problem:

> If by supreme sphere one understands the one that according to faith and the truth of things is immobile, this sphere is not in a place either properly speaking (*per se*) or by accident; in fact, by its own nature it is immobile in its totality and in each of its parts, and no body surrounds it.
>
> On the other hand, if by heaven one understands the last of the mobile spheres of whatever rank, this heaven is in reality in a place, properly speaking, for there exists a body enveloping it and containing it; this sphere is no less in a place than any of the spheres it contains.
>
> Aristotle, however, I understand, did not think that it was in a place, properly speaking, in the same way as other bodies, but that it was only in a place, improperly speaking, by accident, in virtue of an economy with the other bodies. The other bodies are said to be in a place because they are in something that surrounds them and contains them. That is not how heaven is, for according to Aristotle, there is no other immobile body beyond this ultimate sphere; if it is in a place, that is because there is an object at rest by which one can judge the movement of heaven, and because from instant to instant each part of heaven is situated variously with respect to this object.[31]

In order to follow more exactly the thought of William of Ockham, Jodocus of Eisenach should have asked himself what would happen with this place of the supreme heaven if the central body, instead of remaining at rest, were to move. This question, which preoccupied the Parisian of the fourteenth century, seems to have been ignored by the Germans of the fifteenth and sixteenth centuries.

Paul of Venice

Nicoletti sometimes was able to renounce his backward Peripatetic philosophy in order to follow the path traced by the philosophy of the moderns, but often he was divided between these opposite tendencies; he sought to reconcile them, to unite them. From the union of such radically heterogeneous elements emerged odd doctrines that were difficult to characterize. They were too often noticeable only for their mediocrity and absence of logical progression.

These defects are apparent in Paul Nicoletti's theory of place; it is constructed out of pieces furnished by Simplicius, Averroes,

Saint Thomas Aquinas, and the Terminists. Its incoherence allows one to suppose that its author misunderstood the various ideas he forged to one another.

Paul of Venice distinguishes material place and formal place;[32] the material place of a body is the containing body, and the formal place is the extreme surface of the containing body, the surface by which it touches the contained body. We have already indicated that this conception is crude when we spoke about Frederick Sunczel, who adopted it.[33]

Moreover, Paul Nicoletti does not only distinguish material place and formal place; he also considers efficient place and final place:

> The efficient place is a conservational virtue of the contained body residing in the surface of the container; it is the virtue about which Gilbertus Porretanus spoke when he said: Place is a principle of generation.[34]

This virtue is also what was in question in the opusculum *De natura loci*, attributed to Thomas Aquinas.

As for the final place, it is nothing other than the natural place.

These distinctions are concerned with place, properly speaking; but for Paul of Venice there is also place, improperly speaking, and it can also be material, formal, efficient, or final.

Moreover these various kinds of places, improperly speaking, are related to one another in unusual fashions. Here, for example, are the definitions of material, formal, and final places, improperly speaking:

> Material place, improperly speaking, is a volume attributed to an entity that does not occupy space; the Philosopher is speaking about such a place in the first book of the *De Caelo* when he says that heaven is God's place. Formal place [improperly speaking] is the situation that orders the parts with respect to place. Simplicius in his commentary on the *Categories* is speaking of this place when he says that place by its own character is ranked in the category of the situation; by the character of place he understands the form of the place or the order of the parts with respect to each other. . . . Final place is the situation acquired by local movement; in other words, it is the relation *ubi* the Philosopher speaks of frequently when he says that the movement is accomplished with respect to place and that place is the term for movement.[35]

We recognize in this strange mixture of disparate notions the influences of Saint Thomas, Duns Scotus, William of Ockham, as well as those of Burley and Albert of Saxony.

Paul of Venice's opinion concerning the place of the supreme orb is no less confused. The supreme orb is in a place accidentally and by its center. This proposition formulated by our author summarizes Averroes' teaching. But Averroes understood by center an immobile central body of finite dimensions, capable of serving as term of comparison in the study of the movements of heaven. What was appealing about such a theory disappeared in Paul of Venice's summary, which understands by center an indivisible geometric point:

> Even though heaven is divisible, it is in an indivisible place. In the same fashion that permanent beings are in an instant, for their duration is measured by this instant, heaven is in an indivisible point, because its movement is known by this point.[36]

The opusculum *De natura loci*, attributed to Saint Thomas, admitted that the celestial spheres internal to the supreme orb were lodged in two ways; each of them was in a place by its center, like the supreme sphere, and each of them was lodged accidentally by the superior orb containing them. No doubt Paul of Venice wishes to reproduce this theory; but he deforms it to such an extent that it is hardly recognizable. Instead of applying it to the lower spheres only, he applies it to the set of all the celestial spheres; he therefore teaches that this set is lodged by its center and also contained within the orb of the fixed stars, because the orb of Saturn is part of the set.

Following Albert of Saxony, Paul of Venice rejects the authority given to the *De motibus animalium* for the proposition that any mobile body requires the existence of a fixed body;[37] he asserts that Aristotle spoke only of a movement of progression, which, in fact, requires a support. Moreover, he does not reject this authority in order to refute the argument concluding for the immobility of the earth from the mobility of heaven; the argument he proposes to attack is the one by which Campanus and d'Ailly attempted to demonstrate the existence of an immobile supreme heaven, the place of all the mobile orbs.

The movement of heaven requires the immobility of the earth; Paul Nicoletti adopts this conclusion and invokes the reason proposed by John of Jandun in order to establish it: the perpetuity of the generation and corruption of living beings supposes constantly variable celestial influences. Albert of Saxony had shown that this reason, assuming one thinks of it as well-founded, required only a relative movement of heaven with respect to the earth, without requiring anything with respect to the movement or rest of the latter;[38] Paul Nicoletti pays no heed to this remark, which is so clearly true.

Paul of Venice's *Summa totius philosophiae* is a school manual; its defects are those that characterize most manuals, at every time, in every country; formulas of diverse origins are juxtaposed in an artificial order which in no way hides the disparateness and incoherence. In order for them to be more concise, these formulas are emptied of the thoughts that made them live; rigid, dry, and flat, they easily accumulate in the minds of those who believe they have acquired some ideas when they have learned some words. And since these people are numerous, the books that suit them are always assured great success.

Paul of Venice did not wait to conceive some strangely disorganized ideas about place until he wrote his *Summa totius philosophiae*; we can notice a similar incoherence in what he asserts about this problem in his *Expositio super libros Physicorum*.

As in his *Summa totius philosophiae*, he distinguishes eight usages of the word *place*, but these usages are not classed and defined in the same way in the two works.[39]

The word *place* is used for:

1. The lodging body;
2. The ultimate surface of the lodging body;
3. The origin of place; thus according to the Commentator, the center of the world is the place of heaven. In the same fashion Gilbertus Porretanus states that *simple place* is the origin of *compound place*, understanding by *simple place* the position with respect to the center of the world, and by *compound place* the ultimate surface of the ambient body;[40]
4. The *ubi* that stems from compound place;
5. The *ubi* that stems from simple place;
6. The conservational virtue of place;
7. The space that attracts and keeps a number of objects; thus place is where the marketplace is.
8. A space subject to a being that does not itself occupy any place; thus one says that heaven is the place of God.

Paul of Venice is aware that some other authors have attempted to classify the various senses of the word *place* in a manner which is at the same time simpler and more rational. Burley, for example, distinguishes between what the word denotes and what it signifies; what it denotes is simply the ultimate surface of the ambient body; what it signifies is the union of the surface and the action of containing (*continentia*), which is a relation between the lodging body and the lodged body.[41]

Nicoletti rejects this theory; it is contrary to Aristotle's thought in the *Categories*. He opposes it by means of another solution: place implies two things; the first, which it implies directly, which is its subject and matter, is the surface; the second, which it implies indirectly, which is its act and form, is the fact of containment.

Here Paul of Venice seems to be recalling both the teachings of Giles of Rome and Duns Scotus; from one he borrows the distinction of the matter and form of place; from the other, the consideration that the surface of the container and the action of containing are two realities, of which the second is to the first as form is to matter.

A little further on, we can find Giles of Rome's theory about the immobility of place:

> Giles declares that place presents two things to consider, material place and formal place. Material place is the surface of the containing body; formal place is the relative order with respect to the whole universe or, in other words, the distance to the poles and the center of the world. Material place is mobile by accident; formal place is not mobile either by itself or by accident.[42]

Burley gave several arguments against this theory, which Nicoletti reproduces. Here is one: whether by divine power or in thought, let the whole world be displaced in a straight line, except a body contained in air, which is kept immobile. The immobility of this body ought to carry with it the permanence of its formal place. But the distance of this body from the poles and the center of the world, its position with respect to the whole universe, has changed.

Paul of Venice thinks that one can turn against Burley the argument Burley uses against Giles; the immobile body ought to keep an invariable place; however, the ambient medium and therefore the ultimate surface would change.[43]

This reply could have embarrassed Aristotle and the Commentator, but one cannot see how it could bother Burley and any of the Parisian Nominalists; for them the immobility of a body does not require the persistence of the body's place, but only the *equivalence* of the places that succeed each other. No doubt Paul of Venice is at that moment forgetting their doctrine, of which he will soon give a summary.

Be that as it may, Paul Nicoletti attempts to perfect the theory of Giles of Rome. He proposes that the definition of formal place proposed by Giles of Rome does not have to be applied to any place, but to a certain kind of place.

In his *Treatise on the Six Principles*, Gilbertus Porretanus distinguished two kinds of places: *simple place* which is the center of the world, and *composite place* which is the ultimate surface of the ambient body.[44] In the same fashion one must distinguish two *ubi*: the *ubi* that arises from simple place, and the *ubi* that arises from composite place; the former "is the situation of the whole world, coextensive with the world,"[45] while the latter "has the lodged thing as subject; it has no extension in itself and resides there indivisibly."[46]

It would be difficult for us to believe that the two *ubi* considered here by Paul of Venice do not have an affinity in his mind with the two kinds of *thesis* considered by Simplicius, the one corresponding with the *situation* of the body in the world taken as a whole, and the other to the *disposition* of the various parts of this body.

Movement, not movement *per accidens*, but movement *per se*, does not have for its object the acquisition of any *ubi* whatever; the only *ubi* to which it relates is the *ubi* arising from simple place, the situation with respect to the poles and the center of the world. This *ubi* alone cannot be related to an object without a change being produced in the object itself. The other *ubi*, that derives from the ambient surface of the lodging body, is not the object of proper movement; it can change without any change in the lodged body and by the mere movement of the lodging body, for it is a relation of the lodging body to the lodged body.

One cannot see how this distinction can cover Giles of Rome's doctrine against the attacks that Burley and the Paris Terminists directed against it. If God displaced the world by a translation, keeping a single body immobile, there would be a change of *ubi* for this body arising from simple place; however, this body would be without movement. Paul of Venice finds no response to this, other than allowing it as a miraculous effect of divine power. In spite of the weakness of this reply, he holds that the distinction between the *locus situalis* and the *locus superficialis* is capable of resolving difficulties, as we see him returning to the distinction.

Nicoletti says that "Ockham, expounding upon the definition of place given by the Philosopher, says that place is nothing more than the lodging body, considered as layers contiguous with the lodged body, that one can imagine in infinite numbers."[47] Our author objects that this definition of the Venerable Inceptor contradicts Aristotle's theory on many points. Ockham clearly knew this well.

In particular, Paul of Venice makes the observation (which has been made before) that, according to this definition, place would be mobile. "Burley replies that an immobile house within moving air can be in places that are numerically distinct from instant to instant, but that it is always in the same place by equivalence."[48] Our author does not agree with this. He returns to the distinction between two kinds of places: place arising from the situation with respect to the whole universe, which he named *locus situalis*, and which he now names *relative* place; and place consisting in the surface of the ambient body, *locus superficialis*, which he names *absolute* place. From instant to instant, Burley's immobile house is in two different *superficial* places, but its *relative* place remains the same numerically.

But Paul of Venice's doctrine is still open to the objection that the definition of *relative* place, *locus situalis*, can have no meaning unless there exists an absolutely fixed reference; in the Averroist doctrine, a central body, immobile because of its essence, constitutes this reference. Once one considers the earth as capable of being displaced and the whole world as capable of translation, the notion of relative place, as it has been defined, loses it meaning. To their credit the Parisian Terminists recognized the necessity of ridding the notion of local movement of the need for an immobile reference with concrete existence. Paul of Venice is too faithful an Averroist to follow an opinion so radically opposed to the teaching of the Commentator; thus he must fight endlessly against some inextricable difficulties.

> According to Burley, [he says,] since it is certain that the whole world and all its parts move endlessly, and that there is no immobile body outside the world, one must conclude that a body moving locally does not necessarily behave differently from one instant to another with respect to some immobile term.[49]

Our author replies to this, that all local movement corresponds to a change of place, but that the movement producing a change of place is not necessarily the movement of the lodged body; it can be a movement of the lodging body. It is difficult to see the relevance of this response to Walter Burley's observation.

It is impossible to hold the Averroist theory of place if one renounces the proposition: there is a body at the center of the world whose immobility is certain and necessary. Because he did not recognize this as a truth, Paul of Venice has already been confronted by paralogisms; these contradictions become more flagrant when he takes up the important question about the place of the supreme orb.

In order to define the place of the ultimate sphere, Paul Nicoletti first expresses himself almost as a Scotist would:

> The ultimate sphere is in a certain *ubi*, and this *ubi* is engendered by the fact that it surrounds its place; it is in the *ubi* arising from simple place and not in the *ubi* arising from compound place.[50]

Our author thinks this theory, clothed in language borrowed from Gilbertus Porretanus, or the Scotists, or Walter Burley, is plainly in conformity with the Averroist doctrine, which he formulates as follows:

> The ultimate sphere is in a place in one way, heaven as a whole is in another way, and finally the heaven of the planets is in a third way. The supreme sphere is only in a place by accident, because of its center; it is not in a place *per se*, and it is not in a place by the intermediary of its parts. Heaven as a whole is in a place accidentally, because of its center; it is also there by the intermediary of its parts, for it has various parts which lodge one another. Finally, the heaven of the planets is lodged in three distinct manners; it has a place accidentally because of its center; it has a place *per se*, because it is contained in the concavity of the supreme sphere; and it is lodged by the intermediary of its parts, for it has parts that lodge one another.[51]

Paul of Venice not only thinks his doctrine is in conformity with Averroes' doctrine, but he goes further. Because of the identity he admits, following Giles of Rome, between *locus superficialis* and material place and *locus situalis* and formal place, because of his confusion between *locus situalis* as he has defined it and *situation* as considered by Avicenna, and because of another confusion, that he takes material place and formal place as understood by Giles of Rome as identical to place *per se* and place *per accidens* considered by Averroes; because of these plays on words, Paul Nicoletti believes that he can establish an agreement between Avicenna's theory and Averroes's theory:

> Avicenna holds that heaven does not move around a place but in a place, this place being, moreover, a *locus situalis* and not a *locus superficialis*. . . . The Commentator, on the other hand, holds that heaven moves locally around its place, by which he understands the earth; toward this end he distinguishes between accidental or *formal place* and *per se* or *material place*.
>
> It seems to me that heaven moves locally in the manner that the Commentator defined and also in the manner indicated by Avicenna.[52]

Averroes surely would not have upheld this since he fought against Avicenna's theory so strongly. Would he have accepted the concessions that Paul of Venice allowed in his name? That appears extremely doubtful to us. However, let us read the passage that follows; its author has clearly read Albert of Saxony and, above all, Simplicius:

> According to the Commentator, if the terrestrial element and the other elements moved in a circular fashion, as heaven does, heaven itself would no longer have any local movement. Its movement could not be a movement of translation or a movement of rotation. According to the Commentator, in fact, any body moving by a movement of translation changes both its place *per se* and its place *per accidens*, its material place and its formal place; a body moving by rotation suffers a formal change, although it does not move *secundum materiam*. But if the earth turned by a movement of rotation at the same time as the other elements, heaven would no longer have a formal place or a material place; in fact, it would not be moving within a surface capable of surrounding it, and it would not be moving above an immobile surface upon which it is possible to trace circles that allow a reference for its movement either.
>
> However, in the case where the earth turned in the opposite direction of heaven, or in the case where it turned in the same direction as heaven but more slowly, the Commentator would admit that heaven moved by local movement. He would also think this if the earth accompanied heaven in its movement, as long as one of the other elements remained immobile, or that it turned in the opposite direction, or even in the same direction but slower; in this case, in fact, heaven could still describe its various circles above this element.[53]

No doubt the Commentator insisted on the truth of the proposition that no movement could be experienced by us if the term toward which the mobile tends moved in the same direction and at the same speed as the mobile; but he reflected upon the relativity of the movement our senses experience with too much depth to affirm that heaven is or is not in movement (to state what this movement is) before being assured of an absolutely fixed term of comparison; and he wanted—this was the fundamental principle of his doctrine— that this absolutely fixed term be a real and concrete body. He therefore would have rejected the propositions just formulated by Paul of Venice.

On the other hand, he could have accepted the following without contradicting any of his axioms:

> Even when all the elements move with heaven, as long as one agreed that heaven has no local movement, the celestial spheres would have a local movement; in fact, since they do not all move by the same movement, each lower sphere would describe a circle with respect to the concavity of the higher sphere, and the higher sphere would describe one with respect to the convexity of the lower sphere. However, if the earth were in movement, it would be more difficult to experience the local movement of heaven than it is when the earth remains immobile: that is why the Philosopher states, in the second book of the *De Caelo*, that if heaven is in movement, the earth must be at rest.[54]

We believe that the Philosopher understood more by that. Be that as it may, the Averroist theory of place would not be contradicted by the hypothesis that Paul of Venice has just examined, for in that hypothesis, the supreme heaven, deprived of any local movement, would provide the absolutely fixed term that all local movement requires, according to Averroes. One should point out that this hypothesis, which takes the supreme orb as the immobile place to which all celestial and terrestrial movements are related, is precisely the one that Copernicus adopted.

Paul of Venice, pushing forward his hypotheses, attempts the question Duns Scotus formulated and for which he had stated, "seek an answer—*quaere responsum*":

> Even if God annihilates the whole world except the supreme sphere, this sphere would still move by local movement—but not by movement relative to the *locus situalis*, no doubt. The part of heaven that was right becomes left, and the part that was east or south becomes west or north,

or inversely; all that cannot happen unless heaven were animated by a movement consisting of a change of situation.[55]

To state that the part of heaven that was right becomes left supposes that the movement of heaven is contemplated by a being who has a left and a right, and who remains immobile; the proposition formulated by Paul Nicoletti therefore has no sense unless there exists a fixed and extended term that has a left and a right, an east and a west, a north and a south. Where will our author find such a fixed and extended term? Aristotle and Averroes wanted this term to be the earth; but by hypothesis, the earth is annihilated. Damascius, Simplicius, and the Paris Terminists thought of it as an abstract body, a pure being of reason; it would seem that Paul of Venice has to accept their thought. However, he does not. By a strange aberration—whose traces we have already noted when we analyzed the theories of the *Summa totius philosophiae*—this immobile term by which one should distinguish the left and the right of heaven and its north and south is reduced by Paul of Venice to a simple indivisible point at the mathematical center of the universe! Burley had made this mistake, inadvertently, no doubt; Paul of Venice professes it clearly and with insistence:

> The supreme sphere is in a place accidentally, because of its center. . . . One can object as follows to this proposition: if the center were animated by a movement of rotation, as heaven is, the supreme sphere would still be in a place, since it would move by local movement; but in this case it would not be lodged by its center. Therefore it is not so actually. The major premise and its consequents are evident; as for the minor premise, it results from the ultimate sphere moving necessarily, according to the Commentator, around an immobile center.[56]
>
> Here is the response one should give to the above objection: . . . The world has two centers; it has a simple, indivisible mathematical center, and a natural center, the terrestrial element. Even if one were to suppose that the natural center moved by a rotation, the mathematical center would not so move. The movement of the supreme sphere would therefore still be a local movement; this sphere would still be lodged by its center, not by its natural center, of course, but by its mathematical center. . . . However, the Philosopher thought that this natural center cannot move by any movement, for in the book, *De motibus animalium*, he stated that all the gods together could not move the earth.

If the world were homogeneous or if the earth moved
by a rotation, the earth could no longer be the place of heaven
as a whole or of the supreme orb; only the indivisible
mathematical center could constitute this place. In fact, if one
says that the earth is the place of the elements and celestial
bodies, that is because of its immobility, an immobility that
it derives from the indivisible center of the world.[57]

Scotus said *quaere responsum*; Paul of Venice found an awful
response. At least in the passage above Paul of Venice noted the
disagreement between his opinion and that of Aristotle. Elsewhere
he goes further and attempts to have the Philosopher endorse and
take the responsibility for his unacceptable doctrine.

The Commentator draws this distinction: there are two
centers of the world, the natural center and the mathematical
center. . . . By *center* Aristotle understands one or the other
of these two centers. If by *center* he understands the natural
center, the whole heaven moves constantly *secundum formam*,
endlessly describing a new circle around the center of the world;
if by *center* he understands the mathematical center, one can
still admit that heaven moves by formal movement. In fact,
in the same way that it endlessly describes a new straight line
from the circumference to the center, it also endlessly describes
a new circle around the center of the world.[58]

Here, oddly enough, Nicoletti is using an Averroes commentary
on a passage from Aristotle in support of his theory of the place
of heaven.[59] The center for which Aristotle required immobility,
so that the movement of heaven would be conceivable, is no doubt
the natural center, the earth.

The absurd idea that an indivisible point, the center of the
world, can serve as place of the universe, the fixed term for all
the movements produced, is an idea that is particularly dear to
Paul of Venice. We have seen him broach this idea in the *Summa
totius philosophiae* and formulate it with precision in the *Expositio
super libros Physicorum*. But there is a work in which he developed
it with special care, the final work of his career, the *Expositio
praedicamentorum Aristotelis*, completed by March 11, 1428. In this
work the thought is developed to such an extent that it permeates
the whole theory of place. This unacceptable theory is truly the
work and property of Paul of Venice; however, he insists on
attributing it to Aristotle and Averroes:

The Commentator calls the surface enveloping the body, composite place; he says that simple place is the indivisible mathematical center of the world. In the same fashion that fire moves toward the concavity of the lunar orb, toward its natural place, earth also moves toward the center of the world. One can therefore say that, according to the Commentator, the center of the world is not only the place of the terrestrial element, but it is also the place of all the other elements and the place of heaven as a whole. It is not really a containing place, but a contained place; it is not an ambient place which envelops, but a place that serves as measure. In fact, it is by their distance or nearness to the center of the world that one recognizes whether the elements are lodged and situated in a natural manner. The Commentator, in fact, is saying there that heaven and the elements are in a place due to the center.

The *ubi* being an effect of place, there are as many kinds of *ubi* as there are kinds of places. And there are two kinds of place: composite place, which is the enveloping surface, and simple place, which is the center of the world. There are therefore two kinds of *ubi*, composite and simple.

Composite *ubi* proceeds from composite place; the subject that receives it is the lodged body.

Simple *ubi* proceeds from simple place; the subject that receives it is the whole world, which is imbued by it along every dimension. This *ubi* is called the situation (*situs*) of the world and of each of its parts.

The necessity of this simple *ubi* is seen through the movement and rest of natural bodies. Let us suppose that a vase filled with water is moved by a movement of translation; in its movement the water gets nearer or farther from the center of the world. And in the fifth book of the *Physics* it is proved that, properly, movement goes toward a *ubi*; from instant to instant the water must acquire a new *ubi*. But it does not acquire nor lose the *ubi* arising from composite place, for the surface containing it always remains numerically the same. It therefore acquires or loses some other *ubi*, which can only be the *ubicatio situalis* arising from the simple place, which is the center of the world.

Moreover, a tower, a town, or any body whatsoever fixed within moving air remains at rest locally; it must therefore be continually in the same *ubi*; but assuredly, it does not always remain in the same *ubi* arising from composite place, since this composite place changes with the surface upon which it depends *in fieri et in facto esse*; . . . It is therefore in another *ubi*; this *ubi* can only be the immobile *ubicatio situalis* arising from the center of the world.[60]

The center of the world is therefore the term to which one must relate all local movement or rest; to move is to get nearer or farther from the center of the world; to stay at rest is to remain at an invariable distance from the center of the world. If that is so, a celestial sphere is immobile by definition. Paul of Venice does not examine this inadmissible consequence of his theory; but he does examine another for which he appears not to give a very good solution.

> Although there is movement, properly speaking (*per se*), in the category of *ubi*, there is no movement, properly speaking, toward just any *ubi*; there is no movement, properly speaking, toward a composite *ubi* because a new composite *ubi* can be acquired by a body without the body itself changing. But there is movement, properly speaking, toward a simple *ubi* for a new simple *ubi* cannot be acquired by a body if the body does not suffer any change in itself.
>
> That is true with respect to natural power. In fact, God can move the whole world by a translation while maintaining you at rest within some air; you would then receive a new simple *ubi* continuously, for you would be nearer the center of the world from instant to instant. Yet you yourself would not be changing. Thus although movement, properly speaking, is accomplished toward a simple *ubi*, God can make it be that movement, properly speaking, is not accomplished toward this *ubi*.[61]

This amounts to saying that God, because of His omnipotence, can make something be in movement although by definition it is not in movement. There is hardly anything more inane.[62]

There is a constant battle in Paul of Venice's mind between various alternatives, between his Averroist tendencies and his modern Parisian tendencies; sometimes he sides with the former and sometimes he sides with the latter.

Under the influence of Terminist doctrines, Nicoletti renounces the axiom of Aristotle and Averroes, that there exists a body of finite extension in the center of the world, whose absolute immobility is logically necessary, this body being the earth. Our author does not think it absurd that the earth can be animated by a movement of rotation or that the universe as a whole can suffer a movement of translation.

Once one renounces positing in the world a concrete body

serving as term of comparison for the local movements of the heavens and the elements, logic allows only one path to follow; one has to admit that all local movements are defined by comparison to an abstract body, a body that the senses cannot perceive, but about which physical theory can teach us. That is the path followed by Damascius and Simplicius, his disciple, and the path followed by Parisian Terminists.

Paul of Venice does not wish to follow the path traced by Averroes's adversaries completely; he attempts to follow a direction between them and the Commentator. That leads him to a flagrant paralogism: he proposes to relate local movements to a simple mathematical point at the indivisible center of the world. He takes up and develops this doctrine again and again from 1409 to 1428; in this way, he demonstrates that he is a wretched philosopher.

III
Time

7
Time

Time according to John Duns Scotus

The memorable fourteenth-century discussions about the theories of place and movement received their starting point with Etienne Tempier's decision; they were inaugurated by John Duns Scotus, who dared to assert the proposition that even if there exists no immobile term [as reference] a body can still move by local movement. Thus the immobility of the earth ceased to be the essential postulate without which movement would be inconceivable, according to Peripatetic physics.

Duns Scotus added another assertion to the above, one that is no less contrary to the Stagirite's physics: Even if heaven stopped, time would continue to be and to measure the movements of the other bodies. Moreover, even if all movement were to stop, time would still exist and would measure the universal rest. In fact there is a *potential time*. If heaven moves actually, potential time coincides with actual and positive time which measures the movement of heaven; if heaven is immobile, potential time continues to exist— it is then the time that would measure the movement of heaven if heaven moved. We know this time independently from the movement of heaven; therefore, if heaven were immobile, we could, using this potential time, measure the duration of heaven's rest. Such was the doctrine formulated by the Subtile doctor in the following:

Heaven being stopped, Peter could walk after the Resurrection; and this walk would be conceived as existing in our continuous time, not in some other kind of time. Similarly, if the first movement of heaven did not exist, the rest that heaven has, due to the cessation of this movement,

would be measured potentially by this time, which would
measure the first movement, if this movement existed in an
actual and positive manner; every movement existing actually
can be measured by this potential time. Thus the movement
measured in this way does not depend necessarily on the
movement of the first heaven for its existence; it does not exist
necessarily in virtue of the movement of the first heaven—
thus it was for all movement existing while heaven was stopped
during the time of Joshua. Here the measure of a quantity
by another quantity, of a magnitude by an equal magnitude,
is not accomplished by something whose measured magnitude
depends essentially, as it happens, with quidditative measures
[meaning, with measures expressing the composition of an
object by means of the parts constituting it]. It suffices that
in the case when this movement exists, the magnitude can
be known by a distinct knowledge of time, time being either
actual or potential. (*Sed tantum sufficit quod motus iste,
quando est, possit distincta cognosci secundum quantitatem
suam ex cognitione distincta temporis, et hoc vel actualis vel
potentialis.*) Therefore I assert that if the movement of the
first heaven is not, one can still measure all other movements
by means of the time marked off by the movement of the
first heaven—meaning that one can know in which part of
celestial movement the considered movement can be
accomplished—if celestial movement existed; presently it is
accomplished during a part of the celestial rest equal to what
can coexist with such a part of celestial movement. (*Et ita
dico quod, quando iste motus caeli non erit, poterit tamen
alius motus mensurari per tempus hujus motus primi caeli,
une quantum scilicet motus ille posset fieri cum tanta parte
illius motus, si esset; et nunc et cum tanta parte quietis cum
quanta pars motus posset esse.*)[1]

Duns Scotus seems sure of the existence of a potential time known
distinctly in the absence of the movement of any body whatever,
by which we can measure the duration of all movement and rest.
He expresses himself on this matter more formally in one of his
Quodlibets:

> Even if no movement existed, there can still exist a rest,
> properly speaking; in fact, even if no body is in movement,
> a body can always behave in the same fashion, while being
> naturally capable of behaving in one fashion or another.
> . . . There corresponds to this invariable disposition a proper
> measure which is a time. If one imagines any two instants
> in this time, there can be a flow or a movement of some

magnitude between these two instants; since one calls time
the measure of such a flow or movement, this invariable
existence would have a time—this would not be an actual
and positive time, but a potential and privative time. Thus
the intellect, which has a notion of potential and privative
time,[2] can apply the notion to this invariable duration, and
can know its magnitude; in other words, the intellect can
know that this duration would have such a positive
magnitude, if there existed a positive time. (*Nullo motu
existente, potest esse quies aliqua etiam proprie accepta;
quia, nullo corpore moto, posset aliquod corpus uniformiter
se habere et, cum hoc, esse aptum natum aliter et aliter se
habere. . . . Huic etiam uniformi dispositioni correspondet
propria mensura quae est tempus, inter cujus quaecunque
dua instantia imaginata posset tantus fluxus sive motus
intercipi; et ita, si tempus dicitur mensura motus sive
fluxus, illa uniformis existentia habebit tempus, licet non
uniformiter actuale positivum, sed potentiale et privativum;
unde intellectus, habens notitiam temporis potentialis et
privativi, applicando eam ad istam durationem uniformem,
potest cognoscere quantitatem ipsius, scilicet quod tantam
haberet positive, si esset tempus positivum.*)[3]*

What is it that drives Duns Scotus to conceive and formulate the
hypothesis of a potential time which exists and is known distinctly
by the human mind independently of the movement of any body,
and allows the mind to measure the duration of any movement
or rest, and with which time marked off by the diurnal movement
agrees, when the diurnal movement occurs? The example of Joshua's
miracle, chosen by the Subtile doctor, allows us to guess at it.

For Peripatetic physics, time was inherent in the diurnal
movement; if the diurnal movement did not exist, there would be
no time. From this the Averroists derived the following conclusion:
If the diurnal movement stopped, all other movements, all other
changes would have to stop, for there would be no time to measure
their duration. We have seen Robertus Anglicus assert, in his
commentary of the *Sphere* of Joannes de Sacrobosco, that if the
first heaven were to stop rotating, a falling stone would stop falling.[4]
Other Peripatetics held the same thing, and doubtless they drew
from this a reason to deny the miracle of Joshua.

Saint Augustine, on the contrary, derived from the miracle of
Joshua a reason to deny the whole Peripatetic theory of time and
to deny that time has any existence outside the mind.[5]

John Duns Scotus read the *Confessions,* and no doubt his reading contributed to his rejection of Aristotle's doctrine with respect to time; we find the evidence for this in a refutation of Aristotle's doctrine in the fourth book of the *Scriptum Oxoniense.*[6] Here is what the Subtile Doctor responds to those who make of time a passion of the first movement:

> Time is not in the movement of heaven as one quantity is in another quantity; in fact, one must not put, in the same permanent subject, two quantities, of which the first is the subject of the second, so to speak, and the second is a passion of the first. Therefore, time adds to movement only the reason (*ratio*) of measure from the formal point of view, for movement implies a succession which is proper to it; and from the point of view of foundation, it adds the reasons which are required so that the measure may be effectuated, namely, uniformity or regularity and speed: In fact, the first characteristic, uniformity or regularity, renders the measure very exact, and the second characteristic, speed, renders it as small as possible.[7]

In the actual state of things, these characteristics of a good standard for the measure of time are found in the movement of heaven. If heaven were to stop,

> there would no longer be a movement faster than all other movements, or at least there would no longer be a uniform and regular movement; what confers the reason of measure with respect to other movements would no longer be found in the foundation of any movement. Time would therefore no longer exist in the manner presently admitted, as a passion of the first movement.[8]

Must one therefore conclude that there would no longer be any time in any manner, and that there would no longer be any movement, because the movement that must measure the others has ceased to be? Would we invoke the reason, "Where there is no measure, there is no measured object either"?

Duns Scotus concedes that the above proposition is valid with respect to an essential measure, when the existence of the measured objects depends on the existence of the measure itself, when the latter is the principle or the element composing the former.

> But this proposition is not true for an accidental measure which measures by application or coextension, as an ell measures a piece of cloth; in fact, it is evident that the length of a piece of cloth does not depend on the length of an ell.

Moreover, it is only in this manner that the first movement, taken in its successive extensions, and considered as having a relation of measure with respect to other movements, measures these movements; it is their measure by application or coextension, and not as the term for an essential need.[9]

It is therefore not true that stopping the movement of heaven would lead to the stopping of all other movements.

In favor of this reply one can invoke the passage from the book of Joshua where it is said that Joshua fought while the sun and moon were stopped, and that, consequently, once heaven as a whole was stopped, there was a fear that the stopping of the sun and the moon, accompanied by the movement of all the other celestial mobiles, would carry with it too great an irregularity in the movement of celestial bodies.

Saint Augustine said, in the eleventh book of his *Confessions*: While the sun was stopped, the potter's wheel turned.[10]

It seems that we are capturing here the thought that suggested to Duns Scotus his reflections on the theory of time. The example of Saint Augustine pushed him to reject the Peripatetic theory of time, whose falseness was rendered clearly manifest by the miracle of Joshua.

However, the Subtile doctor did not completely follow the path pointed to by the bishop of Hippo. He did not go so far as to deny time any existence outside the mind; he placed potential time in things.

The condemnations of 1277 perhaps give us the reason for this intermediary opinion between Aristotle's and Saint Augustine's.

Etienne Tempier condemned this proposition:

"156 [79]. If heaven stood still, fire would not burn flax, because time would not exist. (*Si caelum staret, ignis in stupam non ageret, quia nec tempus[11] esset.*)" But the bishop of Paris also anathematized this other proposition:

"200 [86]. Time and eternity have no existence in reality but only in the mind. (*Quod aevum et tempus nihil sunt in re, sed solum apprehensione.*)"

The first decree struck at Aristotle; the second touched Saint Augustine. The theory proposed by Duns Scotus had nothing to fear from either of these two condemnations. As we shall see, William of Ockham expounded upon the theory of potential time in the greatest detail.

Time according to Peter Aureol

Peter Aureol appeared as firmly convinced of the existence of the time Duns Scotus called *potential time* as was the Subtile doctor.

Time, considered in itself, before being measured and reduced in number, is a purely successive and continuous quantity. In the same way that dimensions fix the order and establish the continuity between the various parts of any permanent quantity, time is what fixes an order and establishes a continuity between the parts of any successive quantity or any movement whatever. One can say that "time is the succession of movement';'[12] or that "in a formal way, time is nothing more than what has come before (*prius*) and will come after (*posterius*) to which continuity is added."[13]

Properly speaking, time has no parts;[14] it is the formal succession of the parts of movement. Similarly, continuous and permanent magnitude considered in itself has no parts. But that is no longer so when time is submitted to measure; time becomes a determinate quantity—measured time—which is a composite of continuous magnitude and discontinuous or arithmetical magnitude.[15] Moreover, that is the same for permanent magnitudes: a line, taken in itself, is a purely continuous thing; but in a measured line three feet long, the number three, which is a discontinuous magnitude, is implied. What we have just asserted about a line taken in itself and a line three feet long can be repeated about time considered in itself and a duration of three days.

Peter Aureol attaches great importance—with reason—to this discussion; he does not wish that we confuse time, a purely continuous succession without parts, with measured time, which is cut into a certain number of partial durations. He does not wish us to say something of the first which is true only of the second, and vice versa.

Time as continuous succession does not exist outside the mind;[16] that is evident. In fact, time is composed of the past, present, and future; outside the mind the past is no longer, the future is not yet, and of the present there exists only the present instant, the *nunc*; but an instant is neither a time nor a part of time.

> Time consists of something that exists outside the mind and of something that does not exist outside the mind. In fact, whether the intellect considers or does not consider them, the indivisibles of time and of movement exist outside the mind. On the other hand, the past and future, between which the indivisibles establish continuity, have no being if the mind

does not conceive them. Therefore, if one calls something having a certain positive nature (*ratio*) a being existing outside the mind, time and movement are beings only in the mind. The parts of time taken together have no positive nature, except insofar as the mind takes these parts together, conceives them all as in actuality, and concludes from this conception the succession that binds them, that distinguishes what came before from what is coming later. If one understands positive being, or being external to the mind, in this manner, one must say that time and movement are beings only in the mind. They are beings outside the mind only in virtue of being composites of affirmation and negation.[17]

Aureol cites the following Averroes text as support for this doctrine:[18]

Time is composed of past and future; but the past has already stopped being and the future does not yet exist. Time is composed of being and nonbeing. . . .

It is the same for movement; no part of movement is in actuality. Whatever part one designates, it is already distant; therefore, it is also composed of what has already ceased to be and what is not yet.

Such things do not possess a complete existence; these things receive a complete existence from the mind. The mind conceives all the parts and posits them as existing, at the same time that it conceives the indivisible that exists in reality.[19]

Aureol could have cited a number of passages from the eleventh book of the *Confessions* of Saint Augustine along with the passage from Averroes, since the thought of the bishop of Hippo inspired him as much as the thought of the Commentator.

Time receives existence in the soul when it conceives movement and it distinguishes a continuous succession between the various parts of this movement. Does the consideration of these different movements give rise to different times? Aureol's answer is negative.

Each movement will not serve as foundation for a time of particular nature; even if there were an infinity of movements, they would serve as foundation for only a single time. The reason for this is that, even for infinite movements, if there were an infinity of movements, the mind can establish a unique notion (*ratio*) of time; thus it measures all the instantaneous changes (*mutata esse*) by the same present instant (*nunc*), all the movements by a single past and by a single future.

In fact, the mind attributes a single present instant, a single past, and a single future to all the movements it conceives simultaneously.[20]

The present instant (*nunc*) is something absolutely unique in every place for all movements.

At the instant in which I am speaking, in this same present instant, the king of the Tartars sits; therefore there is no present instant for us and another for the Tartars. . . . Even if there existed several heavens, it would be true to say that at the instant I am speaking in this world, surrounded by this heaven, another man speaks in another world, under another heaven.[21]

Once the present instant is unique, time is also unique.

Here is how I prove this conclusion: Time is constituted by the flow of present instants coming one after the other. (*Tempus constituitur per fluxum ipsorum nunc secundum prius et posterius.*) But it is not possible that there is more than one present instant. It is therefore impossible that there is more than one flow, more than one time.

I explain this reasoning with the help of an analogy.

According to what the mathematicians imagine, a flowing point engenders a line, as the present instant engenders time by flowing. But if there can exist only a single point, there can exist only a single flow from this point and, consequently, only a single line.[22]

Thus since the present instant is unique, it can engender a single time by flowing.

This comparison was familiar to Aristotle. He often spoke of the present instant as something that always remains the same along its duration. He also sometimes considered present instants in time differing from each other and succeeding one another. Thus the mathematician sometimes considers a single mobile point on a line that describes the line by flowing, but sometimes he marks on the line some points that are fixed and distinct from one another. Peter Aureol, who willingly places himself in the former camp, as we have just seen, explains his thoughts as follows:

To render all this evident, one must know that time outside the mind does not have the same existence as positive beings outside the mind; in fact the word *time* does not designate anything positive. The only existence that time can have outside the mind is the one establishing continuity between the past and the future; but what is not yet cannot be placed into continuity with what is no longer. Moreover, any continuity between the parts of time stems from the soul which alone continues them one to another.

Aristotle imagines that the soul puts continuity into time in the same fashion as the mathematician imagines the generation of a purely conceived mathematical line by the flowing of a point. . . . Wanting therefore to give the means by which we ought to conceive time, Aristotle takes as example the means by which the mathematician imagines the generation of a line by the flowing of a point. Were I to wish, for example, to render actual a past time, I would imagine that a present instant (*nunc*) flows up to such a marked present (*praesens signatum*); in this fashion I simply imagine that a certain indivisible flows along the changing and successive parts of a movement (*fluens secundum aliud et aliud prius et posterius in motu*), until the final state (*esse*) that the soul wishes to render actual in this movement, until the state where it intends that this movement end. . . . The flow of such an indivisible is time.

Aristotle does not say, as some interpret his writings, that the present instant measures the mobile, but that it follows the mobile. And here is what one should understand by that: In the same way that the mobile engenders movement by its flow, an indivisible, meaning a certain present instant conceived by the mind, engenders time by its flow. This instant follows the mobile in the sense that the mind, conceiving the mobile, conceives the continuity [of the movement of this body]; it finds, in the flow of the mobile, the foundation for a certain indivisible that constantly follows the mobile. Hence, while the mobile flows from one part of its flow to the other, in the same way, the soul imagines that this indivisible flows from one preceding part to one succeeding part of its own flow. When the soul conceives that the mobile is in such a part [which is placed before such another part], the indivisible that it conceives—which is the present instant—is in a region of its flow corresponding to the former part that precedes the one corresponding to the latter part. This is always objective[23] in the mind. (*Et hoc semper in amina objective.*)

Thus, in the same way that the mobile remains always in itself the same mobile, and that what changes in it are only the instantaneous states (*mutata esse*) of its flow, . . . similarly, during the whole duration of its flow, the imagined present instant (*nunc*) remains the same in the mind conceiving it; it differs only in what it is before or after the length of its own flow.[24]

Is that not a very exact description of the way we conceive or imagine time? Is it not true that we consider the present as something indivisible that accompanies a mobile moved of a continuous movement?

Peripatetic philosophy sees in time an attribute of the movement of the first mobile. So far this theory has not been questioned here. Does Peter Aureol reject it? We do not doubt it. The reasons he brings forth to reject the theory are the reasons drawn from the miraculous events reported by the Sacred Scriptures; they are the reasons invoked by Saint Augustine toward the same end.[25]

> As a quantity not measured, time has any movement as support. . . .
> Even if the movement of the first mobile were not to exist, the mind could still capture time in some other particular movement, the movement of a potter's wheel, for example.
> Here is a confirmation of this: The battle directed by Joshua occurred in time. This time, however, was not the movement of the first mobile.
> Every time the mind perceives that there is a changing existence (*apprehendit se in esse transmutato*), it perceives time. Time therefore, considered as a continuous quantity and not as determined by measure, has for foundation any movement whatever; in fact, whoever perceives a movement, perceives successive parts. By taking the parts succeeding each other in this movement, he distinguishes the ones coming before and the ones coming after; and the before and after of movement is time.
> But you might object that the Commentator seems to say the contrary. In fact, the Commentator says that the soul, in perceiving any movement whatever, perceives time, but [it does not grasp this directly], it grasps it indirectly (*per accidens*), since it grasps the movement of heaven.
> I reply that the Commentator here is speaking of time considered as a continuous quantity that has been measured and reduced to number.[26]

This is where Aureol finds a legitimate occasion to use the distinction he established between time taken in itself before any measure, and time measured:

> As measured quantity reduced to number, time is relative to the first movement which is its proper subject.[27]

In fact, the movement of the first mobile

has everything it needs to be a precise measure, to allow the uniform enumeration of the successive parts of the other movements. Therefore, before and after would not play the role of fixed numbers and would not be measured if they did not have as foundation the movement of the first mobile.[28]

The qualities thus rendering the movement of the first mobile alone capable of measuring time are that this movement is circular and uniform, and that all men can perceive it.

> Therefore if you wish to reduce before and after to durations that are fixed exactly, if you wish to reduce them to numbers, it is necessary that you first do this for the movement of heaven, and that you take time as a quantity determined precisely by the movement of heaven.[29]

When William of Ockham began to teach, there were two powerful influences in the order of Saint Francis, that of John Duns Scotus and that of Peter Aureol. And with respect to time, the teaching of Scotus and of Aureol agreed on a number of points; we would not be surprised if these points were to attract the attention of William of Ockham.

Time according to William of Ockham

> Time [says William of Ockham] is the measure of movements whose magnitude is not known to us; in fact, it is by time that we recognize the length (*quamdiu*) of a movement, that a movement moves longer (*diutus*) than another, for we say that the latter moves at greater length when it moves for longer time. Time is also the measure of temporal things; we recognize by time that a permanent thing has a greater duration than another. Finally, in the same way that time measures movements and temporal things, it also measures rest. . . . Those are the principal reasons for which one posits time and for which the knowledge of time is necessary for us.
>
> Moreover the magnitude of a measure must be better known than the magnitude of the measured thing, since we are instructed precisely about the magnitude of the object to be measured by the magnitude of the measure. Therefore, as a result of the preceding, the nature of time, as far as its magnitude is concerned, must be better known than the magnitude of the movements it measures. Therefore, time is not, as some say, something latent and unknowable. Moreover, it is well known to us, and it is not only known to wise men,

but it is known by all who have the use of reason. There is no fool who does not have the knowledge of something temporal, who does not know, for example, which of two bodies measures longer at rest. . . .

If one says that time is not well known, it is because of certain difficulties encountered by those who wish to treat the nature of time using ill-understood texts attributed to certain authors. People who philosophize thus, without understanding manners of speaking and the texts having the authority of the philosophers, have more doubts with respect to time than rustics who use only the common language.[30]

Let us see then how Ockham proceeds to clear the clouds which accumulated around the notion of time. Let us begin with Aristotle's definition.

One must remark that the definition, time is what enumerates the before and after in movement, is not a definition, properly speaking. It is only a definition that expresses the sense of the word (*quid nominis*), as the definition one can give to verbs, conjunctions, adverbs, etc. That is so because time cannot receive any other definition.

In fact, as we have stated several times, the word *time* does not signify something single, distinct in its totality from all permanent things, whose nature or being can be expressed by means of a definition. But we must imagine that this word signifies the first continual and uniform movement, and that it also signifies, or signifies at the same time, the soul that conceives the before and after and what is between the two in this movement. That is to say that this word designates something moving in a continuous and uniform fashion with great speed, and about which the soul says that such a part was first in such a situation and then in such and such other situations. All this is expressed by the word *time*; some of this is expressed directly, and some indirectly—some of it is signified by a verb and some is signified in some way by an adverb.

What the Philosopher expresses by this definition is that the word *time* signifies nothing external to the soul that is not equally signified by the word *movement*. However, besides this, the word signifies or implies the soul that says: the mobile was first here and later there—that says that these two positions are distinct meaning, that the mobile cannot be both here and there at the same time. . . .

Thus the word *time* signifies first and principally what the word *movement* signifies, even though it signifies besides,

not only the soul, but even the act by which the soul knows
the before and after in this movement.[31]

[Ockham also writes,] to say that time is movement comes
to the same as saying that it is by movement that the soul
knows what is the magnitude of any other movement; a
property that cannot belong to a movement without the soul.
It is therefore impossible that a movement be time if it were
not for the soul, in the same way that, without the soul, it
is impossible for time to measure movement.

It is now evident that the operation of the soul must
necessarily have its place in the definition that expresses the
sense of the word (*quid nominis*) *time*. That is why, in
Commentary 88, the Commentator says that time is numbered
in the beings whose actuality is received from the soul; that
is perhaps what he understood when he asserted that . . .
time is one of those hidden beings which would not exist,
except potentially, if the soul did not exist.[32]

We must consider these assertions as literally true: time
is movement and movement is time, for *movement* and *time*
are under the same thing (*pro eodem supponunt*). Yet at the
same time we must, properly speaking, consider this
proposition as true: time is the movement by which the soul
measures how much another movement lasts. We must also
consider this proposition as true: time is movement. Similarly,
here is a true proposition: movement can be without there
being time; but here is a false proposition: time can be without
there being movement. The reason for this is that time implies
(*importat*), besides movement, an act by which the soul
measures in an actual manner; time is, in fact, the movement
by which the soul knows what is the duration of another
movement. It is therefore impossible that a movement be time,
if it were not for the soul. Therefore, in the same way that
the proposition, movement can be without there being a soul,
is true, the proposition, movement can be without there being
time, is true.

And similarly, the proposition, time can be without there
being a soul, as long as one takes the subject [time] for what
it is [in reality], is true. In fact, the movement that is time
can be without there being a soul. This proposition is equally
true: without the soul, time can be movement, as long as one
takes the subject for what it is; in fact, it is equivalent to
the phrase: without the soul, a movement can be a movement.
But this proposition is false: without the soul time can be
time, for although the subject does not have its form as
appellation, the predicate does. Therefore, even though

without the soul time can be movement, time cannot be time
without the soul.[33]

We have here a remarkable example of the subtile precision of
Ockham's logic.

The considerations that the Venerable Inceptor develops with
respect to the instant are similar to those suggested to him by time.

In order to understand what the Philosopher has said about
the instant, he writes, one must know that the instant or the now
(nunc) does not signify anything permanent, so that anything that
can be imagined as existing outside the soul signified by the word
instant or the adverb now is something permanent that can last
some time.

> The instant is not, as the moderns admit, something that
> passes suddenly (raptim) and is distinct from all permanent
> reality in its totality. . . .[34]
>
> The instant is not something that ceases all of a sudden
> (statim) to exist in natural reality; but the instant is nothing
> other than the first mobile itself. . . .
>
> The instant is nothing other than the first mobile whose
> parts exist somewhere where they were not immediately before,
> such that the word instant expresses simply the first mobile
> existing in a place where it was not immediately before and
> where it will not be immediately after.[35]
>
> Thus one sees clearly how it can be that one assigns a
> first now and then a second now posterior to the first. That
> is because one says first, now this part of the mobile is in
> this situation, and one then says, now this part of the mobile
> is in that situation, and later it will be true to say, now it
> is in another situation, and so forth. Thus it is evident that
> now does not signify something distinct, but that the word
> now always signifies the first mobile which always remains
> identical to itself. . . .
>
> We therefore assign a prior present instant (nunc) and
> a posterior present instant; in other words, we first say, this
> present instant is in A and is not in B, and then it will be
> true to say, this present instant is in B and is not in A. Therefore
> it happens that contradictory affirmations are successively
> true.[36]

[In the Système du monde, vol. VII, chap. 4, sec. 4,] we have seen
that Gregory of Rimini took up, with respect to time and movement,
a good portion of what William of Ockham has just asserted; the
theory of the Augustinian philosopher is to be distinguished from

the theory of the Franciscan philosopher, however, in that it insists less than the latter on the attribution of the action of the mind to a definition of time. Gregory, one might say, limits himself to stating that time is a clock; Ockham, on the other hand, formulated the proposition that time is a clock and an intelligence that uses the movement of this clock in order to measure the other movements.

Time according to William of Ockham (*continued*):
The Absolute Clock

This difference can be noticed better in what Ockham says about the choice of the clock, meaning about the choice of the movement that will serve to measure the other movements; the movement that, by definition, will be time.

What is needed for a movement to serve as the measure of the duration of other movements, in order for it to be time, for it to serve as a clock? According to Gregory of Rimini, what is needed, and what is sufficient, is that it be continuous and regular—let us call this uniform, following the manner of speaking now common. Gregory adds no commentary to this assertion; however, it requires some elucidation.

What is signified by the phrase, such a movement is uniform? One can give it two entirely different senses, depending upon the philosophical school to which one belongs.

In its first sense it signifies that, independently of any intervention of the human mind, there are, in nature, movements that are uniform or, at least, a movement that is uniform; there are also movements that are not regular. The role of the human mind wholly consists in seeking among the observable movements the one or ones that are uniform; it then takes up the one (or one of them) in order to define time. There is a clock or some clocks imposed upon man by nature.

But the phrase, such movement is uniform, can take another sense. One can admit that it does not signify anything before the intervention of the human mind. Man is free to choose as he wishes the continuous movement that will serve him to define time, the movement that will become his clock. This movement will therefore be uniform by convention, and it will be the same for all movements regulated by the observer, by means of the first movement. In this second way, nature does not impose upon man the clock he must use; the choice of a clock is the result of an arbitrary convention in which one considers only reasons of fitness and convenience.

It seems that all the ancient philosophers embraced the first view of time.

The Pythagoreans and later the Neoplatonists all admitted, under various forms, the existence of a time transcending the world of bodies, by which all movements and all changes of visible nature could be measured. According to them, a uniform movement was any movement that accomplished equal distances in durations marked as equal by the divine clock. Since they are divine beings, the celestial bodies can only accomplish uniform rotations. Each one of them, and in particular the orb of the fixed stars, is therefore a visible clock regulated exactly by the clock that marks the perfect time in the world of ideas.

Aristotle approached his theory of time in such a way that one could have arrived at the conclusion that the choice of the clock is arbitrary without any great shock: it seems that any movement in which one can enumerate the successive states of a mobile is suitable for defining a time. But Aristotle seemed to have feared that such a consequence might be derived from his teaching; he attempted to affirm the existence of a unique time, one that is the same in all places, on land and on sea, that would be the same in other worlds if, *per impossibile*, there existed other worlds. He seemed not to rest until he had rejoined the Pythagorean doctrines by means of the proposition that the clock that defines the unique and true time is the sphere of the fixed stars.

William of Ockham's writing on the subject of time awakens in us a similar impression as Aristotle's writing had awakened in us. Like the Philosopher, the Venerable Inceptor sometimes seems ready to admit that man could have taken as clock any movement of regular appearance.

Here is, for example, the objection that he formulates and that he resolves in his *Questions on the Physics* against the assertion, time is movement:

> Here a doubt presents itself. In order for a quantity to be the same as another quantity, the parts of the first have to have the same magnitude and the same relation with the parts of the other—in a word, that they be the same. Therefore if movement and time were identical, the parts of time would be equal and identical with the parts of movement, as it has just been stated. But that is false. In fact, if the movement of heaven were twice as fast as it is, heaven would accomplish in an hour the movement that it now accomplishes in two hours. The parts of this movement would have the same

magnitude they now have; they would therefore no longer be equal to the parts of time, since the movement that is now accomplished in two hours would be accomplished in an hour, meaning in half the time. Hence, if time and movement were identical, the movement of two hours would be identical with an hour; the part would be equal and identical to the whole.

I reply to this that time is identical to movement, and that the parts of time are identical with the parts of movement—regardless of whether time is a fast movement or a slow movement. . . .

If the movement of heaven were to become twice as fast, it would be accomplished in half the time it now takes it, since a regular and uniform movement here below is the time and measure of the movement of heaven (*quia aliquis motus inferior, qui est regularis et uniformis, est tempus et mensura motus caeli*); therefore if the movement of heaven became twice as fast as it is now, it would coexist with a twice as small[37] succession of regular and uniform movement by which the movement of heaven is measured. And if the movement of heaven became twice as slow as now, it would coexist with a twice as large succession [of this same regular and uniform movement]. Thus a movement of heaven that is twice as fast would be accomplished in half the time as now. But this does not result in a part being equal to the whole; in fact, the part of the uniform movement considered, that measures the faster movement of heaven, is less than the total succession [of this uniform movement] and also less than the part of this succession by which the movement of heaven is measured, as it is now. It is twice less than it; the part is therefore neither identical nor equal to the whole.[38]

The above passage seems to affirm clearly that there is here below a well-determined, regular and uniform movement, which is true time, and by which the movement of heaven itself is measured (*aliquis motus inferior, qui est regularis et uniformis, est tempus et mensura motus caeli*).

But reading Ockham's reply to the question, "Is there a movement here below which is time?"[39] would impart an altogether different impression.

I reply to this question that one can call time any movement here below, the knowledge of which allows us to have the knowledge of some celestial movement that was unknown to us at first.

A multitude of experiences render this conclusion evident. Thus when the movement of a clock is known, we can, using this movement, measure the movement of the sun and our own doings, especially when we have recognized that the movement of this clock is uniform and regular. He who owns a clock knows how far the sun has traveled on its course, even if the sun is hidden by clouds; but he who makes a clock also measures the movement of the clock by means of the movement of the sun. He who makes a clock can order time in the following way: while the first mobile accomplishes the diurnal movement, the clock moves on its dial. In this way he measures the movement of the clock by means of the diurnal movement. Once the movement of the clock is known, it can then be used to measure the other movements, such as the diurnal movement and the sun's movement; consequently, each of these movements can be called time with respect to each of the others.

Similarly, someone who knows the magnitude of a motive force and who knows what space it has traversed, can, without considering the movement of heaven, know what fraction of its movement has been achieved by heaven and how much time has flowed. If one knows, for example, how far a horse can go in a day—let that be thirty miles—and if a horseman has gone thirty miles on this horse, by evaluating the length of the road traversed by his steed, he can know how much time has flowed; he would not know it if he did not know the length of the road traversed by the horse.

Furthermore, workers often draw the knowledge of what time it is from the knowledge of the work accomplished.

It often happens that one measures the movement of heaven by means of the space traversed by a mobile; if one knows the space traversed by a mobile, one can know what has been the duration of the movement and, consequently, one can know that the time, the measure of the movement, has been long.

Therefore, one sees manifestly that every movement here below by which one measures the movement of heaven is a time; in fact, to measure something by another is simply, by knowing the magnitude of the first, to gain knowledge of the magnitude of the second. . . .

However, the movement of the wandering stars, taken as measure of the other movements and operations, is more properly called time than the movement of anything here below; and that is because the movements of the wandering stars are more uniform, faster, and less able to be impeded than the movements here below. But the movement that is most properly called time is the diurnal movement; in fact, it is the fastest and most uniform of all the movements. It is therefore by means of this movement that the soul can measure the other movements with the greatest certainty.[40]

At first reading this passage suggests the thought that any movement whatever, whether it is accomplished on earth or in heaven, can be taken as the measure of other movements, and therefore be called time. But if we look at it carefully, we see that before taking any movement as measure of other movements, Ockham assumes that we have measured it, recognized its uniformity and regularity, and regulated the clock we are using; that is to admit that there exists a normal movement, imposed in advance, serving to regulate the others, which alone is truly time.

Like Aristotle, William of Ockham believes that the choice of movement serving to define time, to measure other movements, is imposed upon man and is not arbitrary at all. But he presents what he says to explain and justify his opinion as complementing and illustrating what Averroes wrote on this subject. Let us first read the considerations by which Averroes thought to establish the conclusion that the movement of the first mobile is the only movement that allows us to define time and the only one that can serve as clock.

The Commentator attached some value to the system that he was able to construct: "This question [he said] remained a long time [with me] before becoming clear to me; in everything else that I have written about time I followed the commentators, but here I do not."[41]

Averroes first posits as fact that the perception of time can result in us from the knowledge of changes within our soul alone, without our senses revealing any external movement.

> To perceive time [he said] is not to perceive some movement seized by one of our senses, for we have the sensation of time even when we find ourselves in obscurity and no movement reaches our senses; therefore we have this sensation only because we sense any movement whatever in our soul. In fact, as soon as we imagine any movement in our soul, we capture the notion of time.
>
> But there is a difficulty, and not such a small difficulty, in what Aristotle says.
>
> If time is not the consequence of a movement existing outside the soul, if it is a consequence of our imagination, since the imagination does not exist outside the soul, the movement of which time is the consequence also has no

existence outside the soul. How then can Aristotle tell us that time results from celestial movement? . . . If time results only from celestial movement, does it not follow that the blind man would not perceive time, since he has never perceived a celestial movement? Moreover, if time results from any movement whatever, there would be as many times as there are movements, which is impossible.

Evidently then, either time has no existence outside the soul, or else, if it exists outside the soul, it results from every movement, and times are then multiple, as are movements, or else, it is the consequence of a single movement and whatever does not perceive this movement has no sensation of time. All this is impossible.[42]

Averroes sees only one way to escape these difficulties, and it is as follows: it is true that every movement gives us the sensation of time; but it does not give it to us directly and essentially, for there would then be as many times as there are real or even imaginable movements.

The perception of time does not therefore result directly and essentially from the sensation of any movement; it only forms in us in connection with the sensation: "When we perceive any movement, we also perceive the unique movement for which time is an accident."[43] If this were true, the difficulties besetting us would disappear. Let us therefore detail this response.

When we perceive any movement, whether it is external or internal to our soul, we sense that we exist in a mobile fashion, subject to change (*sentimus nos esse in esse moto et transmutabili*), an existence rendered divisible by continual transformation. It is in this way, and only in this way, that we perceive time. If we existed in an existence which the absence of transformation rendered indivisible, this existence would constitute an instant, and there would be no time for us. It is therefore necessary, in order for us to have the sensation of time, that we exist in a changing and mobile existence, and that we perceive that our existence is of this kind.

> Moreover, we perceive directly that we exist in an existence subject to change; we sense, for example, that the instant in which this discourse began is distinct from the instant in which this discourse ended.
>
> Therefore, the movement which, when it is perceived, gives us directly and essentially the sensation of time, is the movement by which we sense that we exist in an existence

capable of change, by which we sense that we change because we are given such an existence.

Thus, as we have said, every time we perceive any movement whatever, we sense that we exist in an existence subject to change; and sensing that we exist in a changing existence is what the sensation of time flows from, for us.[44]

But since Averroes links the sensation of time only to the internal perception of our changing existence, how can he reconcile his thought with Aristotle's thought that makes time the property of the movement of the supreme sphere? We shall see how he does this; the lines we have just cited are followed immediately by these lines:

It is manifest that if we sense that we exist in an existence subject to change, it is only due to the movement of heaven (*manifestum est quod nos non sentimus esse in esse transmutabili nisi ex transmutatione caeli*). If it were possible that heaven stopped, it would also be possible that we existed in an existence incapable of change; but that is impossible. It is therefore necessary that even the one who does not perceive by sight the movement of the celestial body has, nevertheless, the sensation of this movement.[45]

This passage would appear extremely strange and perhaps extremely obscure if we did not have the whole teaching of Aristotle and Averroes to explain and illuminate it. Let us recall Aristotle's assertion in the *Meteorology*, that the whole sublunar world is subordinate to the celestial revolutions; let us also recall that in the *Sermo de substantia orbis* Averroes clearly declares that all the movements of the elementary sphere have the celestial movements as causes, and that if the latter were to stop, the former would also stop. The above passage becomes clearer because of these teachings. The life of our soul, as all the changes here below, flows following a rhythm whose beats are measured by celestial revolutions; to perceive this changing life is therefore to perceive the movement that regulates and governs all the others, the revolution of the first mobile.

Thus Averroes draws the solution of the problem of time from the essential principle of Peripatetic astrology.

That is the theory William of Ockham took up but separated from the astrological principle which is Averroes's real support. Ockham touches upon this theory in his *Summulae in libros Physicorum*,[46] but he exhibits it in greater detail in his *Tractatus de successivis.*[47] These two works are the primary sources for our knowledge of his theory.

The following is how the Venerable Inceptor understands Averroes' theory:

> The Commentator means that anyone who senses any movement can sense time or, in other words, can understand that time exists—but not directly and essentially. In fact, here is the process followed for this.
>
> A man sees a celestial body move or perceives some external movement or even imagines a movement. That done, he can imagine that he coexists with some body moved by continuous and uniform movement (*potest imaginari se coexistere alicui uniformiter et continue moto*); consequently, he can understand this proposition: I coexist with a body moved by a continuous and uniform movement. But by understanding this, he understands something proper to the celestial body, for the celestial body moves continuously and uniformly. Consequently, even though the movement of the celestial body is not perceived by any senses, it nevertheless is understood by the intellect, not by a particular and simple concept, but by a composite concept.[48]

A passage of the *Summulae* develops what we ought to understand by that.

> It is sometimes possible that [one has this kind of composite concept even though] one does not know what the concept is proper to. It is even possible that one has such a concept even though one does not know whether the concept is proper to anything. Thus the concept of a uniform and most rapid movement is suitable for the first movement; but one can have this concept even though one does not know which movement is this uniform and most rapid movement. Similarly, it would be possible to have this concept even when one doubts the existence of such a movement or if one denies it.[49]

This uniform, continuous, and most rapid movement that is purely conceived is the clock by which we compare the various durations among one another.

> He who wishes to determine how much something moves, remains at rest, or lasts, forms this concept naturally and at once; he says in his mind that if there existed absolutely a perfectly uniform and most rapid movement (*si simpliciter esset motus velocissimus et uniformissimus*), the moving thing

would move faster or slower than it. And thus, generally, every time the intellect wishes to measure something by time, it forms the concept naturally. That is how anyone who perceives time conceives a movement, and not only does he conceive a movement, but he conceives the first movement, meaning that he possesses a concept proper to the first movement, namely, the concept of a uniform and most rapid movement.[50]

But we do not immediately know that this concept of uniform movement finds its realization in the first movement—Ockham just asserts this. We could have this concept without knowing that it is realized, while denying that it is realized by any external movement. It is therefore observation, and observation alone, that teaches us that the diurnal movement realizes this continuous and uniform movement whose idea we have just conceived. That is what the *Tractatus de successivis* takes care to affirm.

> Once this conception is formed in our soul, we may form a subsequent perception by which we could know that such body moves continuously and uniformly, meaning we may form a perception from the only movement by which our soul can measure the other discontinuous or non-uniform movements . . . and the soul recognizes thus that time is not suitable except for the movement of heaven, for there does not exist any other body moving uniformly by which the soul may be able to measure the other movements perfectly and certainly.[51]

This theory that began in the *Summulae* and was developed in the *Tractatus de successivis* was taken up in great detail in the *Quaestiones super librum Physicorum*; in order to illustrate it fully, we should cite some passages from the *Quaestiones*.

> Does the Commentator mean that anyone who perceives time perceives the movement of heaven? . . .
>
> With respect to this question, here is our first conclusion: according to the intention expressed by the Commentator in the commentary that we have invoked, it is not necessary for us to perceive that we are really changing in order for us to have the perception of time; it is not necessary that our bodies or our minds be moving. . . .
>
> It therefore appears that we can perceive time without our soul moving, and also without our body moving—without our being moved in any way; the reason for this is, according to the Commentator, that we can perceive that we coexist with a mobile moved continuously and uniformly. The soul can

measure other, non-uniform movements by the movement of this mobile. Moreover, to perceive our coexistence with this mobile is to perceive time; this perception therefore does not require that we be moving either with our body or with our soul. . . .

Second conclusion: we can perceive time without grasping the diurnal movement, either by perception or by a simple concept. . . .

Third conclusion: he who perceives time does not necessarily conceive the movement of heaven by a composite concept proper to the movement of heaven; he can, for example, perceive the movement of a clock and, by this movement, measure other movements without forming any concept [simple or composite] proper to the movement of heaven.

Fourth conclusion: whoever perceives time can, according to the Commentator, grasp the movement of heaven in a [composite] concept which is proper to this movement. Here is the reason for this: Without seeing the movement of any celestial body, a man can perceive the movement of some external body, or even imagine a movement. That done, he can imagine that he coexists with a certain mobile moving continuously and uniformly; consequently, he can understand this proposition: I coexist with a certain mobile moving continuously and uniformly. Once he understands the proposition, he grasps a composite concept which is proper to the celestial body in movement, namely, [the concept of] this uniform, regular, and very fast movement; in fact, no movement is suitable for this except for the movement of heaven. And according to what the Philosopher intends, no other movement is suitable, even though God can make a faster movement, and perhaps a more regular movement, than the diurnal movement of heaven. Thus one sees that even when it is not perceived by any senses, the movement of the celestial body can be grasped by the intellect, doubtless not by means of a particular simple concept, but by means of a composite concept proper to it. That done, a perception can then arise in the soul, a perception by which the soul recognizes that a certain body moves continuously and uniformly; it can then measure the other movements that are not uniform and regular by the movement of this body. Hence that is what is to perceive time essentially, for time is a regular and uniform measure. . . .

Thus one sees how a man who does not see heaven can perceive the movement of heaven, once he perceives himself as existing in an existence subject to change (*se esse in esse transmutabili*), meaning once he perceives his own coexistence with a mobile moving uniformly and continuously, or once he grasps the proposition, I coexist with a certain body moving continuously and uniformly.

Second, as has just been stated, when we perceive that we exist in an existence subject to change, we[52] perceive time essentially, for we perceive then that something moves continually and uniformly—and this is to perceive time essentially.

Third, . . . since there is no body other than heaven that moves in a regular and uniform movement, we can understand that we exist in an existence subject to change without grasping the movement of heaven, at least by a compound concept.

Fourth, we see how time belongs to the movement of heaven accidentally (*accidit*); in fact, there does not exist outside heaven any body moving uniformly by which the soul can measure the other movements by a method that is certain.

Moreover, it is certain that the proposition, heaven moves, is not known by every man who perceives time as existing in things perpetually. A man can grasp the movement of heaven in a compound concept and yet be unaware of the proposition, heaven moves. . . . Someone blind from birth does not know the proposition, heaven moves, for he has never seen the movement of heaven; he can, however, grasp the movement of heaven by a composite concept. It suffices for him to grasp the proposition, I coexist with a certain body moving continuously and uniformly; in fact, that is a concept proper to the movement of heaven.[53]

We can now understand what Ockham meant when he said, "If heaven moved twice as fast, its movement would be accomplished in half as much time as now, for there is here below a regular and uniform movement which is time and the measure of the movement of heaven (*quia aliquis motus inferior, qui est regularis et uniformis, est tempus et mensura motus caeli*)."[54] The regular and uniform movement that allows one to recognize a change of speed in heaven is the movement by which our consciousness tells us that we coexist, once we perceive the changing character of our existence; this movement perceived by our consciousness is the standard clock by which we recognize that the diurnal movement is uniform, and by which we could appreciate the changes in speed of the diurnal movement, if it lost its regularity.

This uniform movement about which we have a consciousness of coexisting, this standard clock, is a concept; Ockham has repeated this many times. The composite concept, he adds, is proper to the diurnal movement because the first mobile is the only body that moves by a uniform movement. But is this uniformity of the diurnal movement necessary for our author, as it was for Aristotle? Assuredly not; Ockham discusses what would happen if heaven moved twice as fast or half as fast, and he even admits the assumption that the diurnal movement might not be completely regular, that God can make a movement that would be more regular. Moreover, that the diurnal movement is regulated exactly on the standard clock about which we have consciousness, and that this clock, this composite concept, is a concept proper to the rotation of the first mobile, are not necessary propositions; for Ockham they are truths of fact. It would not be absurd to suppose that any body external to our intellect can realize this uniform movement that our mind conceives; we would then have another standard clock, which is purely conceived, by which one could measure the duration of various movements.

We should understand in a similar fashion Ockham's reply to the question, if there were several first mobiles, hence several first movements, would there be several times?

> If there were several equally first heavens and several first movements, there would, in reality, be several times; but all these times would be a single time by equivalence (*per equivalentiam*), meaning that these multiple times would make up a single time for measuring.[55]

The locution, *by equivalence*, that Ockham also used in his theory of place, following Scotus, indicates that we should relate Ockham's theory of place with his theory of time; this indication is reinforced in the *Summulae*, since there Ockham borrows an example from his theory of place in order to explain the words, *per equivalentiam*. This relation [between Ockham's theory of place and his theory of time] clears up any obscurity from what we have just cited. Because of it, Ockham's thought becomes clear to us. In the same way that he relates all local movements to an absolutely fixed reference conceived by our reason, he measures time with a purely ideal clock that our mind constructs as soon as it perceives any movement or change whatever. It is by relating time (the time Duns Scotus called *potential time*) to this ideal clock, that man recognizes the uniformity of the diurnal movement, and that he observes or constructs visible clocks.

· But how does man achieve the comparison between the movement of the supreme sphere and the indications of this purely conceived clock? And how is the agreement between the clock our mind can conceive as soon as it has perceived some movement and the corporeal clock located in heaven to be explained? Ockham gives no answers to these embarrassing questions; the Venerable Inceptor seems not to have even thought to pose them.

The Analogy between Time and Place; Franciscus de Marchia

A Franciscan, a contemporary of William of Ockham who sided with the religious opinions of Ockham and Michael of Cesene, clearly affirmed the principle that directed Ockham's theory of time implicitly. According to Franciscus de Marchia, there is the closest analogy between the theories of time and place.

Franciscus de Marchia asks himself, in his *Questions on the Sentences*, "whether time differs from movement."[56] The reply he gives does not differ essentially from the one given by Ockham. The concept of time not only implies the concept of movement, it implies at the same time a certain relation of movement, called time, with another movement measured by the first, so that what is called time is not any movement whatever; it is a certain uniform movement considered as the measure of other movements. Franciscus de Marchia compares time with place in order to clarify this thought.

How do place and the lodged body behave with respect to one another? A place is not a permanent volume (*quantitas*) considered absolutely; it is a quantity playing the role of lodger. Similarly, a thing is not said to be lodged unless it plays the role of contained body. The term *place* expresses, not only the idea of volume, but also a relation of container to contained body; and the terms *lodged body* express in a similar way a certain relation of the body to its place. . . .

I say as much about movement and time. In fact, in the same way that we can consider a greater quantity that contains and a smaller quantity that is contained among the permanent quantities, we can give a quantity that contains and measures and another quantity that is contained and is measured among the successive quantities.

Therefore, in the same way that we do not call any volume "place," but only a greater volume considered in relation to the smaller volume it contains, we do not call any movement "time," but precisely the movement that is uniform and the measure of all others. That is also why time is called the number of movement, whether this enumeration takes place in the reality external to the soul or whether it takes place only within the soul.

Therefore this is what I assert: When we call a permanent volume "place," we do not consider it in relation to itself, but in relation to the other volumes it contains. Similarly, when we call the movement of the first mobile "time," we do not consider it in relation to itself; we consider it in relation to the other movements, in relation to the inferior movements it measures.

Considered with respect to their foundation [composed of two bodies], the containing body (*locus*) and the contained body (*locatum*) are of the same kind; but formally, from the point of view of the relation each presents to the other, they have differing reasons. I say the same for time and movement.[57]

After having set out this doctrine so clearly, Franciscus de Marchia attempts to resolve a classical difficulty that the Scholastics pitted against the proposition, time is a movement. He formulated the objection as follows:

Things that have distinct and opposed properties (*passiones*) are really distinct. But this is true for time and movement, for one says of movement that it is fast or slow, and one does not say that of time.[58]

Franciscus de Marchia's reply is brief, but we believe that it is worth emphasizing.

Time is neither fast nor slow because the movement that is time itself, meaning the movement of the first mobile, is itself neither fast nor slow; *it is regular and uniform, although it can be faster or slower.* In fact, this movement is called time because it measures the other movements; therefore, it is not said to be either fast or slow as long as it plays the role of time; as for the inferior movements, since they do not act as measure, they are said to be faster or slower. (*Dico quod tempus non est velox nec tardum*[59] *quia nec motus qui est ipsum tempus est velox et tardus, scilicet qui est motus primi mobilis; sed est regularis et uniformis, licet posset esse velocior et tardior; quia*[60] *hujusmodi motus dicitur tempus ut est mensurans alios motus inferiores; ideo, ut habet rationem temporis, non dicitur velox vel tardus; alii autem motus inferiores, quia sic non dicunt rationem mensure, ideo dicuntur tardiores vel velociores.*)[61]

Can we be mistaken about the meaning of the above passage? It seems extremely clear, and it appears to assert the following: if we use the diurnal movement of the first mobile to mark time, meaning to measure the other movements, it is not because we know by other means that this movement is regular and uniform. We do not know anything of the sort—we are not even warm; the movement could sometimes be faster or slower and we would still be able to use it as our clock. And once it is used as the clock, it is regular and uniform *by definition*. We can no longer say of it that it is fast or slow; these words have no meaning for the movement that serves as the measure of the other movements. The words *fast* and *slow* have a meaning only when they are used to talk about movements that do not play the role of measure.

If that is the meaning of the passage we have just cited—and we do not think that any other meaning can be attributed to it— Franciscus de Marchia announced formally what William of Ockham perhaps suspected but shied away from.

Is the Absolute Clock Arbitrarily Chosen? Walter Burley, John Buridan, Albert of Saxony, and Marsilius of Inghen

William of Ockham was not the only philosopher of the fourteenth century who thought that man was not free to decree the uniformity of a movement arbitrarily chosen and to decide that this movement will be his clock and will serve to define time for him.

Peter Aureol, for example, taught in his second book of his *Commentary on the Sentences* that time is the gliding, the flow of a certain present instant, unique and identical to itself; he could not therefore admit that time is not something unique and determined, that there are as many distinct times as there are movements, and that man is free to adopt the time that suits him best.

There was nothing arbitrary in time as conceived by Francis of Mayronnes; time, he said, is a relation, "but it cannot be a relation with respect to any creature, for even if there existed only a single creature, there would be a before and an after. I therefore hold that time is the fluxion of a presentness (*praesentialitas*) with respect to God, in the same way that I have asserted of place"[62] that it is a certain presentness with respect to God.

The thought that one could attribute to each movement a time distinct from one movement to another is manifest in the *Questions* of Joannes Canonicus. He distinguishes two kinds of times: a common and general time, which is unique; and particular and specific times, which are multiple. The common and general time, extrinsic measure of all movements, is the first movement that regulates all the others—the movement of the first mobile. As for the particular and specific times, there are as many as there are movements. There corresponds a succession of proper states, of *before* and *after*, to each movement; hence, to each movement there also corresponds a successive being that is particular to it and a time that is just as particular to it.

We are certain, in any case, that there corresponds a time to every movement, because of the following reasoning:

> There cannot be a movement unless there exists positively a time which is its measure; this is evident to any philosopher, for any movement is measured by time. But if the first mobile stopped in such a way that the first time stopped and ceased to be (for it ends with the end of the movement to which it corresponds), some movement here below could still be produced; this movement would have to be measured by a time, but this could not be accomplished by the first time since by hypothesis the first time is no longer. Therefore, it would have to be measured by a time that is proper to it.[63]

Averroes would deny the minor premise that if the first mobile stopped, a movement here below can still be produced. However, Joannes Canonicus proves it by the Scriptures and by reason: "First, by the Scriptures: one sees clearly and literally that, in Joshua, the potter's wheel continues to turn even when heaven is stopped. Then, by reason: God can produce either one or another of two things that are essentially different and that do not depend on each other; and these two movements are this way."[64] It is true that Averroes would have rejected these reasons. There is an antagonism in the theory of time (as there is in every theory) between Catholic doctrine and Peripatetic philosophy. Saint Augustine clearly recognized that with respect to time this antagonism is unreconcilable. What Joannes Canonicus invokes are the examples that the bishop of Hippo himself cited.

The idea that each movement can be taken, if one wishes it, in order to mark the time by which one measures the other movements is barely perceptible in Joannes Canonicus's teaching; it is more clearly declared in Walter Burley's doctrine:

There are four ways of conceiving time: the common way, the proper way, the more proper way, and finally the most proper way.

In the common way, time is a duration having a before and after; this is common to the durations of all movements: the duration of local movement, the duration of movements of alteration, and the duration of movements of dilation or contraction.

Properly, the word *time* is taken for the duration of local movement; the role of time and of the movement from which time results is to measure the other movements. Moreover, the conditions suitable for the movement from which time results, which cause this movement to measure other movements, are more easily encountered in local movement than in other movements; in fact, as the Commentator states toward the end of his treatise on time, the conditions that the movement (from which time results directly) must fulfill are that it be perceived by our senses, that it lends itself to measure more manifestly than the other movements, and that we can recognize its uniformity more readily . . . ; and local movement better fulfills these conditions of being more easily perceived by our senses than the movements of alteration, dilation, and contraction; besides, its uniformity is more easily recognized by us than the uniformity of the other movements.

More properly, one takes as time the duration of a local circular movement accomplished around the center of the world; in fact, by means of the duration of such a movement, the magnitudes of the movements here below can, in the natural course of things, be determined better by the human intellect than by means of any other movement. In fact, the duration of other movements is not the same for all the inhabitants of earth; moreover, the variability of these movements can neither be perceived by the senses nor be known by reason, while we can know it for the circular movement of a celestial body. . . .

Finally, most properly, one takes the duration of the movement of the first mobile as time, because the movement of the first mobile is the first of all movements and the one that is most uniform of all; and those are the essential conditions of the movement from which time results directly. Moreover, the movement of the first mobile is the fastest movement, so that it best plays the role of measure.[65]

The dominant thought in Burley's mind is evidently the following: man can choose, out of a great number of choices, the movement that serves to define time for him. Only one principle should guide his choice, that it is useful to take the movement best suited for measuring others. But when he enumerates the characteristics by which one recognizes that a movement can play the role of clock, he takes care to cite uniformity. He does not seem to admit that the movement chosen to define time will be uniform by definition; he seems to believe that before having chosen the movement that will furnish a clock, we are in a state of examining and deciding whether the movement is or is not uniform.

This thought, which causes problems in Burley's theory, is, as the words of its author teach us, suggested by Averroes's discourse. Thus it deflects Burley from the opinion that appeared to be his own in order to bring him, if not to the Commentator's doctrine, at least to William of Ockham's doctrine. Is not William of Ockham's doctrine what we recognize in the following passage?

> I assert that in perceiving any movement whatever, we perceive the first movement in a confused way; in fact, we perceive that there is a simple and uniform movement which is the measure of the movement we are perceiving. But whether this simple and uniform movement is the movement of heaven or some other movement, we do not perceive. Thus when we perceive any movement whatever, we perceive the first time in some way; in perceiving any movement whatever, we perceive the first movement in a confused way; moreover, in perceiving any movement whatever, we perceive the particular time that results from that movement.[66]

John Buridan's phrasing has great similarities with Burley's phrasing with respect to this matter. Buridan asserts:

> Most properly, time is the first movement; in fact, the role of time is to be the measure of movements. The movement therefore is most properly called time which is the measure, most properly speaking, of the other movements. In every kind, it is reasonable that the first is the measure of the others, more than the opposite would be. In fact, we know the measured thing through measure; the knowledge we have of things that follow the first thing is the most perfect knowledge, most properly speaking, than the one following the inverse order. That is why the science of *propter quid* is superior to the science of *quia*. Moreover, the measure must be regular, and the first movement is the most regular in its succession; the first mobile moves neither faster nor slower today than yesterday. . . . Finally, that is manifest in the case of

astronomers, who, when measuring movements, have recourse to the first movement as the first, most properly speaking, measure of the other movements. . . .

Therefore, if we take the word *time* in the most proper way, we should say that time is the first movement; but only the astronomers, not the common man, make use of this manner of speaking. It is not because of sensible awareness but because of intellectual reasoning that they make use of this movement instead of time in their calculations, when they wish to know the situation that the stars occupy either with respect to other stars or with respect to us. As for other men, although they see this movement, they make use of other movements, known by the senses or the imagination, as time.

For the common man, the movement of the sun, composed of the proper movement of this stellar body and the diurnal movement, is called time more properly speaking than the other movements. . . . This latter movement, in fact, is the movement best known by common men, for it is the most apparent to the senses; they do not know the simple diurnal movement separated from proper movement, so they cannot make use of it for measuring. . . .

Sometimes workers in the mechanical arts use their work as the movement that defines time, because, since the magnitude of the work they are accomplishing is well known to them by force of habit, they often measure other movements by means of this work. Even when they cannot see the sun, from the quantity of work they have accomplished, they conclude that it is three o'clock and it is time to eat.

Ecclesiastics make use of a clock with respect to time; but this is not time, properly speaking, for the movement of the clock must have first been measured by means of the movement of the sun.[67]

John Buridan's thoughts were taken up by Albert of Saxony. When enumerating the characteristics that a measure should have, Albert takes care to declare that it should be "invariable." The movement of heaven is time *"principalissime"* because it has all the characteristics of a good measure and, in particular, "because it is regular."[68] Moreover, our author remarks, as did Buridan, that "for the common man, the movement of the sun is called time, because that is the movement he knows best."[69] He adds that

some call the movement of the moon time, for they distinguish time by lunar years, so that time is for them the movement of the moon. . . . From all this we can conclude that time is said to be *principalissime* only for the local diurnal movement; but in a less *principale* manner, we can call time any movement whatever that we use to measure the other movements—the movement of a clock, for example.[70]

Marsilius of Inghen, attempting to improve upon what his teachers stated, asserts in his *Abbreviationes libri physicorum* that time, most properly speaking, is the movement of the sun, because it is the movement which is known best. He does address himself to the objection that the movement of the sun is "less regular than the diurnal movement, because of the obliqueness of the zodiac and the eccentricity of the solar sphere,"[71] but he replies that "the irregularity of the solar movement is of little importance, for the natural days are all almost equal among themselves, without any sensible difference; therefore this difference does not prevent them from serving as measure, for the measure of movement by time is not a perfectly exact measure, but only a proximate measure of which the difference is not sensed."[72]

Among the *Questions on the Physics* that Marsilius of Inghen composed *according to the method of the Nominalists*, there is one thus formulated: "Is time the movement of heaven?"[73]

This question is evidently inspired by the corresponding question of Albert of Saxony. Marsilius declares that

any movement able to be known to man is a time. . . . If one asked, for example, how much time the lesson lasted, one might reply, the time to accomplish two miles; in this way a movement of progression would be a measure of the duration for him. Someone else could reply: the time to bake bread in an oven; he then would be judging the length of this successive duration by means of a movement of alteration. Others determine the length of a successive duration by means of certain consequences of movements—sound for example. If one asks of them, for how long have you read? They might reply, the time to say a nocturne, or else, the time to say a Pater Noster. . . . It is therefore evident by all this that any movement known by man, of any kind whatever, can be taken as time. However, of all the movements, the movement of heaven is most properly called time, for it possesses, to a greater extent than any other movement, the most numerous and most important conditions of the conditions required for a measure.[74]

And among the conditions required for a good measure, conditions that Marsilius enumerates with great care, is the following:

It should be as invariable as possible; if it were variable, we would be deceived when judging the magnitude of some other thing and using this measure. That is why we do not know evidently and certainly the terrestrial length that corresponds with a degree of heaven, because we do not know whether human feet are today of the same length, larger, or smaller, as when the measurement [of the earth] was accomplished.[75]

At first glance, John Buridan, Albert of Saxony, and Marsilius of Inghen seem to agree with the proposition that any movement can be called time. But we soon begin to understand that their thought presupposes the restriction that the movement must be uniform, or nearly uniform. If it is absolutely uniform, it would give us time, properly speaking; if it is only nearly uniform, it would give us an approximate time.

Our authors therefore admit the following two propositions, more or less explicitly:

1. A movement is or is not uniform independently from the choice of movement serving as time.

2. After having chosen the movement defining time, it is possible to know whether a movement is or is not uniform.

We can determine with certainty, using John Buridan's writings, the meaning they attributed to the first proposition.

Time, as defined by a local movement, says Buridan, is nothing more than the flow or succession that constitutes this movement;[76]

in every local movement, this flow or this succession has a certain magnitude. But we have stated elsewhere that this local movement is revealed to us only by the perception of the changing disposition of the mobile with respect to some place or some immobile body; we are therefore equally constrained to know and to determine the magnitude of the flow by the magnitude of the space to be traversed, or what is traversed, or what we imagine to be traversed. A movement whose successive quantity is greater would traverse a greater space.[77]

Although awkwardly stated, we can easily understand the above thought: the flow constituting local movement is perceived, at each instant, by the speed at that instant. Hence, a local movement whose flow always keeps the same magnitude is necessarily a movement of constant speed, a uniform movement. A uniform movement is therefore characterized by a property intrinsic to it that it possesses before one has chosen the movement that will be called time and that is independent of the choice of this movement. These uniform movements are the only movements that can properly be called time. There are therefore, in nature, some movements that alone can be taken legitimately by man for a clock.

How can man recognize these movements? Is that accomplished simply by observation, by sensible perception?

Ockham and Burley attribute to our soul the faculty to construct, in itself, a conceptual clock that is perfectly regulated and absolutely uniform, and then to compare perceived movements with the indications of this clock. Assuredly, that is not the opinion of Buridan and his students.

For example, Albert of Saxony examines the following objection to the theory that takes the diurnal movement as time: "If the movement of heaven were to slow down or to speed up, time would get slower or faster."[78] He replies: "I agree that if the movement of heaven would slow down or speed up, time would also become slower or faster."[79] No doubt he adds implicitly: but we would not be able to perceive the change.

Albert of Saxony probably considered this complementary thought, although he did not formulate it. Nicole Oresme, his contemporary, was more explicit. He wrote:

> Just as heaven could be stopped in the time of Joshua and could turn backward in the time of King Hezekiah, in the same way its movement could be speeded up or slowed down, if it so pleased God.
>
> Certainly if the movement of heaven were twice as fast or twice as slow or more, however that may be, and if all the other movements or changes here below, which Aristotle attributes to the influence of heaven, were proportionally faster or slower, no one would be able to perceive this mutation; for everything would appear to us humans exactly as it is at present.[80]

It is therefore not sensible perception that indicates to us that a movement—the movement of heaven, for example—is a uniform movement. How can we know this then? The answer is that reason, or physical science, teaches us that no cause can intervene to increase or diminish the successive flow that constitutes the local movement of such bodies.

What had assured the Pythagoreans of the uniformity of the rotation of celestial bodies was their belief in the divinity of these bodies; Aristotle deduced some assurance of uniformity from the unchangeable character of the fifth essence. Buridan and Albert of Saxony asked for this uniformity from principles derived from their physics (today we would call this their dynamics). John Buridan and his disciples—no more than Aristotle or Averroes, William of Ockham or Walter Burley—did not intend to assert that man had the right to choose arbitrarily, with no other guide than usefulness, the movement it would be convenient to call time.

The Atomism of Gerard of Odon and Nicholas Bonet

The assertion that the unique time by which all movements are measured is, if not a conventional time, at least a pure mathematical concept, abstracted from all reality, was an assertion that was clearly affirmed by a fourteenth century philosopher, Nicholas Bonet. But, in addition, Bonet upheld a doctrine of time and movement differing greatly from the ones we have examined until now. This doctrine deserves our attention.

It appears to be intimately connected with the mathematical atomism professed by Gerard of Odon, the general master of the Franciscan orders at the time of Bonet. Unfortunately, we have only an imperfect understanding of the teachings of Gerard of Odon; all that we know is taught to us by its extremely brief summary and its slightly longer refutation in Joannes Canonicus's *Questions on the Physics*.[81]

The exposition of this theory is both brief enough and important enough that we should reproduce its text.

Joannes Canonicus has just given reasons to demonstrate that indivisibles cannot engender a continuum by melding to one another. He pursues this as follows:

> Frater Gerardus autem nititur solvere praedictas rationes. Dicit enim quod, in ista definitione continuorum, in qua videtur quod continua sunt quorum ultima sunt unum, ultimum non debet accipi pro aliqua parte ejus quod continuatur alteri, sed pro differentia loci discretiva, sic quod ante istius et retro illius hoc omnino unum, vel sursum unius et deorsum alterius. Et hoc modo est possibile puncta continuari, et superficies, et lineae ad invicem, et etiam instantia in tempore. Licet ista sint indivisibilia secundum partes quantitativas, sunt tamen divisibilia secundum differentia loci vel temporis. Sicut apparet quod superficies distinguitur per intus et extra, dicimus in quantum quod corpus est intra superficiem et non extra; sed corpus tangens est extra et non intra; et sic superficies, indivisibilis existens, dividitur, et dividitur per intus et extra. Item punctum quod

est in medio potest terminare semidiametrum venientem ex partes dextra non tamen terminando ex partes sinistra.

Hoc premisso, respondet ad rationem Philosophi dicens quod, in illa ratione continuorum qua dicitur quod continua sunt quorum ultima sunt unum, non accipitur ibi ultimum pro aliqua parte ipsius qua continuatur alteri, sed pro differentia respectiva loci, ita quod ante istius et retro illius sunt omnino unum, et sursum istius et deorsum illius; et hoc modo possibile est continuari puncta.[82]

The above text would appear obscure to us, and Joannes Canonicus's discussion would add little to its clarity, if a passage from Nicholas Bonet's *Tractatus de praedicamentis* had not come to our aid; in this passage, we can cite expressions that are almost identical to those cited by Joannes Canonicus. We are therefore certain that Bonet was presenting the doctrine of Gerard of Odon.

Our author has just related Duns Scotus's arguments against those who admit indivisibles within continuous magnitude.[83] He then objects to the demonstrations of the "new philosophers" in this manner:

It is difficult to resolve these demonstrations, for the solution depends partly on metaphysics and partly on physics; the task does not seem impossible, however.

Let us first state that the first reasoning is not conclusive; in fact it admits that things whose extremities (*ultima*) are one are continuous. Indivisibles have no extremities. Therefore, they cannot be continuous with one another.

I reply that indivisibles such as points have no extremities, if by extremity one understands a point; in fact, a point cannot be the ultimate and intrinsic term of another point. But a point has an extremity, if one understands by extremity an ultimate difference of position, meaning a before and after; the before of a point makes up a unity with the after of the preceding point, then, and that is not impossible.

But, one would ask, these differences of situation, the before and after which are in a point, are they really distinct from the point? One replies negatively; but they are distinct by essence (*quidditative*).

In order for that to become evident, let us note that one can distinguish parts (*partes*) in something that is not divisible into parts (*partes*), and that one can assign differences of time and place to it. Let us explain this with respect to each of the five kinds of things that are quantitatively indivisible.

The doctrine is evident with respect to surface, which is indivisible following the dimension that is depth; surface is, in fact, defined and determined by an inside and an outside; the body of which it is the surface is wholly on the inside of the surface; the body touching the preceding along the whole surface is wholly outside the surface. But between the body inside the surface and the body outside the surface there is nothing but the indivisible surface. This surface exists insofar as it is the intermediary between the two bodies, and yet one distinguishes in it an inside and an outside, which are different locations.

One distinguishes a convex side and a concave side for a circular line that is indivisible in width. . . .

Something similar is seen for a point taken at the center of a circle; it terminates the radius that comes from one side, its right, for example, but it does not terminate the radius that goes to the left. . . .

It is the same for the instant; it is at the same time the end of the past and the beginning of the future. The relations designated by the words *beginning* and *end* are distinct from one another.

That is also evident for the instantaneous state (*mutatum esse*), which is an indivisible and the term for movement. It can, however, establish continuity between the preceding movement and the succeeding movement. The instantaneous state (*mutatum esse*) therefore contains in itself that by which it is the continuation and term for the preceding movement, and that by which it is the beginning of the succeeding movement.

Therefore, in each of the five kinds of indivisibles just enumerated, one can distinguish and determine the following differences: before and after, right and left, front and back, and above and below.[84]

The thought that Bonet attributes to Gerard of Odon now appears well defined. Every continuum is composed of things indivisible from the point of view of magnitude; but in each of these things one can distinguish two sides. That is how these indivisible things can constitute continuous magnitudes; for that, it suffices that the second side of each of the indivisibles composing a magnitude makes up a unity with the first side of the succeeding indivisible, that the front of the first is identical with the back of the other.

It is by means of the above principles that Gerard of Odon, and Nicholas Bonet after him, attempted to destroy Duns Scotus's argument against indivisibles.

A point remains unsettled in what we have just asserted. Did Gerard place an infinite number of elements in a finite length, or did he suppose that the multitude was infinite? The text of Joannes Canonicus does not furnish us with a formal response to this question. The discussion to which Joannes Canonicus submits Gerard of Odon's theory appears to indicate that Odon remained undecided between the two alternatives, because some objections that are addressed against him aim at the first hypothesis and some other objections aim at the second hypothesis.

Nicholas Bonet has clearly presented to us the opinion we know to be the opinion of Gerard of Odon. Although he does not explicitly reject this opinion, he does not appear to admit it, for on several occasions he develops a noticeably different theory, a theory that he surely prefers.

According to Gerard of Odon, a continuous line is composed of points. Each point has two extremities, two sides, a front and a back. These extremities, these sides, are not points. Continuity between the various points of a line is established by means of these extremities or sides, the back of each point making up a unity with the front of the preceding point.

According to Nicholas Bonet, it is otherwise.

A line, for example, is composed of indivisible parts. Each indivisible part has two extremities that are points.[85] These indivisible parts are melded to one another because any two consecutive parts have a point in common.[86] This discontinuous structure is the structure that the line has in actual existence, outside the mind. But in conceptual existence, inside the mind, it is otherwise; there the line is continuous. Between any two points whatever, the mind can always mark a third point, so that the conceived line potentially contains an infinity of points.[87] Our author promises to develop these brief indications in his treatise, *De praedicamentis*;[88] it is therefore in this *Treatise on the Predicaments* that one should seek Bonet's doctrine.

In keeping with his usual method, our author exposes in succession various doctrines about the constitution of continuous magnitude; he places last the theory he prefers. He also attributes an author to each of these theories; but, in general, history would not be able to justify these attributions. For example, he attributes to Plato the doctrine that constructs a volume by means of surfaces, a surface by means of lines, and lines by means of points.[89] Similarly, the doctrine that he develops last is "Democritus's doctrine';'[90] moreover, he tells us, toward the end of his book, that "whoever wishes to adhere to the doctrine of the Peripatetics or the Platonists may do so, but Democritus seems to speak more reasonably."[91] It is therefore clearly Bonet's opinion, given the authority of Democritus, that is cited.

What is this theory that "is attributed to Democritus"?

> Democritus thinks that there are two kinds of quiddities.
>
> The quiddity of the first kind has as foundation a simple reality, not capable of being divided into other realities. That is perhaps the quiddity of form or of matter; and that is also the quiddity of intelligences, an absolutely simple (*simpliciter simplex*) quiddity that cannot be resolved into other realities.
>
> On the other hand, the second kind of quiddity is a quiddity that has the agreement of several realities as foundation, such that, if one suppressed one of these realities, there would remain nothing, neither the name, nor the definition, nor the quiddity of the thing we are considering.
>
> The quiddities of substances composed of matter and form are of this kind; humanity, for example, [man's quiddity] does not have a simple reality for foundation. It has for foundation the agreement of two realities, soul and body. If one suppressed one of these two realities from man, there would remain nothing, neither the name nor the definition, for once the soul is separated from the body, there is no longer a man.[92]

Assuredly, the Democritus who holds this doctrine has read William of Ockham to whom this distinction was familiar. What does he teach us about continuous magnitude?

> Continuous magnitude belongs to the second kind of quiddity, to the kind of quiddity founded on the agreement of several realities; if one suppressed one of these realities, one would have both the name and definition of the continuum disappear.
>
> Let us take an example. The line or the quiddity of the line has as foundation the agreement of several realities. Let us suppress one of these realities; neither the name nor the definition of the line remains. None of these realities is by itself the line, although it is something of the line—in the same way that the body is not the man, but something of the man.
>
> Democritus does not think, in any case, that these realities, whose agreement serves as the foundation for the quiddity of the line, are realities of points, because then his opinion would coincide with Plato's opinion. He thinks that these realities are of an entirely different nature and an entirely different order than the reality of points.
>
> One can repeat similar considerations with respect to the

surface and the body, treating each of them in a suitable fashion.

But this continuous quantity, which has as foundation the agreement of such realities of another nature and order than itself, is manifestly an individual in the genus of quantity. It is absolutely impossible to divide it into several individuals of the genus quantity; however, one can dissociate it into several realities, of which none is a quantity, but of which each is a partial reality that agrees to form an individual of the genus quantity—thus a stone is separable into matter and form, meaning into two realities of which neither is a stone.

It is clear that this continuous quantity, the resultant of realities that are not themselves divisible into several realities, has two extremities (*ultima*) that are two points in actuality; it is contained between these two points as between two proper limits.

This quantity can be made continuous with another— meaning another indivisible quantity—as long as the extremities of these two quantities become a single indivisible thing. Then these two indivisible quantities would touch along something that belongs to them (*aliquid sui*) along the same reality; but they would not touch each other in their totality, because they are not simply two realities.

The third quantity, the one resulting from these two indivisible quantities, is something that can be divided into several continuous things, into several quantities. It is a magnitude that has quantitative parts, that has parts outside parts, whose various parts have a determined and distinct situation with respect to one another.[93]

How then must one understand the divisibility of a continuum? Bonet expresses himself with his usual conciseness about this matter.

Here a doubt presents itself. It seems that every part of a continuum, and even all its [successive] parts, are of the same nature. Therefore if one of them is divisible into several magnitudes, into several coexisting continua, it must be the same for each and every one of them.

The solution of this difficulty resides in this remark: there are parts, properly speaking (*per se*), and accidental (*per accidens*) parts in a quantity.

The parts, properly speaking, are indivisible realities to which neither the name nor the definition of quantity belongs.

The accidental parts are otherwise; they possess the name and definition of quantity. It happens (*accidit*) that the line as line is composed of several parts of which each is a line. But the proper nature of the line as line is to be constituted by several realities of which none is a magnitude; one cannot find any line that does not have these parts, although one can find a line without other parts.

But, one might ask, is the line or any other continuous magnitude divisible into parts that are divisible to infinity? Does it have an infinity of parts?

I would reply negatively, according to Democritus's theory. Any continuum actually finite can be resolved until one reaches continua that are no longer divisible into new continua, but that are nevertheless divisible into some other realities; and finally, one would attain absolute indivisibles [that are realities].

This affirmation seems to agree with what Aristotle said about the natural minimum; for in the first book of the *Physics*, he stated that one can reach a quantity of flesh which is as little as possible.

But you might assert that Aristotle teaches that every continuum is divisible to infinity. I would reply that he is referring there to a continuum that is potentially divisible to infinity because there always remains something to be taken beyond what one has already taken. And he states that according to the testimony of the senses. In fact, by natural means, one can barely push the division down to these indivisibles in such a way that there remains nothing to divide; however, in truth, if it were possible to divide a continuum as much as it is divisible, the division would stop at the indivisibles.

Because of what has been asserted, it is clear that points in a line do not follow each other in a continuous fashion and are not continua; however, there is always a divisible or indivisible line between two points.[94]

One must say the same for the surface and the body; one must also express oneself similarly with respect to time and movement.

In fact, movement is composed of movements that are no longer divisible into other movements, but that are each able to be divided into several realities capable of constituting movement (*realitates motus*). All movement, divisible or indivisible, is always enclosed between two instantaneous states (*mutata esse*). There is no movement that is divisible to infinity. There is no infinity of instantaneous states in a movement; there is only a finite number.

It is the same for time. *Per accidens*, a time is composed of several times that are not themselves divisible into several other times. But by itself time is composed of several realities capable of constituting it (*realitates temporis*), of which none keeps either the name or definition of time. One such time, that cannot be divided into other times, is a whole act at once. It is enclosed between two instants (*nunc*). It happens suddenly. Another indivisible time succeeds it; the instant (*nunc*) that establishes continuity between these two times is located between them. The second indivisible time passes suddenly, another succeeds it, and so forth.

Let us therefore conclude, with Democritus, that every continuum is composed of indivisibles, and that it is finally resolved into these indivisibles. It is not composed of an infinity of indivisibles, but of indivisibles in a finite number. Whoever wishes to adhere to the doctrine of the Peripatetics or the Platonists may do so, but Democritus seems to speak more reasonably.[95]

Movement and Time according to Nicholas Bonet:
Although Continuous in the Mind,
Successive Beings Are Discontinuous in Reality

Joannes Canonicus taught us, in his extremely short exposition of the atomism of Gerard of Odon, that the master general of the order applied his theory equally to points, lines, surfaces, and also to instants of time. Among the objections that Joannes Canonicus addresses against this doctrine are some derived from movement;[96] there, what is in question is not only indivisibles or atoms of length, but also indivisibles or atoms of time.

We thus learn that Gerard of Odon did not limit his atomism to the permanent magnitudes of geometry only, but that he extended it to movement and time.

Bonet also declares that one must extend to movement and time what he has stated about the resolution of the line, the surface, and the volume into indivisible elements.

The idea was not novel. It was already proposed by the Mutakallimun, the Arab theologians who, desiring to contradict Aristotle's philosophy in every way, adopted the physics most opposed to Peripatetic physics, namely atomism.

The Latin Scholastics of the fourteenth century knew well the doctrine of the Mutakallimun because of its exposition and refutation given by Moses Maimonides in his *Guide of the Perplexed*.

That is where Gerard of Odon could have read the following:

> [The Mutakallimun] thought that the universe as a whole—meaning every body in it—is composed of very small particles that, because of their subtlety, are not subject to division. Each of these particles is absolutely devoid of quantity; however, when several are aggregated, their aggregate possesses quantity and thus becomes a body. . . .
>
> The third proposition says that time is composed of instants, meaning that there are many units of time that, because of the shortness of their duration, are not divisible. This proposition is also necessary for them because of the first proposition; for they undoubtedly had seen Aristotle's demonstrations, by means of which he has demonstrated that distance, time, and local movement are three things equal as far as existence is concerned (meaning that their relation to one another is the same and that when one of them is divided the other two are likewise divided, and in the same ratio). Accordingly, they were forced to recognize that if time were continuous and infinitely divisible, it would follow that the particles they supposed to be indivisible would be necessarily divisible. Similarly, if distance was supposed continuous, it would follow by necessity that the instant that was supposed to be indivisible would also be divisible—just as Aristotle made clear in the *Akroasis* [the *Physics*]. That is why they supposed that distance is not continuous in principle, but that it is composed of parts at which divisibility ends, and that likewise the divisibility of time ends with the instants that are not divisible. . . .
>
> Hear now what they were compelled to admit as a consequence of these three propositions: They said that movement is the passage of an atom belonging to these particles from one atom to another that is contiguous to it. It follows from this hypothesis that no movement can be faster than another movement. In fact, they said that when you see that two things in movement traverse two different distances in the same time, the cause of this phenomenon does not lie in the greater speed of the movement of the body traversing the greater distance; but the cause of this lies in the movement we call slower being interrupted by a greater number of units of rest than is the case with regard to the movement we call more rapid, which is interrupted by fewer units of rest.[97]

Thus the Mutakallimun professed an integral atomism that was applied to time and movement as well as to distance. Such also was Gerard of Odon's doctrine, if we can judge it by the critique of Joannes Canonicus; such also was Nicholas Bonet's doctrine.

Nicholas Bonet also posited in principle that successive beings, such as movement and time, are in conceptual existence (*in esse cognito* or *in esse conceptuali*), within the mind, other than they are in real existence (*in esse reali*), outside the mind.

What Aristotle and his commentators referred to was movement and time such as they are in conceptual existence: "One agrees that in conceptual existence movement possesses continuity, divisibility, perhaps infinity, and all the other properties described in the sixth book of the *Physics*. But they do not possess these properties in real existence."[98]

In real existence, movement has a discontinuous and atomic constitution similar to the constitution of the real line.

Two things agree in constituting the real line: the indivisible element of the line and the point. The indivisible element of the line has two extremities that are two points; when two indivisible elements constitute one another, the point terminating the first is identical to the point beginning the second.

Similarly, two things concur in constituting a real movement: indivisible movements and instantaneous states (*mutata esse*). In this constitution of real movement, the indivisible movement plays the role that the element of length plays with respect to the line, and instantaneous states play the role that points play.

We should find, in the constitution of time, the equivalent to what we found in the real constitution of movement—the indivisible time corresponding to the indivisible movement, and the instant (*instans* or *nunc*) corresponding to the instantaneous state (*mutatum esse*).

Finite length is a series of a finite number of indivisible lines, each of which is melded into the other by a point. Finite movement is a series of a finite number of indivisible movements, each of which is tied to the next by an instantaneous state. Finite time is a series of a finite number of indivisible times, each of which has a common instant with the succeeding indivisible.

Such is, essentially, the integral atomism of Nicholas Bonet.

Like Joannes Canonicus, Nicholas Bonet admits that the successive beings are constituted differently in the mind than they are in external reality. But except for this affirmation that they share in common, the doctrines of the two authors are, as it were, the inverse of the other. As an imitator of Damascius and Simplicius, Joannes Canonicus places the flowing continuity of successive things—of movement and time—in the reality external to the mind;

it is the mind and only the mind that parcels out and renders discontinuous this flux. On the other hand, according to Bonet, the parceling and discontinuity are inherent in the real nature of successive things; it is only in the mind and in their conceptual existence that movement and time flow continuously.

Is it not interesting that the debate, which before the middle of the fifteenth century had rendered these two disciples of Scotus into adversaries, still rages between the philosophers of our time? One could without injustice make Joannes Canonicus a precursor of Henri Bergson, while Nicholas Bonet might have regarded Evellin as his disciple.

Let us now retrace the details of the doctrine we have just sketched.

Here, first, are some of the passages where Bonet expresses himself about movement:

> Movement belongs in the category of passion; it is nothing other than a relation between the mobile and the motor, or else a relation between the patient and the agent, or else a relation between what is produced and what produces it. . . .
> We should note that movement outside the soul is not a reality divisible into several parts of movement; it is a negatively indivisible reality that passes suddenly (*raptim transiens*); this reality is nothing more than what we have said, an action and a passion, which are indivisible and suddenly occurring accidents.
> Therefore, nothing of movement is produced in actuality unless it is this indivisible which is action and passion; another indivisible succeeds this indivisible, and not something divisible in actuality. If this were not the case, the reality would not be a successive reality, but a permanent reality; in fact, a quantity that possesses parts that are in actuality simultaneous is a permanent, not successive, reality. Thus one commonly says that the existence of a successive thing is reduced to the existence of the indivisible belonging to it, an indivisible that passes suddenly and to which succeeds another indivisible in actuality. In fact, this action is constantly renewed, and new degrees of what is produced also constantly occur. That is how movement comes to be. For example, when fire is engendered in water, the heat, the action that produces the second degree of heat within the water, is not numerically the same as the action that produces the first degree; it is a new action. And thus, new actions are constantly exercised,

in the same way that new degrees of heat occur in the produced heat. Each of these actions is distinct from one another, but each succeeds the other.

It is a result of what has just been asserted that movement outside our mind and in reality is not continuous; in fact, these actions and passions that pass suddenly are not capable of continuing one another. The whole continuity of movement comes from the mind. (*Tota continuitas motus est ab anima.*) When these actions are brought into conceptual existence (*esse cognitum*), the intellect conceives these actions as presenting no interruption, as a single action that exists in a constant flow and that follows a continuous succession. The continuity of successive things therefore exists only when these things are brought to conceptual existence, for it seems impossible that these things possess any kind of continuity in reality. (*Tota ergo continuitas successivorum est ut ipsa redacta sunt ad esse cognitum, quia in re videtur impossibile quod habeant aliquam continuitatem.*)

These consequences result:

First, a movement is not divisible to infinity in reality; in fact, actions and passions that are movements in reality are finite in number; if this were not the case, the part would be equal to the whole.

Second, movement is composed of indivisibles, since it is composed of actions and passions that are negatively indivisible.

Third, there is no infinity of instantaneous states; one can assign a first *mutatum esse*, and a first change (*mutatio*).

Finally, it is not true that one can assign a movement before any *mutatum esse*, or that before any movement one can assign a *mutatum esse*.[99]

The essence (*quidditas*) of movement does not have for its foundation a single, absolutely simple reality, but the agreement of several realities of which none, taken particularly and in isolation, is formally movement. What is formally movement is the total reality, the result of this agreement. But movement is so founded in the agreement of these realities that to suppress one of them suffices to cause movement and its definition to disappear. . . .

Moreover, movement, thus resulting from three realities, is absolutely incapable of being divided into several parts, each one of which would be a movement, although it is divisible into several realities constitutive of movement. Similarly, an element can be subdivided into several realities, but not into several elements.

Movement, taken as a whole in this way, is wholly in actuality at the same time, but it passes suddenly; another movement succeeds it which is also indivisible into partial movements, but which is divisible into several realities, as has been said. And it is the same for other movements consecutively.

In this way one sees clearly that nothing appears of movement outside the mind, unless it is indivisible. In any case, I do not understand by this indivisible an instantaneous state (*mutatum esse*), for this latter thing is absolutely indivisible in itself and in what relates to its measure; in fact, it has for measure the instant (*nunc*) which is absolutely indivisible. This movement, on the contrary, is only indivisible into several partial movements; its measure is [a time] indivisible into several partial times, but divisible into several realities constitutive of time, as we shall assert when we will deal with time. The element of movement (*terminus motus*) therefore does not have for measure an instant (*instans*), but a time indivisible into several times.

Clearly, continuous and successive movement is composed of such movements that are indivisible into several movements, but are divisible in the said manner. However, it is not composed of instantaneous states (*mutata esse*), for these are absolutely indivisible.

You might ask, are these indivisible movements consecutive to one another? I would reply that they are consecutive to one another in such a way that there does not exist any other intermediary movement, either divisible or indivisible, between two indivisible movements. But between two indivisible movements there is always an intermediary instantaneous state, like a point between two lines; that is what we shall state more explicitly in the *Predicaments* in the chapter on quantity.[100]

Let us append the details about time and the instant to these details about movement and the instantaneous state. Bonet first talks about the instant (*nunc*); let us first extract only what he says about the instant in real existence.

It is evident that the first assertion that should be formulated about the instant is the following: the instant is a passive entity that behaves with respect to time as the instantaneous state behaves with respect to movement, and the point with respect to the line.

The instant is the measure of something indivisible, for it is the indivisible of time.

It is a coexistence, and this coexistence is a relation that comes externally. . . .

It is to be distinguished from what it serves as measure, as all measures are to be distinguished from what they measure. . . .

The foundation of the instant (*nunc*) is the instantaneous state (*mutatum esse*) of which it is the measure. . . .

The term of this coexistence that receives the name *instant* is the eternal duration of the First Intelligence, a duration with which all instantaneous states coexist.[101]

It is also a property of the instant to be what establishes the continuity (*continuativum*) between the parts of time, in the same way that the instantaneous state is what establishes the continuity of the parts of movement to each other; and that is established in such a way that there can be continuity (*continuatio*) for successive things.[102]

From the instant, let us turn to time.

Time is a relation that is real and external to the mind; it has its foundation in movement, and the parts of time have as foundation the parts of movement. Time is therefore the set of coexistences of such movement with the First Intelligence; or else a part of time is the coexistence of a part of movement with the eternity of the First Intelligence. Similarly, the instant is the coexistence (which passes suddenly and is indivisible) of the instantaneous state with its First Principle. The various parts of movement in fact coexist, one after the other, with the First Intelligence.

Perhaps you would object that there is nothing of a successive thing existing in actuality outside the mind, unless it is an indivisible thing; in fact, if it were a divisible thing possessing several simultaneous and coexisting parts, we would no longer be dealing with something successive, but something permanent.

The resolution of this difficulty depends upon the following remark:

One must repeat for time what has been asserted about movement. Time is a measure that has as foundation the agreement of two or more realities; each of these realities, taken singly, is not a time, but a part of time. If one of them were suppressed, the word *time* and its definition would disappear; time is the whole that results from these different realities. These two realities can exist simultaneously outside the soul; for such a time is called present time. It passes suddenly, and another present time succeeds it; thus one time continually succeeds another.

The following is evident because of the preceding: when one says that what exists in the reality external to the soul is only an indivisible time, this is not true of the indivisible instant (*nunc*) but of the indivisible which is time. This [indivisible] time, in fact, is something that cannot be divided into several realities, each of which is properly a time; each of these realities is only something of time. This time, which is indivisible into several other times, and which taken all together (*simul*) exists actually, exists simultaneously with the instant, which is the principle and the term for this time. . . . Time and the instant thus exist simultaneously.[103]

The instant and indivisible time are both defined as relations. One of these relations has as foundation, according to Scholastic terminology, the instantaneous state, the *mutatum esse*. The other has as foundation the indivisible element of movement. One and the other have the same term, and this term is the eternity of the First Intelligence. The instant is the coexistence of an instantaneous state with the eternity of the First Intelligence; indivisible time is the coexistence of an indivisible element of movement with the eternity of this Intelligence.

These definitions of the instant and time by Nicholas Bonet are clearly influenced by those of Francis of Mayronnes, who often influenced Bonet's thoughts. Time, said Francis of Mayronnes, is a relation, "but it cannot be a relation with respect to any creature, for even if there existed one creature only, there would be a before and an after. I therefore admit that time is the flux of state of presence (*praesentialitas*) with respect to God, as we have stated about place."[104] This definition of time, which Nicholas Bonet and Francis of Mayronnes share, comes up against an objection that is difficult to resolve. Because of His eternity, God is equally present to almovements, in whatever time these movements are produced: there is no distinction between the past, present, and future for an eternal intelligence. How then can the various parts of movement have, with respect to this Intelligence, various coexistences capable of constituting the succession of time?

By retracing the history of the theory of place, we have said how Bonet attempted to resolve this difficulty. According to him,

the First Intelligence is not included as subject in the line of temporal succession, for it cannot be the foundation of a distance to a time past or a future time; but it can be the term for a distance counted from a time past or a future time. Similarly, we have said, it can be the term for a distance relative to place or to the *ubi*. It can therefore delineate place by designating one place as here and another as there; in the same way it can delineate time by designating one time as before, another as now, and another as later.[105]

Does this distinction suffice to dispel the objection against Bonet's and Francis of Mayronnes's definition of time? One can doubt this. But, so as not to delve into the problem more deeply than our author did, let us be content with what he related to us about the instant and time, as they exist in the reality external to our mind, and let us now see what they are in their conceptual existence.

In what concerns its conceptual existence, the instant (*nunc*) is what engenders (*causat*) time by its flow (*fluxus*), which the mind grasps.

In fact, when the intellect wishes to engender the duration of a day, it first engenders an instant in conceptual existence, at the same time that it posits the first mobile in this same conceptual existence; then, taking this mobile and the instant with it, it carries them from east to west, and brings them back from west to east. Thus by means of this flow, it engenders the duration of a day in conceptual existence. It is therefore one and the same instant that engenders time.

If, on the other hand, one considers it according to its quiddity and its formal reason (*formaliter et ratione quidditativa*), the instant is an indivisible that serves as beginning and end for some divisible [time]; the instant is no longer numerically one. The intellect, in fact, designates an instant as the principle of some time; and then, when it has to mark the term of this time, it does not take the same instant. It posits two different instants in conceptual existence; similarly, the point which is the beginning of the line and the point which is the end of the line are not the same points. And since in the middle of the time in question, the intellect can still designate a new instant, and another instant at the midpoint of the midpoint, and so forth to infinity, there is a potential infinity of instants in a given time, just as there is a potential infinity of points in a given line.[106]

Ever since Aristotle, philosophers have been divided with respect to these two contradictory assertions:

1. A unique instant that always remains the same engenders time by means of its flow.

2. There is an infinity of instants distinct from one another in a given time.

According to Bonet, neither of these propositions is valid with respect to the instant and time as they are in reality, external to the intellect; they are both valid with respect to the conceptual existence of time and the instant. In this way they both can, and must, be affirmed at the same time. The one recalls how we engender a duration in the mind; the other recalls how we subdivide a duration that we assume given.

Although Bonet shows himself to be adept at dispeling the misunderstandings that bother other philosophers, he is perhaps less capable of resolving the objections directed at his own theory of time and movement.

This theory had surely already been given by Gerard of Odon; and it is against the latter's theory that Joannes Canonicus reasoned thus:

> If every continuum were composed of indivisibles, the indivisibles themselves could be divided. This conclusion is false; the premise is therefore also false if one can prove that the reasoning is conclusive. And to render this evident here is what one should assume:
>
> First, one can assign a faster movement and a slower movement in any given time.
>
> Second, given the same time, the faster mobile crosses more space than the slower mobile.
>
> Third, in whatever time given, the space crossed by the faster mobile can be in some relation to the space crossed by the slower mobile. If one supposed that the first mobile is twice as fast as the second mobile, it would cross a double length in the same time; and if the speed of the first is equal to the speed of the second by a relation of 3/2, the traveled lengths would have the same relation in equal time.
>
> That posited, let us reason as follows:
>
> Let us take a mobile which is faster than another by a relation of 3/2; in the same time, the length crossed by this mobile will be 3/2 the length crossed by the slower. Let us divide the lengths upon which these mobiles move, one faster than the other, and let us divide them into indivisibles; if the length crossed by the faster mobile is divided into three indivisibles—A, B, and C—the length crossed by the slower will be divided into two indivisibles only—0 and I—since these two lengths are in a relation of 3/2. The time in which the faster mobile crosses three atoms of length will be divided into three atoms of time, since length and time are divided in the same manner; let a, b, and c be these three atoms of time. But if the faster mobile crosses three atoms of length

in three atoms of time, the slower mobile necessarily crosses two atoms of length in three atoms of time, and crosses an atom of length in half the time. But a time composed of three atoms cannot be divided into two halves, unless an atom can be divided into two equal parts. Combining this final proposition with the first proposition, we find that if every continuum were composed of indivisibles, the indivisible would be able to be divided, which is contradictory.[107]

Nicholas Bonet heard the objection; he even recalls it briefly. Moreover, he seems to appreciate its seriousness. He states: "It is a doubt whose solution is extremely difficult."[108] In fact, one cannot deny that the potter's wheel turns more slowly than the first mobile. Bonet attempts several replies, which amount to agreeing with Joannes Canonicus—resulting in the divisibility of the temporal element relative to the slower movement. Our author seems helpless against the difficulty: "You would object that if I speak in this way I am required to admit that the element of the movement of the wheel is measured by a time [since it is measured by several indivisibles of the movement of the first mobile, to which several indivisible times correspond]. I reply: seek for the answer (*solutionem quaere*)."[109]

Nicholas Bonet does not shy away from bold opinions; his mind, which loves paradoxes, even seems to seek them out and seems satisfied when his reason cannot resolve the objections against which it is pitted. Hence, after having affirmed the radical distinction that he established between movement and time as they exist in the mind, and movement and time as they are in reality, he attempts to draw some surprising conclusions from his principle. Let us take note of two of them.

Outside the mind, a movement is composed of elements that cannot be divided into partial movements. Each of these indivisibles exists for an instant; it then ceases to exist and another indivisible exists, in turn, for an instant. None of these indivisibles can have a permanent duration; none can exist for some time. But can it not be possible that all the indivisibles composing a movement existed all together, at the same instant, in which case the movement itself, taken in its totality, would be accomplished instantaneously? Bonet does not hesitate to declare this possible.

But, you might say that it is impossible; in fact, movement is a successive being, and there is no succession in such an instantaneous movement.

Bonet replies to this objection by distinguishing two meanings of the word *successive*.

He observes that in its proper sense, the words *to succeed* exclude the possibility of coexistence, as when one says that something succeeds another. In this sense, "when I state that things succeed one another, I understand that the affirmation of the second part implies the negation of the first part; in any case, the two assertions cannot be true together, but the affirmation of one and the negation of the other are true together."[110]

In this sense,

> the word *succession* expresses nothing positive above and beyond the parts succeeding one another; all that it expresses is that the affirmation of one of these parts is the negation of the other.
>
> All this is said about succession, properly speaking.
>
> But one also uses the word *succession* in a less strict and less proper sense; one understands by succession an order according to which some things are placed after the other, either in an absolute manner in existence, or in place, or when this series corresponds with a priority of origin, of nature, or of time, but without the affirmation of the one thing implying the negation of the other in such a succession.[111]

Nicholas Bonet resolves the objection against his theory by means of the following distinction:

> The objection, which is called the Achilles [heel] of all the objections, is that movement is essentially successive. Here is how it is resolved:
>
> Does one understand succession in the first sense of the word? By saying that one part succeeds the other, do we understand that the affirmation of the second part implies the negation of the first part? It is not true that such a succession exists *per se* in movement (*per se, perseitate primi modi*). It exists only *per accidens*. It is *per accidens* that in movement, when a second part comes into being, the first part cedes its place and ceases to exist. In fact, it does not seem that these parts, which appear to be of the same nature, are repelled by existing together.
>
> But if you understand succession in the second sense of the word, succession as only an order of priority of origin or of nature . . . we would not deny that such a succession exists in all movement; but such a succession between the parts of movement would not be repugnant to their simultaneous existence in time.[112]

The propagation of light furnishes Bonet an example for explaining his thought. Like all Scholastic physicists, except for Nicholas of Autrecourt, Bonet thinks that this propagation is accomplished in an instant. He therefore sees this as an instantaneous movement, a movement in which succession, improperly speaking, is found, but where succession, properly speaking, is not. He states:

> I take as example the illumination of air; no doubt this illumination is not accomplished in some time, but in an instant; in fact, it is not a successive movement of proper succession, but it could be so, because there are several illuminations in it that precede one another by origin and by nature. Moreover, it does not seem impossible that the first part of air was lighted before the second, and that the first part ceased to be lighted at the moment when the second was lighted, and that it is always thus, successively—in such a way that the affirmation of the second illumination and the negation of the first illumination are true at the same time. If it were thus, the illumination of air would have some duration as measure. It is therefore evident that it is not repugnant for an instantaneous change of state (*mutatio*), as instantaneous state, to become [a movement] which is measurable by a time, as long as this instantaneous state includes several partial illuminations, as the total illumination of air contains several partial illuminations that follow a certain order. However, an illumination that does not absolutely include any partial illumination is absolutely indivisible; such an illumination always occurs suddenly, for one of its parts cannot succeed another.[113]

But if a movement of some duration can be condensed until it became an instantaneous change of state, a time of some duration can equally be condensed into a unique instant. This corollary does not bother Bonet.

> In the same way that all the parts of a movement that can be accomplished in some greater or lesser time can be also accomplished in an instant, because their existence together is not repugnant (as we have previously stated), the various parts of a time can be accomplished successively in greater or lesser duration, and they can also be accomplished instantaneously in the reality external to the mind (*in re extra*); their existence together is no more repugnant than the existence together of the parts of a movement, which are their supports (*subjecta*). In fact, if objects can exist together, so can the passions of their subjects.[114]

But let us note that this coexistence of the parts of a time, which is absolutely possible in the reality of things, is inconceivable for us; in fact, we do not conceive time as it is in its real existence, but such as it is in its conceptual and mathematical existence.

> Let us conclude, then, that, according to its natural existence (*esse naturae*), the various parts of a time can exist all together; but that this is absolutely repugnant with respect to their mathematical existence (*esse mathematicum*).[115]

Time according to Nicholas Bonet: Physical Time and Mathematical Time

Whatever one thinks about the atomic structure Bonet attributes to real movement and real time, this opinion should not dictate what one should think about the theory of the absolute clock proposed by our Franciscan; in fact, the latter theory is entirely independent of his atomism. One can easily reject the one while accepting the other.

The problem of the absolute clock has great similarity with the problem of the absolutely fixed term: What is the absolutely fixed body to which all local movements are referred? What is the particular movement that must serve to mark time for all other movements? These two questions are extremely similar; they are inseparably linked to one another. They are the questions that must be answered at the beginning of a science of movement. One can expect that they have been given similar replies.

But we have not encountered this similarity between the theory of absolute movement and the theory of absolute time when reading the works of Ockham and his successors.

Ockham and his followers admitted that the absolutely fixed body to which all local movements must be referred does not have to be a concrete body, realized outside the mind; it can be an abstract body, a pure concept. But these philosophers did not carry this doctrine to their theory of time; they did not dare declare that the movement intended to mark time for all other movements is also a pure concept not realized in nature—that the absolute clock is an abstract clock, existing only in the mind. They wanted for there to be, outside the mind, a perfectly uniform movement able to mark time for all other movements, and for us to have a means of recognizing it. They stopped midway along the road on which their theory of place led them.

Ockham went farther than midway; one needs only press his theory of time a little to bring his thought almost to the end of the road. But it seems that he did not wish to declare formally that he had reached this end.

Only Franciscus de Marchia seems to have clearly perceived and formally announced the truth that we are free to choose as we wish, in nature, the body in movement to which we would attribute the role of clock (whose responsibility is to measure the other movements and to mark time); the movement of this body would then be regular and uniform by definition. But this thought of Franciscus de Marchia, reduced to the proportion required by the reply to an objection, did not attract the attention of his contemporaries or his successors, and did not show them the path to follow.

However, when Nicholas Bonet enters a particular path, he pursues it to its end. He does not shy away from anything, and he does nothing halfway. That is the character of his genius. We have had several occasions to see this already; we will see it again.

Bonet clearly affirmed that it is useless to seek an immobile body among the real bodies in nature for a fixed place; all bodies move or are capable of moving. This seeking after a fixed place makes sense only for bodies given only conceptual existence (*esse cognitum*). There are degrees of greater and lesser abstraction in conceptual existence; a body can be conceived as formed of this or that substance, as given this or that physical property. It can still have physical existence (*esse physicum*), and it is still capable of movement. An absolutely immobile body is a simple shape that the mind has detached from any particular substance, from any physical property; the immobile place, the absolutely fixed term to which all local movements can be referred, has only mathematical existence (*esse mathematicum*).

Bonet extends this doctrine to his theory of time; he maintains that the absolute clock, the movement that marks time for all the other movements, has no existence other than conceptual and mathematical existence. He even pushes his generalization further; he develops the profound thought that in any order of magnitude whatever, the fixed standard by which one measures has no existence other than conceptual and mathematical.

Bonet therefore insists on the assertion that there is no single time outside the mind and in real existence, but that there are as many distinct times as there are different movements. Moreover, a determinate movement does not necessarily correspond with a determinate time, for this same movement can be accomplished in more time or in less time; that is even true of the movement of the first mobile.

If one says that time is a proper passion of movement, and that consequently it is inseparable from movement, we would be denying the assumption that one and the same movement can be produced in longer or shorter time; each of these times therefore resides in a movement accidentally, because the movement can be separated from each of these times. If, for example, the movement of the firmament became twice as fast as it is, it would clearly be separated from the time that is now its measure, because it would now be measured by only half this time; therefore time is not, by itself, a proper passion of movement, inseparable from this movement.[116]

Thus, not only are there as many distinct times as there are different movements, but also a multitude of various times can correspond to each movement.

Further, if one is referring to a real existence external to the mind, it is not true that there is one and only one present instant (*nunc*); each changing and mobile thing has its own particular present instant.

The property of the present instant, that it is numerically one and that there are not several present instants simultaneously . . . , we assert, should not be granted to the present instant in the reality external to the mind. In fact, in the same way that the different movements of various mobiles have several distinct instantaneous states (*mutata esse*) at the same time, there are also several present instants that have these various instantaneous states as foundation; in fact, where there is a number of distinct subjects, there is also the same number of distinct accidents.

Perhaps you would say that the subject of the present instant is exclusively the instantaneous state acquired in the movement of the first mobile, and not the instantaneous state acquired in another movement; and there cannot be several instantaneous states at the same time in the movement of the first mobile. Therefore there cannot be several present instants either.

I would reply that, first, what you assert is not right; the instantaneous state acquired in the movement of the first mobile is not exclusively the subject of the present instant; the other instantaneous states acquired in the other movements are also the subjects of the present instant.

Here is the reason for this assertion:

Every instantaneous state, in whatever movement it is acquired, has its own coexistence with the eternity of the First Intelligence, just as the instantaneous state acquired by the movement of the first mobile does. The instantaneous state therefore has its particular present instant, which is its proper and intrinsic measure. The present instant, which has the instantaneous state of the first mobile as subject, does not measure the instantaneous states of the other movements except extrinsically. Therefore, beyond this present extrinsic instant, every instantaneous state has its proper and intrinsic measure that resides in it and of which it is the subject.

Moreover, even if the instantaneous state acquired in the movements of sublunary things did not each have its present particular instant, you would still not hold what you held, namely, that there is one and only one present instant at the same time. That there are several worlds, and that there are then several movements that can be properly called first, and that therefore there are several instantaneous states acquired at the same time in these various movements, do not imply a contradiction; but then there are several present instants residing in these various states which have them as subjects. In fact, the multiplicity of the subjects carries with it the multiplicity of the accidents. Therefore, if there are several instantaneous states, there are also several present instants, for there are several coexistences of these instantaneous states with the First Intelligence.

You would then object: hence, in this present instant when I am agreeing with all this, is it not true that another man lives? In this instant when I am speaking, does not the Seine flow? I would reply that these things do not occur in a present instant which is really the same, but in a present instant which is the same only by equivalence—by that I understand a present instant which is an extrinsic measure and not an intrinsic measure.

This is what one must note: the present instant is not the same numerically from the point of view of physical considerations; it is the same numerically from the point of view of mathematical considerations, as we shall later say with respect to time.

If you said to me that I am in contradiction with our ancestors (*progenitores*), I would reply that he who speaks as a mathematician about time and the instant, is he who, using his intellect, abstracts the present instant from this or that instantaneous state taken in this or that movement, and has dealings only with one and the same present state, at least by equivalence (as we have said).

Moreover, our ancestors held as absolutely impossible that there are several worlds and hence several movements properly called first; they concluded from this that there cannot be either several times or several present instants at the same time. But we deviate from them with respect to the principle, and hence we also deviate from them with respect to the conclusion that necessarily follows from the principle.

Let us therefore affirm that the present instant changes with latitude and that the present instant changes with longitude—that one does not have the same present instant on earth, at sea, and in heaven, but different present instants.[117]

It seems that Bonet attempts to give his conclusion the same form as Aristotle gave his; in this way he marks more clearly the contradiction between the new theory and the old doctrine. At the same time he tells us what is the change of principle that was to bring the ruin of the Peripatetic doctrine. Everything that the Philosopher taught with respect to time is, in the final analysis, founded on this dogma: There does not exist and there cannot exist more than one first mobile, and hence more than one first movement; this first unique movement marks one and the same time for all the other movements. By asserting that God can, if He wishes, create several worlds, Etienne Tempier destroyed the foundation that held together the Peripatetic theory of time; in the same way, by asserting that God can impose a movement of translation to the universe, he also deprived the Peripatetic theory of place of any support.

We have seen Bonet distinguish between the multiplicity of present real instants and the unity of present mathematical instants; let us now observe him pursue the same distinction with respect to time. The pages he wrote on this subject are worthy of being reproduced almost entirely; they are certainly among the clearest that anyone has ever written on this difficult subject:

Let us now speak about the properties of time, its unity and its plurality, its simultaneity and its continuity; these subjects, in fact, are worthy of special consideration. . . .

To treat them properly, one must understand that time can be considered in two ways: naturally or mathematically. One must therefore speak about the simultaneity and unity of time in one way if one considers them with respect to their natural existence (*esse naturae*), and in another way if one considers them with respect to their mathematical existence (*esse mathematicum*). That is what Aristotle's Commentator states in commentary 131 of the eighth book of the *Physics*; he notes that the behavior of time outside the mind is similar to the behavior of place. He adds that the study of time is mathematical rather than natural (one manuscript has "mathematical," another "divine," and another "physical").

By the natural existence of time, we understand the existence time has in sensible matter and when it is conjoined to sensible matter; by its mathematical existence, we understand the one it has when it is separated by abstraction from all sensible matter.

Let us take an example: I can consider a line of two feet existing in a piece of wood or a stone—a line to which the wood or the stone serves as subject. I can consider this line from the point of view of natural existence. I can also consider this line of two feet as abstracted from the wood or the stone; I can consider the line without considering the wood or the stone. Such an abstraction is not something deceptive; the study of the line thus abstracted is a study from the point of view of mathematical existence.

Here is another example, drawn from discontinuous quantity and the number ten: if this number includes some sensible matter, as does the number of ten dogs or ten horses, the number is considered from the point of view of natural existence. But I can conceive the number abstracted from horses, dogs, or any sensible matter; such a consideration of number is a consideration from the point of view of mathematical existence.

It is also clear that there are two ways of considering time and place. In the first way, one considers it from the point of view of natural existence; one considers time according to whether it exists in this or that movement. In the other way, one considers time as separated by abstraction from this movement here as well as from this movement there; such a consideration concerns its mathematical existence.[118]

Let us now return to what is in question, and let us speak of time first from the point of view of natural existence and then from the point of view of mathematical existence.

With respect to time considered in its natural existence, taken as existing in every movement, one must repeat what one has asserted about movement; one must repeat this with respect to the simultaneity of particles of time and the unity of time.

In the same way that all the parts of a movement can be successively produced in a greater time or in a lesser time . . . all the parts of a time can be produced in a larger or smaller succession. . . . This can be clearly understood through the example of the line two feet long. Taken from the point of view of its existence in matter, in a mass of air or water, for example, its length can be lengthened or shortened by the effect of the condensation or rarifaction of its subject; the line of two feet can become a line of one foot or three feet— it can become greater or lesser than two feet according to whether the rarifaction of its subject is greater or lesser. Similarly, with respect to its real and passive existence outside the soul, the successive line of a movement or of a time can be shortened or lengthened.

But from the point of view of mathematical existence, consideration of time differs greatly from the preceding. Taken in its mathematical existence, abstracted from all sensible matter, a time cannot be either lengthened nor shortened. . . .

Let us take as example the line two feet long that [really] exists in a piece of wood, but that the mathematician considers without considering the wood [in which it resides]. I say that this line cannot be lengthened or shortened; it cannot become a line that is not two feet long, no matter whether the quantity of the line in the piece of wood contracts or dilates because of a change suffered by its subject.[119]

In other words, the mathematician does not conceive a line that contracts or dilates, but only a line of invariable length, which is shorter or longer than another line of equally fixed length.

The mathematician abstracts from the movement or change suffered by the subject; hence, the line thus considered in no way changes by the effect of a change in its subject. One therefore says rightly that mathematics deals with absolutely immobile things.[120]

Bonet here expresses exactly the thought of Euclid and the Greek geometers who took great care to avoid introducing any consideration of movement in a demonstration of geometry.[121]

One must say the same about the successive line of time. The mathematician considers the duration of a diurnal revolution, and he separates by abstraction this successive line from all matter and all movement; further, with respect to the existence it has in its subject, this duration can also be multiplied, changed, lengthened, or shortened, as can its subject. But considered mathematically, this successive line is absolutely invariable; it can neither be lengthened nor shortened.

Here is another example: A future duration, such as the duration of next year, can be shortened, and so can the revolutions occurring during the year; in fact, it is not contradictory that all the revolutions of a year can be accomplished in a month or in a day, because of an increase in the force of the motor that produces them. But as a mathematician, to conceive this successive line, the duration of the following year, is to separate it by abstraction from all subject. A line thus conceived cannot be either lengthened or shortened; it always keeps the same length.

One therefore rightly says that mathematicians abstract away all change and movement; everything considered from the mathematical point of view is absolutely immobile and invariable. . . .[122]

Let us now deal with the unity of time. Is there one and the same time for all temporal things, or can there be several times at once?

Clearly, we must reply one way from the point of view of natural existence and another way from the point of view of mathematical existence.

From the point of view of the natural existence time has outside the soul, it is evident that there is no unique time for all temporal things; in fact, there is not only one movement, but there are many movements. And the multiplicity of movements carries with it the multiplicity and diversity of times, because in natural existence, the multiplicity of subjects carries with it the multiplicity of the passions of which they are affected.

It is clear that a potter's wheel has a different coexistence with the First Intelligence than has the movement of a ship or the movement of heaven. Similarly, if there existed several worlds, there would be several heavens, and hence several movements properly called first, and therefore several times.

One can also show this by the following reason:

It is the same for any permanent continuous magnitude; that is evident, for the line of two feet existing in a piece of wood is not the same as the line of two feet existing in a stone. It is the same for a continuous quantity; the number of ten horses made concrete in the horses is not the same as the number of ten dogs made concrete in the dogs—in the same way that the unity of horse, consisting of a horse, is not the same thing as the unity of dog, consisting of a dog. Similarly, time taken materially, in its natural existence, is different with respect to various movements. There is no unique time for temporal things; there are several times at once.

Considered mathematically, on the other hand, there is clearly a single time for all temporal things; the multiplicity of movement does not carry for time an equal multiplicity.

Let us take a discontinuous quantity as example.

The mathematician considers the number ten after having separated it, by mathematical abstraction, from all things. He can then apply it, as measure, to several sets he wishes to measure—to several horses, and then to several dogs; several number tens do not result from this. That is what Aristotle and his Commentator asserted: the number of ten horses is the same as the number of ten dogs.

Similarly, for all temporal things, time, taken mathematically, is unique. If, for example, there are several movements properly called first, there would not be several times from the point of view of mathematical existence. In fact, in order to consider time, the mathematician separates it by abstraction from this world as well as from that world; he considers it in itself, as if it were separated from all these worlds in real existence. Thus considered, time is not counted with respect to the number of subjects. Similarly, if the number ten existed separately, as if suspended and floating in air, one could replace the horses by dogs without having to distinguish two different number tens. Therefore, since one can count and measure men as well as horses and dogs by the same number ten, and since the mathematician considers the quantity existing in a thing, whether it is a successive or permanent quantity, as if it were separated from that thing, suspended above that thing (for he does not at all consider anything of the thing in which it resides), as a result, evidently, this quantity, considered mathematically, is not multiplied by the multiplication of the subjects in which it can reside; it is held as immobile in the thought of the mathematician. One therefore says rightly that all things are immobile in mathematics.[123]

The Problem of the Absolute Clock
according to Graziadei of Ascoli

Of all the medieval Scholastic masters, Nicholas Bonet was the only one, as far as we know, to have developed to its fullest the doctrine we have just detailed. It was the logical extension of Ockham's analysis, but the Venerable Inceptor did not proclaim it formally. However, there is a philosopher whose theory is a clear reflection of Bonet's theory, either because he knew the writings of Bonet, or because his own meditations on this matter revealed to him thoughts similar to those of the Franciscan doctor; this philosopher is the Dominican, Graziadei of Ascoli.

Are there not as many distinct times as there are distinct movements? Graziadei proposed a solution to this problem in his *Quaestiones disputatae*.[124] He took it up again, more clearly and with more detail, in his *Quaestiones litterales*; we will therefore refer to *Quaestiones litterales*. There Graziadei wrote:

> Even though time is the measure of all movement, and there is, at the same time, a multitude of movements, there is, however, only one numerically single time, and not multiple times; that is so because the first movement is unique, and time concerns this movement first and properly.

> The above seems to be well stated, but there is a great difficulty with this reply.

> The diversity between the accident and the subject does not carry any multiplicity for existence; but existences are multiplied by reason of the multiplicity of the subjects in which they are immediately and properly applied. Socrates has his own existence and Plato has his own; existence is therefore multiplied because of their multiplicity.

> Similarly, the diversity between the first movement and the other movements might not carry to the multiplicity of times; but, it seems that there would necessarily be multiple times if there were simultaneously several first mobiles, and several first movements.

> In order to gain a clear understanding of this difficulty, here is what one should consider:

> Let something result from another, but without receiving from it its complete existence; this thing receives from the other only its foundation which awaits for its remainder, and this remainder is received from the soul. Sometimes, then, we see that the multiplication of the thing imparting the foundation does not carry to a multiplicity of the remainder [given by the soul]; it would carry to it only if the multiplication of this thing would produce a diversity in the *ratio* of the foundation it provides.

> We see, for example, individual men providing the same thing which is the foundation of the concept of man; and the concept receives its remainder from the soul. Since the various individual men are, according to the same *ratio*, the foundation of this concept, as we see, there can also be as

many as one wishes; however, the soul would form only one numerically single concept of the essence of man.

Now here is what is true of time: it receives its remainder from the soul, as we have said; it receives from the first movement only the foundation upon which this remainder rests. Let us suppose that there are several first movements; the foundation of time would issue from all of them, for the same reason . . . ; from then on time would receive from the soul only one and the same remainder of existence. . . .

We must therefore reply that if there were several first movements, time would be multiple in its foundation; but it would be uniquely one in its remainder, because of the unity of the reason conserved in its [multiple] foundation.[125]

Graziadei's doctrine can be summarized as follows: Time is an abstract concept formed by the intellect from every concrete movement; it remains always the same, whatever the concrete movements are from which it was formed. In its essentials this thought is identical to Nicholas Bonet's thought.

We have noted that the theory of the absolute clock proposed by Bonet was not connected to his atomic theory of time and movement; we now have a clear proof for this assertion, since Graziadei, who does not even mention the latter theory, teaches the former. Far from admitting with Bonet that time, which is discontinuous in reality, receives its continuity from the mind, the Dominican of Ascoli holds with Joannes Canonicus that time is continuous in its real existence and is distinguished into parts by the intellect. But this opposition between the two doctors does not prevent them from conceiving in the same way the existence of a unique time, serving as the measure of all movements.

Conclusion of the Problem of the Absolute Clock

All local movement supposes a term or fixed place to which the positions of the mobile are compared successively. What is this term, then, that cannot be moved and to which all local movement is referred? Aristotle said that it is the earth, for it is contradictory to suppose that the earth, the center of celestial rotations, can be moved. That is an error, responded Etienne Tempier and the doctors of Paris; God can, if He wishes, impose a movement of translation to the whole universe and to the earth which is located at its center. Hence, the fixed place, to which all local movements can be referred, is neither the earth nor any other body actually existing in nature,

for all the bodies of nature are or can be moving, asserted William
of Ockham; this reference is simply an imagined body. Nicholas
Bonet added that this imagined body is a simple concept, a geometric
figure existing only as *esse cognitum*, within the mind of the
mathematician. In this way, the whole Peripatetic theory of place
and movement was upset.

All change is accomplished in time. The determination of this
time requires the existence of an absolutely uniform, privileged
movement that marks the duration of all the other changes. Where
does one find this first clock? This first clock, said Aristotle, is
the diurnal movement of the first mobile. Since the first mobile
is perfect and absolute, it is necessary that its movement be an
absolutely uniform rotation. Moreover, since there can exist only
one world, this unique clock marks the same time for all the
movements accomplished in heaven and on earth. That is an error,
proclaimed Etienne Tempier and his counselors, for God can, if
He wishes, impart a movement of translation to heaven; it is an
error, because God can, if He wishes, create several worlds. Then
Nicholas Bonet and Graziadei d'Ascoli declared that the perfectly
uniform movement, the perfectly regulated clock that marks the
duration of all changes, does not exist in nature; it is a pure concept
that resides in the mind of the mathematician. It does not matter
whether the ultimate heaven actually speeds up or slows down;
the abstract sidereal day, as conceived by the astronomer, would
retain an invariable duration. In this way the whole Peripatetic
theory of time was ruined.

Moreover, Nicholas Bonet pursued his work; one can also
recognize the truth of what he stated about time and place with
respect to the geometer's measures. In order to measure lengths,
one needs a fixed length. Where can one find the standard whose
length remains invariable? Does the ruler made of wood, that one
calls a foot, have the same length it had yesterday? What certainty
do we have about this? Bonet replied that the immutable length
does not exist in any concrete bar of wood or stone, but in a shape,
abstracted of all matter, that the geometer conceives, and about
which he reasons.

Thus in all orders of magnitudes, the unity, the standard, is
not something that exists of real existence outside the mind; it is
an abstraction that has only a conceptual existence within the mind.

All philosophers, ancient as well as modern, agree that
unity is indivisible, for to be one, is to be an individual. But

let us note that this is understood from the mathematical point of view, insofar as unity is abstracted from all sensible matter. If not, it would not be true that unity is absolutely indivisible; in fact, the unity by which a piece of wood is one, which has its foundation in the wood (*quae subjective est in ligno*), can be divided by the division of the wood, as all the other accidents of the wood. In the same way that the piece of wood can be divided into pieces of wood, the unity of the piece of wood can be divided into other unities. But considered from the mathematical[126] point of view, the unity is indivisible, for to be one is to be an individual.[127]

The unity is therefore only a true unity for the mathematician who conceives it separated from all sensible matter; it cannot be actualized without ceasing to be a unity.

And it is the same for whole numbers. The number ten remains always the one and the same number ten, whether it is actualized in ten horses or ten dogs, because it is an abstract number. Once one takes it from conceptual existence to real existence—from *esse cognitum* to *esse realis existentiae*—the number ten ceases to exist.

8
Time in Fifteenth-Century Cosmology

Paul of Venice

Paul of Venice thinks himself a faithful Peripatetic with respect to what he asserts about the nature of time: "Time is not the movement of the supreme orb; it is simply a passion of that movement."[1] This is the proposition our author receives from Aristotle and Averroes. From them he borrows the following argument against those who identify time with the movement of the supreme orb: If there were several worlds, and hence several ultimate spheres, there would be several times—which is inconceivable.

But does not this objection turn against the Peripatetic theory that our author admits? "There corresponds multiple passions to multiple subjects. If there were several heavens, there would be several movements of these heavens; therefore if there were several heavens, there would be several times."[2] This objection has greater force for us than it had for Aristotle; according to Aristotle, whether or not one can imagine several worlds, at least there can exist only one. Christians, on the other hand, believe that God can create several worlds. By recalling this, Paul of Venice is clearly alluding to the decrees of 1277, as he had done many times. Therefore, one can suppose that there are several worlds, but one cannot suppose that there are several times.

This is a serious difficulty; here is how our author attempts to resolve it:

> Even if we agreed to the existence of several worlds and several first movements, one must not agree that there are several times. Time is a passion of the world, whether there is one world or several; similarly, *risibility* is a passion of

man, whether there is one man or several. Already there is a unique time all over this world, for the numerically same time exists in heaven and on earth, both east and west. Similarly, if there were several worlds, a numerically identical time would exist all over these worlds.

In the third book of his *Treatise on the Soul*, the Commentator admits that there is a single intelligence for all men; according to this theory, a numerically same intelligence exists in all the men of this world, or of any other world; the multiplicity of worlds would not carry to the multiplicity of the human intellect, any more than the multiplicity of men does. Thus a time numerically identical to itself—meaning a same hour, a same month, and a same year—would exist in all the worlds, and would be spread all over these worlds.[3]

In order for Paul of Venice's comparison to carry any weight, one would have to admit the Neoplatonic hypothesis of a time separated from the world, subsisting by itself, in the same way that the Commentator considers the active intellect as separated from all matter, subsisting by itself. However, "that time is an eternal substance" is an assumption Paul of Venice knows well and rejects, because "an eternal intelligence cannot have parts and cannot present a continuity."[4]

IV
Void

9
Void and Movement in the Void

The Void and Arabic Philosophy: Ibn Bajja

In 1277, Etienne Tempier, bishop of Paris, condemned the following two errors:

34 [27]. The First Cause cannot make more than one world.

49 [66]. God cannot move the heavens in a straight line, the reason being that He would then leave a void.[1]

The first of these condemnations denied what Peripatetic philosophy taught about the impossibility of infinite magnitude, both potential and in actuality. It required the medievals to take a new tack on the theory of the infinite. And during the fourteenth century the University of Paris split into two schools over this problem; but both the defenders of categorematic infinite magnitude and the holders of mere syncategorematic infinite magnitude equally clamored for the decree brought forth by Etienne Tempier.

The second of these condemnations upset the Peripatetic theory of place. It led the doctors of Paris to establish the philosophy of place and local movement on a new basis, and in order to do so they were required to make a great effort, to work out these problems until they were able to bring to light some thoughts that lay deeply hidden.

These two condemnations also contributed toward ruining the Peripatetic theory of time, enabling a new doctrine to come forth out of the wreckage.

Everything that Aristotle's *Physics* asserted about infinity, place, and time shattered when it was confronted by the power of the condemnations of Paris. New thoughts passed through the breach, many of which can be rediscovered, barely modified, in the writings of our contemporaries who philosophize about the principles of science.

But we have not yet described all the consequences of the decisions brought forth by Etienne Tempier and his counsel. We shall now see how they required Scholasticism to deny Aristotle's objections against the possibility of void.

Philosophers before Aristotle generally accepted, along with bodies, the existence of something that was not a body, but that was homogeneous and indefinite, in which three dimensions could be traced, and in which bodies were placed and moved; the atomists called it void [or vacuum] *Kenon,* and Plato called it space, *Xora.* Aristotle reacted most forcefully against this theory and its variations, which were defended by the majority of his predecessors. He attempted to demonstrate the impossibility of void and Platonic space. Above all, since his predecessors thought they had established that the existence of *Kenon* or of *Xora* was necessary for the possibility of movement, he attempted to demonstrate that rest and movement were inconceivable within the *Kenon* or the *Xora.*

Peripatetic physics did not last long among the Greeks; it was not long before it was eclipsed by Stoic physics. But Aristotle's doctrines about the impossibility of void did not await the triumph of Stoicism before being discarded; Strato of Lampsacus, it seems, had already strayed from the teachings of the master with respect to this point.

The possibility of void was one of the essential dogmas of Stoic physics. Although the disciples of Chrysippus and Zeno did not really believe that there was actually, within the world, spaces devoid of all bodies, as did the disciples of Leucippus and Democritus, at least they believed that an unlimited void did extend above the sphere delimiting the universe.

When the Arabic philosophers first came to know the philosophy of Aristotle, they adopted most of its essential propositions, particularly, those denying the possibility of void within the sphere limiting the universe, as well as outside it.

Soon one could hear al-Farabi formulate the following two propositions which are the echoes of the Aristotelian theses of book IV of the *Physics*:

"The limiting surface of the enveloping body and of the enveloped body is called place."

"There is no void."[2]

The Brothers of the Purity and Sincerity adopted the Peripatetic teachings about the void. In the fifteenth treatise of their encyclopedia, they asserted that "the word *void* designates a free place in which there is nothing; but place is a property of bodies, which cannot reside except in a body and which cannot be found except with a body. . . . Hence the existence of the void is absurd. . . . This rational demonstration proves that there is no void either inside or outside the universe."[3]

Avicenna also rejected the existence of void as formally as Aristotle had. In the description of his philosophy that al-Gazali, his disciple, had given, we can find a reference to the Stagirite's demonstrations from the fourth book of the *Physics* as support for the proposition that there can be no void.[4]

The philosophers of Islam, convinced followers of Aristotle, were therefore unanimous in their rejection of the void. The theologians, the Mutakallimun, on the other hand, saw Aristotle's philosophy as the great enemy of dogma, and thought only to take the opposite of this philosophy at all times; they were atomists, and hence, they believed in the existence of the void, without which the atoms cannot move. Maimonides reported:

> The men concerned with the roots [the Mutakallimun] believe likewise that the void exists, and that it is a certain space or spaces in which there is nothing at all, being accordingly empty of all bodies, devoid of all substance. This premise is necessary for them because of their belief in the first premise [the existence of atoms]. For if the world were full of the particles in question, how can a thing move? It would also be impossible to represent to oneself that bodies can penetrate one another. Now there can be no aggregation and no separation of these particles except through their movement. Accordingly, they must of necessity resort to the affirmation of the void that it should be possible for these particles to aggregate and separate, so that it should be possible for a moving thing to move in this void in which there is no body and none of these substances [meaning none of these atoms].[5]

Maimonides, after citing some of the inadmissible consequences of the hypothesis that continua are composed of indivisibles, adds, "you should not think that these doctrines I have explained to you are the most abhorrent of the corollaries necessarily following from those three premises, for the doctrine that necessarily follows from the belief in the existence of the void is even stranger and more abhorrent."[6]

Although the hypothesis of the existence of a vacuum in nature seems so absurd to Maimonides that he does not bother to discuss it, the judgment of other philosophers with regard to this hypothesis appears to have been less severe. One of the most original thinkers of Islam, Ibn Bajja, whom the Scholastics called Avempace, may have thought that the notion of empty space was not a meaningless notion; at least he rejected one of the objections Aristotle had raised against the hypothesis. In order to do this, he took up, almost verbatim, the reasoning by which Joannes Philoponus had attempted to deny that the fall of a weight would have to be instantaneous in a vacuum.

Averroes reports to us some of Ibn Bajja's concerns "from the seventh chapter of his book." They are as follows:

> In his fourth book, Aristotle evaluated the ratio between the resistance of a medium to a body moving in a medium and the power of a vacuum. But this ratio is not what one might believe it to be following Aristotle's opinion. The ratio holding between the density (*spissitudo*) of water and the density of air is not equal to the ratio of the movement of a stone in air with its movement in water. What is equal is the ratio of the cohesion (*potentia continuitatis*) of water to the cohesion of air with the ratio of the accidental retardation brought upon a moving body by the medium in which it moves, water for example, and the retardation accidentally brought upon it when it moves in air.
>
> In fact, if things are as some believe they are, natural movement would be a movement by constraint. And if there were no resistance, how could there be any movement? It would have to be accomplished instantaneously. Also, what could one say about rotation then? In fact there is no resistance with rotation; there is no division [of the medium], since the place of every circle remains the same—one place does not become vacant while another fills up. Therefore the movement of rotation would have to be accomplished instantaneously. But we can observe as rotative movements both the slowest movement—the proper movement of the fixed stars—and the fastest movement—that of diurnal rotation.
>
> All that is simply due to the difference in nobility between the motor and the moved thing. The more noble the motor, the faster is the thing it moves; when the motor is less noble, it is closer to the moved thing in nobility and the movement is thereby slower.
>
> Such are the words of Avempace.[7]

Averroes adds:

> If one agrees with what Avempace states, Aristotle's demonstration is false. If the ratio between the subtlety of

a medium and the subtlety of another medium is equal to the ratio of the accidental retardation brought upon a mobile by the medium and the accidental retardation brought upon by another medium, and not equal to the ratio of speeds, then it no longer follows that what moves in a vacuum does so instantaneously. Then, in fact, what is taken away from the moving thing is only the accidental retardation that was brought upon it by the medium, and its natural movement would remain, so that its movement would be accomplished in a finite amount of time. Thus what moves in a vacuum, necessarily moves for a certain divisible time, and no impossibility results. Such is the problem posed by Avempace.[8]

In his discussion of this opinion, Averroes multiplies the subtleties:

> Avempace judged that sensible movement is what remains of natural movement. He judged that natural movement is like a quantity (mensura) from which two other quantities can be assigned [in two separate circumstances], themselves proportional to two other quantities. He noted that when one operates with two such quantities, the ratio of the assigned quantities is not equal to the inverse ratio of the others. He judged that, because of the resistance of the medium, natural movement is diminished proportionally with resistance; as we have already stated, he judged that the sensible movement is the remainder of the natural movement after this diminution, somewhat like what is left of a magnitude when one lops off another magnitude from it. He therefore judged that the speed of the sensible movement in one case, to the sensible movement in another case, is not as the resistance in the first case is to the resistance in the other; this ratio is that of the retardations. . . .
>
> No one, before Avempace, had arrived at these answers; he surpassed all the others in depth.[9]

However, Ibn Rushd's admiration for Ibn Bajja did not go as far as to accept the opinion he just detailed; the confidence of the Commentator in the words of Aristotle defied any contradiction:

> Let us assert as a manifest truth . . . that the difference in the density of media is, other things being equal, the cause of the differences in speeds, and that the diversity in speeds whose cause is due to the differences in the subtlety of the media, essentially follows this subtlety. . . . It is therefore evident that the ratio of speeds is equal to the ratio of subtleties or densities of the media. . . . These are propositions which are self evident.[10]

Ibn Rushd rejected the reasoning of Ibn Bajja, but he sincerely admired the author for his reasoning; he said that Avempace had surpassed all others in depth. Never was such praise more justified; in fact, the author had so thoroughly examined this question that he had begun shaking up the foundations of Aristotle's dynamics. But the real innovator was not Ibn Bajja; it was Joannes Philoponus of Alexandria or perhaps Ammonius son of Hermeas, his teacher. Clearly, Avempace had borrowed this reasoning from Philoponus— a most fortunate borrowing. In fact, Western Christianity came to know this doctrine, which carried the seeds of part of Galileo's dynamics, through Averroes quoting Avempace.

We have already stated that it is impossible to establish any relationship between the first principles of this dynamics and the essential axioms of Newtonian dynamics.

Newtonian dynamics distinguishes two elements in a moving body, that which moves and that which is moved; force is that which moves, and mass is that which is moved. The respective magnitudes of these two elements provide the law of motion.

Peripatetic physics distinguished a motor and a moved thing for a moving celestial orb; the motor conjoined to heaven is what Hellenic or Arabic Neoplatonism called the soul of heaven, and the moved thing is the body of heaven.

Aristotelianism also distinguishes a motor and a moved thing for a moving animal—the soul of the animal and its body.

But the Philosopher and his disciples cannot allow this separation between the motor and the moved thing for an inanimate body, such as a weight. And since they want something to be opposed to the motor, something to prevent a thing from attaining instantly that toward which it tends, they seek the cause of this opposition in a resistance extrinsic to the mobile engendered by the medium.

That is the fundamental thought of this dynamics which Avempace shook up when he declared that the role played in the natural movement of a weight by the resistance of the medium is accessory and accidental, when he wanted the simple and essential law for this movement to depend only on the motor and the mobile, and on the comparison one can establish between them.

Averroes recalls the principles of Peripatetic dynamics in order to oppose them to Avempace's theory:

> It is evident that the resistance of the motor is in the moved thing, when the moved thing is in itself distinct; that is the disposition of celestial bodies. But within an element,

the moved thing exists only potentially, while the motor exists in actuality; an element is composed of prime matter and a simple form—the prime matter is the moved thing, while the form is the motor. Since one cannot actually posit a distinction between motor and moved thing for these bodies, it is impossible that they move in the absence of a medium. . . . In fact, if such a body moved without being surrounded by a medium, there would be no resistance toward the motor by the moved thing. Even better, there would be absolutely nothing that would be essentially the moved thing.[11]

By the last statement Averroes means that what essentially deserved to be called "moved thing" is that which resists the gravity of the weight, namely, the medium.

He said that it is true that in a weight the form is the motor and the matter is the moved thing, but he was careful to remark that the moved thing, thus distinguished from the motor, exists only potentially, so that the distinction in question cannot be posited actually.

He repeats this more precisely still in one of his commentaries on the *De Caelo*, where he again takes up the development of these same principles.

In a stone, in fire, and in other simple bodies, the motor and the moved thing are not distinguished actually as they are distinguished in animals. There the motor and the moved thing are one and the same thing with respect to their subject; their difference is only a difference in point of view (*secundum modum*). For example, the motor of the stone is its weight considered as a simple form; and the moved thing is also the weight insofar as it resides in prime matter. The cause is that the prime matter is not an actual being, for the stone is composed of gravity and prime matter. It is otherwise with an animal, which is composed of a body and a soul.[12]

The motive force of a falling thing is its weight; the moved thing is also the weight. The moved thing is therefore really identical to the motive force. That is the thought at the basis of Averroes's and Aristotle's reasonings. It took many centuries and much effort before the human mind was able to distinguish clearly two ideas in this single notion of weight, that of the form [force] which moves and that of the mass which is moved.

Let us affirm, [continues Averroes,] that there has to be some resistance between the motor and the moved thing. In fact, the motor moves the moved thing insofar as it is contrary to it, and the mobile is moved by the motor insofar as it is similar to it. All movement follows from the excess of motor power on the moved thing; any variation in speed or slowness follows the ratio of these two powers.

Resistance stems from the mobile itself when the mobile moves itself, voluntarily, and divides itself into actual motor and thing actually moved; that is the disposition of animals and celestial bodies.

Resistance, on the other hand, can stem from the medium in which the mobile moves; that is the case when the mobile does not divide itself into an actual motor and a thing actually moved. That is the disposition of simple bodies. . . .

The beings in which a self-moved thing divides itself into actual motor and thing actually moved do not necessarily need a [resistant] medium; if there were one, it would be so accidentally.

On the other hand, the beings that move themselves, but do not allow themselves to be divided into actual motor and actual mobile, necessarily require a resistant medium; these are heavy and light bodies. If they were not in a resistant medium, they would accomplish their movement in no time; in fact, there would be nothing actual resisting their motive force, and it would be impossible for them to take their natural movement if they were not constantly hindered. That is why, if we were to put them in a vacuum, they would accomplish their movement in no time, and a heavier body would move with the same speed as a lighter body, which is impossible.

Therefore, what Avempace thought, that without a resistant medium, simple bodies would have their own natural movements, is also impossible.[13]

Ibn Rushd attempted as much as possible to refute Ibn Bajja's theory; in doing so he rendered an inestimable service. He inserted this theory in his *Commentaries*, and with his detailed exposition and the lengthy discussion he devoted to it, he framed it so that a careful reader could not but perceive it. Thus the theory of Ibn Bajja benefited from the extraordinary popularity of Averroes's commentaries in the schools. No Scholastic would be able to examine whether the void is possible without having to think about the reasons of Avempace and his opponent. And the reasons of the latter uncover the roots of Aristotle's dynamics; they demonstrate that it is the roots themselves that one has to sever if one wishes to reduce the role of the medium in the fall of a heavy body to that which Ibn Bajja attributed to it.

And there came to be some Scholastics who believed, as Ibn Bajja did, that the true natural movement of a weight—its essential and simple movement—is the movement it has in a vacuum, and that the movement observed in a resistant medium is a complex movement consisting of the simple movement and the retardation introduced by the resistance of the medium. These Scholastics were the precursors of Galileo, Descartes, and Beeckman, who developed the theory successfully. And by following the trail of Avempace, because of the discussion of Averroes, they would know that they were substituting a new dynamics for Aristotle's dynamics.

The Impossibility of Void and Scholasticism before 1277: Ibn Bajja's Argument; Saint Thomas Aquinas and the Concept of Mass

When the Scholastics were introduced to Aristotle's reasoning against the possibility of void, they freely accepted its conclusions; they even accentuated the rigor of the Stagirite's condemnation. Can one wish, for example, for more formal support than the sentences by which Robert Grosseteste ends his brief chapter on the void from his *Summa* on the eight books of the *Physics*?

> In nature the plenum behaves in such a way that it cannot not be; therefore the void cannot be. . . .
> One can have only an indirect science of the void (*per accidens*); one can in no way have a direct science of it (*per se*).
> The void does not have a real definition, a definition of species. It admits only of a definition in name; it does not result that it is a being from its definition except as a manner of speaking (*nisi secundum vocem tantum*).[14]

In spite of the unanimity of thirteenth-century Scholastics toward the rejection of the possibility of void, Ibn Bajja's objection against one of Aristotle's reasons did not cease to preoccupy some of them. Although Franciscans like Roger Bacon, who often referred to the impossibility of void in his various writings, did not even allude to what Avempace had written, the Dominicans showed themselves to be more attentive toward this restatement of Joannes Philoponus's arguments.

Albertus Magnus gives a detailed exposition and an equally detailed refutation of the objection he attributed to Avicenna and to Avempace (although Averroes had affirmed the priority of the latter with respect to the objection).[15] What Albertus asserts is only a paraphrase of the Commentator's discussion in any case; no new conclusion emerges from his prolix paraphrase.

Albertus Magnus's teaching was long retained by the Dominicans; we can recognize its reflection in what Ulric of Strasburg asserted about movement in the void. After having demonstrated that weight would fall instantaneously in a vacuum, as did Aristotle, he relates, though somewhat obscurely, the objection he attributes to Avicenna and Avempace (as did his teacher). He rejects this objection because, in his opinion, it contradicts the rule that

> Aristotle had formulated in books VII and VIII of the *Physics*: if a motor moves something, for some distance, during some time, the motor would move half of it the same distance, during half the time. [If Avicenna and Avempace are correct,] the motor would be able to move the half more than the whole; it would be able to move half of it the same distance in less than half the time.[16]

No reason is given for this assertion.

Instead of retaining Albertus Magnus's lecture, Ulric would have done better to have studied what Saint Thomas Aquinas wrote about the movement of a weight in a vacuum.

Thomas Aquinas gives a concise analysis, both exact and penetrating, of Averroes's reasonings; specifically he takes up the exposition of the fundamental principles of Peripatetic dynamics (as did the Commentator), whose essential statement is as follows:

> But in regard to heavy and light bodies, when we subtract that which the mobile body has from the mover (meaning the form, which is a principle of movement and which the generator or mover gives), then nothing remains except matter, in regard to which no resistance to the mover can be considered. Hence it follows that in such things the only resistance is from the medium.[17]

But Thomas Aquinas rejects Ibn Rushd's reasons and judges them severely: "*Sed haec omnia videntur esse frivola.*"[18]

> When the form, which the generator imparts, is removed from heavy and light things, a body with magnitude remains only in the understanding. But a body has resistance to a mover because it has magnitude and exists in an opposite site [opposite to where the movement would lead it]. No other resistance of celestial bodies to their movers can be understood.

(*In gravibus et levibus, remota forma quam dat generans, remanet per intellectum corpus quantum; et ex hoc ipso quod quantum est in opposito situ existens, habet resistentiam ad motorem; non enim potest intelligi alia resistentia in corporibus caelestibus ad suos motores.*)[19]

Thomas thinks that this division between motor and moved thing that Peripatetic philosophy had declared impossible can be accomplished, at least in thought; thought distinguishes, on the one hand, a form, the motive force or gravity, and, on the other hand, prime matter given determined dimensions, not prime matter bare and simple, but a quantified body occupying a certain location and resisting the force attempting to bring it elsewhere. Even though this division of a weight into gravity and a body of determined magnitude can only be accomplished in thought, it suffices in order for us to be able to assimilate the movement of heavy or light bodies with the movement of celestial bodies; it also suffices to render inoperative one of Aristotle's objections against the possibility of void.

Thomas's assertion, which we have just quoted, is extremely brief; let us not allow its brevity to make us misunderstand its importance. For the first time we have seen human reason distinguish two elements in a heavy body: the motive force, that is, in modern terms, the weight; and the moved thing, the *corpus quantum,* or as we say today, the mass. For the first time we have seen the notion of mass being introduced in mechanics, and being introduced as equivalent to what remains in a body when one has suppressed all forms in order to leave only the prime matter quantified by its determined dimensions. Saint Thomas Aquinas's analysis, completing Ibn Bajja's, came to distinguish three notions in a falling body: the weight, the mass, and the resistance of the medium, about which physics will reason during the modern era.

A later chapter [*Le Système du monde,* vol. VIII, chap. 10] will demonstrate how John Buridan, during the fourteenth century, discerned the role played by mass in the movement of projectiles; he also identified this mass with prime matter quantified by determined dimensions.

This mass, this quantified body, resists the motor attempting to transport it from one place to another, stated Thomas Aquinas. This thought suggested another to him, the following peculiar thought:

The Commentator replies that the natural movement of light and heavy things requires this impediment from the medium so that there might be a resistance of the mobile body to the mover, at least from the medium. But it is better to say that all natural movement begins from a nonnatural place and tends to a natural place. Hence, until it reaches the natural place, it is not unsuitable if something unnatural to it is joined to it. For it gradually recedes from what is against its nature, and tends to what agrees with its nature. And because of this natural movement it is accelerated at the end.[20]

The science of the fourteenth century will do justice to this unfortunate explanation of the accelerated fall of heavy bodies.

What Thomas Aquinas had fashioned as support for Ibn Bajja's opinion was new and extremely distant from Peripatetic doctrine; moreover, it was expounded in very concise terms. That such a thought remained misunderstood by those one thinks of as the most faithful Thomists, and that it was even rejected by them, should not surprise us.

We ought not be surprised, in particular, by the language of Giles of Rome. He asserts:

In the movement of a heavy or light body, only the resistance of the medium, not the resistance of the mobile, requires time in order to accomplish the movement. In fact, here the reason for movement is the form, and if one were to abstract the form from such a heavy or light body, there would remain only the matter; but the matter does not possess anything in itself by which it can resist such movement. Therefore time cannot be required by the resistance of the mobile, but only by the resistance of the medium.

Perhaps one might say, doubtless when one abstracts the form, the matter does not retain any quality, since all qualities come from the form, but the matter retains quantity, because quantity comes from matter. . . . That would not suffice in order to require time for a movement to be accomplished; as we have already demonstrated, no time is required for a movement by reason of quantity only, without quantity being accompanied by some disposition contrary to movement, or some resistance, or some hindrance.[21]

In order to reject Thomas Aquinas's proposition that a body opposes all motive forces by a resistance due to the quantity of matter it contains, Giles of Rome invokes a reason Giles had previously developed, as we have just heard.[22] The object of this reason is the refutation of the following assertion: That which hinders the movement of a weight in a vacuum is simply the distance the movement must travel. Those who held this opinion did not attribute any role to the quantity of prime matter in the mobile. Their thought therefore had nothing in common with Thomist thought, and what Giles objected to, regardless of its validity, can prove nothing with respect to Thomas's supposition; our author's argument was therefore defective.

Still one can find a reference to Thomas's hypothesis that places in quantified matter the reason by which movement requires time in order to be accomplished; this reference can be found in what Giles of Rome asserts about movement in the void in his commentary oñ the first book of the *Sentences*. There, as in the first of his two arguments presented in his commentary on the *Physics*, the size of the mobile is not in question; only the distance traveled is invoked. It is the reason for the essential duration required by all movement; the resistance of the medium adds to it an accidental duration.

Our author expounds this doctrine with much clarity before attacking it; he refers to it as the doctrine professed by Ibn Bajja.

> In fact, according to the Commentator, Avempace posited two causes by which all movement requires time; one of the causes stems from the movement itself, that is from the distance of its terminations, and the other stems from the resistance of the medium. . . . He thought that if a space filled with water were of the same magnitude as a space filled with air, a mobile would require the same time to travel across one medium or the other, in virtue of the distance of its termination; but the amount of time would vary according to the resistance of the medium.[23]

Our author merely reproduces Averroes's argument against this doctrine, that is, he shows that the opinion of Ibn Bajja cannot be reconciled with Aristotle's dynamics; but to admit the truth of the latter's dynamics is to assume what is in question.

This illogical argument had a serious consequence; the proposition, what prevents movement in the void to be accomplished instantaneously is only the magnitude of the space to be traversed, was thought by some Scholastics to be an expression of Thomas's thought; they did not wait long before discussing it and leaving aside the true Thomist assumption which was profound in other ways and which contained many truths.

And the doctrine that Giles of Rome had fought against, which the discussion had represented as Saint Thomas Aquinas's doctrine, actually expressed Roger Bacon's thought.

The existence of a space devoid of any body, but having dimensions within which a body could exist and move, seems completely impossible to Roger Bacon; he argues against this assumption in his two series of *Questions on Aristotle's Physics*, in the *Opus Tertium*, and in the *Communia Naturalium*. However, after having demonstrated that the void cannot exist and that if it did exist all movement would be impossible, he poses the question, as Aristotle did: If the void could exist, and if a body could cross it, would it cross it in an instant, or would it take some time to cross it?

He first examines this problem in the second series of *Questions on the Physics*, a manuscript which is conserved in the municipal library of Amiens.

It is now an established fact that neither natural movement, nor circular movement, nor violent movement, nor any particular translation at all can be accomplished in the void; but now, with respect to translation in general, we pose the following question: If it were possible that a translation be accomplished in the void, would it be accomplished in an instant or would it take time?

It seems that it ought to take time; Aristotle stated in the eighth book of the *Physics* that the before and after of space are the cause of the *before* and *after* in the translation across the space. But the *before* and *after* in the translation cause the *before* and *after* in time; thus, in this translation there will be a *before* and *after* in time, so that the movement would be successive.

Moreover, any body is divisible and has a distance between its limiting surfaces; it has a prior and posterior part. Therefore it would cross the void by one of its parts before crossing by the other. Hence, there is a *before* and *after* in the parts of magnitude, therefore a *before* and *after* in the movement, and a *before* and *after* in time.

On the other hand, there is no relation between the void and the plenum, and the passage of a body across a plenum takes time; therefore, the passage of a body across the void would take no time. One sees that it was Aristotle's meaning in the text that such movement takes but an instant.

Nevertheless, if a translation were possible, it would be necessary to assume that it would be successive and would take time.

We would therefore reply to Aristotle's authority that it

must be understood as follows: the void is not an accident nor an incorporeal substance; hence, if the void were a separated space, it would be a corporeal substance, so that the void and the plenum would be the same thing; however, the void is nothing at all. Moreover, if for discussion (*gratia disputationis*) we admit that the void is a separated space, we would have to deny the proposition, there is no relation between the void and the plenum. However, since the void truly is nothing at all, Aristotle had reasoned well.

One can still state, if one refers to natural movement, that there is no relation between the void and the plenum, but that is not the case if one refers to movement absolutely. In fact, with respect to natural distinctions, there is no relation, but with respect to distance and dimension, there is a relation once we admit that the void has dimensions.[24]

The above opinion from the second series of *Questions on the Physics* was new for Roger Bacon; in fact, he had asserted the contrary opinion in the first series. Here is the reasoning he had developed in order to prove that all movement is impossible in the void:

All movement exists in becoming and succession. And in succession there is a *before* and *after* stemming from the *before* and *after* of magnitude. Thus the *before* and *after* of time are caused by the *before* and *after* of the magnitude which is moved. Hence the succession of movement and the *before* and *after* stem from the magnitude of the space. But there is no magnitude or corporeal space in the void. Therefore, a movement cannot exist in the void in any way—which I agree with.[25]

Bacon's thought therefore changed between the composition of his first series of *Questions on the Physics* and the second series. Afterward his opinion did not change.

The argument from the second series of *Questions on the Physics* is summarily reproduced in the *Opus Tertium*;[26] Bacon gives it the following conclusion:

I concede that the reasons given prove that a movement in an empty space cannot be accomplished in an instant; but it does not result that the movement will last for some time, for there is a third proposition between these two propositions, namely, that the void cannot allow passage to a body. If a body is allowed passage, it would result that some movement can be accomplished in void, as long as it were not a natural movement.[27]

Similar considerations can be found in the *Communia Naturalium*:

> Here, there surfaces a more important question, namely: If one admitted that the void allowed passage, would a change of place be accomplished instantaneously or would it require time? . . . If the void allowed passage, there would be a dimension of space that would create a *before* and *after* so that the void would not cede passage suddenly, and so that it would not be crossed in an instant, but little by little and in some time.[28]

We do not find anything of the fertile thought of Saint Thomas Aquinas—of the first perception of mass—in this theory to which Bacon returns several times with favor; what prevents a movement from being produced instantaneously in the void is, according to Bacon, the geometric divisibility that equally affects the motor and the mobile. Giles of Rome confused this theory with the theory of the *Doctor Communis* in his refutations of it; assuredly, this was not the means of publishing and clarifying the latter.

It came to be that one borrowed a portion of Aquinas's reasons in order to justify Roger Bacon's theory. We cannot furnish examples of the masters, contemporaries of Aquinas, who furnished the elements of such a synthesis, but we can present the union of these two doctrines in the teachings of the Dominican, Graziadei of Ascoli, at the University of Padua during the fourteenth century.

The care taken by Graziadei to follow exactly, in his *Lessons on the Physics*, the order that Thomas Aquinas had imposed upon his own commentary, shows the admiration that the professor of Padua felt for his illustrious brother of Saint Dominic; further, this section and the ones following will show us the extent of the influence of Roger Bacon's teaching—especially those on the void and weight—on Graziadei.

Graziadei twice discussed the possibility of movement in the void. One of the questions he disputed at the University of Padua has as title: "The existence of the void having been assumed, can movement be accomplished in it? Assuming that a weight fell in the void, would it have fallen instantaneously?"[29] Later, when he wrote his *Lessons on the Physics*, he devoted the ten questions of lesson 11 to the examination of these problems.[30]

Our author develops the reasoning that interests us in almost the same fashion in both expositions; therefore, we cite the later writing:

> According to what we said in our *Quaestiones Disputatae*,

in order to decide the question we have to consider the following:

One finds a certain resistance to the agent in the transmutation that leads matter to some form, so that there resides within the matter a disposition opposed to the disposition that the force of the agent must introduce, and that repulses the new disposition; similarly, it must be necessary that in the change in location of a mobile, we find some resistance to the motive force. It stems from the fact that the mobile has a certain opposed disposition that repulses the disposition which must be brought about because of its change of place. In fact, the disposition that the mobile acquires directly by local movement is called the *ubi*; also local movement is called movement toward a *ubi*. The disposition rejected by the mobile is a *ubi* opposed to the one which must be introduced, and that repulses the latter. The reason for this opposition is as follows: Two things are said to be opposed to one another and repugnant to one another when they cannot exist simultaneously in the same subject, but can only exist one after the other; and the *ubi* that stem from different parts of space can exist in the same mobile one after another, but they cannot exist simultaneously there. . . . It must therefore be that the *ubi* that stem from different parts of space are opposed to one another and repulse one another. Consequently, a resistance to the motive force must necessarily be encountered during the change of place of a mobile; that is because the mobile must be conceived as existing in a part of space other than the one which the motive force must bring it to.

But in a change of the form of the matter . . . what necessarily requires time is the resistance stemming from the contrary disposition existing within the matter; and the time required is as great as this disposition is distant from the disposition to be introduced—which it repulses. Similarly, one must assert that for a change in place of a mobile, time is required in virtue of the resistance, and that this resistance stems from the *ubi* existing in the mobile and which repulses the *ubi* intended to be introduced by the local change; other things being equal, this time is as great as the distance is great.[31]

The start of this reasoning drew its inspiration from what Saint Thomas Aquinas had stated; he had justified the existence of an intrinsic resistance by the mobile from the fact that it is in a situation opposite from that which it must take. But the conclusion brings us to Bacon's theory: If the movement of a body requires some time—even in a vacuum—it is because of the distance it must traverse, and because a mobile cannot be in two different places at once.

This deduction, allowing us to go from Aquinas's theory to Bacon's theory, conceals what was valuable in the former; it masks the first notion of mass which Aquinas had identified with quantified prime matter, the *corpus quantum*.

Graziadei, in any case, would have formally refused to attribute to the *corpus quantum* the resistance without which the motive force would lead the mobile to the termination of its movement instantaneously. Let us note the firmness by which he rejects Thomas Aquinas's doctrine:

> Prime matter cannot play the role of motor; that is self-evident. It is the same with respect to matter understood under corporality only and given magnitude and dimensions. In order to understand this well, let us take note of the following: One says that mathematical beings, taken according to their own essence, are immobile and abstracted from movement. Why? Because when considering them according to their own essence, one abstracts away natural qualities, and these are the principles of the various movements. On the other hand, one says that natural bodies are mobile, that they are related to movement because they are related to the natural qualities. But it is not so, neither for matter taken by itself nor for matter submitted to the form of corporality and quantitative dimension only. It is not related to any quality when it is this way, for the body, insofar as it is body and insofar as it is given magnitude only, is not a natural body, but rather a mathematical body. In order for it to conserve a natural quality, it must be taken under some natural form, such as the form of a heavy body or the form of a light body; it is under these forms that the natural quality of gravity or lightness results. Therefore, it is only when it is under gravity or levity that a body can play the role of mobile, of moved body; that is the principle and essence of its mobility.[32]

That a body can have a mass in virtue of its quantity of prime matter only is Thomas Aquinas's inspired perception, but it is a premature perception. The minds contemporary with the *Doctor Communis* were not prepared to recognize its truth; none had understood it. It is a thought that the physicists of the following centuries had to discover a second time, with much labor.

The Impossibility of Void and Scholasticism before 1277:
The Void and the Plurality of Worlds

For thirteenth-century Scholasticism, the proposition, the void is impossible, appeared as a kind of axiom whose negation would constitute a real absurdity. We have seen this declared by Robert Grosseteste. This axiom seemed able to serve as a major premise for some deductions. That is how the impossibility of void served to justify—by a method Aristotle had not used—the Peripatetic proposition, several worlds cannot exist.

We first find this argument in the commentary of the *Sphere* of Joannes Sacrobosco that Michael Scot had composed for the emperor Frederick II.

One of the first questions examined by Michael Scot is: Does there exist one or several worlds?

In order to prove the impossibility of several worlds, the noted astronomer summarily reproduces Aristotle's reasoning, but he precedes it with the following argument:

> Between the convex surfaces delimiting the differing worlds, there necessarily exists a certain amount of space. Therefore, either a body exists occupying this space or not. But there cannot be a body filling this space; this body, in reality, would be estranged from all worlds since it would be outside the spheres delimiting the worlds. If there is no body filling this space, it is then a void; and there can be no void in nature, as Aristotle has demonstrated in the fourth book of the *Physics*; therefore there cannot be a plurality of worlds.[33]

This demonstration had not been given by Aristotle; was it Peripatetic in spirit? Toward the end of the chapters of the *De Caelo* where the plurality of worlds is refuted, the Stagirite had written:

> It is evident that there is no place or void or time outside heaven. For in every place body can be present; and void is said to be that in which the presence of body, though not actual, is possible, . . . and outside heaven, as we have shown, body neither exists nor can come to exist.[34]

If Aristotle had admitted the simultaneous existence of several worlds, and had he also asserted that no body exists or can exist which does not belong to one of these worlds, it seems that he would not have had to alter the sentences we have just quoted in any way. He would have continued to declare that there is no void outside these worlds. The argument formulated by the astrologer of Frederick II would not have been taken as completely convincing by a strict Peripatetic. But has the weakness of an argument ever kept it from being in vogue?

The above argument appears to have seduced a number of philosophers of the thirteenth century, and not just the lesser ones. Among them we can cite William of Auvergne.

Let us suppose, said the bishop of Paris, that there are several worlds or an infinity of worlds outside one another. In addition to these worlds will there exist a body outside them and a stranger to them? Assuredly not. The existence of such a body is impossible; it is so for reasons similar to the ones invoked by the people who want to prove that nothing exists outside this world. "In fact, a world necessarily contains only the absolute totality of bodies or the totality of bodies that suit it; but we cannot imagine something that suits neither this world nor any other world."[35]

Since nothing can exist (whatever its nature) between these various worlds, then the various spherical surfaces containing them would need to touch each other, not only at a point, but along certain areas; in fact, no distance can separate the spheres from each other—"only the presence of an intermediary body can allow for a distance between the two bodies."[36]

Would one say that there is a void between these two worlds that nothing separates? No. William of Auvergne has established by means of various arguments that the void is an impossibility. Therefore the supporters of a plurality of worlds are driven to the absurdity: two spheres can touch not only at a point, but along a whole surface.[37]

This argument is not the only one William of Auvergne opposed to the plurality of worlds; he formulated others which we will detail in a later chapter devoted to this question and to the discussions it provoked [part V, chap. 12]; as for now, the problem of the plurality of worlds interests us only in its connection with the problem of the possibility of void.

We should not be surprised to recognize the influence of Michael Scot and of William of Auvergne in the writings of Roger Bacon. Bacon cites Aristotle's translator on several occasions, although he treats him harshly. As for the bishop of Paris, Bacon relates that he was taught by William of Auvergne in his youth.

There is, however, another influence on Roger Bacon, one he felt most strongly, that drove him to regard the void as an impossibility and authorized him to take this impossibility as an axiom capable of supporting a demonstration. This influence was Robert Grosseteste's.

From the beginning of his teaching, Bacon demonstrates that he wanted to admit the impossibility of the void. The *Questions on the Physics* seems to be a record of Roger Bacon's thoughts while he was master of arts at the University of Paris. And in the second series of these questions, dealing with the fourth book of the *Physics*, we see him respond to the query: "Must one admit the existence of the void?"[38] His reply is as follows:

> It appears that the void cannot be. In fact, if it were, it would be a substance or an accident. But the void is not an incorporeal substance, for it would then be a soul or an intelligence. Neither is it a corporeal substance, for then it would occupy a place. Finally, it is not an accident, for an accident cannot exist separate from a substance, and the void is a separate dimension. It is therefore nothing at all (*ergo nihil est*)—which is a truth I share with Aristotle, since he said that the void is nothing at all.[39]

Is that to say that even God's omnipotence is prohibited from the production of an accident separate from all substance—hence the void? Bacon does not grant this without some precautions. In the fashion of all Scholastics, he distinguishes between God's absolute omnipotence, for which nothing is impossible except that which is contradictory, and His restricted omnipotence, which cannot produce that which divine wisdom would disapprove. The production of void, prohibited from restricted omnipotence, is not prohibited from absolute omnipotence however. That is what Bacon teaches in the second series of his *Questions on the Physics* after giving the response we have reported.

In the passage that presently interests us, Bacon asks "whether one has to admit the existence of a vacuum below heaven."[40] He begins by clarifying the question, declaring that one could consider the void as a space separated from any body, but having dimensions (*dimensio separata*). He first enumerates the reasons one can invoke for the possibility of void understood in this fashion. The first of these reasons is as follows:

> The power of the First Being surpasses any finite act. And the existence of such a separate dimension is a finite act. Therefore, the power of the First Being can make such a void dimension exist actually.[41]

Bacon objects to this argument as follows:

> I reply that it is not true. In fact, a dimension is an accident, and an accident cannot exist without a subject. Making an accident without a subject is an act that is not in the order of things (*actus inordinatus*). This separate dimension can exist in virtue of the absolute power (*potentia abstracta*) of the First Being, for thus understood this power surpasses any finite act. But if we are speaking about what the power does, about the ordered power (*de debito potentie et ordinatione potentie*), then the power of the First Being does not surpass any act; it is equivalent to acts and effects in the order, which are possible according to the possibility of things (*ordinatis et possibilibus fieri secundum possibilitatem rei*); in this way it would appear that to produce the void would be to produce a substance, which entails the contrary of what is assumed.[42]

Here Bacon does not consider the existence of void as a pure absurdity, since he admits that God can create a separated dimension with His power taken absolutely. But everywhere else he expresses himself in a more formal, and more cutting, fashion. In the *Opus Tertium*, as in the *Communia Naturalium*, he repeats that separated dimensions cannot exist, for they would be accidents isolated from any substance: "*Accidens non potest per se stare.*" "*Accidens non potest esse sine subjecto.*"[43] The existence of the void appears as a pure impossibility in these formulations.

One can therefore make use of this impossibility in the fashion of an axiom and deduce, for example, the impossibility of the existence of several worlds from it.

In his *Opus Majus*, Roger Bacon devoted a chapter to the examination of these two questions: Can there be several worlds? Does the matter of the world extend to infinity? Here is what one can read in this chapter:

> Aristotle stated in the first book of the *De Caelo* that the world collects all its own matter into a single individual of a single species, and that it is the same for each of the principal bodies making up the world; in this way the world is numerically unique, and there cannot be several distinct worlds belonging to the same species (and there cannot be several suns, nor several moons either, even though many have imagined such things).
>
> In fact, if there existed another world, it would be spherical like this one. These two worlds would not be distinct from one another, since if they were, there would be an empty space

between them—which is false. They would have to touch; but by proposition 12 of Euclid's third book of the *Elements*, they could only touch at one point (as it has been previously demonstrated by means of circles). Therefore, at every place other than this point, there would be an empty space between them.[44]

In the *Opus Tertium*, Bacon summarily takes up Aristotle's argument against the plurality of worlds;[45] he does not appeal to any reasoning derived from the impossibility of void. But he again takes up this reasoning when he writes his *Communia Naturalium*, or better, when he writes the treatise, *De Caelestibus*, which the famous manuscript of the Bibliothèque Mazarine has as the second book of the *Communia Naturalium*. After having summarized Aristotle's argument, he writes:

> We can add a mathematical demonstration to this argument.
> If there were several worlds, one would have to admit the void—which has already been refuted generally. This consequence is evident (by the twelfth proposition of Euclid's third book) to anyone who knows the purity of geometry; I assumed this purity of geometry throughout the *Naturalia*, since I have expounded it previously. In fact, if these worlds are distant to one another everywhere, there will be a void. If they are conjoined, they would be so only at a point (by the twelfth proposition), and their convexity would separate them from one another.[46]

In any case, the chapter of the *De Caelestibus* in which Bacon denies the plurality of worlds is soon followed by a long chapter which develops Aristotle's thought, that outside the world there does not exist and there cannot exist any body, so that outside the world there is no filled or empty space. Bacon writes:

> We have previously demonstrated, in our general discourse, that the void cannot exist in nature, but we have only demonstrated that it cannot exist below heaven. Until now we have reserved the special consideration about the void above heaven. When during the several occasions I assumed the nonexistence of the void, I assumed it in virtue of the general consideration by which it has been proven that the void cannot exist. I now assert that in a similar fashion it cannot exist outside heaven.[47]

A lengthy argumentation, often confused and sometimes faulty, supports the above assertion.

Because of Michael Scot, William of Auvergne, and Roger Bacon, the proposition, the world is necessarily unique, became so intertwined with the proposition that the void is impossible, that the anathema that condemned the former also struck at the latter.

The Condemnations of 1277 and the Possibility of Void

The impossibility of void was considered by many authors, such as Grosseteste, as an axiom whose negation would imply a contradiction. This axiom seemed to them capable of serving as the major premise of a deduction; we have seen it used in this fashion in the demonstration of the proposition, there cannot be several worlds.

God cannot make anything contradictory; He cannot therefore make an empty space. Consequently, any effect that would necessarily entail the production of an empty space is prohibited even for God's omnipotence. That is how the authors—unknown to us—who formulated the proposition, God cannot give the universe a rectilinear movement because the world would then leave a void behind it, must have reasoned.

In 1277, Etienne Tempier condemned the above proposition, and at the same time he struck down the assertion that God cannot create several worlds. Obliged to think these two theses erroneous, several Paris doctors believed that they had to hold the production of the void as a possible thing, at least with respect to God's omnipotence.

Godfrey of Fontaines

There are those, however, who in their lectures on the *Sentences* or in their quodlibetal discussions, while agreeing that God could create several worlds, attempted to safeguard their belief in the impossibility of void; Godfrey of Fontaines can be numbered as one of them:

[The existence of two worlds] does not require one to admit the void. The void, in fact, is an empty place (*locus inanis*); it is the surface of a body, a surface capable of containing another body, but containing none. If there existed another world, it would have its own place as does this one. There would be no void between them, for there would be nothing there capable of containing something, and containing nothing; such a thing can only exist because of our imagination, in the same way that we imagine the void outside heaven.[48]

Henry of Ghent

As we have already stated, the above remark appears to conform with Aristotle's doctrine.

But it is valid only if we assume that the two worlds coexist from all eternity; it ceases to be valid if we were to pose the question as the Scholastics did: Can God actually create another world outside this world? To answer yes is to deny Aristotle's proposition that there is no body and no body can be engendered outside the world. And that is the proposition that allows us to conclude that there is no place or void above heaven.

The question that Henry of Ghent attempts to answer in one of his *Quodlibets* is, "Can God create, above heaven, a body not contiguous with heaven?"[49]

The Solemn Doctor answers that "God can create a body or another world beyond the ultimate heaven, in the same way that He has created the earth within the world or within heaven, and in the same way that He has created the world itself and the ultimate heaven."[50]

But where would this new body, or this new world, be created? Is there an empty space outside the world, some *separated dimensions*, as the Stoics would have it? Must we say that the new body or the new world is created in this void or in this space? When expressing himself with respect to these problems, Henry of Ghent clings to the teaching of the Stagirite; he is still firmly convinced—perhaps too much so—that there is no place, no void, outside the world.

> [The body or the world that God can create outside the world] would not be produced in something, but in nothing (*in nihilo*). We must not understand these words materially as if nothingness were something. We must understand that the body succeeds the nothingness because it is created where there was nothingness before; that is not to say that there was something there like a separated dimension (*dimensio separata*) and that nothingness was in this something—that there was something there, something in which the dimensions of the body could have been received after having chased away the nothingness existing in this something. One must understand the proposition completely negatively, as if one said there is not something there, understanding by that that

one is denying both the existence of place (*ubitas*) and the existence of something (*aliquitas*). It is in a similar fashion that we assert: this body or this world has been made from nothing.[51]

Thus God, when creating a new world, would not create it where there was an empty space before; the existence of the new world would no more be preceded by the void than the existence of this world was. Henry of Ghent was certainly in agreement with Etienne Tempier with respect to the condemnation of the following error: 201 [190] "That He who generates the world in its totality posits the void, because place necessarily precedes that which is generated in it; and so before the generation of the world there would have been place with nothing in it, which is the void."[52]

God can therefore, above the ultimate heaven, create a new body or a new world. Can He create this body or world in such a fashion that it does not touch the ultimate heaven? Roger Bacon and all of Peripatetic physics denied this. Between the two worlds or between our world and the new body there is no other body; therefore, there is no distance between them, for the distance between two bodies is an attribute, an accident of the bodies interposed between the two bodies. The existence of a distance between the two worlds, though there is no body between them, is equivalent to the existence of an empty space between the two worlds; according to Peripatetic philosophy, the two existences are expressed by one and the same proposition, and this proposition implies a contradiction.

That is not the conclusion of Henry of Ghent; he introduces a subtle but essential distinction.

> I claim that two bodies can be distant from one another in two different ways.
> In the first way, they can be distant properly speaking (*per se*); that is what happens when there is an actual distance (*positiva*) between them because of the dimension of an interposed body.
> In the second way, they can be distant accidentally (*per accidens*). In this case, there does not exist any actual distance (*positiva*) between them, but beside them or outside them, there is an object in which a dimension is realized, and this dimension allows the recognition of distance between the two bodies.
> Let us suppose, for example, that there is a void between two bodies, and that these two bodies touch a wall three feet

high, one touching the top and the other touching the bottom; we would then say that three feet is the distance between the body above the void and the body below the void.

Then if there is nothing between two bodies, but if a body of some dimension can be received between them, we would judge that the interval between these two bodies has precisely the same dimension, but that it has it *per accidens*.[53]

In this way the Solemn Doctor defined how it is possible to attribute existence to the void: the void is nothing other than the dimension or distance between two bodies, between which there exists no other body,

a distance which, as we have said, exists only *accidentally*, either because a dimension may be actualized (*positiva*) along these bodies, or because an actualized dimension (*positiva*) is capable of being placed between these two bodies or along them.

The void itself has no other existence than an *accidental* existence, in that the bodies between which it exists are disposed in such a manner that the dimension of a body is capable of being placed between them.[54]

Let us cite an example Henry of Ghent furnishes. In another quodlibetal discussion, he asks whether God can bring the void into being.[55] He replies affirmatively to this question and, in order to justify his response, he takes up, though with fewer developments and less depth, the considerations we have just analyzed.

To this end he imagines that God destroys all the elements between the earth and the lunar orb, without changing the magnitude and location of the two bodies. There will then be a void between the two bodies, but it will exist only *accidentally*. This purely accidental existence will consist in the fact that God could make the destroyed elements exist actually, and that water, air, and fire could be placed between the earth and the lunar orb. The thickness of the spherical layers formed by the three elements susceptible of being lodged between the terrestrial element and the lunar orb would be the *accidental* distance between the two bodies.

The Solemn Doctor attempts to distinguish between the void (*vacuum*) as just defined and the nothingness (*nihil*) outside the world. Above heaven the void does not exist, even *accidentally*: "In fact, there is no distance *per accidens* there, because there is no body capable of being received in some intermediary void."[56] Therefore, above heaven, there is no plenum and no void, as thought the Philosopher.

After a new body or a new world had been created by God, above the ultimate heaven and not in contact with it, we would assert that between the [new] body or the [new] world and the ultimate heaven there is a void, and that this void would have a determined dimension, that is, the dimension of a body which could have been received between the ultimate heaven and the newly created body; but we could not assert that there is a void in places other than between the heaven and the new body. Similarly, we cannot presently assert that above the ultimate heaven there is a plenum or a void, but only that there is a pure nothingness.

Therefore, if God were to create now, above heaven, a body tangent to heaven, this body would not have been created in the plenum or in the void, but in the pure nothingness; and on the side not related to heaven, the body would continue to subsist in absolute nothingness, nothingness being taken as a pure negation. Similarly, heaven was created in pure nothingness, and pure nothingness was once where this body is now located—all that must be understood in a purely negative sense, in the manner we have indicated.[57]

The body newly created by God would then border the void on one side and nothingness on the other side. And "if the elements contained by heaven were destroyed, we would have to admit that the void exists in the concavity of heaven, but we would not have to admit it in any way above heaven; there would be only pure nothingness there."[58]

The corollaries that Henry of Ghent deduced from his theory themselves condemn it. This body created above the ultimate heaven is in the void on the side near the ultimate heaven and in nothingness on the other side; how could one designate on the surface of the body the boundary between the plane bordering the void and the plane touching nothingness?

The attempt by Henry of Ghent both to attribute to God the power of creating a new body outside the world and to accept the Philosopher's doctrine that there is neither plenum nor void outside the world was condemned to failure in advance; the first assertion entailed the ruin of the second.

Richard of Middleton

The doctrines of Richard of Middleton are generally a reflection of the teachings of Henry of Ghent. With respect to the possibility of void, this reflection is so weak that it almost seems to disappear. The lecture of the English Franciscan on this subject is not, however, devoid of interest; it can demonstrate the decisive role that the condemnations, brought forth by the "articles of Paris," played in the discussions occupying us.

Richard agrees that God could have created another world. Among the many reasons he invokes for this conclusion, let us cite only the following:

> I state that God could have and can still now create another universe. There is, in fact, no contradiction in attributing this power to God.
>
> Such a contradiction cannot originate from that which the universe might have been fashioned, since God did not make the universe from anything.
>
> It does not originate from the receptacle of this universe, since the world in its totality is not in some place (*spatium*). The Philosopher stated in the first book of the *De Caelo* that there is no place, no void, no time outside heaven; and that is how we ought to understand the ultimate heaven. . . .
>
> In order to establish this opinion, we can invoke the sentence of Lord Etienne, bishop of Paris and doctor of sacred theology; he has excommunicated those who teach that God could not have created several worlds.[59]

Richard of Middleton, like his teacher, Henry of Ghent, wishes therefore to reconcile Etienne Tempier's decision with Aristotelian dogma: there is no place, no time, and no void above heaven.

A more difficult task is the attempt to reconcile Peripatetic dogma with the proposition that it is not impossible for God to impart a movement of translation to heaven.

Our author remarks that "in fact, any movement of translation transports a body from one place to another. But according to the Philosopher in the fourth book of the *Physics*, the ultimate heaven is not in a place; and according to him in the first book of the *De Caelo*, there is no place, no plenum, and no void above the ultimate heaven. It is therefore impossible for God to move the ultimate heaven by a movement of translation."[60]

However "the following article has been excommunicated by Lord Etienne, bishop of Paris and doctor of sacred theology: God cannot move heaven by a movement of translation."[61]

The contradiction between the teaching of the Philosopher and the teaching of Etienne Tempier is difficult to resolve.

Richard allows that God can move a portion of the firmament, or any body whatever above the firmament. The remainder of the firmament, remaining immobile, would furnish a fixed reference for the local movement of the body. But the difficulty remains with respect to the movement of the ultimate heaven or the universe taken in its totality.

> [God can impose on the universe this movement of translation] provided that He creates a space (*spatium*) outside the universe. In fact, it is impossible for any power whatsoever to move an object in its totality by a movement of translation unless there were a space external to it. If there were no creature except for an angel, God could not move this angel by a movement of translation except insofar as He creates a space outside this angel or around this angel.[62]

Richard adds that "there is another defect in the argument [condemned by the bishop of Paris]. If God gave a rectilinear movement to the ultimate heaven, a void would not be produced because heaven is not in a place."[63] Richard forgets what he has just taught; before imparting such a movement to heaven, God would necessarily have created a space outside heaven. It is therefore true that a void would be created by the displacement of heaven.

Richard in any case admits that the existence of the void is not contradictory. His thought on the subject appears to be in conformity with Henry of Ghent's, although he expresses himself less clearly.

> God can destroy the substance created between heaven and earth without moving heaven and earth. Afterward, heaven would not be distant from the earth; in fact, there has to be a dimension between two things that are distant locally, and any dimension is something created. But afterward, heaven would not touch the earth either, because, without changing either one or the other, God can create a distance between them. Thus, to be distant and not to be distant are contradictory and to be joined and not to be joined are contradictory, but, by means of what has been asserted, not to be distant, and at the same, not to be joined does not imply a contradiction. . . . God can make it be that there is a void; it does not result from this that He can make two contradictory things coexist.[64]

Ramon Lull

Henry of Ghent and Richard of Middleton attempted to maintain Aristotle's assertion that there is no plenum, no place, and no void above the ultimate heaven. They resolutely opposed those who, following the example of the Stoics, wished to extend outside the world a space (*spatium*) or separated dimensions (*dimensio separata*). But this put Henry of Ghent in an embarrassing position when he wished to allow that God can create a body contiguous to the ultimate sphere outside the world; and, in order to explain how God can give the whole universe a movement of translation, Richard of Middleton had to concede that God would first create a space around the universe.

Ramon Lull, on the other hand, seems to admit the existence of a space existing before the world, having been created by God before the world so that the world would have a place. It is because of the existence of this place separated from the world that the world can be moved by God with a rectilinear movement. Such is the doctrine that Lull has his interlocutor, Socrates, expound (in a rather obscure manner) in order to refute the proposition condemned at Paris: God cannot impart a movement of translation to the world because it would leave a void behind it.

> Let your reason rise above your imagination, Socrates, and consider this: When God created the world, He created place so that the world could be lodged in a place. In the same fashion, He created the principle by which the world could have a principle (*esse principiatus*); similarly, He created time so that the world could exist in time, and similarly for quantity, movement, and things of that kind, so that the world could have quantity and movement. God therefore has created a place in which the substance of the world resides (*in substantia mundi sustentatum*). The world is lodged in this place. In the same way that your body can move from one place to another without abandoning its essential place, its color, or its surface, God can move heaven by a movement of translation without the world leaving its essential place.[65]

William Varon

The hypothesis of a space distinct from the world in which the world has its place appears to have the support of William Varon.

William Varon devotes a whole question in his commentary on the *Sentences* to establishing that God can create a world outside this world.[66]

Among the objections that can be raised against this conclusion, Varon discusses the following:

> It is impossible for God to make a void, because the existence of the void implies a contradiction; and if there existed two worlds, there would exist a void. If there existed

two spheres, they could only touch absolutely at a point, for they would be round shapes, like two apples; since there would be nothing between the surfaces that delimit the worlds, there must then be a void.[67]

This objection is the one to which our author gives the greatest development. He details the reasons that philosophers hold the proposition that two bodies touch when there is no positive distance between them, a distance actualized in a third body. He refutes the objections one can give to this opinion, which he seems to make his own. Nevertheless, we hear him formulate the following conclusion:

> Outside this world, which is spherically shaped, God can make another spherical world that does not touch the first world; God can do this because it does not imply a contradiction. The reason by which He can make it be that the parts of one heaven are distant from the parts of another heaven is also the reason by which He can make it be that a whole heaven is distant from another whole heaven according to His will. In fact, the creation of this world has not diminished His power.
>
> Before the creation of this world, there was absolutely nothing where this world is, and God created this world; therefore, He can do the same outside this world. One can in fact imagine a quasi-infinite space in which, however, there is absolutely nothing (*contingit enim ymaginari spatium quasi infinitum in quo tamen penitus nihil est*); in the same way that He created a world where there was nothing, He can create an infinity where there is absolutely nothing—a potential infinity, by which I mean that He can never have created so much that He cannot create more.[68]

John Duns Scotus

Duns Scotus's doctrine on the void is the same as Henry of Ghent's doctrine. In spite of Aristotle's objections, he allows that God can produce an empty space.

> One does not see that it must be contradictory to admit a concave surface without admitting any relation between the surface and the body that would be contained, provided, however, that there exists a body naturally capable of being contained by this surface.[69]

The Subtile Doctor, who has just declared possible the existence of a body contained by nothing—such as the final celestial sphere—

rightly remarks that this possibility must extend to the body containing nothing.

> God can easily realize a concave surface containing nothing:
>
> God can absolutely annihilate the elements without changing anything with respect to the existence of heaven; that posited, the inner surfaces of heaven would not be reunited instantaneously, for nature cannot accomplish such a change in an instant; the concave surface of heaven would then subsist, and this surface would contain no body.[70]

What then is the force of Aristotle's argument against the possibility of the void?

> This demonstration holds only if one thinks of the void as a space devoid of natural qualities, but actually endowed with dimensions (*spatium actu dimensionatum, licet non habeat qualitates naturales*). . . .
>
> But the void that we are asserting possible with respect to God is not a space that has positive dimensions; there is only the possibility of receiving positive dimensions of some magnitude there, and at the same time the absence of all dimension in actuality (*possibilitas ad tantas dimensiones positivas, cum carentia cujuscunque dimensiones in actu*). . . .
>
> By an interval (*medium*) one can understand either a positive and actual interval or a privative and potential interval. In either sense there is no interval between two bodies touching, or between two surfaces that coincide. Between the inner surfaces of a vacuum there is no interval in the first sense, but there is an interval in the second sense; a body can be included between these two surfaces, a body equal to the body included when the space is actually full. Therefore, there is a potential intermediary, and as a result, it is a privative intermediary (in which there is an intermediary in a privative fashion), since these inner surfaces are missing an intermediary equal to the one which can be included between these two extremities. . . .
>
> Formally, distance consists in a relation between two extremes; this relation resides in one of the two extremes and has the other as term. Here we can assign two positive terms [to the distance between the inner surfaces of an empty space], even though there is no positive interval. If a relation can be called positive when only the two terms between which it is established are positive, then we can allow that there

is a positive distance there; but if [in order to call the relation positive] we also required that the interval be positive, we would have to say that there is a privative and potential distance there, and not a positive and actual distance.[71]

Duns Scotus relates his theory of time to this theory of space (in which the thought of Henry of Ghent so clearly appears).

If the first heaven moves, there is a positive and actual time; between two given instants, a positive and actual duration flows— meaning, a real portion of the diurnal movement. If God were to stop the first heaven, there would no longer be positive and actual time, but there would still be privative and potential time. If one can conceive two instants in the movement of any body or even in the rest of heaven, there would not be a positive duration between these two instants because no portion of the diurnal movement would actually be accomplished between these two instants; but a potential duration of determined magnitude would flow between them, meaning, if the first heaven had continued to move, there would have been accomplished a determined portion of the diurnal movement.

> It is the same in the case we are concerned with and in the case of the positive and privative distances about place. In the same fashion that privative duration measures the successive parts of something, privative distance measures the permanent parts. . . . Thus one can report the following argument in support of what we wish to establish: It is possible that there is a distance in time, so to speak, between two things, even though there is no positive intermediary between these two things, meaning, no positive measure of the [celestial] movement; in order for there to be a kind of temporal distance between them, it suffices for there to be a potential time taken in a privative manner; that suffices in order to say that one comes before the other or after the other at such a temporal distance. By similitude, we can conclude that it is the same for place and local distance.[72]

Aristotle, in his argument against the void, often attacked those who thought of the void as a body endowed with magnitude, with dimensions, but devoid of all other physical properties. He took care to remark that his argument was valid against the *Xora* of Plato as well as the *Kenon* of the atomists. Duns Scotus concedes to Aristotle that such a void cannot receive a body. He no more allows the compenetration of bodies than does the Stagirite; he does not even conceive the possibility of a total mixing similar to the one the Stoics considered—thinking of place as a penetrable body, in the fashion of Syranius or Proclus, does not even come to his mind. According to him, the void is not actually separated dimensions—a real body although one having no properties other than magnitude—the void is only potential dimensions—the possibility to receive a body of determined magnitude and shape. Like potential time, it is not a movement; it is only the possibility to receive a movement of determined duration.

What are these two possibilities exactly?

Are they two concepts? Surely yes; Duns Scotus explicitly declares that we have the distinct awareness of potential time. Are they only two concepts? Certainly not; they are things which, if they are not realized, at least they are realizable. Potential time will be realized after the end of the world, when the diurnal movement will be stopped; it was realized when Joshua bade heaven to stop. If God wished to annihilate the elements without changing heaven in any way, then void, or potential space, would be realized.

What is the nature of these realizable possibilities? To this question Duns Scotus says no more than what we have reported.

Joannes Canonicus

Because of the influence of Richard of Middleton and Duns Scotus, the possibility of void, at least with respect to divine power, appears to have been generally accepted by the Franciscans.

Gerard of Odon, for example, clearly believed that the existence of the void was not contradictory; however, his atomism did not require that the void be actually realized, for according to him the indivisibles were welded to one another in a continuous manner.

Joannes Canonicus borrows almost everything he writes about the void from Gerard of Odon, as he takes care to remind us. He expresses himself as follows:

> Unless shown otherwise, I believe that God can destroy all intermediary bodies below the ultimate sphere without the ultimate sphere being affected. . . .
>
> In fact, heaven is an absolute thing, essentially distinct, not dependent in any way on the lower spheres. It therefore does not seem impossible that God can make it be—although it is against nature—that there is no fire or air, or any other body from earth to heaven. As a result, there can exist a space separated from all bodies, but capable of receiving a body. God can, in fact, create anew an earth, some air, and the other elements, and these bodies can place themselves below heaven

as before. Any reasoning that can be raised against this conclusion stems from purely natural principles.[73]

Petrus Aquilanus

While admitting that God can, if He wishes it, produce an empty space within the world, the Scotists did not seem to have thought it necessary, in general, to admit the actual existence of any empty space outside the world. It did not seem to them that this assumption is necessary in order to safeguard God's power to create a new world outside the limits of this world.

We can find an explicit assertion of this point of view in the commentary on the *Sentences* written by the Franciscan, Petrus Aquilanus, surnamed *il Scotello*.

Petrus Aquilanus teaches that "everything that implies no contradiction is feasible for God, and producing a second world of spherical shape implies no contradiction."[74]

One can object as follows against the conclusion that can be drawn from the above:

God cannot produce the void; and if He were to make another spherical world, the void would necessarily result, for the two worlds would touch only at a point.[75]

Il Scotello replies as follows to this objection:

It is not absurd to suppose that God can make a void; moreover, it appears necessary to admit it. If one supposed, in fact, that God annihilated the totality of air and everything that is located between the inner surfaces of a container, the annihilation would be accomplished in an instant; if afterward nature had to fill this space, it could only accomplish it during some time. Therefore, the space would have to remain empty for at least an instant.

I assert, in any case, that admitting the existence of two spherical worlds is not the same as supposing the void; in fact, even though these worlds touched only at one point, there would be outside them neither plenum nor void.[76]

Robert Holkot

The above did not suit all Scholastics; it did not suit Robert Holkot in particular. That the power accorded to God to create a world outside this one implied not only the possibility but the real existence of an empty space outside the world is what our Dominican formulates with a clarity that had not been attained by anyone else.

If God had the power to create another world, it must be that He would create it somewhere (*alicubi*), as this world was, in such a way that there are distances between the various portions of that world. But, I ask, what is actually there, where the world was created, nothing or something? If there is something, there is then, in fact, something outside the world. If there is nothing, one can reason thus: outside the world there is nothing, and outside the world there can be something; therefore, outside the world, there is a void. For, where a body can exist and where there is no body, there is a void. Hence the void must now exist.[77]

Holkot does not reply to this argument; but since a little further on one can hear him declare that there is no contradiction in supposing that God can create a second world,[78] we ought to think that he upholds the position that he has just formulated: the void now exists.

In any case, elsewhere Holkot explicitly proclaims the possibility of an empty space.[79]

Walter Burley

During the fourteenth century, then, the theologians appeared unanimous in declaring that faith in the omnipotence of God required one to believe in the possibility of a void inside the world and its actual existence outside the limits of the universe. It is a point at which the consequences of revealed dogma are in absolute contradiction with the teachings of the philosophers.

In this conflict, what was the attitude of those who reasoned according to the principles of natural philosophy, and who, since they were not theologians, did not have to take revealed truth as principles for their deductions? What did the masters of arts say?

Faithful to the system he had formulated many times, John of Jandun simply exhibited the teachings of Peripatetic physics concluding for the impossibility of the void; he did not even allude to the theological reasons concluding otherwise.

But the majority of the masters of arts were not able to maintain this serene impassivity which seems to ignore even the existence of the conflict. Burley, in what he stated about the void, showed himself to be the faithful disciple of Aristotle and Averroes; however, he did not allow his readers to ignore the fact that Catholic theology required them to admit conclusions contrary to those which had been formulated by the Philosopher and his Commentator. He indicated what the decisions taken by Etienne Tempier required of them.

> Those who admit the creation of the world must hold the following:
> In the same fashion that God has created a discontinuous world, formed of various parts, a discontinuity in virtue of which the various parts of the world are properly in a place, God could have created a body absolutely continuous in all its parts and have created nothing other than this continuous globe. Let us then suppose that when God created the world, instead of this world He created an absolutely continuous spherical body. Since every body is in a place, this spherical body would also be in a place; it would not be in a place because of its parts, however. In fact, none of its parts would be in a place, since place is a divided containing body, and the body in question is absolutely continuous. It remains then that the body is located in the void. Thus, those who admit the generation of the world must also hold the existence of the void.
> Let us assert with respect to this that the theologians of various religions (*loquentes cujuslibet legis*) affirm that God can create a similar, absolutely continuous, spherical body that would fill the space occupied by this world. Once this is assumed, those who speak from the point of view of physics (*loquentes physice*) must hold that this body would not be in a place, for it cannot be so either because of its parts or because of the ultimate part of the containing body, since there is nothing outside it, nothing containing it. They would conclude that it is not part of the nature (*ratio*) of body to be necessarily in a place.[80]

In fact, this conclusion is the one which Duns Scotus had formulated.[81] Burley pursues this as follows:

> But someone might say that God can move this body by local movement, either giving it a movement of rotation or a movement of translation transporting the mobile to another region of space; moreover, any local movement requires a place. Once we admit the existence of this isolated body, we must grant a place to it, and we cannot grant any place to it other than a place previously empty. We assumed that God created this body and nothing else. He therefore did not create a place for this body; then the place existed before, deprived of a body.
> Here is what we ought to reply to this: If we assume that such a continuous body exists and nothing outside this

continuous body exists, God cannot move this body by a movement of translation without creating a new place to which He would transport it. . . . He cannot give it a movement of rotation either, or else, if He were to give it a rotation, it would not be a local movement but rather a movement of situation (*motus situalis*).

It seems to me that it is difficult to avoid this consequence: those who speak suitably about our religion and who admit the creation of the world must suppose that there is a void outside the world. They in fact admit that God who created this world can create another as well. Let us therefore assume that God creates a second world. I ask then the following question: between the convex surfaces delimiting the two worlds, is there or is there not any distance? If there is something between these surfaces, it is a void, for it is a divisible space, holding no body, and capable of receiving a body. If, on the contrary, there is no intermediary between these spherical surfaces, is it that they touch at one point, or that they touch along a whole plane? They cannot touch only at a point, since in fact, between a point of the first surface and a point of the second, there will be something divisible which could only be a void—from which our conclusion follows. Could we hold that they touch along a whole plane? No, a spherical body cannot touch another spherical body along a whole divisible plane. If a surface touches a convex surface along a whole plane, it is because the surface is concave along the region of contact. And it is impossible for the spherical surface delimiting a world to be concave. We see then that those who adhere to our religion are obliged to admit the existence of a void. We have treated this question at greater length in the first book of the *De Caelo*.[82]

Required to admit the possibility of the void because of his faith, Burley examines how this void can be. We can understand by void, he asserts,

an empty space capable of receiving a natural body, although deprived of the presence of such a natural body. But that can be understood in two ways. In the first way, there is no natural body nor separated dimensions in this space. In the second way, there are separated dimensions in this space.

The first sense implies a contradiction; to assume that something capable of receiving a body has no volume, is to admit that a volume has no volume, for only a volume is capable of receiving a volume.

The second way, considering the void as a volume having length, width, and depth, but separated from all sensible volume, is less impossible. According to the theologians, it is possible for God; in the same fashion that in the sacraments of the altar there remains a volume without any corporeal substance supporting it, God could make it be that a volume existed without any quality. A similar volume separated and capable of receiving a body is what the ancients called the void.[83]

That is also the way in which John Buridan conceived the possibility of void.

John Buridan

Walter Burley did not dissimilate the seriousness of the conflict between the teaching of the Catholic theologians and Peripatetic physics with respect to the void; he gave the impression that this conflict was not beyond disturbing him. John Buridan also defined the terms of the conflict, with no less precision than Burley, and confessed his embarrassment about it in a clearer fashion than Burley had done.

John Buridan treats the void according to natural principles—meaning according to Peripatetic doctrines—in one of the first questions of his *Physics*.

He distinguishes two ways of understanding the word *void*.

In the first way, the void is "a space distinct from the magnitudes of natural bodies, which does not have to give way in order to receive natural bodies; each natural body occupies some part of this space equal to itself."[84] The void is then the place of all bodies. "One sees that this void is a volume (*dimensio corporea*) equal in length, width, and depth to the natural body that would fill it up if one placed it in this void."[85]

In the second way, the void is defined as a place without body, and place itself is understood in the Aristotelian fashion:

Place is the surface of the containing body. Hence, if there is a void, one has to imagine it thus: One removes the body contained in a filled space or else one destroys it while the place keeps its shape—the inner surfaces of the place do not get nearer to one another. Let us imagine, for example, that the sublunar world were totally annihilated while heaven kept the shape and size it now has; the concave surface of the lunar orb, which is presently a place filled by the world below, would then become an empty place, for no body would be contained in it. Moreover, it would no longer contain any space or any volume (*dimensio*)—it would then contain nothing at all.[86]

However one understands it, the word *void* designates something that no natural power can realize. But is God's omnipotence prohibited from accomplishing what nature's powers cannot accomplish? That is what Buridan examines in another question. Here is what he asserts about this topic:

> Some of my lords and masters of theology have reproached me for mixing up theological considerations within questions of physics, which is not something artists ought to do. I humbly reply to them that I would like not to be compelled to act in this fashion. And all the masters, when they begin in the arts, swear that they will not discuss, in a determined manner, any purely theological question—the Incarnation, for example; they also swear that if they happen to discuss or to settle any question concerning both faith and philosophy, they would settle it in conformity with faith and would resolve their objections in the manner in which they ought to be resolved. And if there is a question concerning faith and theology, assuredly it is the following: Can the void exist? If I wish to discuss this question, I must, in order not to perjure myself, assert what seems to me must be affirmed according to theology and avoid as much as possible the reasons that seem to conclude in an opposite sense. But I cannot resolve these reasons without exhibiting them; therefore I am forced to exhibit them.
>
> I therefore say that we can conceive the void in two different ways, as we have explained in the preceding question. There are two ways, then, that the void can exist by divine power. That is for me an item of faith and not a proof based on natural reason. I therefore do not intend to prove this, but merely to state how this seems possible to me.
>
> With respect to the first way of conceiving the void, I concede that God can make an accident without a subject, and that He can separate the accidents from the subjects that have them and conserve them after having separated them; He can therefore create a simple volume (*dimensio*) without there being any substance coexisting with this volume. . . .
>
> With respect to the second way of conceiving the void, I believe that God can annihilate the world below and conserve heaven as it is now with regard to its size and shape; hence the cavity of the lunar orb would be empty.[87]

Albert of Saxony and Marsilius of Inghen

With respect to this question about the possibility of the void, the teachings of Albert of Saxony and Marsilius of Inghen are only the faithful echoes of the teachings of Buridan.

Albert does not admit that the void can exist in the first way: "a separated dimension must not be allowed, for one must not admit that an accident can exist separated from its subject."[88] Doubtless, our author understands this under the condition "by some natural means," since the considerations suggested by the Eucharist transubstantiation had led the theologians to admit that God can separate an accident from all subjects; William of Ockham, for example, taught this proposition formally.[89]

"In the second way, the void can exist by supernatural means since God can annihilate everything existing between the inner surfaces of heaven; that done, heaven would be empty."[90] In that case, would the inner surfaces containing the void be distant from one another? "The inner surfaces of heaven would not be conjoined; heaven would remain a spherical globe—its inner surfaces would not immediately touch each other. However, there would remain no distance between them. . . . Thus it can happen supernaturally that some bodies are not distant and that they are not contiguous or neighboring."[91] We can perceive a reflection of Duns Scotus's teaching here.

Marsilius of Inghen does not discuss the possibility of void in his *Questions on the Physics*; he merely examines whether a heavy body would move successively in the void. On the other hand, he broaches the problem of the existence of the void in his *Abbreviationes libri Physicorum*. Marsilius's work on this topic is, as it is generally, a simple summary of Buridan's *Questions on the Physics*:

> The void can exist supernaturally in two ways.
>
> That is evident with respect to the first way; it is possible that God conserves a volume (*dimensio*) separated from other bodies. He can make this volume receive a body without place ceasing to be. That is the void understood in the first way.
>
> Similarly, it would be possible for God to annihilate everything below the concave surface of air while this surface remained in the same situation and disposition as now; one would then have the void in the second sense of the word, for we would then have a place which no body would fill.
>
> It is evident [that God can do both] since neither implies a contradiction.[92]

Nicole Oresme

Neither John Buridan nor his disciples, Albert of Saxony and Marsilius of Inghen, refused God the power to produce the void; but they did not refer to the considerations presented by Robert Holkot and Walter Burley about the actual existence of the void outside the world and what these considerations attempted to demonstrate. However, this actual existence of the void outside the world was affirmed in a most formal manner by the illustrious contemporary of Albert of Saxony and Marsilius of Inghen, Nicole Oresme, the great adversary of Aristotle.

The problem of the plurality of worlds is what leads Oresme to expound upon this topic.

Oresme affirms that "in truth, God could create *ex nihilo* new matter and make another world; but Aristotle would not admit this."[93]

After recalling Aristotle's various objections against the plurality of worlds, he continues in this fashion:

> He argues again in chapter 24 that outside this world there is no place or plenum, no void, and no time; but he proves this statement by saying that outside this world there can be no body, as he has shown by the reasoning above to which I have replied. So it is unnecessary to answer this argument again.
>
> But my position can be strengthened or restated otherwise; for if two worlds existed, one outside the other, there would have to be a vacuum between them, for they would have to be spherical in shape; and it is impossible that anything be void, as Aristotle proves in the fourth book of the *Physics*.
>
> It seems to me and I reply that, in the first place, the human mind consents naturally, as it were, to the idea that beyond heaven and outside the world, which is not infinite, there exists some space, whatever it may be, and we cannot easily conceive the contrary. It seems that this is a reasonable opinion, first of all, because, if the farthest heaven on the outer limit of our world were other than spherical in shape and possessed some high elevation on its outer surface similar to an angle or a hump, and if it were moved circularly, as it is, this hump would have to pass through space which would be empty—a void—when the hump moved out of it. Now, we may assume that the outermost heaven is not thus shaped or that nature could not make it thus, nevertheless, it is certainly possible to imagine this and certain that God could bring it about.

From the assumption that the sphere of the elements or all bodies subject to change contained within the concavity of heaven or within the sphere of the moon were destroyed while the heavens remained as they are, it would necessarily follow that in this concavity there would be a great expanse and an empty space. Such a situation can surely be imagined and is definitely possible, although it cannot arise from purely natural causes, as Aristotle shows in his arguments in the fourth book of the *Physics*, which do not settle the matter conclusively, as we can easily see by what is said here.

Thus outside heaven is an incorporeal space quite different from any other plenum or corporeal space, just as the extent of this time called eternity is of a different sort than temporal duration, even if the latter were perpetual, as has been stated earlier in this chapter.

Now this space of which we are talking is infinite and indivisible, and is the immensity of God and God Himself, just as the duration of God called eternity is infinite, indivisible, and God Himself, as already stated above.

Also, we have already declared in this chapter that, since our thinking cannot exist without the concept of transmutation, we cannot properly comprehend what eternity implies; but, nevertheless, natural reason teaches us that it does exist.

Likewise, since apperception of our understanding depends upon our corporeal senses, we cannot comprehend nor conceive this incorporeal space which exists beyond heaven. Reason and truth, however, inform us that it exists. In this way the scriptural passage, Job 26, that speaks about God can be understood: *Qui extendit aquilam super vacuum.*

Therefore, I conclude that God can and could in His omnipotence make another world besides this one, or several like it or unlike it. Nor will Aristotle or anyone else be able to prove completely the contrary. But, of course, there has never been nor will there be more than one corporeal world, as was stated above.[94]

To admit that there is an infinite empty space outside the limits of the world is to take up the teachings of Stoicism against Peripatetic thought. But to admit that the empty space is none other than the immensity of God is to propose the doctrine that Newton later discovered and that Clarke later maintained against Leibniz.

Graziadei of Ascoli

That one ought not refuse God the power to create an empty space was what the decrees of Etienne Tempier and his counselors were the first to proclaim, at least indirectly. These decrees forced the University of Paris and the schools under its influence to adopt the above proposition; but soon it was received even in the universities where the condemnations of Paris carried no weight. Thus we can hear the proposition professed by the Dominican Graziadei of Ascoli at Padua.

Graziadei states that "the First Cause can introduce an empty space in the world."[95] The First Cause does not act necessarily, but by a completely free will. "Then it can take out of existence absolutely everything located in the sphere of passive things, without doing so for the things in the sphere of the celestial orbs; in fact, it gives existence to what it wishes to maintain not by necessity, but by an absolutely free will."[96]

Graziadei knows the objections commonly raised against the above proposition. Some assert with the Commentator that if the void existed, there would be a magnitude that no body would carry, meaning an accident without a substance. Others write that the three dimensions stem from the considerations of the mathematician; therefore, one would have mathematical beings existing separated from any physical properties. In order to avoid these objections, others support the following opinion:

> If the void existed, not only would a separated dimension not result, but there would not be a distance between the extremities of an empty space; if the bodies included between heaven and earth were suppressed, the distance between heaven and earth would also disappear, so that heaven would find itself conjoined to the earth.

> How frivolous all this seems, [replies Graziadei,] the void does not suppose real dimensions, but only conceived dimensions (*imaginatae*); if it existed, there would not be real accidents without substance, nor would these be a mathematical reality without sensible matter. The mathematical being would merely be conceived.

> But the distance between the extremities of an empty space would not be suppressed, for if it were, the void would suppose no dimension, no real or conceived extension—which is false.

In fact, there can be no dimension or extension, either real or merely conceived, in a point. The void, on the other hand, supposes three conceived dimensions awaiting a body possessing three real dimensions, which this space can receive.[97]

Graziadei's thought clearly approximates Richard of Middleton's thought on this subject.

10
The Void in Fifteenth-Century Cosmology

Nicholas of Orbellis

In what he says about the void and movement in the void, Nicholas of Orbellis follows John Duns Scotus.

He believes in the possibility of an empty space with Duns Scotus and against Aristotle:

> Even though there is no void in the reality of nature, however, [the volume] heaven encloses can become empty by divine power. In fact, God can annihilate everything contained within the concave surface of heaven.[1]

Our author then recalls how Aristotle attempted to establish that all movement would be instantaneous in the void in order to demonstrate the impossibility of the void.

> With respect to this subject, here is what Scotus says in the second distinction of the second book of the *Sentences*: If one placed a weight in the void, the weight would not move, according to the Philosopher, because the void cannot give way to the weight. . . . If one admitted that the void can give way, the movement of the weight would be successive, for the mobile would have to cross one part of space before the other. In the same way that the space is then divisible, the movement is then divisible into successive parts, just as it is in the plenum; there is, therefore, an essential succession in this space. [In a plenum] some additional speed or some diminishing of speed is added in virtue of the accidental condition whereby the medium favors some movement or prevents it. . . . A movement can therefore be produced in a successive manner in the void; one can compare this movement to a movement in the plenum as long as it is with respect to its essential succession, and not with respect to the

added speed and slowness; the mobile possesses no added speed or slowness in the void, although it does possess some in the plenum. . . . What the Philosopher has established against his adversary who affirms the existence of movement in the void, then, is that there can be no movement in the void possessing a speed or slowness added to the essential succession.[2]

John Hennon

Is the void possible? Our author does not doubt that the supernatural power of God is capable of creating an empty space; he invokes an argument that had long been classic at Paris in order to support his opinion.

That the void exists does not imply a contradiction. . . . In fact, there is no contradiction in that the surface of a concave body subsists without having any relation with another body contained by it. Therefore, etc.

Moreover, the elements are corruptible in their parts, and since in them the whole is of the same nature as a part, there does not seem to be any contradiction in that the totality of the elements can be destroyed, at least by divine power. That done, there would be a void between the inner sides of the lunar orb.[3]

Our author again takes this up with the help of a postulate which Ockham used constantly.

If two abstract things are essentially distinct, and if one of these two things does not depend essentially on the other, God can separate the two things and conserve them separated from each other; He can also destroy one while keeping the other. But heaven is something essentially distinct from sublunar things, and it does not depend essentially on these things; God can therefore, by annihilating that which is contained in the lunar orb [and conserving heaven], make the void exist.[4]

If the void were thus produced in the cavity of the lunar orb, can one still say that there is distance between the inner walls?[5]

If by "distance between two surfaces" one understands a relationship implying the existence of a positive intermediary medium, there would no longer be such a positive distance between these inner walls. But the word *distance* can be taken in another sense. Doubtless there is no positive medium between these inner walls, but between these walls a body can be received exactly equal to the body received when the interval is actually filled. Our author calls this capacity (*potentialitas*) to receive a body of such magnitude a privative medium; he then asserts that there is a negative distance between the inner walls of an empty receptacle.

After having affirmed that the existence of an empty space does not imply any contradiction, that God can therefore realize such a space, John Hennon asks whether an empty volume can be found in nature. He answers that

> it is not possible for the void to exist naturally, and that is evident, for nature abhors a vacuum (*quia natura abhoret vacuum*). . . .
>
> If that were not the case, then one cannot see why the pleats of a bellows cannot be separated from one another when all the orifices are shut; that does not seem to have any cause other than avoiding the void that would exist between the inner surfaces of the bellows if one separated its pleats in this manner.
>
> Similarly, one cannot see why water cannot flow out of a clepsydra when the air does not enter from any hole, if one does not hold that nature abhors a vacuum and cannot suffer its existence.[6]

Our author who believes that ice is denser than water also declares that the water in a vase hermetically sealed "cannot congeal, or else the vase would be broken by the force of the universal nature that abhors the existence of the void."[7] If one can separate two smooth plates, it is because the interposed air becomes more rarefied and fills the space which opens up between the plates; "otherwise, no natural force could separate them; universal nature, which abhors a vacuum, would prevent it."[8]

The doctrine Roger Bacon had developed in the Faculty of Arts at Paris during the middle of the thirteenth century was still in favor there two hundred years later.

George of Brussels and Thomas Bricot

George of Brussels professes the same opinion as Nicholas of Orbellis and John Hennon with respect to the possibility of void:

> The void can be realized by a supernatural power, but it cannot be realized by natural power. . . . A lodged body

can be annihilated while the containing body keeps its size and shape; it is therefore possible that the void is produced by a supernatural power. . . . As for the second part of our assertion, it is evident for nature abhors a vacuum; and we see weights ascending and light bodies descending in order to avoid a vacuum. It is therefore not possible that a void be realized by a natural power.[9]

If the void were realized, movement in it would be produced successively; for example, the fall of a weight would not be instantaneous; some time would be required in the fall of a weight. In order to establish these conclusions, George, in some long and confused discussions, dilutes what his predecessors, particularly Nicholas of Orbellis, had borrowed from Duns Scotus.[10] These complicated considerations contain only one short passage worthy of being cited. It concerns the refutation of Thomas Aquinas's opinion, which attributed the successive character of movement in the absence of all resistance from the medium to a limitation of the motive power. According to our author, one can reason in the following manner against this limitation, in which one guesses resides the intuition toward the first notion of mass:

> One possibility would be that there would be a greater limitation in a larger body; yet a heavier body would not fall faster than a lighter body. In fact, as we add to the weight we add to the limitation, but it seems that they maintain the same proportion between them. Another possibility would be that there would not be a greater limitation in a larger body; but that is impossible. In fact, this limitation is a quality or substantial form of the body, and the larger a body is, the greater is its quality or substantial form, and therefore the greater the limitation. (*Vel in majori corpore esset major limitatio, et sic magis grave non deberet descendere velocius quam minus grave; quia quantumcunque additur de gravitate, tantum additur de limitatione, et sic semper videtur manere eadem proportio. Vel non esset in majori corpore major limitatio; sed hoc est impossibile, quia illa limitatio est qualitas vel forma substantialis, et quanto corpus est majus, tanto major est qualitas vel forma et, per consequens, tanta major est limitatio.*)[11]

Clearly, our author glimpses the truth that, at least for bodies of the same nature, the limitation—the mass—must be proportional to the magnitude of the body, yet there would be the same relation between the weight and the mass for all these bodies.

Clearly also, he recognizes that the above proposition entails the consequence that in the absence of all extrinsic resistance, in a vacuum, a large body would fall with exactly the same speed as a small body. But this corollary, that we recognize as a great truth, strikes him as an absurdity. That a large body must fall faster than a small body, even if the resistance of the air were abstracted away, is a proposition which appeared as certain as an axiom; from the time of Joannes Philoponus, the false belief in this law prevented the notion of mass from entering science and prevented the formulation of the true principles of dynamics.

However, George of Brussels believes in this "limitation of the motive power." If the fall of a weight requires some duration, even in a vacuum, he believes that it is due at least in part to the existence of this limitation; in order to avoid his own objection, he takes recourse with the worst subterfuges.[12] How else could he reason about this? He renders mass into the falsest of ideas. As a faithful Peripatetic, he cannot imagine this cause of slowness under some aspect other than as an antagonistic force. He says, "the mobile resists the motive force in the fashion of a weight resisting the motive virtue that would raise it from the earth to heaven, even if the space between heaven and earth were empty."[13]

There was no notion more essential in order to constitute mechanics than the notion of mass; and there perhaps was none more difficult for the human mind to abstract from experience.

That which we have just reported about movement in the void was only the opinion of George of Brussels. Would Thomas Bricot have been of the same opinion? We think not; in fact, toward the end of the fourth book of the *Physics*, after having treated time, the *Cursus* again takes up the following two doubts already examined and resolved: (1) Can the void exist naturally? (2) Would a weight move instantly in the void? Although the reflections about the first doubt agree with those exhibited before, that is not the case with the reflections about the second doubt. The author states:

Let us note that there are two opinions about this.

The first is the one we have previously reproduced. It maintains that if a weight were placed in a vacuum and if it moved, it would move successively and not instantaneously. This succession would stem from the fact that the terms of space are not compossible; the weight cannot be in both places at the same time.

But another opinion seems more probable and conforms better with Aristotle's thought; it is that, in local movement,

succession stems wholly from the resistance of the medium.
. . . In conformity with this opinion, then, if a weight were
placed in the void, we say that it would move instantaneously
and not successively.[14]

Further, among the reasons for the *"opinio prius recitata"* was this
confused allusion to the notion of mass as conceived by Saint
Thomas Aquinas: "If a weight placed in a vacuum would move
in an instant, it is above all because it would not encounter any
resistance; but that is false because the moved body would still resist
gravity (*adhuc corpus motum resisteret gravitati*)."[15]

Our author replies that "the weight does not resist gravity,
but on the contrary it inclines toward downward movement."[16]

Clearly, the notion of mass has not yet been apprehended.

Parisian Doctrines in Germany

Is the existence of an empty space possible? All the German
masters we have been able to consult have been in agreement with
respect to the answer to this question.

The existence of an empty space does not imply a contradiction;
therefore nothing can prevent God from realizing such a space if
He so wished it.

On the other hand, natural forces always put obstacles into
play against the production of a vacuum, which nature abhors.

Let us read from Frederick Sunczel's work. He replies to the
question, "Can the void exist by the effect of some power?"[17] by
providing two conclusions. The first conclusion is as follows:

The void cannot exist naturally. . . . All things desire
their mutual contiguity in order to receive better the influences
conserving them; and the void would be a drawback in nature,
a kind of disorder. Nature then, meaning common nature,
abhors the void, so that things are conserved in their being.
That is why one can prove by means of many experiences
that natural things even go against their own natures in order
to avoid a vacuum.[18]

And our author continues by describing several experiences
that had been the custom to cite in this circumstance, from the
time of Albertus Magnus and Roger Bacon.

But he appends the following to his first conclusion:

The existence of the void does not imply any contradiction;
that is, the void can exist supernaturally. That is evident:

something is possible if its actual existence does not entail any impossibility, and no impossibility results if one admits the existence of the void.[19]

Let us now read from Conrad Summenhart's work.

The existence of the void is impossible naturally. Nature in effect does not admit what it abhors; and nature abhors a vacuum, which it therefore does not admit.

Here is proof of the minor premise: Nature prefers to operate against its natural inclination rather than to admit a vacuum; therefore it must abhor it.

The antecedent of this proof is evident given the example of a clepsydra full of water; if one plugs up its upper orifice, water does not escape through the lower holes, even though its natural inclination would be to fall out of the clepsydra; it does so because a vacuum would be formed. The clepsydra to which I am referring is an instrument used to water one's garden, similar to the clepsydras used at Paris; that is what one calls them there.[20]

One has to add another assertion to this first assertion:

A vacuum can exist by the effect of divine power. One can prove this: There is no contradiction in supposing that God has annihilated all the bodies located below the ultimate sphere, while the ultimate sphere remains. That accomplished, there would remain an absolute void, for there would be a place—the concave surface of the ultimate sphere—which no body fills up.[21]

Gregory Reisch says the same.

Can there be a body without place or a place without a contained body, [asks the disciple]?

Assuredly, [replies the master,] that can be accomplished by divine virtue; it would occur if God created some body outside heaven, where there is no place, or else if He annihilated some body within heaven, in the latter case an empty space would remain. The word *void* designates a place no body fills, but which can be filled by some body.

However, created nature abhors this; it allows heavy bodies to move up in order to avoid the formation of the void; one sees this in the behavior of water. In a clepsydra whose upper orifice is obstructed, water does not flow from the lower holes; it remains suspended until air can enter and fill up the place left by the water when the upper orifice is opened.[22]

The writings of Jodocus of Eisenach echo those of Gregory Reisch.

One might ask whether the existence of the void is possible by means of some power. One must reply that the existence of the void is possible by means of supernatural divine power, but it is not possible by any natural power. . . . No person of faith can doubt that the void can be produced by divine power. . . . By this power a contained body can be destroyed while the place of this body would keep the same size and shape; the existence of the void is therefore possible. . . .

Here is proof of the second assertion:

Nature abhors the compenetration of dimensions, the subsistence of accidents deprived of subjects, and the void. This abhorrence of the void can be demonstrated by extremely evident experiences; it cannot, in any case, be demonstrated otherwise. We observe, in fact, that in order to prevent the formation of a vacuum, a heavy body climbs in order to fill up the empty space.[23]

Our author then recalls the properties of the syphon and of the clepsydra; the phrase that he devotes to the latter instrument is taken word for word from the *Margarita Philosophica*. He adds,

It is the same for innumerable other experiences; further we can relate the saying of Themistius about this: Rather than permitting the formation of the void in the upper region of air, nature would require a millstone to leave its natural place and climb up in order to ward off this effect.[24]

Under the avowed influence of Peter of Abano, Jodocus of Eisenach places this abhorrence of void under the heading of universal nature,[25] which he conceives exactly as the Paduan Doctor had done before him, and Roger Bacon before him.

Paul of Venice

What Paul of Venice asserts about the void in his *Summa totius philosophiae* little resembles what he asserts about it in his *Expositio super libros physicorum*. Let us first cite the latter work, then discuss the former work.

The *Expositio super libros physicorum* supports the doctrine of Aristotle and the Commentator with respect to the void; it maintains that the fall of a weight in the void cannot last for some time, no matter how small the time is. The fall would be instantaneous. It reproduces against this opinion the doctrine that Averroes attributed to Avempace who derived it from Joannes Philoponus.[26] It sets out at great length Averroes' objections to this doctrine; it concludes that the cause, in virtue of which such a movement is successive, is due totally to the resistance of the medium. It takes care, in any case, to borrow the following objection from the Commentator:

[If Avempace's theory were true] there would result the absurdity (*inconveniens*) that two bodies, of which one is heavier than the other, all other things being equal, would move with the same speed; that is clearly what would happen if two bodies with unequal weights were to move in the absence of a medium. (*Et propter hoc sequitur aliud inconveniens, dicit Commentator, quod aliqua duo corpora aequivelociter moventur, quorum unum est altero gravius, caeteris paribus; patet de duobus inaeque gravibus motis sine medio.*)[27]

The truth, in the void two unequal weights would fall at the same speed, was thought by all to be an inadmissible proposition; it was sufficient to condemn any theory of which it was a corollary. Already Joannes Philoponus had to defend his theory against having to admit a similar consequence.

However, Paul of Venice rightly stated elsewhere why the proposition was not in any way absurd, once one admits the theory of Ibn Bajja and Joannes Philoponus.

Here is what Avempace would reply to this objection: It is not absurd that two unequal weights move with equal speed in the void; there is, in fact, no resistance other than the intrinsic resistance due to the application of the motor on the mobile, in order that its natural movement be accomplished. And the relation of the motor to the mobile, with respect to the heavier body and the lighter body, is the same. They would then move with the same speed in the void. In the plenum, on the other hand, they would move with unequal speed because the medium would prevent the mobile from taking its natural movement. (*Non est inconveniens in vacuo inaequaliter gravia aequevelociter moveri quia non habent resistentiam nisi intrinsecam ex applicatione motoris ad mobile facientem motum naturalem; et quia eadem est proportio motoris ad mobile in graviori et leviori, ideo ambo aequevelociter moventur in vacuo; in pleno autem inaequaliter moventur ratione medii impedientis mobile a motu naturali.*)[28]

The above glimpses what we, from the time of Newton, have expressed as follows: Unequal weights fall with the same speed in the void because the relation between their weight and their mass has the same value.

Aristotle did not invoke experience against the possibility of void; Paul of Venice does so on several occasions.

These experiences, of the pipette or the suction cup, for example, are, according to Paul of Venice, commonly attributed to the pull of the vacuum, *tractus vacui*. This pull, how are we to understand it?

Paul first reproduces Giles of Rome's explanation:

> Giles states that this pull of bodies is not accomplished by an intrinsic principle, but by an extrinsic principle in virtue of heaven. In the same fashion that the magnet attracts the iron in order to unite itself with it, heaven, possessor of a regulative virtue toward the whole sphere of the elements and toward all the parts which constitute it, attracts natural bodies so that the void would not separate them from one another.[29]

Paul addresses various objections leading him to reject this explanation:

> Here is what one should reply to this: the pull one says comes from the void stems from an intrinsic principle; it is a natural movement belonging to any body that can move in any direction. It is true that the Philosopher asserts in the first book of the *De Caelo* that in each simple body there is only one simple movement. One must understand that it is so for a simple body considered in itself (*per se primo*) from the point of view of its specific nature, but when one considers it not in itself (*per se non primo*), from the point of view of its generic nature, one can attribute all movements to a simple body. One can in fact consider a simple body insofar as it is an element of a determined species; in this way, only a single movement is proper to it. One can also consider it as a natural body, then all movements can be proper to it. They are in it naturally by an intrinsic principle in order to suppress the void.[30]

One understands thus why water does not flow from openings of a clepsydra when one places a finger over its upper orifice.

> In that case, water encounters a prohibition which is not external, but internal; in fact, water tends to descend because of its specific appetite, but it tends to conjoin itself with some bodies because of its generic appetite; and since it is of a genus before it is of a species, it tends to conjoin with a body more than it tends to descend. And if it did descend, an empty space would be produced between the water and the container. Hence, in order that the void not be produced, the generic appetite prevents the descent.[31]

That is Roger Bacon's doctrine; we should not be surprised to discover it in Paul of Venice's writings. In fact, we know that it was traditional at Padua; Peter of Abano taught it at the start of the fourteenth century, and Graziadei of Ascoli later developed it.

This borrowing from Roger Bacon's theory is an addition to the Peripatetic doctrine about the void; however, Paul of Venice earlier asserted something about the void that neither Aristotle nor the Commentator would have accepted at any cost; taking God's omnipotence as a premise, he concluded that there exists an infinite empty space outside the world.

Against the Philosopher, one can prove that there exists some space above heaven; once that is proven, it must be necessary, according to the Commentator, that this empty space be infinite.

In fact, God can create a stone outside the world and move it away from heaven; from instant to instant this stone would get farther from heaven. There would then have to be some space between the stone and heaven. However, this space already exists, since our assumption admits that God has created nothing except the stone.

God can still create three other worlds outside this world touching this world at three points. The surfaces of the four worlds would then have distances between them; there would then be space and distance between them. But this space and distance has not been created by God, for I am assuming only that God has created the three worlds; hence the space already exists.

From now on we must admit that there is some space above heaven, in the same way as there would exist one between the inner walls of heaven if God were to annihilate all the bodies contained in the concavity of the lunar orb; for after such a destruction there would be space and distance between the inner walls of heaven because these inner walls would not touch and would remain concave.

All these propositions must be admitted, for they are true if one takes divine power into account. Aristotle did not admit them; he did not know the infinite character of divine power. . . .

If the sphere of things that are active and passive were destroyed, one would still have to admit that the two halves of heaven are distant from one another; this distance would not be the one about which we commonly speak. It would be a rectilinear distance; that is not to say that there would exist some straight line between the two halves of heaven, but only that there could exist one. The relation between the two points is said to be rectilinear whether there exists a straight line between them or not, as long as one could draw a straight line. Even though we agree that there is a distance between the inner walls of heaven, we must not think that there really exists a thing between them which is a space or an interval; that is not the sense that the word *between* takes in the proposition in question. We must understand only that the two inner walls of heaven differ from one another by some distance. Similarly, when we say that there is a difference between two men, we do not think that the difference is something interposed between the two men, but simply that there is a difference by which the two men differ from one another.[32]

To posit God's omnipotence in principle and to declare that this omnipotence can produce in numerous circumstances what Aristotle's physics declares impossible is to draw upon the same inspiration which directed the decisions of Etienne Tempier during 1277. To conclude that a new physics must be constructed that does not put limitations on the divine power is to follow the example given by the physicists of Paris during the fourteenth century. We see that Paul of Venice felt the effects of the inspiration and was seduced by its influence.

But he often returned to an intransigent Peripatetic philosophy. Thus he withdraws, in his *Summa totius philosophiae*, the conclusions he has just reached in his *Expositio super libros physicorum*.

There he recognizes that the existence of the void does not entail a formal contradiction, a contradiction in terms, but he declares it impossible nevertheless:

If, in conformity with the ancients' opinion, the void existed, there would exist an accident without a subject to carry it. In any case, the void would be a body because of its length, width, and breadth; however, it could receive a lodged body without having to withdraw from the place the body would occupy. There would then be some compenetration of bodies.

If God were to annihilate all the elements except the terrestrial element, heaven and earth would not be distant from one another any more; they would be neither nearer nor farther, for proximity and remoteness exist only because of magnitude (*quantitas*) and there would be no magnitude between them, mediate or immediate, since they would be touching.

The following corollary results from this: two bodies can be distant from one another and then not distant from one another immediately after without either of them suffering any change carrying one of them toward the other. That is evident. Let us suppose, in fact, that God begins to annihilate all the elements except the terrestrial element. In that case, immediately after this operation, heaven and earth would no longer be distant; there would be neither rectilinear nor curved distance between them.[33]

Paul of Venice, as we know, is not afraid of contradicting himself.

V
The Plurality of Worlds

11
The Problem of the Plurality of Worlds in Peripatetic Philosophy

Aristotle and the Plurality of Worlds

The concepts of natural movement and natural place are the foundation of Aristotle's reasoning about weight and levity, and about the shape, position, and immobility of the earth. These concepts play an equally important role with the problem the Stagirite attempts to resolve about the plurality of worlds. Perhaps there is no problem in all his physics better capable of demonstrating the precise meaning he attributes to these two concepts.

"Since we habitually call the whole or totality heaven (*ouranos*)," states Aristotle, "we give the name to any body included within the extreme circumference."[1] In the *De Caelo* Aristotle first demonstrates that the universe is limited, then immediately broaches the question: Can there be more than one heaven, that is, several universes?[2] He resolves this question negatively. To justify his solution he makes use of two principles.

We have seen Aristotle invoke the authority of the first principle many times. This principle distinguishes natural rest and natural movement from rest by constraint and movement by constraint. Nowhere else in his writings does he formulate in as clear a fashion the two axioms that he uses freely in his deductions; they are as follows:

1. If a thing can rest naturally without constraint in some place (which is therefore its *natural place*), when one puts the thing outside the place, its movement toward the place will be natural; and reciprocally, if a thing moves naturally toward a certain place it is because the place is its natural place, where it will reside at rest without constraint.

Thus the natural place of fire is the region located immediately below the sphere of the moon; if one puts fire somewhere else,

431

on earth for example, it will climb naturally toward the sphere of the moon. In the same way, a fragment of earth would move toward the center of the world; hence, that is where its place of natural rest is located. We have indicated that this corollary serves as the point of departure for the explanation of the shape, location, and immobility of the earth.

2. If one has to exercise a constraint in order to maintain a thing at rest in a certain place, then that thing cannot move toward that place without violence.

A fragment of earth, for example, would not reside around the sphere of the moon unless it were constrained to do so; if it were put on the surface of the earth, it would not climb without being pushed by some power foreign to its nature.

The second principle upon which Aristotle rests his demonstration is the following: if a world exists outside the one we know, since this world is similar in nature to ours, it must be composed of the same bodies as ours. It would not be composed of elements that one would call earth, water, air, and fire, having a mere verbal similarity with ours, but being essentially different from our earth, our water, our air, and our fire. If it were so, this world would have only a verbal analogy with our world; it would not, in reality, be another world. If must be that the earth of the other world has the same substantial form (*idea*) as the earth of our world; one can say the same for fire, air, and water.

Each of the elements of the second world, having the same substantial form as the corresponding element of the first world, would also have the same power (*dynamis*); for example, since the earth of our world naturally seeks the center, its natural motion in the second world will also tend toward the center of that world. In the same fashion, the nature of fire will always carry it away from the center of the world within which it is located.

On the strength of his two hypotheses, of which at least the second does not seem to follow necessarily from his *Physics*, Aristotle undertakes to prove that the simultaneous existence of two worlds is an absurdity.

The earth of the second world would have the same substantial form as the earth of the first world, hence the same power, hence the same natural place; if one puts it at the center of the first world, it would reside there immobile without constraint. Consequently, placed without constraint elsewhere, within the other world, for example, it must move toward the center of the first world naturally; but in order to so do, it must move away from the other center, and this implies a contradiction, for we have seen that the natural movement of the earth within the second world consists in moving nearer the center of this world.

One can put forth similar considerations with respect to fire. They would lead to the same conclusion: the coexistence of the two worlds is an absurdity.

Aristotle's argument may be opposing a doctrine that would appear much more plausible to our modern conceptual sensibilities: a fragment of earth has the tendency to move at the same time toward the center of the first world and toward the center of the second; in either of the two centers it would occupy its natural place, but the tendency bringing it to a center varies with the distance from this center. When this distance increases, the intensity of the tendency decreases; of the two tendencies bringing the fragment of earth toward the centers of the two worlds, the greater is the one connected with the nearer center, and it is the one that transports the fragment.

Undoubtedly this doctrine was current during Aristotle's time, for without bothering to expound it, he takes care to refute it. Let us linger for a moment with this refutation; it touches upon an essential feature of the subject we are considering.

It is unreasonable to maintain that a heavy thing moves toward the center of the world with respect to its distance from that center; what makes it tend toward this spot is its nature itself (*physis*). We would have to admit that the nature of a heavy thing varies with the distance that separates it from its natural place. But how can this distance have any relevance to the nature of a thing? Two heavy objects at different distances from the center of the world are different for us, but with respect to their substantial forms they are identical.[3]

Aristotle's reply, although contrary to our modern conceptual sensibilities, follows logically from the principles of Peripatetic physics. A body is heavy when it is by nature potential with respect to the center of the world, which is its natural place; whether it is near or far from the center of the world, it always has the potential to lodge there, and this potential cannot be spoken of in degrees; it can only be ended when the body is actually at the center of the world.

In any case, it also makes little sense to maintain that an element, earth, for example, can have two natural places of the same kind, but numerically distinct, that a heavy body can tend toward the center of one world and the center of another. To the unique substantial form that characterizes the earth in one world and in another must correspond a unique natural place, not only of kind, but of number.

This principle brings with it a new consequence.

Outside the extreme circumference containing our world, can there be any body whatsoever? The Stagirite replies negatively to this question.[4] A body cannot reside either naturally or unnaturally outside the last sphere.

An element cannot have a natural place outside the supreme sphere since it already has a natural place inside this sphere, and, as we have already seen, the same element cannot have two natural places. Further, since mixtures are composed of elements, no mixture can be naturally situated where no element has a natural place.

Neither can a body be outside our world unnaturally. If a body is somewhere unnaturally, the external place will be natural to some other body; but we have just proven that no body has a natural place outside the extreme circumference.

Hence there is no bodily mass whatever outside the circumference of the world. What is there then, a void? No; void is said to be that in which the presence of a body, though not actual, is possible; but a body cannot reside outside the eighth sphere. Above the heavens, then, *there is no place or void.*

There is no time either, for there is nothing corporeal, nothing capable of alteration or change. Where no change is possible, there is no passage from possibility to actuality—there is never any movement. With the absence of movement, time, the measure of movement, disappears. Everything outside the last sphere occupies no place; so it is immaterial. Time does not age it, it is not corrupted, and it does not change; so it is eternal.

The world thus includes within it all the actually existing matter.[5] Further, it includes all matter that has ever existed as well as all possible matter, for matter is capable of transformations but it may not be created or destroyed. The world is not only unique in actuality; it is also unique temporally. No other world has preceded it; no other world will succeed it. Heaven is one, permanent, and perfect.[6]

Aristotle's argument can serve to refute some doctrines he does not mention explicitly, but about which he might have been thinking.

Stobaeus tells us that "Heraclides Ponticus and the Pythagoreans held that each star constituted a world, that each contained an earth surrounded by water, and that the whole is within the luminiferous ether; the same doctrines are in the Orphic hymns, for they make a world out of each star."[7] By affirming that the earth has a unique place, Aristotle contradicted the doctrines according to which each star contained an earth; his refutation of the plurality of worlds pitted itself against the opinions that the Copernicans took up later.

The Plurality of Worlds according to Simplicius and Averroes

Aristotle's arguments against the hypothesis of the plurality of worlds brought forth innumerable commentaries; we will not analyze them here in their entirety. But we will report on the interpretations that Simplicius and Averroes gave on the Peripatetic argument against the plurality of worlds. These two interpretations, so different one from the other, will serve to clarify what Aristotle intended by "natural place" and by "tendency toward a place." Further, they will enable us to know better the diverging doctrines from which the doctors of Christian Scholasticism made a choice.

It is by its own nature that a heavy thing tends toward the center of the world, stated Aristotle. This nature does not change as the distance between the heavy thing and the center of the world changes; therefore, distance does not influence the tendency pushing a heavy thing toward its place. In other words, the weight of a body does not vary in magnitude when one places the body nearer or farther from the communal center of heavy things. That is how, it seems, we should understand Aristotle's thought; and that is how it was interpreted by various commentators.

Simplicius seems to have attributed another sense to Aristotle. Here is what he wrote in his *Commentaries* on the *De Caelo* about the passage we are considering:

> The author [Aristotle] expounds and refutes an objection that someone might put forward; it consists in claiming that the earth of another world would not move naturally toward the center of this one because of its great distance from it. In that case the contradictions standing in the way of the supporters of the plurality of worlds would fail; the earth of the other world would no longer move up, and neither would fire move down. It is unreasonable, replies Aristotle, to consider that distance is capable of suppressing the very powers of bodies. Whether simple bodies are more or less distant from their natural places, their nature would not

change, and neither would their natural movement. In fact, in this world here, what different property can a body possess when it is separated from its natural place by some distance or another? This one only: it begins to move more weakly toward its natural place when it starts from a farther position. There is a constant relation between the weakness of the movement and the extent of the distance; but whether the distance is greater or smaller, the movement remains of the same kind. If simple bodies existed in another world, they would start to move more slowly, in proportion to their greater distance, than the ones in this world. But the type of movement natural to them would not be altered, for this kind results from their substance itself, and it would be unreasonable to accept the amount of distance as a cause of substantial generation or corruption.[8]

Simplicius, ordinarily so perspicacious when it comes to discerning and explaining Aristotle's thought, does not appear to us to have grasped either his thought or the objection against which it was directed.

The Athenian commentator believes that at any distance from the center of the world a heavy thing would direct itself toward the center, whereas a light thing would direct itself away from it. Neither the existence of this tendency nor its direction vary with distance, but the intensity of this tendency is inversely proportional with distance. Assuredly, the latter proposition is denied by Aristotle.

If we accept Simplicius's opinion, might we not reason thus: If a world exists outside ours, a fragment of earth placed within the other world would continue to move toward our center even though this tendency would be extremely weak. Two tendencies would impel this mass, the first, a weak one toward the center of our world, the second, a strong one toward the center of the other world. The second one carries it away; the mass of earth would be moved toward the center of the world within which it is located, not ours. It seems that this is the objection Aristotle attempted to refute. It is intended to be applied to the principle accepted by Simplicius and rejected by Aristotle, that gravity decreases as the distance of the mobile weight increases from the center. We find no reason capable of denying this objection in the treatise of the Athenian commentator.

It seems to us that Simplicius misjudged Aristotle's doctrine here. On the other hand, Averroes seems to have understood its meaning. Because of what he asserts about this doctrine, he seems to have merited the title of Commentator that Christian Scholasticism gave him.

In his commentaries on the *De Caelo*, the philosopher of Cordoba expounds lengthily Aristotle's argument against the plurality of worlds. When he arrives at the passage we are concerned with, he expresses himself in these words:

> Aristotle then examines another objection. . . . One can, in fact, assert that the earth of the other world does not move toward the center of this world (and inversely), even though the earth is of the same nature in both worlds; and that is the same with respect to the other elements. If someone were to take a body formed out of one of these elements and place it somewhere not equidistant from the two similar natural places that suit it in the two worlds, even though it remains the same at all times, it would move toward one of the two natural places it is nearer. For example, the earth of our world is nearer the center of this world than the center of the other world, and it moves toward the first center, not the second. But if it were located in the other world, it would move toward the other center. Thus, even though its nature remains the same, this earth would be subject to two contrary movements depending upon its proximity or remoteness from the two specific similar places which are differently situated. It can move either from the first center toward the second, or from the second toward the first, even though the two movements are opposite one another. Without a doubt, the element, insofar as it is simple, cannot move with contrary movements; but this becomes possible by the effect of its proximity or remoteness, because proximity or remoteness add something to the simplicity of its nature. By virtue of the composition resulting, the body can at different times move with two opposite movements.
>
> Aristotle replies that this discourse is not reasonable. The natural movements of simple bodies differ among themselves only by dint of the differences existing between their substantial forms; the differences issuing from relation, from quantity, and from all other predicables would not be able to change anything with respect to these movements. In other words, a change in proximity or remoteness does not reach the substance.
>
> We should note on this subject that proximity and remoteness have no influence except for the movements of bodies under the action of an external cause, for then these bodies would be more proximate or more remote from their

motive source. Thus it would be relevant to prove here that the movements of the elements do not have their cause outside these elements. This proposition may seem selfevident; Aristotle, nevertheless, makes use of it in reflections intended to counter what the ancient philosophers said about the rest and movement of the elements, earth in particular. In fact, with respect to the rest and movement of the earth, these philosophers assigned as cause a mutual attraction between the whole earth and its natural place. But it is evident that a fragment of earth does not move toward the whole earth regardless of the position of the terrestrial globe; in fact, if a fragment of earth moved toward the whole earth, this motion would be like the motion of iron toward a magnet. Thus it could happen that the earth might move naturally upward.

Since the motion of the earth toward the center is not the effect of an attraction produced by the nature of the place itself, nor by the nature of the body which occupies this place, nor is it impelled by the motion of heaven, it is clear that Aristotle's reasoning is conclusive.[9]

Aristotle's reasoning revolves around the proposition the Commentator so clearly demonstrates, that weight is neither the effect of an attraction emanating from the center of the world nor the effect of an attraction emanating from the heavy body occupying the center. This principle dominates everything that Aristotle wrote about the natural movements of sublunar bodies.

In order to denote clearly that the weight of earth is not an attraction, Averroes contrasts it with the attraction of iron for a magnet. It would be useful, in order to understand the force of this contrast, to know what the Commentator of Cordoba taught about magnetic actions. It would be inappropriate to attempt to derive his opinion from Aristotelian texts, but one can state at least that the opinion is in conformity with the spirit of Peripatetic physics.

An action by which the attracted body moves, and in which the attracting body remains immobile, as happens in the case of iron and magnet, is not properly speaking an attraction; it is an attraction only metaphorically. In reality the magnet does not attract the iron, but the iron moves toward the magnet like a heavy body moves toward its place, which is the center of the world.

However, there is a difference between the natural movement of a heavy body and the movement of iron toward a magnet: "The body tending toward its own place does so whether it is near or far from it."[10] Averroes thinks, on the other hand, that the tendency of iron to move toward the magnet diminishes as its distance increases, and even that the distance can be so great that all action disappears. He thinks this because, "the iron does not move toward the magnet, unless it is affected by a quality coming from the magnet. . . . It is by this quality that the iron becomes capable of moving toward the magnetic stone."[11]

Moreover, the iron receives the quality through the intermediary of the air in between; the magnet first alters the air, communicating to it a particular quality, and the air then communicates a similar quality to the iron.[12]

It is interesting to note the affinity between Averroes' opinion on magnetic attraction and those favored by contemporary physicists. As soon as a magnet is brought to some place, it begins to determine the appearance of a property in the air surrounding the place—magnetic polarization. The region polarized gradually extends itself into regions where air is not polarized; the surface separating these two regions from one another propagates like a luminous wave and at the same speed. When this magnetic wave reaches a piece of iron, the iron is polarized, and soon its various parts become subject to forces moving it toward the magnet.

Averroes thinks that all actions where a body seems to move toward another at a distance, with a force decreasing in intensity as the distance increases, are accomplished in the same manner as magnetic action. Twice he likens them to the action by which rubbed amber attracts pieces of straw—with which modern science would agree.

Many contemporary physicists would like to accept Averroes's opinion about actions at a distance at its fullest. They want to put all these actions, particularly universal gravitation, under the heading of electromagnetic attractions, but their wish is still far from being realized. On the other hand, the Commentator, like Aristotle, wishes to remove weight and levity from under that heading since he does not consider them as attractions. He affirms that they do not depend upon the distance that separates the mobile from the place toward which it tends.

12
The Problem of the Plurality of Worlds in Scholastic Philosophy

Scholasticism and the Plurality of Worlds before the
Condemnations of 1277. The Plurality of Worlds and the Void:
Michael Scot, William of Auvergne, and Roger Bacon.
The Plurality of Worlds and the Change of Weight
according to Its Distance from the Center of the World:
Albertus Magnus and Saint Thomas Aquinas

When Copernicus, instead of leaving the earth at rest in the center of the world, gave it not only two rotations on its own center, but also an annual revolution around the sun, astronomers were able to maintain that these hypotheses are not given as realities, that it suffices for them to be fictions by which the phenomena are saved in a simpler and more exact manner than is possible using Ptolemy's devices. But physicists did not willingly use this loophole; they not only saw in the system of Copernicus a model enabling them to construct new tables of celestial movements, they also imagined something of an entirely different nature, something that claims to reveal a truth. They imagined that the earth is a planet of the same nature as Venus, Mars, or Jupiter. The problem that the new astronomy laid down for them is as follows: can each of the bodies we call wandering stars be a world similar to the world in which we are living, having at its center an earth covered by water, surrounded by air?

Aristotle had answered this question negatively. Let us imagine that outside our world there exists another world having an earth of the same kind as our earth at its center. This earth would have its natural place at the center of its world, in the same way our earth has its natural place at the center of our world. But this other earth, being of the same kind as ours, would also have its proper place at the center of our world. Hence this earth has two proper places toward which it must tend by nature; this is an absurd assumption.

An intellect trained by modern physics would immediately propose a reply to Aristotle's objection: Evidently, this earth would tend toward its neighboring center with greater force than toward its farther center; therefore it would go toward the former, not the latter. It is only in the case where it is placed at a distance equidistant from the two centers that it would remain in equilibrium.

Such a reply implies the following axiom: The force by which a mass of earth tends toward the center of the world, which is its proper place, varies with the distance from this center; it diminishes as the distance increases. Would such an axiom have been accepted by Aristotle? It is extremely doubtful.

Simplicius thought it valuable; he attempted, moreover, to negate the force of the reply that one could pit against Aristotle's argument. But Averroes seems to have been the more faithful interpreter of the thought of the Stagirite, when he maintained that proximity and distance have no influence on the movement of the heavy body toward its proper place.

Without insisting on the two opposing doctrines of Simplicius and Averroes that we have previously examined, let us limit ourselves to remarking that they posed an important problem for the masters of Scholasticism: Does the weight of a heavy body depend upon its distance from the center of the world?

But the writers of the thirteenth century who first disputed the problem of the plurality of worlds did not latch on to this problem at first—that there cannot exist several worlds seemed to them to result from the impossibility of the void. Aristotle had stated that outside the world there cannot be any place because there are no bodies; there cannot be a void either, because a void is a place where there is no body but where there can be a body, and outside the world there cannot be one. This argument would fail once the existence of another world outside the world were thought actual or merely possible. Hence, one can argue against the plurality of worlds from the impossibility of void.

We have previously mentioned how this argument was developed by Michael Scot, William of Auvergne, and Roger Bacon.[1] We shall not here repeat these ramifications; we shall only relate what they asserted against the plurality of worlds, above and beyond their proofs from the impossibility of void.

After having summarily recalled Aristotle's reasoning against the plurality of worlds, Michael Scot adds: "There are those who maintain that God, who is omnipotent, could and can still create, in addition to this world, another world, or several other worlds, or even an infinity of worlds, creating them out of elements similar to those making up this world, or out of different elements."[2]

The astrologer of Frederick II replies to this objection that "God can do this, but nature cannot withstand it. The impossibility of the plurality of worlds results from the nature of the world itself, from its proximate and essential causes; God, however, can make several worlds, if He so wishes it."[3] In fact, one must distinguish between the power of God taken absolutely and His power relative to the subject upon which it is acting. There are things that God's power, taken absolutely, is capable of doing; but these things cannot be realized by His power, taken relatively, because nature is not able to receive the divine power. It is in this way that nature cannot receive a plurality of worlds.

Ernest Renan has called Michael Scot the founder of Averroism.[4] The passage we have just analyzed is not of such a nature that it forces us to reform this judgment. The God of Michael Scot is One whose creative power finds before it a predetermined nature that puts limits and preconditions upon Him; this God (who cannot act except within some limits that nature is able to withstand) is closer to the God of Averroes than the God of the Christians.

The problem of the plurality of worlds is therefore linked to a problem Greek philosophy did not foresee, the problem of the creative omnipotence of God; it is in the name of God's omnipotence that Christian Scholasticism rejected the Peripatetic solution to this problem.

William of Auvergne, like Michael Scot, thinks that the impossibility of void provides an argument against the existence of multiple worlds; we have previously reported the argument he gave along these lines [part IV, chap. 9]. But he invoked other proofs against the plurality of worlds than those we analyzed.

First, one might propose a loophole against the objection drawn from the impossibility of void. It would consist in the assumption that another world extends above the sphere delimiting our world, the other world completely encircling our world. This world would be contained within a sphere far from the one encircling our world. "But then, since the ultimate sphere of that world contains the heavens of that world and our heavens also (those manifest to our senses), it is clear that this sphere and everything enveloped by it form a unique world, containing everything in it."[5]

One can find many arguments against the thesis that the world is unique. Here is an example: A unique world would not be able to contain all existing things. But, William replies, either one assumes that God has created an infinity of worlds or that He has created a finite number. If the number is finite, a single world would be able to contain as many things as several small worlds, and the unique creation is better suited to God's majesty. In his discussion, the bishop of Paris forgets the second horn of the dilemma he posed.[6]

The above difficulty is not unique. Here is another example:

> God created the world out of His pure and free benevolence; He could just as easily have created a great number of others. Therefore He has created them. The cause that created one by its benevolence, will, for the same reason, create a great number of others. . . .
>
> His generosity has no end, neither do His riches. How could the effect of His generosity and liberality be limited? If the world were finite, then the gifts of God would be finite, divine generosity would be narrowed and restricted. . . .
>
> You see that this reasoning appears to conclude not only against the creation of a unique world, but also against the creation of a finite number of worlds; however many worlds are created, they would not equal God's benevolence and generosity, for everything existing outside God, far from being equal to Him, is nothing in comparison with Him.
>
> I therefore state that God could not have created a finite number nor an infinite number of worlds, and that He cannot create them in actuality; this impossibility is not a defect in God, nor a defect issuing from God, rather it is a defect on the part of the world, which cannot exist in multiples (as I have previously demonstrated). . . . In the same way, God does not know the relation of the diagonal of a square to its side, not that there is a defect in God's knowledge, but because the relationship cannot be known.[7]

The impossibilities that Peripatetic physics discovers against the plurality of worlds are considered by William as mathematical absurdities; God cannot create several worlds, not because His omnipotence is limited by it, but because such a work would imply contradictions.

Bacon read Michael Scot (whom he judged rather harshly); and he studied at Paris under the tenure of William of Auvergne. We should not be surprised if we recognize in his thought a reflection of the thoughts of these two authors.

Roger Bacon devotes a chapter of his *Opus Majus* to the examination of these two questions: Can there be several worlds?

Does the matter of the world extend to infinity? Here is what he writes in this chapter about the plurality of worlds:

> Aristotle stated in the first book of the *De Caelo* that the world collects all its matter into a single individual of a single species, and that it is the same for each of the principal bodies making up the world; in this way the world is numerically unique, and there cannot be several distinct worlds belonging to the same species (and there cannot be several suns, nor several moons, even though many have imagined such things).[8]

In fact, Aristotle wrote, at the end of his argument against the plurality of worlds, that "the world as a whole includes all its appropriate matter."[9] Bacon is therefore giving us an accurate commentary of this thought. Within a species, individuals are numerically distinct, according to Peripatetic philosophy, when their common form affects the different parts of matter in each of them. If the form of an individual is united to all matter capable of receiving this form, this individual is necessarily unique within its species. Thus it is, according to Aristotle, not only for the whole world, but also for each of the stellar bodies and spheres. It is an essential proposition in the Aristotelian doctrine of the unity of the world; and it is one of the doctrines that Christian Scholasticism firmly denies.

After having reported and commented upon the Stagirite's proposition, Bacon, in his *Opus Majus*, draws from the impossibility of the void the proof that there cannot exist several spherical worlds external to one another.

In his *Opus Tertium*, Bacon summarily takes up Aristotle's argument against the plurality of worlds without alluding to the impossibility of the void.[10] But he takes up this principle again, along with the reasoning based on the impossibility of the void, in his *Communia Naturalium*, or better, in the treatise *De Caelestibus* which the manuscript of the Bibliothèque Mazarine gives as the second book of the *Communia Naturalium*.

Bacon now adds the following reflections to Aristotle's reasons and the proof taken from the impossibility of the void:

> Neither can one maintain that a second world encircles the first, for the center of one would be the center of the other, so that there would be only one earth for both. It could be the same with the other parts of the world; hence there could be only one world.
> Further, if there existed a reason for there being two

worlds, for the same reason there would be three, four, and so forth to infinity, for everything about the world is indifferent to any one number. There has to be, then, an infinity of worlds or one only; but there cannot be an infinity of worlds. Therefore, there is only one world.[11]

The author of these lines had read the *De Universo* of William Auvergne.

Michael Scot, William of Auvergne, and Roger Bacon, wishing to prove that there cannot be several worlds, drew their principal argument from the impossibility of void, which Aristotle had not used toward this end. They appear not to have bothered with the reasoning that the Stagirite so carefully developed, the reasoning concerning the movement of the earth toward its natural place. Michael Scot and William of Auvergne did not refer to it, and Bacon was satisfied to cite it in passing while he was enumerating the various Peripatetic reasons against the plurality of worlds.

However, Aristotle's reasoning revolves about an essential problem. First, it poses the question: Does the weight of a heavy body vary with the distance the weight is from the center of the world? Averroes has already shown that within this question lies another, extremely important question: Is weight the effect of a sympathetic attraction seeking to reunite the various fragments of the same element, as the Pythagoreans would have it, and as taught by the *Timaeus*? Or is it, as Peripatetic doctrine would have it, a tendency by which the form of the heavy thing strives toward the place where it will attain its perfection?

The importance of Aristotle's principal argument against the plurality of worlds did not escape Albertus Magnus or Saint Thomas Aquinas, even though the teacher and his disciple took differing viewpoints with respect to it.

Albertus Magnus follows closely the commentary of Averroes. Let us quote a passage from his lengthy exposition:

> Perhaps some contrary person will claim that the nature of elementary bodies, when these are situated in different worlds, is modified with respect to the greater and lesser distance that separates them from their natural places. For example, earth placed outside our world would be farther from the center of our world and closer to the center of the other; it would then be influenced by the nature of the latter center and not by the nature of the former, in such a way that it would move toward the latter center and not toward the former. Thus we see that the magnet attracts a neighboring piece of metal, since the piece of metal acquires a certain property

from the attracting stone; but the magnet does not attract a distant piece of metal, since the virtue of the stone does not reach up to that piece of metal.

We will reply that this discourse does not conform to the rules of reason, and that, consequently, it is erroneous. The movement of the elements is not the effect of an attraction, for if the elements moved by attraction, each of them would be pulled by its own kind in such a way that if we placed a larger earth above a smaller earth, the smaller would necessarily climb toward the larger. In this way, a movement depending on proximity or remoteness is a movement produced by an extrinsic motive power; whereas, the movement of the elements is due to an intrinsic motive power.

In fact, as we have already stated in the eighth book of the *Physics*, when an element is engendered, what engenders it gives it not only its form, but everything resulting from this form; it gives it, in particular, its natural movement and its natural place, which are the consequences of its intrinsic form. If proximity or remoteness from its natural place has influence on the substantial form of the element, it would have to be that the element is composed of two forms having opposite properties; one of these forms would pull the body toward what is nearest to it. This would be a form emanating from the attracting body similar to the form the magnet produces on the iron. The other would be the natural form given by the generator; without any attraction intervening, it would determine the movement of the body toward its natural place. It would be comparable to the form that gives weight to the iron which the magnet attracts. The elements would therefore be composites. All movements of such an element would be composed of two distinct movements, like the movement of an earth approaching the center of a world while getting farther from the center of another. . . .

The coexistence of two such forms is impossible. One must then conclude that a body can be more distant or less distant from its natural place without its form experiencing any change. . . . Whether it is near or far from its natural place, it always moves with a simple movement.[12]

Albert's inspiration was due to Averroes alone; William of Moerbeke had not yet translated the commentaries on the *De Caelo* composed by Simplicius. However, Aquinas did read these commentaries; their influence is often felt in his work, and it is particularly evident here.

The Angelic Doctor follows Simplicius's opinion; he concedes that the distance of a heavy body from the center of the world, without changing the form that carries the heavy body toward its natural place, can vary the intensity of this form. He renders this opinion more precise and relates it to another assumption Simplicius presented as Aristotle's in other circumstances—the increase of weight a heavy body undergoes as it is nearer or farther from the center of the world causes the acceleration noticeable in its fall.

Saint Thomas expresses himself in this fashion:

> According to Aristotle, we must consider unreasonable the opinion that the nature of a simple body would differ according to whether the thing is more or less distant from its natural place, so that it would move toward its natural place when it is near it, but not when it is far from it. In fact, it does not seem that the greater or lesser distance that separates a thing from its place can determine a change in its nature. . . . It is reasonable that a thing move more rapidly when it gets nearer its natural place, even though the kind of movement and the kind of mover remain unchanged; for difference in speed is a change of magnitude and not a specific change, as is the change of distance.[13]

The problem of the plurality of worlds, which Michael Scot and those following him had linked with the problem of the impossibility of void, finds itself linked by Saint Thomas with another debated question, the explanation of the accelerated fall of heavy bodies. We have previously studied the solutions the the Middle Ages proposed for this problem;[14] hence it would not be useful to repeat them here.

In his discussion on the plurality of worlds, Saint Thomas Aquinas could not have been content with the borrowed opinions of the Stagirite and his commentators, such as Simplicius and Averroes. In favor of the opinion that the existence of several worlds is possible, Christianity fashioned an argument based on the creative omnipotence of God, an argument that pagan antiquity could not have foreseen. Here is how the *Doctor Communis* expounded and refuted this argument:

> Know that many are trying to demonstrate the possibility of the plurality of worlds by other means.
> Here is a first argument: God made the world, but the power of God is infinite. The production of this single world does not attain its limits. It is then unreasonable to maintain that the Creator is not able to produce another world. To this argument we must respond thus: If God were to make other worlds, either He would make them similar to this world,

or He would make them different. If He were to make them completely similar to this one, He would be making something in vain, which is not in keeping with His wisdom. If He made them otherwise, it would be that neither of them in itself contains the totality of the nature of sensible bodies; neither of them would be perfect, and it would be their totality which would constitute a unique and perfect world.

A second argument is as follows: The more noble a thing is, the more its kind has the power to be realized. And the world is a more noble kind than any of the natural beings it contains. If the genus of such an object, of horse or oxen, for example, is capable of making up a number of individuals, then *a fortiori* the genus of the universe can contain several individuals. To this we shall respond that it takes a greater power to produce a single perfect individual than to produce a great number of imperfect individuals; and the individuals, the natural kinds in the world, are all imperfect. None of them contain in themselves everything proper to its kind. On the other hand, the world possesses this kind of perfection; it suffices to show that its kind is more powerful than all the rest.

Third, one could object thus: It is better to multiply the better things than the worse things; it is better, then, to create many worlds than many animals or many plants. To which we shall reply: It is relevant to the world's goodness that it is unique; its unity is itself the reason for its goodness. We see that, for certain things, division suffices to forfeit the goodness proper to them.[15]

The Plurality of Worlds and the Condemnations of 1277: Godfrey of Fontaines, Henry of Ghent, Richard of Middleton, and Giles of Rome

The question about the plurality of worlds, like many other problems, seems to place in opposition the impossibilities decreed by Peripatetic physics and the creative omnipotence Christianity recognized in God. Michael Scot, William of Auvergne, Roger Bacon, and Saint Thomas Aquinas attempted to prove by various means that this limitation of power is only apparent, that the lack of power to achieve what Aristotelianism declared impossible is only an effect of divine perfection. Christianity did not accept these subtle explanations; it considered the assertion that a second world cannot be produced as an impious pretension of philosophers, placing a limit on God's power. This belief finds expression in the decree brought forth on March 7, 1277, by Etienne Tempier, bishop of Paris, and his advisors. Among the errors that the decree condemns is the thirty-fourth: "That the First Cause cannot make several worlds. (*Quod Prima Causa non posset plures mundos facere.*)"[16]

This condemnation required the Parisian masters to shift the focus of their teaching with respect to the problem of the plurality of worlds; it was no longer possible for them to claim that this plurality is impossible. Moreover, since this impossibility was deduced from several essential Peripatetic physical theories, they had to reject these theories or to submit them to profound transformations. We have already seen how the condemnations of 1277 completely changed the ideas about the void that had currency until then; we shall demonstrate the consequences it had for other Peripatetic principles.

The problem of the plurality of worlds was clearly one of the problems freely debated in the theological discussions following the decree of 1277.

Godfrey of Fontaines, for example, in one of his *Quodlibets*, discusses the following question: "Outside our world, can an earth of the same species as the earth of this world be made?"[17] In his affirmative reply, our author attacks primarily three arguments invoked by Peripatetic physics. The first, which Bacon formulated most clearly, is that a new world cannot be produced because all the matter proper for the construction of a world is already in this one. The second is drawn from the nature of natural movement. The third is based on the impossibility of void. We have already reported on what Godfrey asserted about the last argument; we therefore limit ourselves to the first two.

> The Philosopher posits that God cannot do anything without the movement of heaven acting as intermediary, and that He cannot do anything unless it is a change imposed upon matter. In all new production (*factio*), he assumes that the matter which is the subject of this production preexists. According to him, the production of new matter is impossible. Hence, another world or another earth in this world is something whose production is impossible, for this world here contains all nature, both actual and potential.
>
> But God, who has already produced some matter, can produce some new matter, and from it produce something else. The fact that this world here is formed from all its matter does not render impossible the existence of another world.[18]

Aristotle's principle is valid for natural causes whose actions are limited to forming preexisting matter in some new way; these causes would not be able to produce a new world, because there is no matter capable of becoming a world, outside this world. But this argument does not hold for God, whose creative power is not limited to forming preexisting matter, but whose omnipotence can even create new matter.

As for the objection derived from natural movement, Godfrey provides a solution based on principles of Peripatetic and Arabic Neoplatonist philosophy.

If our earth has a proper place at the center of the world, where it remains naturally at rest, and toward which it would move naturally if it were separated by violence, it is due to its disposition with respect to heaven and to the influence it receives from heaven: "In fact, [according to Aristotle,] the first movement is the cause of all the other natural movements."[19] Similarly, in another world, the heaven surrounding the earth of the other world would hold it in its proper place; it would therefore move toward the center of the other world by natural movement: "Since this other earth would have no relation with respect to our heaven, and since it would not receive any influence from it, if it came to move toward the surface that encloses our world, it would be by violent movement, and not by a natural movement that would be in virtue of a relation with our earth."[20]

Henry of Ghent professes a doctrine similar to Godfrey of Fontaines's on the topic of the plurality of worlds; he expresses himself on two separate occasions about this topic.

He writes in his twelfth *Quodlibet*:

> The sun contains all its matter, that is, all the matter capable of receiving the form of the sun; however, it does not contain all that will be made or can be made by God. That is why God can make a new matter which is capable of receiving the form of the sun, matter which is the same as that which now exists under the form of the sun; moreover, He can, if He so wishes, make a new sun.[21]

And the Solemn Doctor declares in his thirteenth *Quodlibet*:

> God can create a body or another world beyond the ultimate heaven, in the same way that He created the earth within the world or the heaven, and in the same way that He created the world itself and the ultimate heaven.[22]

In this question he also attempts to refute the objection derived from the impossibility of the void; we have already reported what he asserted about this argument.[23]

Richard of Middleton's thought on the subject of the plurality of worlds conforms with Godfrey of Fontaines's and Henry of Ghent's thought.

"I understand by universe a set of things a single surface contains, including the surface, and on the condition that this set is not itself bounded by any other surface surrounding it."[24] By means of this condition, Richard of Middleton avoids the assumption of worlds within worlds, an assumption William of Auvergne, and others influenced by him, found extremely appealing.

> I therefore state that God could have and can still now create another universe. There is, in fact, no contradiction in attributing this power to God.
>
> Such a contradiction cannot originate from that which the universe might have been made, since God did not make the universe from anything.
>
> It does not originate from the receptacle of the universe, since the world in its totality is not in some place. The Philosopher stated in the first book of the *De Caelo* that there is no place, no void, no time, outside the last sphere; this proposition applies to the ultimate heaven.
>
> This contradiction cannot be because of divine power, since God's power is infinite. Since the universe is finite, it is impossible that it equals divine power.
>
> Finally, the contradiction cannot be derived from the nature of the beings contained by the surface of the second universe, since God made them of the same kind as those of this universe. In the same fashion that the earth of our universe rests naturally in the center of the first universe, the earth of the second universe would rest naturally in the center of the universe to which it belongs. If the earth of the second universe were put in the center of our world, it would rest there motionless; and if the earth of our universe were put by God in the center of the other, it would find its natural rest there. If, with respect to its natural behavior, a thing were indifferent between two places, it would remain at rest in either of the two places where it was first put; it would not tend toward the other.
>
> In order to establish this opinion, one can invoke the sentence of Lord Etienne, bishop of Paris and doctor of sacred theology; he has excommunicated those who teach that God could not have created several worlds.[25]

We have just heard professed the profound change that the decree of 1277 produced with respect to the problem of the plurality of worlds on the thought of the Parisian masters.

Richard of Middleton is not satisfied to admit that the plurality of worlds is not something contradictory and that therefore God's omnipotence would be able to accomplish it. He goes further; he undertakes to counter the main objection that Peripatetic philosophy puts forward against the possibility of several worlds.

As for the objection derived from the impossibility of the void, our author indicates in passing that the world is not in space, and reminds us of the teaching of the Stagirite, that there is no place, no void, and no time outside heaven. In any case, we have heard him declare, in another circumstance, that the production of void is not an impossibility for God.[26]

Giles of Rome, without formally siding with the doctrine of the plurality of worlds, takes care not to contradict the condemnation of 1277.

In his *Opus Hexaemeron*, he teaches formally that "heaven and each part of heaven is formed from all its matter."[27] But he interprets his proposition as follows:

> In every eternal thing, insofar as it is eternal, and in every incorruptible thing, insofar as it is incorruptible, it is not possible to distinguish between what it is (*esse*) and what it can be (*posse*). Everything that can be in such a substance actually is. In fact, if such a substance can possess something and does not actually possess it, then with respect to that thing, it would not be eternal, but subject to corruption.
>
> And the whole heaven and each of its parts can have another *ubi* than that which they presently have. . . . With respect to its *ubi*, one can distinguish what a portion of heaven is from what it can be. . . . But with respect to its essence (*esse*), there is no difference between being and being able to be; heaven and each of its parts have all the essence that they can have. Hence, we say that they can change relative to their *ubi*, but they cannot change relative to their essence.
>
> Heaven therefore contains all its matter; and it is the same for each of its parts. In fact, nothing can be under the form of heaven or under the form of any part of it whatever that is not so actually; otherwise, there would be generation and corruption of heaven relative to its essence and form.[28]

Thus the axiom, heaven contains all its matter and so does all its parts, is equivalent to, in its totality and in each of its parts, heaven is not subject to generation, corruption, and change. Evidently, this proposition does not exclude the possibility of multiple heavens. Giles is careful to point this out, for he adds:

> Moreover, if there existed two suns, one can assert that each of them contains all its matter; there would be no change concerning the form and essence of each of them. All matter that can be under the form of the first sun is entirely in actuality, and it is the same for all the matter that can be under the form of the second sun. With respect to this, one cannot establish a difference between what is and what can be, between what it has and what it can have. Hence each of the two suns contains all the matter that it can possess in actuality; it would then be correct to say that it contains the totality of its matter.[29]

This interpretation, which seems to be in conformity with Peripatetic philosophy, weakens one of the objections commonly cited against the plurality of worlds.

Giles of Rome's doctrine from his *Opus Hexaemeron* was only a restatement of what he held in his *Quodlibets*:

> To speak of the production of a new species with respect to simple bodies is to refer to the production of a new heaven or a new element; both are impossible.
>
> A new heaven cannot be produced by natural means (*via naturae*), for each heaven contains the totality of its matter; a heaven cannot therefore be corrupted into another heaven. One heaven cannot be converted into another either. There cannot be any novelty of form or movement to the heavens. Therefore, a new heaven cannot be produced by natural means.
>
> Neither can the element of a new species nor an element that is taken in its totality be new. If there is some innovation in an element, it is always in a partial manner; such an element is partially destroyed and another element is partially engendered. But that an element is engendered or destroyed totally, or that an element is produced which would never have existed, or that an element or a body is made which is neither fire, nor air, nor water, nor earth, is not possible. Taken in their totality, the elements are not capable of generation or destruction. . . .
>
> Thus by natural means, it is never possible for a new element to be engendered. It is not possible that there was once three elements or fewer. The elements are susceptible to generation and corruption in their parts, but not in their totality. By divine power, an element could, in its totality,

be changed into another element, or be engendered from other elements, but not by natural means.[30]

Giles of Rome therefore argues only about what can or cannot be produced by natural means; he does not intend his arguments to impose a limit on the sovereign power of God.

William Varon, John of Bassols, and Thomas of Strasburg

The decree of 1277 therefore marks a complete reversal in the opinion of the Parisian masters about the plurality of worlds. Before the decree, they accumulated reasons derived from Peripatetic physics in order to establish that the existence of several worlds is an impossibility; therefore they refused God the power to multiply worlds. They endeavored to prove that this refusal was not a limitation on God's creative omnipotence. After the decree, all theologians held for certain that God can create multiple worlds, if He wishes to. They endeavored to destroy the reasons given from physics that were pitted against this proposition, or at least to interpret them in such a way that they were no longer objections.

After the ban formulated by Etienne Tempier, the masters of Oxford also accepted this decision and elaborated a doctrine in which one could recognize the thoughts of Godfrey of Fontaines, Henry of Ghent, and Richard of Middleton; it is the doctrine expounded in great detail though sometimes with a confused verbosity in a question on the *Sentences* of Peter Lombard composed by William Varon:

> One can consider the world from two points of view.
> By *world* one can understand the universality of creatures taken all together; therefore, a world other than this world, once this world has been created, would not contain the universality of creatures. Hence, it would not be another world, but only a portion of the universe.
> One can also understand by *another world* another celestial sphere, within which would be four elements ordered in the same way that our four elements are ordered under our heaven. It is in the second way that one understands the question.[31]

Among the reasons supporting an affirmative answer, the following passage as an argument from authority can be cited:

> The possibility to produce two worlds does not imply a contradiction on the part of the Producer, since He is all-

powerful. Neither does it imply one on the part of what is produced. Matter which has been produced was fashioned from nothing and possesses the totality of form proper to it; since it exists by creation, we can see no reason keeping another matter from being fashioned from nothing and also having all its own form. Finally, this is not repugnant to the world already created; similarly, if there were only one man, and he had all the matter of man, it would not be repugnant to him that the creation of another man be possible.[32]

Among the objections against the existence of several worlds, William does not omit the one derived from the impossibility of void commonly given after Michael Scot; in fact, it is the one about which he comments most lengthily. We have previously reported the essential features of his thought;[33] we now report the conclusion of his lengthy discussion:

We would therefore assert that outside this spherical world, God can make another spherical world that does not touch this world at any point. He can do it, because it does not imply a contradiction. In the same way that He can make it be that there is a distance between one portion of heaven and another, He can make it be that two wholes are as distant from one another as His will has ordered; moreover, His power did not diminish because of the creation of our world. Before the creation of our world there was nothing here, and God created this world; thus He can create another world outside our world. In fact, we can imagine a quasiinfinite space in which there is nothing; and in the same way that He created a first world where there was nothing, He can create, where there is absolutely nothing, other worlds whose multitude is potentially infinite—that is to say that He cannot create so many that He would not be able to create more (*id est non tot quin plures*).

A proof of this is that, if it were otherwise, the creation of this world would have equalled, would have exhausted, God's power. This consequence is false, for that which is actually created is finite, and nothing finite equals God's infinite power. As for the reasoning, it is clearly conclusive. In fact, if a source of heat can produce only one warm body, the active power of this source would be completely exhausted by the production. In the same way, the production of the Son exhausts and depletes the power of the Father; thus He would be able to produce only one Son. It would be the same here.[34]

Varon is not satisfied to refute only the objection derived from the impossibility of the void; he also endeavors to destroy those objections Aristotle formulated.

> The Philosopher asserts that the world is unique because it contains all its matter. We would reply: That is conclusive as long as another heaven cannot be produced by a created agent. In fact, the action of such an agent presupposes some matter; it can act only by dividing some matter. But it is not conclusive for an uncreated agent who can produce another heaven and whose action does not presuppose any matter. Therefore, in spite of this reason, another heaven may be created by an uncreated agent whose action does not presuppose some matter, but produces the matter; such a being, who acts by creation, can, in fact, create the matter and the form. Since he can act thus, there is no longer a contradiction in the assumption of another heaven produced by creation. Similarly, if there existed only one man who had the totality of human matter, and if this man could not engender another man, it would not be repugnant to this man if God were to create another man of the same species, not by dividing the [already existent human] matter of the first man, but by creating new [human] matter. It is the same for the case in question.[35]

Let us now examine how Varon replies to Aristotle's other objection—the earth of the other world, being of the same kind as the earth of this world, would have the same proper place as this earth, and would therefore have to tend toward the center of our world by a natural movement.

> If the new earth created outside our world were of the same kind as the earth contained by our heaven, here is what one ought to say:
> The bodies of the same species that are connected to one another and are brought together by the same containing body have the same natural place; but that is not true for bodies of the same species not connected and not brought together by a single body. Thus the vital spirits of two separate men are of the same species; however, they have separate places and, as it were, separate domiciles. Similarly, for the case in question, we ought to say that all earth contained by our [celestial] sphere has a single natural place; but the earth of the other world, not being connected or contained under our sphere, does not descend toward the center of our world. It is at rest in its own center.[36]

This response is merely a detailed repetition of Godfrey of Fontaines's response. Varon continues as follows:

> If one says that the other world is not of the same species as this one, we would respond that the argument is not conclusive, then, because the center of the other world can be other than the center of this world.[37]

But perhaps our author does not think that Godfrey of Fontaines's response is absolutely conclusive because he adds:

> One might also say that that argument is even more conclusive if the created earth in the other world remains there by constraint and against its own inclination; it does not result from this that God cannot make a second center, even though the earth would remain there by constraint and against its own inclination.[38]

William of Varon's meticulous discussion cannot have failed to attract attention; it seems to have had many readers. For example, what John of Bassols asserts about the plurality of worlds resembles greatly the opinion of the English Franciscan. Bassols teaches that,

> God can make a universe other than ours, either of the same species as ours, or of another species.
> Second, I see no objection to God having created an infinity of worlds of the same species as ours.
> Third, I see no objection to His having created a great number of worlds differing in species from ours.[39]

These conclusions run counter to several objections, some formulated by Aristotle; let us cite a few with the replies by which John of Bassols attempts to resolve them. Here is the first: If there existed another world, it would necessarily be of the same nature as ours; therefore, the earth of each of the two worlds would move toward the center of the other. John of Bassols replies that:

> It is not necessary that the earth of one of these worlds move naturally toward the earth of the other world, nor that it be able to move thus toward the other earth. In fact, the natural tendency of the earth toward its center would not exceed the limits of its own world. It goes without saying that divine power would be able to move it. If you say that, in this case, the earth of the other world would not be of the same species as the earth of this world, I would reply that it is not necessary that it be of the same species. Allowing that the second earth is of the same species as ours, still, the earth of each of the

two worlds would not move toward the center of the other
world, but only toward the center of the world of which it
is a part, so that the natural appetite of this earth would not
extend above the whole of which it is a part.[40]

One might object that "that which is made from the totality of
its proper matter cannot be multiplied, for it is by matter only
that there is multiplicity. And as we see, in the first book of the
De Caelo, the world is made in this fashion."[41] To which Bassols
replies, "I assert that God can produce another matter numerically
distinct or of the same species as that which exists, and that the
world does not contain all possible matter."[42]

With respect to the plurality of worlds, Thomas of Strasburg
upholds a doctrine similar to William Varon's and John of Bassols's;
he seems to have been influenced directly by the former.

To the question, "Can God make another universe while this
universe continues to exist?" Thomas replies:

I hold the affirmative—that is, God can make another
universe while this universe continues to exist. He can make
another universe similar to this one, or a better one; He can
make another similar if He forms it out of parts similar to
the ones constituting this universe and a better one if He forms
it out of better parts. I support this assertion as follows:
Let there be a cause that can produce all the parts of
a whole and produce them according to the order needed to
constitute this whole, that can produce them with a perfection
equal to the perfection that they possess in the whole, or with
a greater perfection. This cause will be able to produce a whole
having as much perfection or more perfection than that which
has been produced. And God can produce all the parts necessary
in order to constitute a universe; not only can He produce
them as perfect, but He can produce them even more perfect
than they have been. He can produce them in the order
necessary to constitute a universe; therefore, He can produce
this universe.[43]

Peripatetic physics addresses a number of objections against
the above thesis, objections formulated by Aristotle himself, or
objections later deduced from his principles—such as the objection
derived from the impossibility of the void. Thomas of Strasburg
sets forth all these objections and carefully refutes them. He believes
that "one should not limit God's power because of these and similar
sophisms."[44] For example:

It is true that the virtue of no natural agent would be able to multiply that which is constituted by the totality of its matter, for the action of such an agent presupposes some matter; however, divine virtue can multiply it, for this virtue creates the matter at the same time as it creates the thing formed out of this matter, and its action presupposes nothing.[45]

One can also object as follows:

Either the earth of the other world would be perpetually kept by constraint outside the center of this world, or else it would descend to the center of our world by virtue of the natural inclination carrying it there; these two assumptions are equally impossible.[46]

Our author escapes this dilemma in the following manner:

Neither of the two impossibilities would hold. In order to prove this, we would need to assert that the earth of the other world would not have a natural inclination toward the center of this world; it would have a natural inclination toward the center of the world within which it was created by God. Therefore, it would not be by constraint but by nature that it would remain at rest at the center of its world.[47]

John of Jandun

In order to refute Aristotle's objection that the earth of the other world would have to fall toward the center of our world, William Varon, John of Bassols, and Thomas Strasburg all repeat Godfrey of Fontaines's reply: The earth of each world tends exclusively toward the center of the world it belongs to; it has no inclination driving it toward the center of another world.

The Peripatetics cannot have been content with this reply. That which gives a body its substantial form also assigns it its natural place; the identity of substantial form entails the identity of natural place, and two bodies cannot have two distinct places if they do not have distinct substantial forms—that is, to assert that an earth has as its proper place the center of a world, and that another has as its proper place the center of another world, is to assert that these two earths do not have the same substantial forms, that they are not of the same species. Aristotle's argument therefore retains its force and demonstrates that there cannot be two worlds of the same species.

Already John of Bassols foresaw this objection; in fact, it was formulated during his time by John of Jandun.

John of Jandun firmly held, against almost all the doctors of his time, that there cannot be several worlds. "That is evident by the authority of Aristotle and of the Commentator, and it can be proven by the reasoning of Brother Thomas."[48] His argument is directly influenced by Saint Thomas:

> There is a possible evasion here. . . . It seems possible for there to be another world and that the earth of the other world not move toward the center of this world because the two worlds are not of the same species. . . . But that goes against those who affirm the existence of several worlds, for all these worlds are of the same species according to them . . . the earth of the other world and the earth of this world have to be of the same species; moreover, it must be that the earth of the other world moves naturally toward the center of this world.[49]

Jandun also refuses to accept the reason, that "the earth of the other world would not move toward the center of this world because it is too distant from it—thus the iron is not attracted to the magnet at any distance."[50] Like Averroes, whose authority he invokes here, our author rejects any assimilation between the tendency of a heavy body toward its proper place and the tendency of iron toward a magnet; he takes up, on this occasion, the theory of magnetic actions the Commentator proposed.

This theory of magnetic actions, that opposes the movement of iron toward the magnet with the fall of a heavy thing, seems to be extremely important for John of Jandun. If one allows the comparison between the two kinds of movements, if one allows that a heavy body falling "is moved by a natural force exerted by the natural place (*virtute naturali loci*), then the earth of the other world would not fall toward the center of this world; but this proposition is false and it contradicts Aristotle."[51] It would, however, be the consequence of the assumed principle; a force exerted by the place (*virtus loci*) is, in fact, a corporeal virtue acting only up to some determined distance.

Does not John of Jandun risk the condemnation of Etienne Tempier because of his doctrine of the impossibility of the plurality of worlds? No, because he also declares the following:

> All that does not say anything about divine power; one always safeguards its infinite freedom and infinite power to create several worlds, even though this reasoning cannot be derived from sensible things; and Aristotle derives his

reasoning from sensible things. However, it is a thing we have to believe firmly, giving our respectful assent to the doctors of the faith.[52]

We know that for John of Jandun the above is not a precaution against the condemnation of 1277; it is a corollary of a general system on the relation between Peripatetic philosophy and Catholic faith.[53]

William of Ockham and Robert Holkot

Those who hold the possibility of several worlds also hold that the earth of each world tends toward the center of its own world; it would therefore not fall toward the center of another world. To which the Peripatetics reply: To attribute to two earths two different natural places, is to attribute to them different substantial forms, to rank them as two different species.

This objection seems decisive against the opinion of the early physicists. But although it can be derived from the works of Aristotle, Ockham showed that it implied too narrow a conception of natural place. Doubtless, masses of earth of the same species must have a *specifically* unique place, but it is not necessary that this place be *one*, that it be a unique point. If the Philosopher believed that the natural place of the earth can only have been a point, it is because he considered only the movements of heavy bodies. If he had considered the movement of fire, he would have recognized that a natural place, though specifically unique, can be formed out of geometrically distinct parts, and that an element tends toward one or the other part according to its location in the world. This remark destroys the principal objection of the Peripatetics against the plurality of worlds, in the name of Peripatetic philosophy correctly interpreted.

William of Ockham devotes an entire question on the problem of the plurality of worlds in his commentary on the *Sentences* of Peter Lombard. He holds this plurality as possible:

> I say that God can make a world better than this one, which would be only numerically distinct from this one. My reason is as follows: God can make an infinity of individuals of the same species (*ratio*) as the individuals existing today; He can therefore make as many or more individuals than those having been already produced, and He can make them of the same species. But God is not constrained to produce them in this world; He can produce them outside this world, and

thereby make another world in the same fashion that He made this world from individuals already produced.[54]

The assertion that several worlds are possible cannot hold unless one counters the Peripatetic arguments against it; Ockham therefore endeavors to counter them.

The Stagirite asserted that the various parts of an element all tend necessarily toward a unique natural place—that therefore there cannot be two worlds whose centers would be two distinct natural places for the earth. Here is how the Venerable Inceptor replies to this:

> All the individuals belonging to one species will move naturally toward the same numerical place, if they were put successively in the same position outside this place; it does not follow that they would always move toward the numerically same natural place. It is possible that they would move toward two numerically different places.
>
> If someone put two different fires of the same species into two different regions of space, they would both climb toward the circumference of heaven, but they would not both tend toward the same place; they would move toward two numerically distinct places. However, if someone took the first fire and put it where the second fire was, the first fire would tend toward the place where the second tended.
>
> It would be the same with respect to the question now occupying us. If someone were to take the earth belonging to the other world and put it inside our heaven, it would tend toward the same place as our earth. But when it is outside this world, when it is inside the other heaven, it would no longer move toward the center of our world—no more than the fire at Oxford would tend toward the same place where it would tend if it were put in Paris—but it would move toward the center of the other world.
>
> It is not only because the two earths are numerically distinct that they move toward two distinct places, as maintained by the objection Aristotle was refuting; they would move toward the two numerically distinct places because they occupy two different positions in the heavens—in the same way as the two fires move toward different portions of heaven because of their different locations.[55]

Would Peripatetics be convinced by this argument? Certainly not, for they would reply with their teacher that the natural movement of the earth within the second world would carry it toward the center of the second world; therefore it would get farther from the center of the first world. Therefore the earth gets farther from the center of our world by a natural movement; however, when it falls toward that center, it falls by means of a violent movement by virtue of the following axiom: If a body gets farther from a place by a natural movement, it can only approach this place by a violent movement. William of Ockham does not hesitate to deny this axiom, or better, to correct it:

> If a thing gets farther from a place naturally, *regardless* of its initial position it will tend toward this place only by a violent movement. But if it gets farther from this place *from certain initial positions* only, it is not necessary that it always approaches it by a violent movement.
>
> The fire located between the center of the world and the circumference of heaven imparts an example; when it tends toward the nearer portion of this circumference, it strays away from the opposite side. If, however, one puts it between the center and this opposite side, it would tend toward it naturally.[56]

The Stagirite therefore must have held principles that were too narrow; he must have been led to them because he considered only the movement of heavy bodies. If he had analyzed the movement of light bodies, he would have been forced to allow different principles, and to have set aside some of his arguments against the plurality of worlds.

However, he would have retained the objection that there cannot exist several heavens, because heaven contains all the matter pertaining to its nature. But Ockham then replies that "heaven is made of all already existing suitable matter, but not of all matter that can exist. In fact, God can create again celestial matter in the same way that He can create a new quantity of matter of any kind whatsoever."[57]

Robert Holkot often shows himself to be the faithful disciple of Ockham—which is particularly true with respect to the plurality of worlds.

His commentary on the second book of the *Sentences* of Peter Lombard furnishes Holkot the occasion to discuss the problem: Can God have known from all eternity that He would create the world?[58] It is during the course of this problem that he treats the plurality of worlds.

> We must here speak in conformity with faith; I assume, in fact, that although God's power is infinite in intensity,

since He freely created the world, the goodness and perfection of the world are finite.

This posited, I formulate three propositions:

First, God can make a world other than this one, more perfect than this one, of the same species as it, and having only a numerical difference from it.

Second, God can make a world other than this one, more perfect and of another species.

Third . . .

Here is how I prove the first proposition—by admitting, with respect to God's power, the commonly accepted assumption that God can do anything that does not imply a contradiction; I supplement this assumption with the following:

God can create anything that does not imply a contradiction; and there is no contradiction in supposing the existence of another world only numerically distinguishable from this world; therefore . . .

Here is a proof of the minor premise: There is no contradiction in the assumptions that there are two suns, there are two moons, and there are two earths. God can therefore create celestial bodies of the same species as ours and, consequently, He can create a second world of the same species as ours having only a numerical difference from it.[59]

Holkot does not neglect the reasons given by Peripatetic philosophy against the plurality of worlds; he reduces these reasons to three primary ones:

First, the world is formed out of all the matter proper to it; therefore it cannot be multiplied and there cannot exist multiple worlds. . . .

Second, there would be no more reason for a heavy body to tend toward the center of this world than toward the center of the other world.

Third, any body that strays from a place by a natural movement, can only tend toward it by a violent movement. But a heavy body put in another world would move by a natural movement toward the center of its world; this heavy body would therefore tend toward the center of our world by a violent movement.[60]

Our author responds as follows to these objections: First, the reasoning that heaven contains all its matter so that it may not be multiplied is not conclusive. "In fact, it can be multiplied by an agent whose action does not presuppose any matter, but which creates everything, the matter at the same time as the form."[61]

"To the second objection I would reply: A heavy body placed within a world would move toward the center of the world within which it is located; another heavy body, placed in another world, would move within the other world toward the center of the other world."[62]

Finally, one has to reject the following Aristotelian principle: Everything that naturally gets farther from a place would approach it only by constraint.

> A heavy body can get farther naturally from the place toward which it naturally tends. . . . A heavy body placed at the center of the world would naturally approach a magnet placed above it. It would do the same to prevent the production of a void.
> We can easily see that it is the same for light bodies. Let us conceive a line passing through the center of the earth and extending to heaven. Let AB be the diameter of the sphere of elementary bodies, the diameter whose extremities are two points of heaven. It is evident that a mass of fire placed on the surface of the earth directly below A will climb naturally toward A; by doing so, it will get farther naturally from point B. However, if this mass of fire were placed on the other side of the earth, directly below B, it would approach B by a natural movement and would get farther from A. Thus a light body can get nearer to or farther from a place by natural movement, depending upon where one places it. I say the same with respect to a heavy body if it were first placed within one world and then within another.[63]

In this last argument we can rediscover the perceptible influence of William of Ockham on Robert Holkot. Like his teacher, the disciple takes Aristotle to be in contradiction with his own theory of natural place; nothing was more fitting in order to contradict this theory, and nothing could hasten its destruction more.

John Buridan and Albert of Saxony

> Although carefully reproduced by Holkot, Ockham's criticism of the Peripatetic theory of natural movement did not fare as well with some of the Parisian masters. John Buridan, who could not have been ignorant of it, talks as if he did not know it. As for Albert of Saxony, he asserts that it is not well founded.

One of Buridan's *Questions on the De Caelo* is, if there exists several worlds, would the earth of one world move naturally toward the earth of the other?

Buridan writes:

> Know that although the existence of another world is not possible naturally, it is however possible absolutely (*simpliciter*). Because of faith we hold as certain that God who made this world here, can make another or many others. We must therefore believe that the following is not a good consequence: If there existed several worlds, the earth of one world would move naturally toward the earth of another. However, Aristotle does attempt to prove this consequence.[64]

Buridan first recalls the hypotheses upon which Aristotle's argument lies:

> First, this world and the other would have the same nature (*ratio*) and would be composed from principles that are the same specifically. . . .
>
> But in truth, it is not necessary to concede this assumption; in fact, God can produce dissimilar actions through His omnipotence and His free will.[65]

The discussion of Aristotle's arguments furnishes Buridan the occasion to make several interesting suggestions about the acceleration in the fall of heavy bodies, as we have previously reported.[66] He concludes this discussion as follows:

> Such is the path Aristotle explicitly follows.
>
> But it seems that it is not demonstrative. Doubtless, that is the nature of the heavy body moving it, but its motive action is dependent on the celestial bodies and God. [. . .] Hence, let us assume that divine omnipotence annihilates all the bodies except the air of this world and some earth; this earth would remain at rest within the air. It would not move; there is no reason, in fact, for it to move to one side or the other. One portion of the air would not be higher or lower than another; there would be no virtue in one portion of the air that would not be in another. All this is because the coordination originating from heaven has been sidestepped. In the same way that all things in this world receive their way of life from heaven (*moderantur a caelo*), it would be the same in another world with respect to the heaven of the other world; the earth of the other world would not descend naturally toward the center of this world.[67]

Buridan therefore simply returns to the doctrine of Godfrey of Fontaines.

Albert of Saxony seems to be influenced by the teaching of John of Jandun. Like John of Jandun, Albert thinks that the plurality of worlds is impossible according to the principles of physical science; he does not deny that God can create a world other than this one, but if He were to do it, it would be a miracle that physics cannot explain.

Albert of Saxony knew the arguments favorable to the plurality of worlds that had been set forth by Thomas Aquinas: "It is better to multiply that which is good and perfect than not to multiply it; and the world is good and perfect. It is better, then, that many worlds exist rather than one only. And since God can make it so, and since among the possibilities God always actualizes the best, it follows that several worlds exist."[68]

The above argument is refuted by Albert: "It is not always true that the multiplication of a good thing is better than its unity; if it were so, it would be better that there be several Gods than only one. And this is false, since it is impossible."[69] This reply was already given by John of Jandun.[70]

Albert of Saxony knew equally well the objections by which Ockham attempted to destroy the reasoning of the Stagirite, but he did not assign to them the value that the Venerable Inceptor assigned to them.

According to William of Ockham, the various portions of an element are not compelled to a unique natural place. "We see, in fact, that a fire can tend toward its natural place by climbing toward the north pole, and another by climbing toward the south pole, so that they are tending toward these two numerically distinct places." To which Albert of Saxony replies, "these two fires move toward a place which, taken in its entirety, is numerically unique (it is the concavity of the lunar orb), even though the various portions of the fire tend toward partial places which are numerically distinct."[71]

The following objection is also borrowed from Ockham:

It would seem that distance has some influence on gravity and levity. In fact, if a mass of fire was in the center of the world, it would move toward heaven, which is the place of fire, in such a way that a portion would direct itself toward the north pole and another toward the south pole; whereas, if one placed this mass of fire between the center of the world and heaven, it would all move toward a single part of heaven [namely, the part nearest].[72]

But Albert is not embarrassed by this objection: "Distance can make different portions of an element tend toward their place by differing paths, but it cannot make a body cease to tend toward its natural place."[73]

Another consideration might be that the weight of a thing depends on its distance from the center of the world:

> When the earth is at the center, it no longer has weight; it seems to have lost all tendency to move toward its natural place.
>
> On the contrary, [replies Albert of Saxony,] when it is in its place, its tendency is to reside there, whereas when it is not in its place, its tendency is to return. . . . Therefore, it is not true that the earth no longer has weight when it is in its natural place. Since it has gravity when it is outside its place, it would not lose this gravity when it arrives there. Hence it has weight when it is in its natural place as when it is not; but this gravity has a different function when the earth is in its natural place and when it is not. In the first case, it inclines the earth to move toward its natural place, and in the second, it inclines it to rest.[74]

What Albert is alluding to here is linked to one of his favorite doctrines (which we have touched upon elsewhere): the gravity of a thing is invariable, but it can exist *actually or potentially.*[75]

Another doctrine of Albert of Saxony—one of the more important of those attributed to him—consists in asserting that a mass of earth remains at rest when its center of gravity is at the center of the world.[76] Then, if an earth formed a layer within two concentric spheres whose center was the center of the world, this earth would be in its natural place, even though each of its parts would be very far from the common center of things. From this Albert's odd conclusion follows:

> If there existed many concentric worlds, the earth of one world would not tend toward the center of the other; each of these earths would, in fact, have the same center. And we are able to conceive that an earth shaped like a spherical layer whose center coincides with the center of the world would be resting naturally, just like our earth. Aristotle's reasoning, based on the thought that the earth of a world would naturally move toward the center of the other, does not conclude against

the plurality of concentric worlds. It is not necessary to prove the proposition we take to be a secondary conclusion: there cannot exist a plurality of worlds eccentric to one another, at least naturally.[77]

What do these last few words, "at least naturally," signify?

Albert of Saxony fully admits with Aristotle that the coexistence of several worlds is an impossibility; but, doubtlessly in order to cover himself against the condemnations brought forth by Etienne Tempier, he admits that this natural impossibility may be surmounted in a supernatural fashion by divine omnipotence. Still the coexistence of the two worlds created in this fashion by God would constitute a permanent miracle, a continuous contravention of the natural laws.

> Following Aristotle's doctrine, we conclude that the existence of several nonconcentric worlds is impossible naturally. It is no less true that God could create many worlds, since He is omnipotent.[78]

> A last conclusion in accord with the preceding conclusions: By supernatural means, there can exist several worlds, simultaneous or successive, concentric or eccentric, by the will of God.[79]

"If by a miracle there existed several worlds eccentric to one another," what would happen to the elements contained in these various worlds? With respect to this matter, we could give free rein to our imagination and put forth any assumption we wished "in keeping with the rule that one can conclude anything about the impossible."[80] We could, for example, assert that God has given to each earth the tendency to move toward the center of its own world only.

Among the conclusions we are allowed to formulate once we admit the miraculous coexistence of several worlds eccentric to one another, Albert of Saxony mentions the following:

> If there existed two worlds, the earth of one of the worlds would not tend toward the earth of the other but toward the center of its own world, because it would tend toward the center it is nearest. But if it happened to be equidistant from the two centers, it would remain at rest between them like a piece of iron between two magnets attracting it with equal force.[81]

Without a doubt William of Ockham would have agreed with this conclusion; Albert of Saxony sees it only as a fantastic consequence of an impossible hypothesis: *"Ad impossibile potest sequi quodlibet."*

Oxford University and the Assimilation of
Weight to Magnetic Attraction

The conclusion that Albert of Saxony treats as a fantasy alludes to a theory that considers weight as an attracting force from the center of the world to the heavy body, and compares it to the attraction that the magnet exercises on a piece of iron. This theory, in fact, defuses an important Peripatetic objection against the plurality of worlds. But Averroes fought it with persistence, and the majority of the masters of Latin Scholasticism espoused the opinion of Averroes on this matter. There was, however, a master who formally rejected the doctrine of the Commentator of Cordoba and who recognized in the action of the magnet on iron an attraction exercised at a distance without any intermediaries. Those who received the opinion of William of Ockham had no difficulty in perceiving weight as an attraction exercised by the center of the world on the heavy body. We should not be surprised that such an assumption seemed seductive to some masters of Oxford University.

Among the teachings of Master Clay to his Parisian students about Oxford doctrines are various considerations about the actions of the magnet. These considerations begin with a phrase that is worth noting. "If the center of the world was a point, as some think, and if it was in movement, it is certain that any heavy body, no matter how large it is, would follow this point with a speed equal to its displacement, for this point is the universal place of heavy bodies."[82] The location of this thought [in magnetic theory] indicates that the holders of this opinion assimilate the tendency of the heavy body toward the moving center with the tendency of iron toward a moving magnet.

Although this opinion was well known at Oxford, it was not universally accepted; John of Dumbleton takes care to reject it. He sees a profound difference between the movement of heavy bodies to the center of the world and the movement of iron toward a magnet. "These bodies do not follow that toward which they move, as the iron follows the magnet that one moves. If the point which is the center of the world moved, the earth would not follow it."[83]

When he was issuing or reporting this opinion, master Clay could not have foreseen the fate it was going to have. Forced to renounce the Aristotelian theory of gravity, Copernicus one day conceived that within each stellar body was a point that moved with the body; he had to conceive that every part of a stellar body would tend constantly toward this point. Later, when Copernicus's conception was admitted by many physicists, William Gilbert assimilated the tendencies of the parts of stellar bodies toward a point within the stellar body with the tendency of iron toward a magnet. Thus he constructed his magnetic theory, which was destined to carry the approbation of Francis Bacon and Otto von Guericke. And all of this magnetic philosophy was latent in the reflection of master Clay.

The Return to the Platonic Theory of Weight: Nicole Oresme

Aristotle established a close relation between his argument against the plurality of worlds and his theory of natural place. It was then difficult to maintain this latter theory after the decree of Etienne Tempier affirmed that God can create several worlds. Actually, even after this decree, several masters, like John of Jandun and Albert of Saxony, remained confident in the teachings of Aristotle on the tendency of heavy bodies toward the center of the world; it is true that they allowed that God has the power to create and conserve several worlds, but by supernatural means, by a permanent miracle, constantly contradicting the most stable laws of physics. However, those who held this doctrine were not numerous; the majority of the masters of Paris and Oxford followed another thought. They attempted to correct, to retouch the theory of natural place so that it became compatible with the existence of several worlds.

The various modifications brought from all quarters to Aristotle's teaching did not seem sufficient for the more audacious minds; they did not hesitate to abandon completely Aristotle's doctrine about weight and to return to the one which the *Timaeus* seemed to propose, and which Plutarch had developed so magnificently in his pamphlet, *De Facie in Orbe Lunae*.[84]

According to this doctrine, if the various elements move by natural movements, it is not because they tend to occupy a determinate *position* with regard to the center of the world; what they aspire to is a certain *disposition* which, without accounting for anything alien to their nature, coordinates them one to the other; they move so as to distribute themselves into spheres or spherical concentric layers, superimposed according to decreasing densities. When the earth thus forms the inner sphere, it becomes covered by water, then by air, and finally, fire envelops the whole; the four elements remain at equilibrium, no matter where in the universe this ensemble is located. Consequently, there is no reason to oppose the coexistence of several sets of similarly disposed elements; the Aristotelian argument against the plurality of worlds no longer holds.

In the Parisian school of the fourteenth century, there was a master who could take and develop this doctrine with greater firmness and fullness than Plutarch could have achieved. This master was Nicole Oresme. Did Oresme read the treatise, *De Facie in Orbe Lunae*? He did not refer to it, and it seems as though he did not know it. Probably the *Timaeus* and his own meditations were sufficient to suggest to him thoughts similar to those of Plutarch. Certainly, the objections of William of Ockham against the arguments put forward by the Peripatetics about the plurality of worlds urged him to seek another theory of weight.

Ockham's influence is easily recognized in the chapter of the *Traité du Ciel et du Monde* that bears the title, "In Chapter Sixteen he proposes to find out whether there are or can be several worlds, and by two arguments he proves that there cannot."[85]

The canon of Rouen cites Aristotle's argument against the possibility of there being two worlds (one of the earths would move toward the center of the other), but he continues it with the following:

> But earlier I expressed a strong disagreement with this principle.
>
> Let us imagine a portion of the fiery element at the very center of our world, so that one-half of this portion lies on one side of the center and the other half lies on the other side. Let a be the center and b one half and c the other half. Now I posit that everything that would hinder the natural movement of fire should be removed.
>
> Each of the two portions of fire move upward toward opposite parts of the circumference and will separate from each other.
>
> Now if these two portions of fire were joined together in a sphere so that they could not be separated or divided from each other, and all encumbrances were removed, this little sphere or portion of fire would not move, for there would be no reason for it to be moved more to one part of the

circumference than to the other. But if it were outside the
center of the earth, then it would go to that part of the
circumference nearest to it.

That is in full agreement with Aristotle's philosophy.[86]

We recognize in this the thought of William of Ockham; we
also recognize, by the way these thoughts are stated, the doctrine
against which Albert of Saxony directed his replies.

The passage we have just cited is immediately followed by
another where we discover the conclusion about which Albert said,
with some disdain, *"ad impossibile potest sequi quodlibet"*:

> In like manner, one can say that, if a portion of earth
> were equidistant between two worlds and if it can be separated,
> one part would go to the center of one world and the other
> portion to the center of the other world.
>
> If the portion cannot be divided, it would not move at
> all because of the lack of inclination, being like a piece of
> iron halfway between two magnets of equal [strength].
>
> If it were nearer one world than the other, it would move
> in the direction of the center of the nearer world.[87]

In any case, regarding the states of equilibrium he has just
considered—the equilibrium of a sphere of fire whose center is the
center of the earth; and the equilibrium of a portion of earth
equidistant from the centers of two worlds—our author recognizes
clearly that they will be unstable:

> I think this to be true, if the case is as stated in the proposition
> above; but it cannot exist or endure in this manner by nature
> because of the variations, or changes, or other movements that
> commonly occur, as in the case of a heavy sword, which would
> not stand for any length of time upright on its point.[88]

As Oresme stated, all these remarks are "in full agreement with
Aristotle's philosophy," a philosophy he will abandon in his final
chapter on the plurality of worlds, "In Chapter Nineteen he refutes
the opinions contrary to that which is stated in the preceding
chapter."[89] Here is how, toward the end of this chapter, he announces
to us that he will be detailing his own thoughts:

> Now we have finished the chapters in which Aristotle
> undertook to prove that a plurality of worlds is impossible,
> and it is good to consider the truth of this matter without
> considering the authority of any human but only that of pure
> reason.

I say that, for the present, it seems to me that one can imagine the existence of several worlds in three ways.

One way is that one world would follow another in succession of time, as certain ancient thinkers held that this world had a beginning. . . .

But this opinion is not touched upon here and was reproved by Aristotle in several places in his philosophical works. It cannot happen in this way naturally, although God can do it and could have done it in the past by His own omnipotence, or He can annihilate this world and create another thereafter. And, according to St. Jerome, Origen used to say that God will do this innumerable times.

Another speculation can be offered which I should like to toy with as a mental exercise. This is the assumption that at one and the same time one world is inside another so that inside and beneath the circumference of the world there is another world similar but smaller. Although this is not in fact the case, nor at all likely, nevertheless, it seems to me that it would not be possible to establish the contrary by logical argument.[90]

After establishing this odd proposition "to toy with as a mental exercise," Oresme continues as follows:

But also I submit that there is no proof from reason or experience or otherwise that such worlds do exist. Therefore, we should not guess nor make a statement that something is thus and so for no reason or cause whatsoever against all appearances; nor should we support an opinion whose contrary is probable; however, it is good to have considered whether such an opinion is impossible.

The third manner of speculating about the possibility of several worlds is that one world should be entirely outside the other in space imagined to exist, as Anaxagoras held. This solitary type of other world is refuted by Aristotle as impossible.[91]

Oresme now sets forth his new theory of the elements's natural places and weights in order to show the lack of firm basis for Aristotle's arguments. Here is how he continues his discourse:

But it seems to me that his arguments are not clearly conclusive, for his first and principal argument states that, if several worlds existed, it would follow that the earth in the other world would tend to be moved to the center of our world and conversely. . . .

To show that this consequence is not necessary, I say in the first place that, although up and down are said with several meanings, as will be stated in book II, with respect to the present subject they are used with regard to us, as when we say that one-half or part of heaven is up above us and the other half is down beneath us.

But up and down are used otherwise with respect to heavy or light objects, as when we say the heavy bodies tend downward and the light tend upward.

Therefore, I say that up and down in this second usage indicate nothing more than the natural law concerning heavy and light bodies, which is that all the heavy bodies so far as possible are located in the middle of the light bodies, without setting up for them any other motionless [or natural] place. . . .

Therefore, I say that a heavy body to which no light body is attached would not move of itself; for in such a place as that in which this heavy body is resting, there would be neither up nor down because, in this case, the natural law stated above would not operate and, consequently, there would not be any up or down in that place. . . .

From this it follows clearly that, if God in His infinite power created a portion of earth and set it in heaven where the stars are or beyond heaven, this earth would have no tendency whatsoever to be moved toward the center of our world. So it appears that the consequence stated above by Aristotle is not necessary.

I say, rather, that, if God created another world just like our own, the earth and the elements of this other world would be present there just as they are in our own world.

But Aristotle confirms his conclusion by another argument in chapter seventeen and it is briefly this: all the parts of the earth tend toward a single place, one in number; therefore the earth of the other world would tend toward the center of this world.

I answer that this argument has little appearance of truth, considering what was said in chapter seventeen. For the truth is that in this world a part of the earth does not tend toward one center and another toward another center, but all heavy bodies in this world tend to be united in one mass such that the center [of the weight of this mass] is at the center of this world, and all the parts constitute one body, numerically speaking. Therefore they have a single place. And if some part of the earth in the other world were in this world, it would tend toward the center of this world and become united with the mass, *and conversely.*

But it does not have to follow that the portions of earth or of the heavy bodies of the other world, if it existed, would tend to the center of this world because in their world they would form a single mass possessed of a single place and would be arranged in up and down order as we have indicated.[92]

Nicole Oresme formulated with perfect clarity the principle of this new theory of weight: "the natural law concerning heavy and light bodies . . . is that all the heavy bodies so far as possible are located in the middle of the light bodies *without setting up for them any any other motionless [or natural] place.*" The consequences of such a principle are obvious. The earth's weight does not require it to remain stationary at the center of the universe, as it does for Aristotle's physics; surrounded by elements, the lighter ones containing the heavier ones, it is free to move about space in the manner of a planet. In any case, nothing prevents each planet being formed out of a heavy earth surrounded by water, air, and fire analogous to ours. This new doctrine allows one to compare the earth and the planets while the Peripatetic theory absolutely prohibited it. Thus Oresme's opinion was adopted by those who wished to number the earth among the planets; it was adopted by Nicholas of Cusa first, then later by Leonardo da Vinci, then by Copernicus, and finally by Giordano Bruno who made of it one of his favorite theses.

In any case this theory of weight, so strongly opposed to Peripatetic theory, was not new to physics; it is the theory Plato supported in the *Timaeus*. And Plato derived from it a different definition for natural movement than Aristotle's. Natural movement is not the movement directed toward the center of the world or the movement away from it, depending upon whether the body is heavy or light; it is the movement by which an object attempts to rejoin the set of elements it belongs to and from which it was violently separated (finding itself with an element of a different nature). Thus air descends naturally when it is in the sphere of fire, as it naturally climbs when it is surrounded by water, for in either case it seeks to get nearer the sphere of air; these two contrary movements, centrifugal and centripetal, are equally natural for air as they are equally violent for it. In order to be able to choose which of these opposites one must attribute to it, one must know in which medium the air is located.

This opinion, deducible from the principles advanced in the *Timaeus*, is in contradiction with Aristotle's physics; for, according to this physics, there is only one natural movement for each element, a movement which is always circular, always centrifugal, or always centripetal. But Oresme plainly accepts the Platonic doctrine; he carefully develops it, and he seems to enjoy exposing the opposition that it provides to the Peripatetic theory of natural movement.

Oresme expresses himself as follows:

> But I still doubt, and I imagine the case of a tile or copper pipe or other material so long that it reaches from the center of the earth to the upper limit of the region of the elements, that is, up to heaven itself.
>
> I say that, if this tile were filled with fire except for a small amount of air at the very top, the air would drop down to the center of the earth for the reason that the less light body descends beneath the lighter body.
>
> And if the tile were full of water save for a small quantity of air near the center of the earth, the air would move upward to heaven, because by nature air always moves upward in water. From these examples it appears that air can, by reason of its nature, descend and move upward to the distance of the semidiameter of the sphere of the elements. Now, these two movements are both simple and contrary, and thus a simple body is by its nature capable of moving in two simple contrary movements.
>
> I reply to this that perhaps we may say that the downward movement of this small amount of air, in the case above, is natural up to the point where the air is directly above the region where the proper sphere of this element, air, is located.
>
> And afterward the air descends again by violence as it meets the proper sphere of fire which is lighter and which mixes it up and casts it down beneath it. So, the descent is in part natural and in part violent.
>
> In like manner, the upward movement of the air rising in the water is natural while it rises from the center of the earth to the point where it meets the region of air, its natural place.
>
> After this the air is moved up by violence, because the water lifts it up and pushes under it by reason of its heaviness.
>
> Thus, insofar as the up and down movements of this air are opposed to each other, one movement is natural and the other violent.[93]

That a simple body cannot have two natural movements distinct from one another was, for Aristotle, one of the reasons why the diurnal rotation of the earth was inadmissible. Oresme is well aware that to discard the principle allows one to discard its consequences; and doubtless it is in order to discard the one that he undermines the other. We have seen how he responded to Aristotle's argument for the immobility of the earth:

> As for the first argument where it is stated that every simple body has a simple movement, I say that the earth, which as a whole is a simple body, has no movement, according to Aristotle in chapter twenty-two.
>
> Against the interpretation of anyone who maintains that Aristotle means that this body has a single simple movement not proper to itself as a whole, but applying only to its parts when they are out of their proper place, we can cite the case of air moving downward when it is in the region of fire and upward when it is in the region of water, both being simple movements. . . . If any part of such a body is out of its place or outside the main body, it returns to it as directly as it can, once the hindrance is removed.[94]

By supporting the hypothesis of the rotation of the earth and by destroying the Peripatetic arguments opposed to it, Oresme was a precursor of modern science; he also helped by formulating a theory of weight that made the Copernican revolution possible. Audaciously innovative (for it imposed axioms to celestial mechanics that are identical with the axioms of sublunar mechanics), this theory became the theory of the new school astronomers until the theory of universal gravitation proposed by Kepler supplanted it.

The Spot on the Lunar Disk (The Man on the Moon)

In a booklet entitled *On the Face That Can Be Seen on the Lunar Disk,* Plutarch upheld the plurality of worlds and upset the whole theory of natural place. That he did so with respect to this topic was no mere coincidence. Although Galileo, using his telescope, showed that there were spots on the sun and thereby gave the final blow to Aristotle's celestial physics, this physics was already confronted by a perpetual contradiction with the spot on the moon. It was impossible to observe the spot on the moon without thinking that it denotes a certain heterogeneity in the structure of the moon, a certain irregularity incompatible with the geometric purity of celestial essence as defined by Peripatetic philosophy; it requires that one considers the moon as a body comparable to our earth.

Plutarch seized upon and developed this comparison to its fullest; but many others before him thought of it. Already, if one is to believe Stobaeus, "Heraclides and Ocellus thought of the moon as an earth surrounded by clouds."[95] Even Aristotle, who so clearly distinguished between celestial substances and elementary substances, seems to have wanted to modify the sharpness of this opposition in the case of the moon; that is what he appears to indicate in this passage from *The Generation of Animals*: "But the form that fire assumes never appears to be peculiar to it, but it always exists in some other of the elements, for that which is in fire appears to be either air or smoke or earth. That kind of substance must be sought for in the moon, for it appears to participate in the element removed in the third degree from earth."[96]

Aristotle therefore conceded that around the moon the celestial substance mixed with fire, the most subtile of the elementary substances.

Did Aristotle go further? Did he go so far as to suppose a kind of affinity between the lunar substance and the terrestrial element? Averroes, on several occasions, indicates that this was the thought of the Stagirite;[97] according to the Commentator, the Philosopher wrote in his *Histories of Animals*, that the nature of the moon has a similarity and a relationship with the nature of the earth, understanding by that that the moon is not luminescent by itself. The *Index Aristotelicus*, appended by the Berlin Academy to its edition of the works of Aristotle, makes no mention of this text, and we have not been able to discover it. But whether it is authentic or not, this text had, through Averroes, a free rein in the science of the Middle Ages; it quieted the scruples of the most rigid Peripatetics regarding the assimilation of the moon with sublunar bodies. In any case, according to Averroes,

> Aristotle stated in the *Treatise on Animals* that the nature of the moon has a relation with terrestrial nature, because it is not luminescent. Everything that is luminescent by itself has a nature related to the nature of fire; as for the parts of the moon that are translucent, that do not glow by themselves and do not have the power to light, they possess a nature which has a relationship with the nature of water and air.[98]

Averroes does not think that there is an identity of substance for this relationship between the natures, he merely wishes to designate an analogy: "Insofar as they are bodies, the celestial bodies have in common with the elements the properties that consist in being translucent, luminescent, and obscure; that is why Aristotle stated in the *Treatise on Animals* that the nature of the moon is similar to the nature of the earth because of its obscurity; in the same fashion, the luminescent portion of the celestial sphere is similar in nature to fire."[99]

The Commentator understands that the analogies between celestial and elementary bodies derived from their behavior with respect to light do not diminish the irreducible opposition that Peripatetic philosophy places between the eternal bodies and the perishable and changeable bodies. Thus he does not believe that he is straying far from the philosophy of Aristotle by using these analogies in order to explain the details of the spot on the moon.

> Here is what is most rightly said about this subject: The spot is a portion of the surface of the moon that does not receive the light of the sun in the same way that the other portions do. That is not something which celestial bodies are prohibited from doing; in fact, in the same way that we discover something luminescent in some way we can also discover something obscure in these bodies. Such is the moon; thus Aristotle stated in the *Treatise on Animals* that the nature of the moon is similar to the nature of the earth. By that he understands that the moon is not luminescent by itself, that it derives its luminescent character from others, like earth from fire. That is not so for the other stars, as it is evident. Since the various parts of the celestial body are distinguished with respect to whether they are translucent or not, or luminescent, it is not impossible that the various parts of the moon receive the light of the sun differently.[100]

But clearly, the first question to be answered is, in what way does the sun enable the moon to light up?

> It has been demonstrated that if the moon acquires the power of lighting up from the sun, it is not from reflection. That has been proven by Avenatha [that is, by Abraham ben Meir ibn Ezra] in an interesting treatise. If it illuminates, it is by becoming a luminous body itself. The sun renders it luminescent first, then the light emanates from it in the same way that it emanates from the other stars; that is, an infinite multitude of rays are issued from each point of the moon. If its power of illumination issued from reflection, it would illuminate some determined places on earth depending upon its circumstances; reflection is only produced for some determined angles.[101]

Abraham ben Ezra and Averroes are right in declaring that the light transmitted by the moon does not behave like the light of the sun reflected as if by a mirror; they did not appear to have considered that the moon might be a rough body transmitting the light in every direction by diffusion. Doubtlessly, that the surface of a celestial body was not perfectly smooth and polished seemed contrary to its perfection for them. They therefore attributed to the moon the power to transmit light, but only after the sun's illumination predisposed it; the property they attributed to the moon is not unlike what we call fluorescence.

A reason similar to the one above avoids the hypothesis that the spot on the moon is only the image of terrestrial objects, mountains or seas reflected on the surface of the moon. If that were so, the shape of this spot would change according to the change in the relative position of the earth and moon.

The spot therefore admits no explanation other than this: When the light of the sun predisposes and excites them, the various parts of the moon become luminescent; but they do not all become luminescent in the same way.

In these considerations about the nature of the lunar light and of the spot on the moon, Averroes endeavored not to fault Peripatetic philosophy in any way; he attributed a heterogeneity to the substance of the moon, but this heterogeneity is restricted to the qualities designated by the words, dense or thin, opaque or translucent, obscure or luminescent. And in his *Discourse on the Substance of the Orb*, the Commentator professes that these words can be said of the celestial substance as well as for sublunar bodies,[102] although they do not have the exact same meaning, but only similar meanings.

Scholastic Christianity before the thirteenth century did not have to burden itself with such precautions. It was not Peripatetic, it was Platonic. It did not believe that the celestial bodies were formed from a substance absolutely separate from sublunar substances; it saw there a mixture of the four elements where fire predominated. In order to explain the spot on the moon, it attributed to it a structure similar to the structure of the bodies around us.

The book *On the Constitution of the World*, falsely attributed to the Venerable Bede, asserts the following:

> The moon is formed out of the four elements. Three of these elements are well mixed and finished, for they are naturally transparent and give back light from themselves.

On the other hand, where the spot is found, earth has not been mixed well with the other elements; there it is rough and does not transmit light.[103]

In his opusculum entitled *The Images of the World*, Honorius Solitarius teaches that the ether within which the planets move is identical to pure fire. The moon is

> igneous in nature, but its mass is mixed with water. It does not have its own light, but it is lighted by the sun like a mirror. . . . One can perceive a kind of small cloud on it because of its aqueous nature. One says that if it were not mixed with water it would light up the earth like the sun does; and because of its nearness to the earth, it would devastate it by its excessive heat.[104]

Around 1270, the astrologer of Baudoin de Courtenay reproduced, in his *Introduction to Astronomy*, the ideas that were current in ancient Scholasticism. The opinion of the pseudo-Bede can be seen through what he asserts about the spot on the moon.

> There are two opinions about the moon; the first is Aristotle's, which is held to be heretical, and the second is a common one, which almost all philosophers hold—that the body of the moon is aquatic and of a different kind than the other planets because of its proximity to water and earth; and because it is near cold things, that is, water and earth, it has no heat or light of itself. Thus it is proper that it receives them from the sun. For it is a polished and smooth body like glass or crystal, and when the rays of the sun hit it, it shines just like a mirror. And although it is very polished as I said, however, there are parts of it which are rough and flawed where there is an accumulation of the water and earth. And that part has a natural obscurity and darkness, although the moon is a body full of light.
>
> Aristotle said that the body of the moon was of the nature of fire, but it has much of the nature of water and earth.[105]

Our author, in keeping with the customs of the time, cited Aristotle but had not read him.

When the physicists of the thirteenth century read Aristotle and Averroes they became perplexed about the explanation of the spot on the moon; the Aristotelian doctrine about celestial essence did not seem reconcilable with the existence of the spot. They saw in the spot the evidence of a kinship between the moon and the elementary substances.

That Albertus Magnus read Averroes is beyond doubt. The influence of the Commentator is evident in what the future bishop of Ratisbon states about the light emitted by the planets; in fact, it is only an extension of what Abraham ben Ezra and Averroes stated about the light of the moon.

According to Albertus, who was influenced by the *Liber de Elementis* of the pseudo-Aristotle, neither the wandering stars nor the fixed stars emit their own light; for each of them the sun is the primary source of the light they emit. But if they illuminate by means of the light received from the sun, they do not do it by reflection. If a star reflected the light of the sun like a mirror, it would reflect the light in only one direction, and would not transmit rays throughout all of space.

> Doubtless, one must concede that the light does not come to us by reflection; rather, as we have already stated, the light is incorporated in the stars. . . . They are like spherical receptacles of light; once they are touched by a solar ray, they are filled immediately with light throughout their whole body, the only exception to this being the moon, which is the least noble of all the stellar bodies.[106]

Like Averroes, Albertus rejects the opinion that sees the spot on the moon as a reflection of our mountains and seas. He adds, "if it were thus, the light of the moon would result from a reflection of this body, and not from an inhibition of the solar light by the thickness of the body; that is the opinion we reject. But we state that this shape is the result of the nature of the moon, which is of terrestrial nature. (*Sed dicimus quod haec figura est de natura Lunae quae est naturae terrestris.*)"[107] Then, with no further explanation, our author describes the shape of the moon in detail.

The moon is of terrestrial nature—that is a statement a faithful Peripatetic could not take literally. Saint Thomas Aquinas therefore attempted to soften its blow.

Following Averroes, the *Doctor Communis* begins by arguing that the spot cannot be explained by the interposition of some foreign body or by the reflection of some of the accidents of the terrestrial surface; he continues in this fashion:

> Others state more rightly that the reason by which the moon seems more varied is the disposition of its substance and not the interposition of some body or some reflection. But they are divided into two camps.

Some maintain that the forms of the effects preexist within the causes in some way; the higher the cause, the more uniform are the various forms of the effects, and the lower the cause, the more the forms of the effects are distinct from one another. The celestial bodies are the cause of the bodies here below, and among the celestial bodies, the moon is the lowest; therefore, a kind of exemplary heterogeneity of bodies capable of being generated can be found within the moon and on its surfaces. Such was the opinion of Iamblichus.

Others assert: It is true that the celestial bodies are of another nature than the four elements; nevertheless, they preexist within them, but they do not preexist in the same way that they exist within the elementary bodies; they exist in a more excellent manner.

And among the elements, the highest is fire, which possesses the most light; the lowest is earth, which possesses the least light. However, the moon, which is the lowest of all the celestial bodies, has some relation (*proportionatur*) with the earth and some resemblance with its nature; thus the sun cannot render it totally luminescent. That is why there is an obscure portion of the moon although the sun illuminates it perfectly.[108]

Thomas does not tell us which of these two opinions he prefers; however, if he is following the Scholastic customs about exposition, he would be indicating that he prefers the latter.

A true Peripatetic, a reader of Averroes, cannot be satisfied by this; the opinion still has too much resemblance to the Platonic theories of early Scholasticism. Although the elements it places within the celestial bodies are superior to those here below, it strays too far from the Aristotelian teaching about the fifth essence. It therefore has to be rejected; such is the thought of Robertus Anglicus.

There is another oddity in the moon whose truth I have not found discussed by any author as fully as it should be, namely, that obscurity in the form of a man which appears in the moon, and which, according to the rustics's stories, is said to be a certain peasant who stole thorns and he, with a load of these on his back, was stellified in the moon and that darkness is his image.

There is another opinion about the spot on the moon, that the moon, as a body halfway between celestials and terrestrials, shares the nature of both. And as it shares the nature of celestials, clearness appears in it; and as it participates in the nature of terrestrials, obscurity and darkness appear in it.

Another more probable explanation is that the moon is like a polished, smooth, and pure body, in which the form of things facing it shine as in a pure mirror. Therefore, the parts of the earth covered by water appear in the moon clearly, and the parts of the earth uncovered by water appear in the moon obscurely, and according to the shape of the site of these parts on earth appears the image on the moon.[109]

Our author continues as follows:

Another possible explanation—and I believe a better one— would be to assume that in the body of the moon there is a greater rarity in some parts and greater density in some others. Then those parts which are rarer receive more sunlight and more deeply, wherefore they appear brighter. The denser parts receive less and so appear darker, and such a figure appears corresponding to the location of that density.

But should anyone raise an objection and say that I am wrong to posit density and rarity in heaven, I answer that this is not impossible according to Averroes in the book *On the Substance of the Orb*, who holds that rarity and density may be posited in heaven just as here below, though perhaps only equivocally (*aequivoce*), as is there stated.[110]

Thus we see in 1271 the Averroist explanation for the lunar spot, derived from Peripatetic principles, preferred over the explanations of the early Scholastics that likened the nature of the moon and sublunar beings, the explanations toward which Albertus Magnus and Saint Thomas Aquinas still leaned.

One of the manuscripts conserving the *Tractatus super totam astrologiam* of Bernard of Verdun carries a marginal note that holds a peculiar explanation of the spot on the moon. According to this explanation, the moon is a perfectly transparent spherical body containing an obscure body within it; the latter appears as a dark spot when seen through the former.[111]

This strange assumption was formulated in order to show that the moon can describe an epicycle without having to rotate on itself and without the spot having to change shape.

This strange assumption did not pass unperceived; in 1310 Peter of Abano referred to it in his *Lucidator Astronomiae*. Unfortunately, this reference is in the most undecipherable passage of the generally unreadable manuscript that holds the *Lucidator*. There the Paduan doctor states that "it is not within the depths of the moon, but at the surface that one thinks the spot to be; that is the most common opinion."[112]

Peter of Abano also discusses the opinion according to which the spot on the moon is the reflection of terrestrial objects. Reproducing almost word for word an objection by Albertus Magnus, he asserts that the moon would then be illuminated "by a reflection which would make a mirror out of it and not by an inhibition of the solar light by its thickness."[113]

It is true that he foresees a difficulty for this theory of the lunar light:

> Perhaps someone will refute it by observing that during an eclipse of the sun, the moon is completely dark, which cannot happen if it can receive light throughout its depth and if it retains some of this light in itself; it seems, then, that it receives this light because of its surface.[114]

In spite of this reason, the opinion that seemed best able to explain lunar light around 1310 was Averroes' explanation.

It is Averroes' theory that Giles of Rome expounds with extreme precision:

> All the stars and the moon receive their light from the sun.
>
> The sun, therefore, is the source of light. The luminous orbs are the medium through which this light reaches the stars and the moon; as for the stars and the moon, they are dense bodies, smooth and polished, that reroute the light they receive from the sun toward the other bodies. In fact, if the stars shine, it is because of their density and because they are the densest part of their orbs; that is how they transmit the light they receive from the sun to the other objects. Therefore, if a star was less dense in one of its parts, it would not shine in this part; that is what one sees with the spot on the moon; where the spot is, the moon does not shine. We believe that it is so because the moon is not as dense there as it is elsewhere.[115]
>
> There is density and rarity in the celestial bodies, greater and lesser density, just like in the things here below. Without having to invoke action or passion . . . we can reconcile what we see in heaven with what we perceive by sight. In fact, we see color here and light there; elsewhere we perceive neither color nor light, but translucencies and transparencies. Translucency and transparency stem from rarity; color and light depend upon greater and lesser densities.
>
> We can distinguish in heaven two colors and three kinds of light which all stem from greater and lesser densities.

There are two colors in heaven, [azure] and a grey color (*maculosus*); that is the color we observe in a portion of the moon we call the lunar spot. In fact, the moon does not glow in this portion where the spot is located, but it displays an obscure color, a cloudy and grey color; thus Averroes states in the second book of the *De Caelo et Mundo* that the moon participates in the terrestrial nature, but without any terrestrial characteristics, and that we can reconcile this spot using only weakness of density. . . .

We would therefore assert that the spot on the moon has some density because it is not transparent; when it is interposed between us and the sun, it eclipses the sun. Since the sun is eclipsed by the totality of the lunar disk, by the part containing the spot as well as by the brilliant part, the moon is dense throughout. However, the spot on the moon has a density, but it is the weakest density in heaven; and since there is a weak density there, there is also a least accumulation of light. . . .

As for what the Philosopher seems to hold, that the moon appears to participate in the terrestrial nature, he did not assert that the moon does not belong to the fifth essence, he merely asserted it because this spot seems to display a color similar to the earth's.[116]

John of Jandun wrote in his *Commentary on the Discourse about the Substance of the Orb*:

We do not have to seek here the reason for the spot on the moon; we will examine this in the second book of the *De Caelo et Mundo*. What we should remember, as the Commentator stated, is that its cause is in the diversity of the parts of the moon with respect to density and rarity; one portion of the moon is so rarified that it cannot receive the light of the sun in the same fashion as the others. This part draws a kind of figure that looks dark on the surface of the moon.[117]

John of Jandun did not keep his word; he did not take up the explanation of the spot on the moon in his *Questiones super libri de celo et mundo*, but what we have just read is sufficient for us to know his opinion; it is plainly in conformity with Giles of Rome's thought.

Averroes's influence can be clearly felt in what Buridan asserts about lunar light.

The moon does not reflect like a mirror; if it did so, it would not scatter light in all directions.

But some wish to save this reasoning by stating that the moon is like a wall; when the rays of the sun hit a wall, it becomes lighted completely, and not only following the lines where the incident and reflected rays form equal angles. Thus it is with the moon.

But this solution is not sufficient. If there is a reflection from all parts of the wall to our eyes, it is because of the roughness of the wall; it is because of this roughness that the rays are directed in all directions. If, on the contrary, the wall were perfectly smooth, like a steel mirror, one would not see a maximum clarity throughout the whole wall, one would see it only where we have already said.

That is what we can clearly see in still waters; only a small portion of this water reflects with intensity the light of the sun or a star. But if we were to trouble the water so that the surface is no longer still, the same light would spread over a greater expanse of water.

And we assume that the moon is perfectly smooth, and presents no roughness to us; Aristotle thought that all the celestial bodies were made this way.

Others assume, with greater probability, that the moon is not actually luminescent, so that it cannot disturb a transparent medium; but it is in a proximate potentiality to become luminescent by its own natural disposition. And when the solar light falls on it, it is constrained to shine actually (*reducitur ad actum lucendi*).[118]

John Buridan rendered more precise the kind of fluorescence Averroes attributed to the moon, and in a better fashion than had his predecessors.

As for the spot on the moon, here is what the rector of Paris thinks:

The Commentator stated with greater probability that this spot stems from the diversity of the parts of the moon with respect to rarity and density. The parts that display the spot are more rarified; thus they are less capable of shining and of delimiting the light of the sun. One can say the same about the Milky Way. The portions of the stellar orb are denser there than elsewhere, thus they can retain and delimit the light of the sun up to a certain point, although they do not do it perfectly; in this manner, this region appears whiter than the rest of heaven.[119]

It is the Averroist explanation of the lunar spot that Nicole Oresme endeavors to explain, in French, to the "*gens de noble engein.*"

Let us abandon these and other opinions that offer no semblance of fact. To understand the more reasonable opinion, we must first know that the moon's light comes from the sun—a readily apparent fact because the portion of the moon turned toward the sun shines and the other does not; also, when the moon is eclipsed, the shadow of the earth deprives it of its light.

Nor do we see the sun's light on the moon as in a mirror, for we should not then see the moon as we do; rather the sun would appear in only a small portion of that part of the moon lighted to us, and at times it would appear in no part at all; and it would be seen in different parts at different times, and not from every point from which the illuminated portion can be seen.

It would be exactly as though we were looking at the sun in a mirror or in water; we do not see it from every position from which we can see the mirror, nor from every angle, but only at a certain position and at a certain distance, and from another distance we see it at another place.

The cause is easily understood; in conformity with the laws of perspective and our own experience, the line passing from the eye to the mirror and the one returning from the mirror to the sun by reflection make two equal angles above a point on the surface of the mirror where the sun appears. It must follow then, that from one position or distance the sun will appear in one place or one part of the mirror, and from another position or distance not in the line passing from the eye to this point the sun will appear in another part of the mirror, as can be shown by example or in a diagram.

According to the remarks of Averroes, a certain Avenatha wrote a special treatise to show that the moon does not receive light from the sun by refraction or reflection.

Some bodies are not diaphanous and not transparent or they are dark like iron, black peas, or such things, and neither the rays of the sun nor of anything else can pass completely through them if they are not very thin. In such bodies light permeates little or not at all, being turned back by reflection or refraction.

If such bodies are highly polished, the light rays are returned or are pushed aside in a perfectly regular order so that such bodies act like mirrors.

If they are not polished, the reflection or refraction is not orderly, but on the contrary, with some rays moving and turning about in one way and some in another. Such a body is not a mirror for reflecting a shape, although it may reflect color or light.

Other bodies are diaphanous, transparent, or clear like glass, crystal, and water, and if such bodies are not too wide or thick, the light penetrates, pierces, and passes completely through. Thus the light penetrates these bodies in proportion to their degree of transparency and makes them conspicuously visible.

The moon is a perfectly polished spherical body, as will be stated in chapter twenty; so from what we have said, if it were a nontransparent dark body like iron and steel, it would reflect the sun's light like a mirror, which we have already shown not to be true; consequently, it follows that the moon is a transparent, clear body such as glass or crystal, at least in those parts near the surface. However, such bodies are somewhat dark.

It follows, moreover, that the sun's light penetrates the moon to some degree, but does not pierce nor pass completely through because of the great size and depth of the lunar body. The sun's light does not penetrate very far in relation to the size of the moon, for, as we can see in very clear water, if the the water is quite deep, the sun's light does not reach the bottom. If the moon were equally clear or transparent in those parts receiving the sunlight, it would be evenly and equally illuminated in one part as another; and the contrary is evident from the presence of the dark spot or shadowy figure of which we are speaking.

Therefore the parts of the moon by their very nature cannot all be uniformly transparent and clear, but rather in different degrees, as we observe certain differences in other parts of heaven. This is the explanation for the appearance of the spot mentioned above.

But it should be noted that, just as in the case of the alabaster stone, those veins and sections that are most clear and through which one can see almost as clearly as through crystal seem darker and less white than the other parts; and the same is true of the parts of the moon. Thus the clearer some parts are, so that the sun's penetration is deeper, the darker those parts appear, and the others proportionally lighter. The shape of the spot on the moon, then, is of this kind just explained.[120]

Albert of Saxony's lecture on this matter can teach us nothing new; we have already read in the works of John Buridan and Nicole Oresme what is to be found in the *Quaestiones in libros de Caelo et Mundo*; however, we ought to detail and reproduce Albert's words because of the importance of the German master's treatise over that of his two French predecessors; while Buridan's and Oresme's works are unedited, Albert's works were printed many times during the end of the fifteenth and the beginning of the sixteenth centuries. It is through Albert that the Renaissance understood the Parisian teachings about lunar light and the spot on the moon. He wrote:

> There is some doubt about the process by which the moon receives its light from the sun. There are several opinions about this.
>
> Some state that the surface of the moon is perfectly polished without any roughness, so that it reflects the light of the sun toward us, in the same way that the various colors are reflected by a well-burnished and well-polished mirror; the moon appears luminescent to us because of this reflection of the solar light.
>
> But this opinion is not admissible; doubtless a smooth and polished body would reflect rays toward the eye, but this reflection does not issue from every part of the smooth body. The mirror is an obvious example. When my face is in front of a mirror, every part of the mirror reflects a ray from my face; but it is not true that any part of the mirror transmits to my eye any ray whatsoever. One part transmits one ray and another part another ray. In fact, in order for a part of the mirror to transmit a certain ray, it must be that this ray from my face falling on the mirror and the ray attaining my eye form equal angles of incidence and of reflection on the surface of the mirror. . . . Then, if the moon reflected the light of the sun toward us in the said manner, that is, like a mirror, doubtless, the whole surface of the moon would offer us a weak light; and we would not perceive an intense light except in some small portion such that the angle of incidence would be equal to the angle of reflection to our eyes.
>
> But one might object to this reasoning. If the light of the sun strikes a wall, the wall seems lighted on all its surface and not only at a point corresponding to an angle of reflection equal to an angle of incidence. This objection is worthless: The moon is not like a wall. Because of the roughness of its surface, a number of parts of the wall can reflect rays to our eyes; hence, a large extent of the wall appears lighted to us. But if the wall were perfectly smooth like a mirror or like the body of the moon, the solar rays would not light up all its surface when striking it, but only at a point where

the incident ray from the sun and the reflected ray toward the eye give equal angles of incidence and reflection. That can be seen easily with still water. Only a small part of the surface of the water represents the light of the sun or of a star with intensity. But if one troubles the surface of the water a little, it no longer remains perfectly smooth, and the light of the sun is sent to us with intensity throughout a greater region of the surface.

One must therefore utter another opinion. That is why I state that the light of the sun is incorporated in the moon. The moon is a translucent and transparent body, at least on its surface, and perhaps in its totality, even though the size of the moon's body does not allow the light of the sun to cross its whole length, so that this light cannot be as intense on the side of the moon that does not face the sun as on the side of the moon that does. Thus the light of the moon we see is not simply the light of the sun reflected off the body of the moon, but the light of sun that the moon soaked up and that became incorporated with it.

One can also express oneself as follows:

The moon is not luminescent actually; it cannot itself disturb a transparent medium. However, by its natural disposition, it has a proximate potential for luminosity (*luciditas*); this potential is brought to actual luminosity by the incidence of the solar light on the moon.[121]

One can recognize Buridan's thought throughout the whole quote; the final passage is even borrowed almost verbatim from the philosopher of Bethune.

Must we hold a similar theory for stars? Albert does not profess a categorical opinion about this matter. He notes that the *Book of the Elements*, which he attributes to Aristotle, holds that all the stars receive their light from the sun, like the moon. On the other hand, Avicenna holds the contrary, that the stars have their own light; and six reasons support this opinion. Our author adds,

the question, "Do the stellar bodies other than the sun and moon receive their light from the sun?" can be considered neutral; the reasons one gives for one side can be as easily refuted as those one gives for the other side. Therefore, for the love of Aristotle, the prince of philosophers, I will refute the six opinions formulated against Aristotle's opinion (in favor of Avicenna's opinion), and I will assert that all the stellar bodies other than the sun and the moon, whether they are planets or fixed stars, receive their light from the sun.[122]

Avicenna's first objection is formulated by Albert as follows:

> According to whether they get nearer or farther from the sun,
> the stars would take on the shape of a crescent, like the moon
> does; and this appearance would be marked in the case of
> Venus and Mercury which are below the sun.[123]

Avicenna did not know this, since the discovery that Venus
and Mercury have phases like the moon was accomplished through
the use of spyglasses; Albert, no less ignorant of the fact, replied
to the objection that "Venus and Mercury are so transparent that
the light of the sun becomes incorporated with these stars and gets
soaked up in all their parts, which does not happen for the moon."[124]

Another objection of Avicenna also gets resolved because of
the transparency of Venus and Mercury:

> Let us suppose that Venus and Mercury, which are less
> elevated than the sun, do not have their own light, but receive
> their light from the sun; when Venus or Mercury are interposed
> between the sun and our eyes, they would eclipse it, as does
> the moon. But we do not see this.[125]

That is what one sees with the help of a spyglass, but that
is what the naked eye has never revealed.

The explanation for the lunar spot is derived from the
considerations we have just read. Here is how Albert of Saxony
develops it:

> One wonders . . . if the spot on the moon issues from
> the diversity of the parts of the moon or whether its cause
> is extrinsic to the moon.
>
> One attempts to prove that it does not issue from the
> diversity of the parts of the moon.
>
> First, the moon is a simple body in fact; and the parts
> of a simple body, considered under a single relation, are similar
> to one another. That is apparent for water, air, and all other
> simple bodies.
>
> Second, the parts of the sun or those of any other stars
> are similar and uniform in rarity and density; it is therefore
> the same for the parts of the moon. And, consequently, this
> appearance of a spot cannot stem from the diversity of the
> parts of the moon.
>
> Third, if it had such a cause, it would be because various
> parts of the moon would be more rare and others less rare;
> but one can prove that it is not so, for during the eclipses
> of the sun, the rays of the sun would reach us by traversing
> the rarest parts of the moon. And that is clearly false.

Finally, one can prove that the appearance of the spot issues from an extrinsic cause. The body of the moon is a smooth body—well polished and like a mirror; the earth finding itself with respect to the moon as if it were in front of a mirror, produces its image and likeness on it. Hence, when we look upon the moon, we see the earth by reflection and receive the appearance of a spot from this.

I will first examine this question itself and develop the various opinions held on this subject and refute them. Second, I will develop the opinion I hold to be true.

First, there was an opinion according to which the spot on the moon was caused by a vapor stirred up by the moon itself; interposed between the stellar body and ourselves, this vapor would obscure some parts of the moon. The Commentator adds that according to some, the moon would attract such a vapor toward itself in order to nourish itself.

Others assert that the moon has a great power on the waters and humidity; its nature is to attract such a vapor about it. Hence, all these authors agree not to account for the spot on the moon by the diversity of lunar parts, but to account for it by some extrinsic cause.

But this opinion is not valid. These exhalations and vapors would not be equally attracted at all times; they would not always have the same shape, but would be essentially changeable. And, on the contrary, the spot always appears constant and always keeps its shape; consequently, it is not caused by a vapor or an exhalation interposed between the moon and us.

Above all, one cannot consider as valid the first opinion according to which the moon attracts vapors toward itself in order to nourish itself; celestial bodies do not have to nourish themselves, for they are not subject to generation, destruction, or alteration.

Another opinion holds that the spot is the representation of some object from this world below, be it some earth, some mountains, or some analogous thing; these bodies would be seen on the moon as bodies can be seen by reflection in a mirror, because, according to this opinion, the moon is smooth like a mirror.

This opinion is not valid. In fact, as the moon moves, the portion of the moon where the spot appears would have to change from movement to movement, exactly as images change positions in a moving mirror. And that does not happen.

Furthermore, if the moon had the power to reflect the images of bodies, the whole image of the earth would appear

on the moon. And that is false, for the earth does not leave
the shape of this spot.

Second, the Commentator issues a third opinion, which
I believe to be true. The spot issues from the diversity of the
parts of the moon; these parts being more or less rare and
more or less dense than one another. The parts in which the
spot is seen are the rarest, which renders them least capable
of glowing. The parts next to them are the densest, and because
of it, they glow most. This is to be understood by analogy
with alabaster; the portions of alabaster that are very dense
and nontransparent appear very white; those that are
transparent like glass are obscure and tend toward black. If
one asks why the moon exhibits such differences between its
various parts, one must reply that this is its nature. . . .

Replies to the arguments from the beginning: I would
reply to the first that the moon is simple in substance in fact;
but that does not prevent it from exhibiting differences in
density and rarity between its various parts.

I would reply to the second that there is no comparison
between the sun and the stars, on the one hand, and the moon,
on the other. There is no cause that can be assigned for this
dissimilarity; it stems from the nature of the bodies.

As for the third, I would say that it is true that a portion
of the moon can be somewhat rarer than another, but it is
not so rare that the solar rays can traverse the thickness of
the moon.

What one should respond to the final argument follows
from the refutation of the second opinion.[126]

Albert of Saxony's treatise summarizes most completely and most
clearly what Scholastic physics taught about lunar light and the
spot on the moon.

The spot on the moon suggested to Plutarch another thought:
The moon is heterogeneous. He did not believe that such a
heterogeneity can be discovered in a body formed out of the
Peripatetic fifth essence; he concluded that the makeup of the moon
was similar to that of the earth. From there, there was only one
step to take in order to believe that there are four elements on the
moon as there are four here, and that the moon is a world similar
to ours. Plutarch took that step. But he found himself engaged
in a dispute against the arguments by which Aristotle attempted
to prove that there cannot be more than one world. In order to
wage this dispute, Plutarch needed to upset the whole Peripatetic
theory of natural place and substitute an entirely different doctrine
for it, which the *Timaeus* suggested.

Parisian physics during the fourteenth century, with Oresme as its spokesman, proposed a theory of weight similar to Plutarch's, but came to it by an entirely different route than that followed by the Platonic philosopher—an evident indication that the Scholastics did not read the treatise *On the Face That Can Be Seen in the Lunar Disk*.

The proposition that the Scholastics attempted to justify was the same one that preoccupied Plutarch: several worlds can exist, each one of which is composed out of the four elements. But what convinced them of the truth of this proposition was a dogmatic condemnation brought forth by the bishop of Paris in 1277 and not some reflections on the lunar spot.

With respect to the lunar spot they all believed, with some insignificant variations, what Averroes asserted; and the Commentator took care that his teaching could be reconciled with the Peripatetic theory of celestial substance.

One day Leonardo da Vinci read and meditated on the *Quaestiones in libros de Caelo et Mundo* of Albert of Saxony; he became most interested in what these questions asserted about the spot on the moon. The Averroist explanation they proposed did not satisfy him. He sought another reason for the mechanism by which the moon transmits the solar light; this reason was suggested to him by a passage that Albert had received from Buridan: if the moon diffuses the solar light in all directions, it may be due, he thought, to its being partially covered by an ocean whose surface is blown by the wind; the dark spots are lands. Hence, Leonardo placed some earth, water, and air on the moon to make of this stellar body a world similar to ours. He saw as natural the extension of this assumption that includes stars. He therefore proposed a theory of the plurality of worlds strongly analogous to Plutarch's, which a meditation similar to Plutarch's suggested to him. But this meditation was brought about by the Parisian physics of the fourteenth century as expounded by Albert of Saxony.

13

The Plurality of Worlds in Fifteenth-Century Cosmology

John Hennon

What John Hennon asserts about the plurality of worlds shows no trace of originality; he borrows almost verbatim from the *Quaestiones in libros de Caelo et Mundo* of Albert of Saxony. Like Albert, and for the same reasons, he believes the coexistence of several worlds to be naturally impossible, but he concedes that God can produce this coexistence in a supernatural fashion.

If we cite master Hennon's answer to this problem, it is because of the following passage:

> It is not contradictory that a world not be constituted from the totality of the matter of sensible things. In fact, it is stated in a Parisian article, *Quod Deus non posset movere Caelum motus recto, error.* It is therefore evident because of this, that God can put the world in a place other than the one in which it is presently.
>
> I ask you then, that done, can God put a man or some other body where the world is presently [from which God has removed the world by hypothesis]? If the answer is affirmative, then the world is not formed from all the matter of sensible things. And one cannot answer negatively, for it is manifest that God does not have less power on a man or on a stone than on the whole world.[1]

It has been asserted about the condemnations of 1277 that "the condemnations first troubled some timid minds, since they were maintained by those with vested interests, but they were soon confronted with indifference, and the masters unacquainted with the Thomist doctrines did not hesitate to blame them and to declare them worthless."[2] But, on the contrary, we have just seen that the decree of Etienne Tempier did not cease to furnish the masters of

499

Paris with weapons against Aristotle's physics during the whole fourteenth century; the above quote demonstrates that there remained some strength in it still, and that it continued to play its role during the second half of the fifteenth century. We shall find still other examples of this.[3]

We shall not pursue further the analysis of John Hennon's treatise. Doubtless we can find in it treatments of many Parisian doctrines. We can find our author's statement that God can produce an infinite multitude, since He created the world from all eternity, and the multitude of actually existent souls would be infinite; we can also find that God can produce an actually infinite magnitude, "which is to be proven thus: No contradiction is implied in the supposition that a magnitude is actually infinite; in fact, infinity does not, in any way, suppress the notion (*ratio*) of magnitude nor any of the necessary consequences of this notion, and magnitude does not suppress the notion of infinity either."[4] These conclusions are in conformity with those of Gregory of Rimini; but Hennon does not make use of the daring and rigorous logic of the illustrious Augustinian in order to establish them.

George of Brussels and Thomas Bricot

John Hennon was content to follow Buridan and Albert of Saxony's doctrine on the plurality of worlds—the simultaneous existence of several worlds does not imply a contradiction, so it can be achieved in a supernatural fashion by God's omnipotence, but two worlds cannot exist naturally since the earth of one of the worlds would tend toward the center of the other world.

George of Brussels and Thomas Bricot were more daring; their allegiance was not to Buridan's doctrine but to Oresme's.

Doubtless the existence of several worlds can only be realized in a supernatural manner; it was also in a supernatural manner that this world was created. The production of a world is not within the power of any natural cause; but if several worlds, external to one another, were created, these worlds could subsist without nature being violated, in spite of the "reason of the ancients."

Worlds can be eccentric in two ways, depending upon whether they touch one another—each of them being completely outside the other, however—or whether they are completely separated from one another; either way, several worlds can exist supernaturally.

The Philosopher's authority should not be an obstacle to this proposition, for his side should not be upheld. He states that heaven is composed from all possible matter, from all matter capable of receiving the celestial form; but that is not true. It is not true that no natural body can be produced outside this world either; such a body, in fact, would reside there naturally in the same fashion that natural bodies, whether simple or mixed, reside in this world. In the same fashion that in this world the simple bodies order themselves, some up and some down, the simple bodies of the other world, if it existed, would group themselves with the simple bodies of similar species, and there would be an up and a down.[5]

Aristotle attempted a *reductio ad absurdum* against the possibility of several worlds; if there existed two worlds, we can demonstrate either of these two contradictory propositions, as we wish: (i) The earth of one world would move toward the center of the other world, and (ii) the earth of one world would not move toward the center of the other world. The coexistence of two worlds is therefore an impossibility.

But it is possible, according to the truth of things, for faith assures that God can create several worlds. Therefore we do not have a conclusive reasoning here: if there were several worlds, the earth of one would move toward the center of the other. In fact, while the conclusion is false, the premise can be true.

Even when the earth of this world is of the same species as the earth of the other world, since it is numerically different, it is not necessary that it moves in virtue of its proper nature toward the center of the other world. Moreover, if a supernatural power were to take all or part of the earth of this world and place it in the celestial concavity of the other world, it would move toward the center of the other world.[6]

What we have just read is clearly a summary of the thoughts by which Nicole Oresme rediscovered the doctrine of Plutarch (about which, undoubtedly, he had no knowledge). From the wealth of audaciously innovative ideas the Parisian physics of the fourteenth century accumulated, George of Brussels and Thomas Bricot knew enough to preserve a few precious slivers.

Paul of Venice

The various places in which Paul of Venice treats the plurality of worlds are occasions for us to witness the influence that the teachings of Paris exercised on his intellect.

Our author devotes an entire chapter of his *Summa Philosophiae* to the problem of the plurality of worlds. This chapter does no more than summarize, somewhat faithfully, what Albert of Saxony said on this topic. Like Albert of Saxony, Paul Nicoletti concludes that there can only be one world.

> Let us suppose, however, that there are two worlds. Even though the earth here is of the same kind as the earth of the other world, it cannot move toward the other earth; the heavens would be an obstacle to its path and would prevent it from moving from one world to another. Yet if we imagined that we took a fragment of our earth and put it inside the other world, it would move toward the earth of this other world. In the same fashion, in our hemisphere, fire moves toward the arctic pole, but it would move toward the antarctic pole in virtue of the same inclination, if it were put in the other hemisphere. Consequently, if there existed several worlds, the fire of the first world would move toward the concavity of the lunar orbit of the second, and inversely; and the air of the first would move toward the concavity of the igneous sphere of the second, and reciprocally.[7]

In the passage we have just cited, Ockham's influence seems to temper the conclusions of Albert of Saxony. But in his book *On the Composition of the World*, Paul of Venice reasons otherwise:

> There is only one world, not several; we shall prove this.
>
> If there existed several worlds, either they would be contained one within the other, or each one would touch the next one at an indivisible point.
>
> The first assumption is inadmissible, for if there were a world containing this world, by the same reasoning there would have to be a third world containing the second within it, and so on to infinity; this cannot be, for one would have, in this way, an infinite series of motive forces and moved objects. The existence of such a series has been shown to be impossible in the eighth book of the *Physics*.
>
> There cannot be a second world touching this world at one point either, for using the same reason, there would exist a third world touching the second, and so on to infinity.[8]

What an awful argument, one might think. It does not even have the merit of being original; it only summarizes the thoughts developed by Ristoro d'Arezzo in his treatise *Della composizione del mondo*.[9]

"This assumption is also wrong for another reason," adds our author; "it would require that there be an infinite void outside the world, and it has been shown in the fourth book of the *Physics* that this cannot be."[10]

That is Michael Scot's reason, which Bacon and several others took up; but Paul of Venice repudiated it in his third book of his *Expositio super libros Physicorum*. He conceded that God can create other worlds outside this one, and he deduced from it the existence of an infinite void above the ultimate sphere.

It is true that he added that such a reasoning took as principle God's infinite power and that Aristotle had not known this power. Here also he remembered this power capable of contradicting all Peripatetic physics, for he ends his argument with this proviso: "Yet God who is omnipotent and infinite could, against the tendencies of nature, make it be that there is a void, and create an infinity of worlds touching at a point, two at a time."[11]

Perhaps this declaration is an act of deference toward the Parisian decree of 1277, but above all it is an imitation of the ending Ristoro d'Arezzo gave to his chapter on the plurality of worlds and to his whole work: "Ma importanto la potenza di Dio altissimo, sublime e grande, lo quale regga e conserva lo mondo, e puo fare tutte le cose che piacciono a lui colla sua potenza, la quale e infinita."[12]

John Major

John Major, master of the Collège de Montaigu in Paris, was not convinced by Aristotle's argument, nor, for a better reason, by those of Paul of Venice; in the very first question of his treatise, he asserted his belief not only in the plurality of worlds, but also in the existence of an infinite number of worlds. He said:

> Speaking from the natural point of view, there is an infinity of worlds; one cannot give any convincing reason for the opposite of this opinion. It is easy to refute the objection formulated by Aristotle that the earth of one world would tend toward the center of another; it is equally simple to refute any other objection. This opinion was, in any case, that of

Democritus, the distinguished philosopher whom Aristotle praised so highly in the first book of *De Generatione*.[13]

John Major does not tell us by what means it is easy to refute Aristotle's objection; no doubt he intends to allude to the path traced by William of Ockham.

In any case, he does cite the exception pointed out by Albert of Saxony against the reasoning of the Stagirite: "Aristotle's reasoning is not conclusive against a plurality of concentric worlds."[14]

It is no longer Aristotle, but rather Saint Thomas Aquinas that seems to be the target of this passage:

> Speaking in a purely natural way, it does not seem that one is able to prove in an entirely satisfactory manner the opposite opinion to ours, that there is only one world; in keeping with normal usages, I understand by world the set of celestial spheres and that which they contain.
>
> If you say, all these worlds are one world, it is that you do not understand your own words; if it is thus, Aristotle would not have bothered to have discussed this.[15]

Here is a reply which he evidently addresses to Michael Scot and Paul of Venice: "If you say, there will be a void between these two worlds, I will reply that your argument is equally applicable toward Aristotle's doctrine, for there will actually be a void outside heaven."[16]

And John Major ends his argument with a kind of defiance: "If you ask me by which arguments I concluded for the plurality of worlds, I ask you those by which you maintain the contrary opinion; and what I am stating, I am stating it from a purely natural point of view."[17]

Thus, even at the end of the fifteenth century, the problem of the plurality of worlds engendered impassioned debates in the schools.

Gaetano of Thienis

We have seen that during the fifteenth century, Paul of Venice restated the argument of Michael Scot, Roger Bacon, and Walter Burley, and deduced from it a conclusion against the plurality of worlds; we have also seen that the Scot, John Major, refused to admit the validity of this argument. He was preceded in this by Gaetano of Thienis.

Here is what we can read in the commentaries on Aristotle's *Physics* by Gaetano of Thienis.

> Burley thinks that Christians, because they admit the creation of the world, must equally admit the reality of the void outside the world. God could, in fact, engender another world outside the confines of this world. Let us assume that He has made one; one would then ask whether these worlds are distant from one another or whether they touch. If they are distant, there will be a void between them, for there will be between them a divisible space capable of receiving a body yet not containing one. If they touched, it would not be by the length of a plane, since they are terminated by a perfect convex sphericity; they would then touch at a single indivisible point. Therefore, there will still be a void within the divisible space existing between them, as before.
>
> But none of this is necessary. . . . One can say that these two worlds are certainly not separated by a mass, since there is no mass between them. They are not separated by a void either; a void is a place deprived of bodies, and between these worlds there is no place, no void, no plenum. The distance separating them is purely formal; it consists of certain relations caused within these two worlds. And this remains true even when they are touching. In any case, there may be grounds for asserting that two worlds can be entirely external to one another without being able to assert either that they are separated or that they touch.[18]

Nicholas of Cusa

If Nicholas of Cusa deserves to be numbered among the precursors of Copernicus, it is due more to his reflections on the plurality of worlds than his doctrine on the movement of the earth.

To remove the diurnal movement from the ultimate heaven in order to attribute it to the earth was neither extremely daring nor extremely useful.

It was not extremely daring. We can find, in almost every period of the history of science, from Heraclides Ponticus to Nicole Oresme, some thinkers who preferred to attribute the diurnal movement to the earth rather than ascribe it to heaven. This was particularly true during the second half of the fourteenth century when Nicole Oresme, John Buridan, Albert of Saxony, and Pierre d'Ailly ranked themselves for or against this hypothesis after having carefully discussed it.

It was not extremely useful. The mere substitution of the rotation of the earth for the rotation of heaven in order to explain diurnal movement did not simplify the astronomical theories in any way; it left the problem of the movement of the wandering stars in the same state in which it was found. That is what prevented a good number of Parisian masters from rallying for the hypothesis of the diurnal movement of the earth.

The great accomplishment of Copernicus was not making the earth rotate on itself instead of making the ultimate sphere rotate, it was renouncing the geocentric hypothesis and returning to the heliocentic system proposed by Aristarchus of Samos; that is how he was able to account for the movement of the wandering stars more simply than Ptolemy could have. And that is his principal claim to glory.

Moreover, the substitution of the heliocentric system for the geocentric system upset the commonly received doctrines of the physicists more deeply than the mere exchange between the rotation of the earth and the rotation of heaven; most Peripatetic doctrines remained undisturbed by whether the diurnal rotation is attributed to the earth or kept by heaven. That is not the case with the attribution of a movement for the earth similar to the movement of the planets. To do that leads one naturally to think of the earth and the planets as bodies having the same nature, and no longer to consider the wandering stars as spheres made of a fifth essence distinct from the four elements, but to think of each of these stellar objects as possessing its earth, water, air, and fire like our globe.

Such an assumption would require that Aristotle's theory of natural place and weight be altered. The natural tendency of the elements would no longer be to occupy a certain place, determined absolutely, lying between the center of the universe and the sphere of the moon. The various elements intended to make up a single body would tend to take a relative position, to form spherical layers on top of each other according to an order of decreasing density, without bothering about the absolute place that their set occupies in the universe.

This new theory of weight was presented, with much precision and clarity, by Plutarch in his treatise, *On the Face That Can Be Seen in the Lunar Disk*.

Nicole Oresme also expounded the same doctrine; he concluded that "if God created another world just like our own, the earth and the other elements of this other world would be present there just as they are in our world."[19]

Nicholas of Cusa asserted that the earth, the moon, and the planets are similar to the stars that move around the pole with respect to their movements around the center of the world.[20] He asserted that "the moon moves from west to east less than Mercury, Venus, or the sun, that it is so for the various stellar bodies by various degrees, and that the earth moves still less than the others."[21] It is obvious that he ranked the earth as one of the moving stellar objects, at least with respect to its movement.

But this analogy is not limited to movements; according to Nicholas of Cusa, the earth is really a star like the sun or the moon because of its makeup.

> If someone were on the sun, he would not see the brightness we see; we find that the mass of the sun has a kind of earth which is its most central part; at its circumference is a kind of luminescence whose nature reminds one of the nature of fire, and between the two there is an aqueous cloud and clearer air.
>
> The earth also has these elements. If someone were outside the region of fire, at the circumference of this region which depends on our earth, the earth would appear to him as a bright star because of its fire. In the same way, we who are located at the circumference of the region of the sun, see the sun as a bright object.
>
> The moon does not appear as bright because we are located within its circumference [which limits the region of its elements]; we are located toward its central parts—in the aqueous region of the moon, for instance. Thus its light cannot be seen by us, even though it does have its own light; this light would be seen by those outside the extreme circumference of the lunar region, while we only see the light of the sun reflected by the moon. Similarly, the moon certainly produces some heat because of its movement, and this heat is greatest at the circumference of its region where the movement is greatest. And this heat is not communicated to us as the heat of the sun is.
>
> Hence our earth seems located between the region of the sun and the region of the moon; with the sun and moon acting as intermediaries, it is influenced by the other stars—which we do not see because we are located outside their proper regions. That which we see of the other stars is only their regions [it is not their bodies]; that is why they scintillate.
>
> The earth therefore is a noble star—*est igitur terra stella nobilis*—it possesses light, heat, and influence. This light, heat, and influence differ from that of any other star; similarly,

> every star differs from every other star by its light, heat, and influence.[22]

The above passage affirms most strongly that the earth is analogous to the sun and the other stars in every detail.

Such a conclusion demands that the Aristotelian theory of weight be abandoned. By what theory might it be replaced? Nicholas of Cusa says little about that, but the broad outline he sketches reminds one of the thoughts that Plutarch and Nicole Oresme had formulated explicitly.

> The movement of a part has as object the perfection of the whole; that is why heavy bodies tend toward the earth and light bodies tend upward. It is why earth tends toward earth, water toward water, air toward air, and fire toward fire. As much as possible the movement of the whole tends toward circularity and the shape of the whole tends toward sphericity.[23]

If each element of a star tends to form a unique mass and if these various masses tend only to dispose themselves into concentric spherical layers, then neither the parts nor the set of these elements have the least disposition to seek the center of the world or to escape from it; such a body would no longer be either heavy or light. Plutarch most clearly affirmed this proposition, which was well suited to seduce the author of the *De docta ignorantia*, according to whom the universe has no center.

Nicholas of Cusa does not seem to have perceived the above corollary of his theory of weight, which is so favorable to it. He does indicate some thoughts which are obscure because of their brevity, but which seem to require the following interpretation:

Of the various elements composing a star, such as the earth, some are heavy and tend toward a certain point, while others are light and tend away from this point. The whole star does not tend toward or away from the point—it is neither heavy nor light—because the weight of some of its elements is compensated exactly by the lightness of the others; due to this compensation, the star remains suspended in space. In order to create the world, God called forth the four mathematical sciences—geometry, arithmetic, astronomy, and music; the exact equilibrium to which we are referring was the work of geometry.

That is the sense we attribute to the following passages:

> God calculated by geometry the proportion of the elements in such a way that firmness, stability, and mobility

flow from this proportion as He wished it. . . . The elements
then were constituted by God following an admirable order;
He created everything with number, weight, and measure.
Number was due to arithmetic, weight to geometry, and
measure to music. . . .

Gravity, in fact, is upheld in space because levity
constrains it; earth, which has weight, is suspended in space
by means of fire—levity fights gravity as fire fights earth, for
example. . . .

How can we prevent ourselves from admiring the Artisan
who utilized so perfect an art when He made up the celestial
spheres, the stars, and the various regions of the stars? Because
of His precision, we find variety everywhere, yet everything
is in harmony. . . . He has set the mutual relations of the
various parts of stars in such a way that, in each, the parts
move toward the whole, the heavy bodies tend toward the
center below, the light bodies climb, tending away from the
center, and the whole set describes the rotary movement around
the center that we notice in the stars.[24]

Someone like Nicole Oresme would have allowed that space
can contain several systems, each of which can be composed of
an earth surrounded by water, air, and fire; for each of these various
worlds, he would have applied a theory of weight similar to the
one Plutarch proposed. But less daring than Plutarch, he would
not have been tempted to place inhabitants on these worlds. As
far as we know, no one defended such an assumption during the
Middle Ages, when Nicholas of Cusa, whose imagination knew
no bounds, proposed it.

The future bishop of Brixen declares that the earth is not the
most vile of all the celestial bodies, and that the humans, animals,
and plants inhabiting it are not inferior in nobility to those
inhabiting the sun or the other stars.

However, our philosopher recognizes that we cannot know
much about the beings inhabiting the various stars.

It may be conjectured that the inhabitants of the sun are
more solar, more bright, clear, and intellectual; we assume
that they are more spiritual than those who inhabit the moon—
who are lunatic. Those on the earth are more material and
rough, so that the intellectual beings found on the sun are
more in actuality and less in potency, while the inhabitants
of the earth are more in potency and less in actuality; the
inhabitants of the moon are somewhere in between these two
extremes.

These opinions are suggested to us by the influence of the sun, which is of igneous nature, by that of the moon, which is both aqueous and igneous, and by the material heaviness of the earth.

The regions of the other stars are similar to this, for we believe that none of them is deprived of inhabitants.[25]

The first time in Western Christianity that one heard someone speak about the plurality of inhabited worlds, it was proposed by a theologian who had spoken before an ecumenical council a few years before. The person who sought to reflect upon the characteristics of the inhabitants of the sun and moon in a book that became well known had the confidence of the popes; the highest ecclesiastical honors were bestowed upon him. There can be no greater proof of the extreme liberality of the Catholic church during the close of the Middle Ages toward the meditations of the philosopher and the experiments of the physicist.

Notes

Preface

1. Alexandre Koyré, Review of Duhem's *Le Système du monde*, *Archives Internationale d'Histoire des Sciences* 35 (1956): 252

2. Pierre Duhem, "Notice sur les titres et travaux scientifiques de P. Duhem," *Mémoires de la société des sciences physiques et naturelles de Bordeaux*, ser. 7, vol. 1 (1917-27): 159.

3. Marshall Clagett, *The Science of Mechanics in the Middle Ages* (Madison: The University of Wisconsin Press, 1959), p. 583.

4. Aristotle *Physics* III, c. 6 (206a 18-24).

5. Ibid., c. 7 (207b 16-21).

6. William of Sherwood, *Treatise on Syncategorematic Words*, trans. N. Kretzmann (Minneapolis: University of Minnesota Press, 1968), p. 41

7. Aristotle *Physics* IV, c. 5 (212b 8-10).

8. Ibid., (212b 12-14).

9. Ibid., c. 14 (223a 21-19).

10. Ibid., c. 7 (213b 31-35); c. 8 (214b 18-20).

11. Ibid., c. 8 (215b 24).

12. Ibid., (215a 19-22).

Chapter 1

1. This phrase has been coined by someone who has meditated most on the progress of Christian thought during the Middle Ages, my friend and colleague, Albert Dufourcq.

2. See Duhem, *Le Système du monde*, vol. 1, chap. 4, pp. 177ff.

3. [The text has "infini" incorrectly, I think.]

4. Aristotle *Physics* III, c. 6.

5. Averrois *Destructio destructionem philosophiae Algazelis*, Disputatio prima, reply to 6th [7th?] Ait [van den Bergh 1:14].

6. Avicenna *Metaphysica* II, tract. VI, c. 2.

7. Ibid., tract. VIII, c. 1.

8. *Philosophia* Algazelis I, tract. I, sexta divisio de ente in finitum et infinitum, capitulum undecimum, fol. d2, col. d; fol. d3, col. a. The text of this passage is extremely flawed—one is often required to guess at its meaning instead of translating it.

9. Algazelis *Destructio philosophiae,* in Averrois Cordubensis *Destructio destructionem philosophiae Algazelis,* Disputatio prima, 6th Ait Algazali [van den Bergh, 1:8-9].

10. Averrois Cordubensis *Destructio destructionem philosophiae Algazelis,* Disputatio prima, reply to 6th Ait Algazali [van den Bergh, 1:9-12].

11. Maimonides *The Guide of the Perplexed* I, c. 73-74 [pp. 194-222].

12. Sancti Thomae Aquinatis *De aeternitate mundi contra murmurantes opusculum,* in fine [p. 25].

13. Sancti Thomae Aquinatis *In libros Physicorum Aristotelis expositio,* librum III, lectio 9, in fine [actually lectio 10, p. 175].

14. Sancti Thomae Aquinatis *Quaestiones disputatae de Scientia Dei,* quaest. II, art. 10: Num infinita Deus efficere possit; conclusio [p. 108].

15. Gratia Dei Esculani, seu ab Esculo *Quaestiones in libros Physicorum Aristotelis,* in *Studio Patavino disputatae.*

16. J. Quétif and J. Echard, *Scriptores Ordinis Praedicamentorum* 1:603, col. a.

17. Leandro Alberti *De Viris illustribus,* fol. 153, col. b.

18. See *Le Système du monde* 6:89.

19. Gratia Dei Esculano *Quaestiones litterales,* lib. III, lect. 9, quaest. 8, fol. 37, col. b.

20. Sancti Thomae Aquinatis *Quodlibeta,* quodlib. IX, art. 1: Utrum Deus possit facere infinita esse actu [p. 336].

21. Sancti Thomae Aquinatis *Summa Theologica,* pars prima, quaest. VII, art. 3 [pp. 57-68. I cannot find the third paragraph from Duhem's quote; standard editions give the paragraph as: "It is manifest that a natural body cannot be actually infinite. For every natural body has some determined substantial form. Therefore since accidents follow upon the substantial form, it is necessary that determinate accidents should follow upon a determinate form; and among these accidents is quantity. So every natural body has a greater or smaller determinate quantity."]

22. Aristotle *Physics* III, c. 7, in fine.

23. Aristotelis Stagiritae *De Physico auditu libri octo cum* Averrois Cordubensis *variis in eosdem commentariis,* lib. III, commentarius 72 [Minerva facsimile, p. 119].

24. Ibid., comm. 59 [p. 113].

25. Aquinas, *Summa,* ad 3m [p. 58].

26. Ibid., ad 1m [p. 58].

27. Ibid., Pars prima, quaest. VII, art. 4 [p. 59].

28. Ibid., Pars prima, quaest. VII, art. 3, ad 4m [p. 59].

29. Fratris Rogeri Bacon, Ordinis Minorum, *Opus majus ad Clementem Quartum, Pontificem Romanum,* p. 93 [p. 173].

30. Joannis Duns Scoti *Scriptum Oxoniense in II librum Sententiarum,* distinctio II, quaest. IX: Utrum angelus possit moveri de loco ad locum motu continuo. Maurice du Port indicates, in his edition of the *Scriptum Oxoniense,* that he views the passage we are analyzing as authentic [*Opera Omnia,* vol. 6, pt. 1, pp. 230-33].

31. Joannis Canonici *Quaestiones in libros Physicorum Aristotelis*, lib. VI, quaest. unica.

32. Guilhelmi de Ockham *Quodlibeta Septem*, quodlib. I, quaest. IX: Utrum linea componatur ex punctis [Franciscan Institute ed., pp. 50-61].

33. Gregorius de Arimino *In primum Sententiarum nuperrime impressus*.

34. Maximillian Curtze, *Ueber die Handschrift* R. 4. 2, *Problematum Euclidis explicatio der Konigl. Gymnasialbibliothek zu Thorn (Zeitschrift fur Mathematik und Physik*, Supplement, p. 65).

35. See *Le Système du monde* 6:704.

36. Alberti de Saxonia *Quaestiones super libros de Physica auscultatione Aristotelis*, lib. VI, quaest. I.

37. *Quaestiones subtilissimae* Johannis Marcilii Inguen *Super octo libros Physicorum secundum nominalium viam*, lib. VI, quaest. I.

38. Joannis Duns Scoti *Scriptum Oxoniensis super Sententias*, lib. II, dist. II, quaest. 9 [vol.6, pt. 1, 256-57].

39. [Ibid., vol. 6, pt. 1, 257.]

40. See *Le Système du monde* 6:619-44.

41. Bibliothèque Nationale, fonds latin, ms. no. 16130, fol. 129, col. a; capitulum I, conclusio IIa.

42. *Tractatus Venerabilis Inceptoris* Guilhelmi de Ockam *de sacramento altaris*, quaest. I: Utrum punctus sit rest absoluta distincta realiter a quantitate; quaest II: Utrum linea et superficies realiter distinguantur inter se et a corpore [pp. 7-93].

43. [Ockham, *De Sacramento Altaris*, p. 9.]

44. [Ibid., pp. 31-32. I cannot find Duhem's text, which seems to be a paraphrase instead of a quote. The closest to Duhem's citation, from standard editions, is as follows:

> Moreover, a line is not the cause of a point and vice versa, as is obvious by a survey of all the causes; therefore, God can, of His absolute power, make a line without any point.
>
> Then I ask, whether the line is finite or infinite. If it is finite and without any point, it is consequently needless to posit a point to terminate a line; and yet it is not posited by those who thus posit it. If, however, the line is not terminated or finite, it is in the nature of things *per causum*; it, therefore, is infinite, which obviously is manifestly false, for that line will not by virtue of this be greater or longer because that point is separated or destroyed. Therefore, in no mode will there be that infinite, no matter to what extent all the points may be destroyed. Moreover, if all the points are destroyed by God and the line were saved, I ask whether or not that line is continuous. If it is, it is not through a point (for it is not *per causum*); there is, therefore, a continuous line without any point, and consequently a point is needlessly posited here. If it is not continuous, I ask whether any part of it be continuous, in which case the preceding argument applies, or else no part is continuous; from which it follows that there is some line which is not composed of continua, which is impossible.
>
> Similarly, that line is continuous or discrete; if continuous, the thesis is established; if discrete, then each part is divided into an infinite quantum as regards all its parts, which is impossible. But this reason proves that a line is sufficiently continuous and finite through its proper nature without any other thing added to it; and consequently, since a point ought not be posited by virtue of another, a point is needlessly posited to be such an indivisible thing.]

45. [Ibid., p. 39.]

46. Guilhelmi de Ockham *Quodlibeta Septem*, quodlib. I, quaest. IX: Utrum linea componantur ex punctis [pp. 50-61].

47. [Durandi A Sancto Portiano *Super sententias theologicas Petri Lombardi commentarii*, in fine, vol. 2, fol. 423, col. 6.]

48. Ibid., lib. II, dist. II, quaest. 4, art. 2. [vol. 1, fol. 133, col. c].

49. [Loux, *Ockham's Theory of Terms*, pp. 142-47.]

50. *Preclarissimi viri* Gualtery Burlei *anglici sacre pagine professoris excellentissimi super artem veterem Porphyrii et Aristotelis expositio sive scriptum feliciter incipit.* Liber Praedicamentorum on the topic: Propriae autem quantitates hae sunt solae quas diximus; fol. sign. d, cols. c, d.

51. Burleus *Super octo libros Physicorum*, lib. I, tract. II, on the topic: Melissus autem quod est infinitum dicit esse; fol. sign. b3, cols. b et seq. [which is the same as fol. 11, cols. b et seq.].

52. Gregorius de Arimino *In secundum Sententiarum*, dist. II, quaest. 2, art. 1, fol. 35, col. d [fol. 37, col. d].

53. [Ibid., fol. 37, col. d; fol. 38, col. a.]

54. Ibid., fol. 37, cols. b, c, d [fol. 38, col. d; fol. 39, cols. a, b].

55. *Acutissimi Philosophi reverendi Magistri* Johannis Buridani *Subtilissime quaestiones super octo Phisicorum libros Aristotelis diligenter recognite et revise A magistro* Johanne Dullart De Gandavo *antea nusquam impresse*, lib. VI, quaest. IV, fol. 96, col. c.

56. Ibid., fol. 97, col. a.

57. Ibid., cols. a, b.

58. Ibid., col. c.

59. Ibid., cols. b, c. The text reads "condiciones"; it seems to us that it ought to read "conclusiones."

60. Joannis Buridani *Quaestiones in Metaphysicam Aristotelis*, lib. XII, quaest. X, fol. 72, col. c.

61. Johannis Buridani *Quaestiones super octo libros Physicorum Aristotelis*, lib. VI, quaest. IV, fol. 98, col. a.

62. Joannis Buridani *Quaestiones in Metaphysicam Aristotelis*, lib. XII, quaest. X, fol. 73, col. b.

63. Guilhelmi de Ockam *Tractatus de sacramento altaris*, De distinctione puncti, linae . . . Primo: Utrum punctus sit res absoluta, fol. sign. B, col. d [p. 81].

64. Alberti de Saxonia *Quaestiones super libros de Physica auscultatione Aristotelis*, lib. VI, quaest. I.

65. *Incipiunt subtiles docrinaque plene abbreviationes libri Physicorum edite a pretantissimo philosopho* Marsilio Inguen *doctore parisiensi*, lib. VI, fol. sign. g4, col. d and the three following columns.

66. Johannis Marcilii Inguen *Quaestiones super octo libros Physicorum*, lib. VI, quaest. III.

67. [Ibid.]

68. Aristotle *Physics* I, c. 4.

69. Aristotelis Stagiritae *De Physico auditu libri octo cum* Averrois Cordubensis *variis in eosdem commentariis*, lib. I, comm. 38.

70. *Emptor et lector aveto.* Sancti Thomae Aquinatis *In libros Physicorum Aristotelis interpretatio sum et expositio.* Divi Roberti Lincolniensis *Super octo*

libris Physicorum brevis et utilitis summa feliciter incipit, fol. after the fol. sign. Q2, cols. c, d.

71. [Ibid.]

72. Beati Alberti Magni *Liber Physicorum,* lib. I, tract. II, cap. XIII: De destructione opinionis Anaxagorae in eo quod posuit principia esse finita et quodlibet esse in quodlibet.

73. Sancti Thomae Aquinatis *In libros Physicorum Aristotelis expositio,* lib. I, lect. IX [p. 35].

74. Sancti Thomae Aquinatis *Summa Theologica,* Pars prima, quaest. VII, art. 3: Utrum possit esse aliquid infinitum actu secundum magnitudinem [pp. 57-58].

75. Sancti Thomae Aquinatis *Quaestiones disputatae de potentia Dei,* quaest. IV: De creatione materiae informis, art. 1: Utrum creatio informis praecesserit duratione creationem rerum [p. 13].

76. Egidii Romani *In libros de Physico auditu Aristotelis commentaria accuratissime emendata,* lib. III, lect. XIV, text comm. 59-60, dubium 1a, 2a, fol. 59, cols. a, b, c.

77. Ibid., lib. VI, lect. IV, text comm. 15, dub. 1a, 3a, fol. 121, col. d [fol. 139, cols. b, c].

78. Egidii *cum* Marsilio *et* Alberto *de generatione. . . . Quaestiones super primo de generatione fundatissime doctoris domini* Egidii *Ordinis fratrum Heremitarum sancti Augustini,* quaest. X: Utrum corpus continuum sit divisible in infinitum, fol. 56, col. a [p. 115, col. b to p. 116, col. b].

79. *Quodlibet domini* Egidii Romani *Theoremata eiusdem de corpore christi,* Guiliermus de Ockam *de sacramento altaris,* quodlib. IV, quaest. VI, fol. 44, cols. b, c; theoremata X, fol. 93, col. a.

80. *Quodlibeta Doctoris eximii* Ricardi de Media Villa ordinis minorum, quodlib. III, quaest. V: Utrum magnitudo naturalis sit divisibilis in infinitum, pp. 91-93 [vol. 1, fol. 31, col. b].

81. Joannis de Janduno *Super octo libro Aristotelis de Physico auditu subtilissimae quaestiones,* lib. VI, quaest. I.

82. Joannis de Janduno *Quaestiones super* Averrois *sermonem de substantia orbis,* quaest. VIII: An forma naturalis ad maximum et minimum determinatur.

83. Burleus *Super octo libros Physicorum,* lib. III, tract. II, cap. 4, fol. 71, col. b.

84. Guilhelmi de Ockam *Super quatuor libros Sententiarum annotationes,* lib. II, quaest. VIII: Utrum mundus potuit fuisse ab aeterno.

85. Johannis Buridani *Quaestiones super octo Phisicorum libros Aristotelis,* lib. I, quaest. XIII: Utrum entia naturalia sint determinata ad minimum, fol. 17, col. a.

86. [Ibid., cols. a, b.]

87. Alberti de Saxonia *Quaestiones super libros de Physica auscultatione Aristotelis,* lib. I, quaest. X, quantum ad 3m [quaest. IX, fol. 9, cols. b, c].

88. Marsilii Inguen *Abbreviationes libri Physicorum,* 6th folio (unnumbered), cols. c, d.

89. Marsilii Inguen *Quaestiones super octo libros Physicorum,* lib. I, quaest. XIII.

90. Aristotle *Physics* III c. 6 [206a].

91. Ibid., c. 7 [207b].

92. Aristotelis *De Physico auditu libri octo cum* Averrois Cordubensis *variis eosdem commentariis*, lib. III, comm. 60 [p. 113].

93. Burleus *Super octo libros Physicorum*, lib. III, tract. II, cap. 4, fol. 71, col. c.

94. Fr. Rogeri Bacon *Opera quaedam hactenus inedita*, *Opus tertium*, cap. 39, pp. 132-33.

95. Ibid., pp. 134-35.

96. Ricardi de Mediavilla *Quaestiones in quatuor libros Sententiarum*, lib. I., dist. XLIII, art. 1, quaest. 6, vol. 1, p. 386, col. b.

97. Joannis Duns Scoti *Scriptum Oxoniensis*, lib. II, dist. II, quaest. 9 [VI, pt. 1, 250-52].

98. Burleus *Super octo libros Physicorum*, lib. VI, tract. I, cap. 1, fol. 155, col. d.

99. Petri Hispani *Summulae logicales cum* Versorii Parisiensis *Clarissima expositione. Parvorum logicalium eidem* Petro Hispano *ascriptum opus nuper in partes ac capita distinctum*, tract. VIIi: parvorum logicalium tract. VIIus; cap: De infiniti quinque acceptionibus et propositionibus en ipso formatis, fols. 259, 260 [pp. 119-20]. [De Rijk, in his edition of Peter of Spain's *Tractatus*, in which he argues that Pope John (erroneously crowned Pope John XXI) is the author of the *Summulae logicales*, treats the *De exponibilibus* in which our passage appears as an inauthentic tract.]

100. [Ibid., pp. 121-22.]

101. Burleus *Super octo libros Physicorum*, lib. III, tract. II, cap. 4, fol. 70, col. c.

102. Gregorius de Arimino *In secundum Sententiarum*, dist. II, quaest. II, art. 1, fol. 33, col. a [fol. 35, col. b].

103. [Ibid.]

104. [Ibid.]

105. Johannis Buridani *Quaestiones super octo libros Physicorum Aristotelis*, lib. III, quaest. XVIII, fol. 63 (incorrectly marked 62), col. b.

106. Ibid.

107. Aristotelis Stagiritae *De Caelo et Mundo libri quatuor, e graeco in latinum ab* Augustino Nipho *Philosopho Suessano conversi, et ab eodem etiam . . . aucti expositione*, lib. I, fol. 31, col. d.

108. Anonymi *Tractatus*, cap. I, conclusio 6a. Bibliothèque Nationale, fonds latin, ms. no. 16130, fol. 121, col. a.

109. Gregorius de Arimino *In secundum Sententiarum* [fol. 35, col. b].

110. Johannis Buridani *Quaestiones super octo libros Physicorum*, lib. III, quaest. XVIII, fol. 61, cols. c, d.

111. Ibid., fol. 62 (incorrectly marked 61), col. b.

112. [Ibid.]

113. Ibid., cols. b, c.

114. Ibid., lib. VIII, quaest. III, fol. 111, cols. b, c.

115. Alberti de Saxonia *Quaestiones super libros de Physica auscultatione Aristotelis*, lib. III, quaest. X [quaest. IX, fol. 37, col. c].

116. Ibid.

117. Egidii *cum* Marsilio *et* Alberto *de generatione. Quaestiones . . . super primo de generatione*, D. Egidii, quaest. XI, fol. 57, col. a [p. 117, col. b].

118. Burleus *Super octo libros Physicorum*, lib. III, tract. II, cap. 4, fol. 70, col. b.

119. Gregorius de Arimino *In secundum Sententiarum*, dist. II, quaest. II, art. 1, fol. 33, col. d to fol. 35, col. d [fol. 33, col. a to fol. 40, col. a].

120. Ibid., fol. 34 (incorrectly marked 28) col. c [fol. 36, col. c].

121. Johannis Buridani *Quaestiones super octo libros Physicorum*, lib. III, quaest. XVI, fol. 59, col. b.

122. Ibid., quaest. XVIII, fol. 63 (incorrectly marked 62), col. d.

123. Ibid., cols. c, d.

124. Alberti de Saxonia *Quaestiones super libros de Physica auscultatione Aristotelis*, lib. III, quaest. XIV, art. 1 [fol. 41, col. c].

125. Aristotle *De Caelo* I, c. 11.

126. Averrois Cordubensis *Commentarii in quatuor libros Aristotelis De Caelo et Mundo*, lib. I, summa decima, cap. II, pars 2, comm. 116.

127. Sancti Thomae Aquinatis *Expositio super libros De Caelo et Mundo Aristotelis*, lib. I, lect. XXV.

128. Joannis de Janduno *Philosophi acutissimi super octo libros Aristotelis de Physico auditu subtilissimae quaestiones*, lib. VI, quaest. I, fols. 85 (incorrectly marked 74), 86 [the quote is from fol. 86, col. b].

129. Johannis Buridani *Quaestiones super octo libros Physicorum*, lib. I, quaest. XII, fol. 16, col. a.

130. Ibid., col. c.

131. Johannis Buridani, rectoris Parisius, *Expliciunt quaestiones super libris De Caelo et Mundo*, lib. I, quaest.: Quaeretitur utrum potentia debeat diffiniri per maximum in quod potest, fol. 79, cols. a, b [p. 97].

132. [Ibid.]

133. Ibid., quaest. XXI: Quaeretitur utrum sit dare maximum in quod potentia potest, fol. 79, col. c to fol. 81, col. a [pp. 98-112].

134. Ibid., fol. 79, col. d [p. 101].

135. [Ibid., p. 102.]

136. Ibid., fol. 80, col. d [p. 102].

137. [Duhem's complaints are not without foundation. Although there is additional evidence that the *Quaestiones super libris De Caelo et Mundo* is Buridan's work (that is E. A. Moody's position on pp. xxiii-xxiv), the passage Duhem is analyzing is not to be trusted entirely; the text differs substantially between the two manuscripts to which Moody refers, the Bruges 477 Ms. having twice as much text as the Codex Latinus Monacensis 19551, to which Duhem referred. See Moody, pp. 101-2.]

138. Wood, *History of Antiquities of Oxford* 1:448. C. L. Kingsford, *Swineshead (Richard)* in *Dictionary of National Biography* 55:231.

139. Bibliothèque Nationale, fonds latin, ms. no. 16621.

140. Ro. Swineshead *De primo motore*, differentia VIIIa, cap. I, fol. 81r.

141. Ibid., fols. 85r-92v.

142. Ibid., fols. 87r-88v.

143. Ibid., fol. 88v.

144. R. L. Poole, *Dumbleton (John of)* in *Dictionary of National Biography* 16:146.

145. Bibliothèque Nationale, fonds latin, ms. no. 16146.

146. Joannis de Dumbleton *Summa*, pars VI, cap. I, fol. 57, col. a.

147. Ibid., cap. II, fol. 59, col. a.

148. Bibliothèque Nationale, fonds latin, ms. no. 16621, fol. 159v.

149. Magistri Roberti Holkot *Super quatuor libros Sententiarum questiones* [preface to Roberti Holkot *Determinationes quarundam questionum*].

150. *Quaestiones subtilissimae* Alberti de Saxonia *in libros De Caelo et Mundo*, lib. I, quaests. XIV, XV. According to J. Aschbach (*Geschichte der Weiner Universität* 1:365), Albert of Saxony composed a treatise, *De maximo et minimo* which is in manuscript form at Venice. If this treatise actually exists, we would have to believe that it is about the subject occupying us [quaests. XII, XIII, fol. 95, col. c to fol. 99, col. b].

151. Alberti de Saxonia, *Quaestiones in libros De Caelo et Mundo*, lib. I, quaest. XIV, quantum ad primum articulum [quaest. XII, fol. 95, col. d to fol. 96, col. a].

152. Ibid., quaest. XIV, quantum ad secundum articulum [quaest. XII, fol. 96, col. c].

153. Alberti de Saxonia *Quaestiones super libros de Physica auscultatione Aristotelis*, lib. III, quaest. XIII [quaest. XII, fol. 40, col. c].

154. [Ibid.]

155. *Quaestiones subtilissimae* Alberti de Saxonia *in libros De Caelo et Mundo*, lib. I, quaest. XV, quantum ad secundum articulum [quaest. XIII, fol. 97, col. d].

156. Ibid., quaests. XIV, XV et passim.

157. Bibliothèque Nationale, fonds français, ms. no. 1083, fol. 24, cols. b, d [p. 193].

158. R. L. Poole, *Heytesbury (William)* in *Dictionary of National Biography* 26:327-28.

159. *Tractatus* Gulielmi Hentisberi *de sensu composito et diviso.—Regulae eiusdem cum sophismatibus.—Declaratio* Gaetani *supra easdem.—Expositio litteralis supra tractatum de tribus.—Questio* Massini *de motu locali cum expletione.* Gaetani—*Scriptum supro eodem* Angeli de Fosambruno.—Bernardi Torni *Annotata supra eodem.*—Simon de Lendenaria *supra sex sophismata.—Tractatus* Hentisberi *de veritate et falsitate propositionis.—Conclusiones eiusdem.*

160. Gulielmi Hentisberi *Tractatus de sensu composito et diviso*, quartus modus, fol. 2, col. d.

161. Ibid., quintus modus, fol. 3, col. a.

162. *Regulae solvendi sophismata preclarissimi* Magistri Gulielmi Hentisberi *De maximo et minimo*, fol. 29, col. c.

163. [Ibid.]

164. Gulielmi Hentisberi *Probationes conclusionem in regulis positarum, Regulae observandae de maximo et minimo*, art. 2, fol. 194, col. a.

165. *Quaestiones subtilissimae* Johannis Marcilii Inguen *Super octo libros Physicorum secundum nominalium viam*, lib. I, quaest. XIV

166. Marsilii Inguen *Abbreviationes libri Physicorum*, 6th folio (unnumbered), col. a.

Chapter 2

1. This error is the thirty-fourth in Etienne Tempier's decree; it was given as the twenty-seventh in Mandonnet's thematic grouping; Pierre Mandonnet, *Siger de Brabant* 2:178.

2. Burleus *Super octo libros Physicorum*, lib. III, tract. II, cap. V, fol. 75, cols. b, c.

3. *Quodlibeta* Magistri Henrici Goethals A Gandavo Doctoris Solemnis, quodlib. XIII, quaest. III: Utrum Deus possit facere corpus aliquod extra caelum quod non tangat caelum, fol. 423v.

4. Ibid., quodlib. XI, quaest. I: Utrum Deus possit facere sub una specie specialissima angeli aliquem alium angelum aequalum in natura et essentia specici angelo jam facto sub illa, fols. 438v [incorrectly marked 439], 439r.

5. Ibid., quodlib. V, quaest. III: Utrum in Deo sit ponere aliquam infinitam idearum vel cognitarum, fol. 155r.

6. [Ibid., fol. 155v.]

7. [Ibid., fol. 156r.]

8. [Ibid., fol. 150v.]

9. [Ibid., fol. 156r.]

10. [Ibid.]

11. Petri Lombardi Episcopi Parisiensis, *Sententiarum libri quatuor*, lib. I, dist. XLIII.

12. Magistri Ricardi de Mediavilla *Super quatuor libros Sententiarum Petri Lombardi quaestiones subtilissimae*, lib. I, dist. XLIII, art. 1, quaest. IV, vol. 1, pp. 382-83.

13. Ibid., quaest. V, vol. 1, pp. 383-86.

14. [Ibid., p. 384, cols. a, b.]

15. [Ibid., p. 385, cols. a, b.]

16. [Ibid., p. 385, col. b; p. 386, col. a.]

17. Ibid., quaest. VI, vol. 1, p. 386 [col. a].

18. [Ibid., p. 386, col. b.]

19. Ibid., lib. II, dist. II, art. 3, quaest. IV, vol. 2, p. 17 [col. a].

20. Gulielmi Varonis *Quaestiones super libros Sententiarum*, lib. I, quaest. LXIX, fol. 57, col. b.

21. *Commentariorum in primum librum Sententiarum Pars Secunda*, auctore Petro Aureolo Verberio, dist. XLIIII, art. 1, pp. 1041-45.

22. Ibid., p. 1042, col. a.

23. Ibid., p. 1043, col. a.

24. Let us recall that the Scholastic term, *objective*, is almost equivalent to our modern term, *subjective*.

25. Petri Aureoli, *Commentatorium*, dist. XLIIII, art. 1, p. 1044, col. b.

26. Ibid., art. 2, p. 1045, col. a.

27. Instead of *possibile* the text erroneously had *impossibile*.

28. Petri Aureoli *Commentatorium*, dist. XLIIII, art. 2, p. 1046, col. b; p. 1047, cols. a, b.

29. Guilielmi de Ockam *Super quatuor libros Sententiarum annotationes*, quaest VIII: Praeterea quaero, de terminis augmenti charitatis, utrum sit dare summam cui repugnet augmentari [Franciscan Institute ed., III, 553-54].

30. [Ibid., III, 554-56.]

31. [Ibid., III, 567-68.]

32. [Ibid., III, 563-64.]

33. The Jesuits of the University of Coimbra, referring to the texts we are studying, classify Ockham as one who holds the following proposition: *Potest infinitum actu divinae virtutis produci*—which is to understand Ockham poorly; (*Commentarii Colegii Coninbricensis, Societatis Jesu, in octo libros Physicorum Aristotelis Stagiritae*, lib. III, cap. VIII, quaest. II.)

34. *Quodlibeta septem una cum tractatus de sacramento altaris* Venerabilis inceptoris fratris Guilhelmi de Ockham, quodlib. II, quaest. V: Utrum Deus potuit mundum fecisse ab aeterno [pp. 128-35].

35. Joannis Canonici *Quaestiones super octo libros Physicorum Aristotelis,* lib. III, quaest. III, art. 2, fol. 38, col. b [fol. 43, col. b].

36. Durandi A Sancto Portiano *Super sententias theologicas Petri Lombardi commentariorum libri quatuor,* lib. I, dist. XLIII, quaest. II, fols. 86-87 [fol. 113, col. b].

37. Joannis Duns Scoti *Scriptum Oxoniensis,* lib. II, dist. I, quaest. III: Utrum possibile sit Deum producere aliquid aliud a se sine principio durationis [vol. 6, pt. 1, p. 52].

38. T. A. Archer, *Baconthorpe, Bacon or Bacho (John)* in *Dictionary of National Biography* 2:379-81.

39. Doctoris Resoluti Iohannis Bacconis *Anglici Carmelitae, liber secundus super Sententias,* dist. I, quaest. VII, vol. 2, fol. 19, col. d; fol. 20, cols. a, b.

40. Ibid., fol. 21, col. b.

41. *Illuminati doctoris fratris* Francisci de Mayronis Ordinis minorum *in Primum Sententiarum scriptum Conflatus nominalium,* dists. XLIII, XLIV, quaest. X: Utrum Deus potuit producere aliquid actualiter infinitum, arts. 1, 2, fol. 128, col. d, fol. 129, cols. a, b, c [fol. 139, cols. b, c, d; fol. 140, cols. a, b; the quote is from fol. 139, col. d].

42. [Ibid., fol. 140, col. a.]

43. Instead of *majora* the text has *minora.*

44. [Francisci de Mayronis *In Primum Sententiarum,* fol. 140, col. a.]

45. [Ibid.]

46. [Ibid., col. b.]

47. [Ibid., fol. 139, col. d.]

48. [Ibid.]

49. [Ibid, fol. 140, col. a.]

50. [Ibid., fol. 139, col. d; fol. 140, col. a.]

51. [Ibid., fol. 140, col. b.]

52. Instead of *finitam* the text has *infinitam.*

53. [Francisci de Mayronis *In Primum Sententiarum,* fol. 140, col. b.]

54. [Ibid., col. a.]

55. [Ibid., col. b.]

56. *Opera* Joannis de Bassolis . . . *In Quatuor Sententiarum libros . . . Quaestiones in Primum Sententiarum,* dist. XLIII, quaest. unica, fol. 211, cols. b, c.

57. Ibid., fol. 210, col. d.

58. Ibid., fol. 211, col. d.

59. Ibid., fol. 212, col. d.

60. Ibid.

61. Ibid., col. c.

62. Ibid., fol. 213, col. b.

63. Ibid., col. c.

64. Ibid., fol. 212, col. c.

65. Ibid., fol. 213, col. c.

66. Ibid.

67. Magistri Roberti Holkot *Super quatuor libros Sententiarum questiones,* libro secundi, quaest. II: An Deus potuit producere mundum ab aeterno.

68. [Ibid., 26th col. from the end of quest. II.]
69. [Ibid., 25th col. from the end of quest. II.]
70. [Ibid., 26th col. from the end of quest. II.]
71. [Ibid., 22d col. from the end of quest. II.]
72. [Ibid.]
73. [Ibid., 26th col. from the end of quest. II.]
74. [Ibid., 23d col. from the end of quest. II.]
75. [Ibid., 21st col. from the end of quest. II.]
76. Thomae ab Argentina *Eremitarum Divi Augustini Prioris generalis Commentaria in quatuor libros Sententiarum*, lib. II, dist. I, art. 4: Utrum Deus sua virtuta infinita aliquod infinitum producere possit.
77. Ibid., art. 2: Utrum mundus ab aeterno esse potuerit.
78. Fratris Nicolai Boneti *Physica*, lib. VII, cap. II, fol. 167r [fol. 132, col. b].
79. [Ibid.]
80. Ibid., lib. II, cap. IX, fol. 124r,v [fol. 95, cols. b, c].
81. [Ibid., fol. 95, cols. c, d.]
82. Nicolai Boneti *Theologia Naturalis*, lib. II, cap. I, fol. 192, cols. c, d; fol. 193, col. a.
83. Ibid., lib. VII, cap. XI, fol. 287, cols. a, b.
84. Ibid., col. d.
85. Ibid., cap. XII, fol. 288, cols. a, b.
86. Gregorius de Arimino, *In primum Sententiarum*, lib. I, dists. XLII, XLIII, XLIV, quaest. IV: Utrum Deus per infinitam suam potentiam posset producere effectum aliquem actu infinitum, art. 2; Gregorius de Arimino *In Secundum sententiarum*, dist. I, quaest. III: Utrum per aliquam potentiam fuerit possibile aliquam rem aliam a Deo fuisse ab aeterno, art. 2.
87. Gregorius de Arimino *In primum Sententiarum*, lib. I, dists. XLII, XLIII, XLIV, quaest. IV, art. 2, fol. 179, col. b. [fol. 173, col. d].
88. [Ibid.]
89. Ibid., [fol. 173, col. d; fol. 174, col. a].
90. Ibid., fol. 179, col. d [fol. 174, col. b].
91. Ibid., fol. 177, col. d [fol. 172, col. c].
92. Gregorius de Arimino *In secundum Sententiarum*, dist. I, quaest. III, art. 2, fol. 13, col. d [fol. 14, cols. b, c].
93. Gregorius de Arimino *In primum Sententiarum*, dists. XLII, XLIII, XLIV, quaest. IV, art. 2, fol. 179, col. a [fol. 173, col. c].
94. Gregorius de Arimino *In secundum Sententiarum*, dist. I, quaest. III, art. 2, fol. 13, col. d; fol. 14, col. a [fol. 14, col. c].
95. Ibid., fol. 14, col. b [fol. 14, col. d].
96. Gregorius de Arimino *In primum Sententiarum*, dists. XLII, XLIII, XLIV, quaest. IV, art. 2, fol. 178, col. d [fol. 172, col. a].
97. Ibid., art. 1, fol. 177, col. c [fol. 173, col. b].
98. Ibid., art. 2, fol. 190, col. c [fol. 175, col. a].
99. [Ibid.]
100. Ibid.
101. Ibid., fol. 177, col. c [fol. 173, col. b].
102. Ibid., fol. 181, col. a [fol. 175, col. b].
103. Ibid., fol. 180, col. d [fol. 175, col. b].
104. Ibid., fol. 181, col. c [fol. 175, col. d].

105. Ibid., [fol. 175, col. d].

106. Gregorius de Arimino *In secundum Sententiarum*, dist. II, quaest. II: Utrum angelus sit in loco divisibili aut indivisibili, art. 1: An magnitudo componatur ex indivisibilus, fol. 35, cols. c, d [fol. 37, cols. c, d].

107. Gregorius de Arimino, *In primum Sententiarum*, dists. XLII, XLIII, XLIV, quaest. IV, art. 2, fol. 179, col. a [fol. 173, col. c].

108. Johannis Buridani *Quaestiones super octo Physicorum libros Aristotelis*, lib. III, quaests. XIV, XV, XVI, XVII, XVIII, XIX.

109. Ibid., quaest. XVI: Utrum aliqua linea gyrativa sit infinita?

110. Ibid., fol. 58, col. d.

111. Ibid., fol. 59, col. b.

112. Ibid., cols. b, c.

113. Ibid., col. c.

114. Ibid.

115. Ibid., quaest. XVIII, fol. 61, col. b.

116. Ibid., fol. 62, col. b.

117. Ibid., quaest. XIX, fol. 65, col. a.

118. Ibid., col. b.

119. Alberti de Saxonia *Quaestiones super libros de Physica auscultatione Aristotelis*, lib. III, quaest. XIII [quaest. XII, fol. 39, col. c].

120. [Ibid., quaest. XI, fol. 39, col. b.]

121. Johannis Buridani *Quaestiones super libris De Caelo et Mundo*, lib. I, quaest. XVI: Utrum possibile est esse unum corpus infinitum, fol. 77, col. d [pp. 79-80].

122. [Ibid., p. 80.]

123. [Ibid.]

124. Louis Couturat, *De l'infini mathématique*, vol. 2, chap. 2: Du nombre infini concret.

125. Jules Tannery, *Introduction à la théorie des fonctions d'une variable*, p. VIII.

126. Baire, *Bulletin de la Société Mathématique de France* (1905), p. 263.

127. Nicole Oresme, *Traité du Ciel et du Monde*, livre I, fol. 11, col. d [pp. 109-11]; also Nicole Oresme, *Tractatus de difformitate qualitatum*, pars III, cap. 13, fols. 265v, 266r.

128. Oresme, *Traité du Ciel et du Monde*, livre I, fol. 13, cols. b, c [pp. 119-21]; also Oresme, *Tractatus de difformitate qualitatum*, pars III, cap. 12, fol. 265r.

129. Oresme, *Traité du Ciel et du Monde*, lib. I, fol. 14, col. b [p. 125].

130. [Ibid., p. 127.]

131. *Abbreviationes libri Physicorum edite a prestantissimo philosopho* Marsilii Inguen, fol. sign. d, col. d; fol. sign. d2, cols. a, b.

132. [Ibid.]

133. [Ibid.]

134. *Quaestiones subtilissimae* Johannis Marcilii Inguen *super octo libros Physicorum secundum nominalium viam*, lib. III, quaests. IX, X.

135. Ibid., quaest. IX.

136. Ibid.

137. [Ibid.]

138. [Ibid.]

Chapter 3

1. Pauli Veneti *Expositio super octo libros Physicorum*, lib. III, tract. II, cap. IV, fol. sign. r ij, col. d.
2. Ibid., fol. sign. r iij, col. a.
3. [Ibid.]
4. Ibid., pars I, 3d folio after fol. sign. r iij, col. c.
5. [Ibid.]
6. Pauli Veneti *Summa Totius Philosophiae*, pars secunda, cap. VI [fol. 26, col. a].
7. [Ibid., cols. a, b.]
8. [Ibid., col. b.]
9. Ibid., cap. VII [fol. 26, col. b].
10. Ibid., cap. XIII [fol. 29, col. a].

Chapter 4

1. Bernard Carra de Vaux, *Avicenne*, p. 85.
2. Ibid., p. 191.
3. [Ibid.]
4. Averrois Cordubensis *Commentaria magna in octo libros Aristotelis de physico auditu*, lib. IV, summa prima: De loco, cap. IX, comm. 45.
5. Ibid., comm. 43.
6. Ibid., cap. VIII, comm. 41.
7. Ibid., cap. IX, comm. 43; Averrois Cordubensis *Commentarii in quatuor libros Aristotelis de Caelo et Mundo*, lib. II, summa secunda, quaest. II, comm. 17.
8. Averrois *Commentarii in quatuor libros de Caelo et Mundo*, lib. II, summa secunda, quaest. II, comm. 17.
9. Ibid., quaest. V, comm. 35.
10. Averrois Cordubensis *Expositio in XIIII Metaphysicae Aristotelis*, lib. XII, summa secunda, cap. IV, comm. 45.
11. Averrois *Commentaria magna in octo libros Aristotelis de physico auditu*, lib. IV, summa prima, cap. IX, comm. 43.
12. Ibid., comm. 45.
13. *Questiones supra librum Phisicorum a magistro dicto* Bacuun, fol. 46, col. d to fol. 47, col. b. [XIII, pp. 216-22. Quaest. I: It is asked about the place of heaven, but first it is asked whether heaven has a place. Quaest. II: It is then asked whether heaven has a place *per se* or *per accidens*. Quaest. III: Given that heaven has a place *per accidens*, it is asked in what way this place *per accidens* can be reduced to place *per se*. Quaest. IV: It is asked whether heaven has some kind of place *in quo*. Quaest. V: It is asked whether the whole universe has a place. Quaest. VI: It is asked about the orbs of the planets and of the place of these orbs, whether they, namely the planets, have a place *per se* or *per accidens*.]
14. Ibid., quaest. I, fol. 46, col. d [XIII, p. 216].
15. Ibid., quaest. II, fol. 46, col. d [XIII, p. 217].
16. Ibid.
17. [Ibid., p. 218.]

18. Ibid.
19. [Ibid.]
20. [Ibid.]
21. [Ibid.]
22. Ibid., fol. 47, col. a [XIII, p. 218].
23. [Ibid., p. 219.]
24. [Ibid., pp. 218-19.]
25. [Ibid. pp. 219-20.]
26. Ibid., quaest. V, fol. 47, col. b [XIII, p. 221].
27. Ibid., quaest. VI, fol. 47, col. b [XIII, pp. 221-22].
28. [Ibid., XIII, p. 222.]
29. Albertus Magnus composed an opusculum under the title, *Liber de natura loci ex longitudine et latidudine ejusdem proveniente*, in which there is nothing concerning the general theory of place and movement, but only some interesting developments concerning the theory of natural place.
30. Beati Alberti Magni, Ratisbonensis episcopi, Ordinis Praedicamentorum, *Parva naturalium . . . Liber de motibus progressivis*, tract I: De modo motus progressivi, caps. II, III, IV, pp. 509-11.
31. Beati Alberti Magni *Parva naturalia . . . liber secundus de motibus animalium*, tract. I: De ipsis motibus et proprietatibus ipsorum, cap. III, p. 125.
32. *Expositio super librum de motibus animalium secundum* Petrum de Alvernia, lect. I, fol. 35, cols. b, c [fol. 47, col. c].
33. Alberti Magni *De Caelo et Mundo*, liber secundus, tract. I, cap. VIII: De causa finali propter quod caelorum motus oportet esse plures [Westfalorum Monastery ed., p. 123].
34. Alberti Magni *Physicorum*, lib. IV, tract. I, caps. XI, XII.
35. [Ibid.]
36. [Ibid.]
37. [Ibid.]
38. Sancti Thomae Aquinatis *Expositio super libros De Caelo et Mundo*, lib. II, lect. IV.
39. Sancti Thomae Aquinatis *In libros Physicorum Aristotelis expositio*, lib. IV, lect. VII [p. 216].
40. [Ibid., pp. 216-17.]
41. Divi Roberti Lincolniensis *Super octo libris Physicorum brevis et utilitis summa*, lib. IV [fol. 281, col. a].
42. Sancti Thomae Aquinatis *In libros Physicorum Aristotelis*, lib. IV, lect. VI [pp. 210-11].
43. [Ibid., p. 213.]
44. [Ibid.]
45. Sancti Thomae Aquinatis *Opuscula*, opusc. LII: De natura loci. Christi sub specie panis et vini realiter contineri? [Marietti ed., VIII, p. 501].
46. P. Mandonnet, "Les écrits authentiques de Saint Thomas d'Aquin," *Revue Thomiste* (1909-10).
47. [Aquinas, *De natura loci.*]
48. A clear and concise treatise on logic entitled *Totius logicae Aristotelis summa* (opusc. XLVIII) is attributed to Thomas Aquinas. This *Summa* treats in succession Predicables or Universals, Predicaments or Categories, Syllogism, and finally Demonstration. The author refers to place and movement in the second part, devoted to the Categories. We do not report what he says here; not only has the apocryphal character of the *Summa*, recognized by Prantl (*Geschichte der Logic*

in Abendlende [Leipsig, 1867], p. 250), been formally established by P. Mandonnet, but the theories expounded there have only a remote kinship with the authentic doctrines of Saint Thomas—they carry the recognizable influence that the debates caused by the teaching of Duns Scotus would have exercised. (See P. Duhem, *Le mouvement absolu et le mouvement relatif; idem,* "Note: Sur une Somme de Logique attribuée à Saint Thomas d'Aquin," *Revue de Philosophie* 8, no. 5 [May 1909]).

49. Sancti Thomae Aquinatis *In libros Physicorum Aristotelis,* lib. V, lect. VI.

50. Egidii Romani *In libros de Physico auditu Aristotelis commentaria accuratissime emendata,* lib. IV, lect. VII, text. comm. 41, dubitatio 2, fol. 72r [fol. 81, col. a].

51. [Ibid., fol. 81, col. b.]

52. Ibid., lect. VIII, text. comm. 46, dubitatio 2, fol. 73v [fol. 82, col. d].

53. Ibid., lect. VII, text. comm. 41, dubitatio 2, fol. 72v [fol. 81, col. b].

54. Ibid., lect. VIII, text. comm. 46, dubitatio 4, fol. 74r [fol. 83, col. a].

55. [Ibid., col. b.]

56. Ibid., dubitatio 2, fol. 74r [fol. 82, col. d].

57. [Ibid.]

58. [Ibid.]

59. *Incipiunt preclarissime quaestiones litterales edite a fratre* Gratia Deo Esculano sacri ordinis predicatorum *super libros Aristo. de Physico auditu: secundum ordimem lectionum Divi Thome Aquinatis* lib. IV, lect. I, quaest. II, fol. 39, col. b.

60. Ibid., lect. VI, quaest. II, fol. 43, col. b.

61. Ibid., fol. 42, col. d; fol. 43, col. a.

62. Ibid.

63. Ibid., fol. 43, col. b.

64. [Ibid.]

65. Ibid., quaest. V, fol. 44, col. b.

66. Ibid., quaest. IV, fol. 43v; fol. 44, col. a.

67. [Ibid.]

68. Ibid., quaest. III, fol. 43, col. c.

69. Fratris Rogeri Bacon *Communia naturalium,* liber primus, partis tertiae, dist. 2a: De loco et vacuo, habens capitula octo, capitulum primum est de distinctione modorum loci, fol. 52, col. a to fol. 54, col. a [pp. 182-89].

70. Ibid., fol. 52, cols. a, b [p. 183].

71. [Ibid., p. 184.]

72. Ibid., fol. 53, col. a [p. 185].

73. Ibid., cols. a, b [p. 186].

74. [Ibid., p. 187.]

75. [Ibid., pp. 187-88.]

76. Ibid., col b [p. 187].

77. [Ibid.]

78. Ibid., col. c [p. 188].

79. Ibid., cap. 4m: De vacuo quantum ad ejus necessitatem propter locata et propter motum augmenti et nutrimenti, et propter motum localem, fol. 59, col. d [p. 209].

80. See *Le Système du monde,* 1:341-42.

81. Celebratissimi Patris Domini Bonaventurae Doctoris Seraphici *In secundum librum Sententiarum disputata,* secunda pars, libri secundi, distinctionis XIV, pars quarta, quaest. III: Utrum conveniat alicui orbi moveri absque stellis?

82. *Theorica planetarum* Campani, the second chapter after the Proemium [p. 183].

83. [Ibid.]

84. Divi Roberti Lincolniensis *Summa*, cap. 216, pp. 546-47.

85. Joannis Duns Scoti *Quaestiones quodlibetales*, quaest. XI [XII, p. 263].

86. Joannis Canonici *Quaestiones super VIII libros Physicorum Aristotelis perutiles*, vol. IV, quaest. II [fol. 51, col. a].

87. Alberti de Saxonia *Quaestiones in libros de Caelo et Mundo*, lib. II, quaest. VIII [fol. 107, cols. a, b].

88. Reverendissimi Domini Petri de Aliaco, Cardinalis et Episcopi Cameracensis, Doctorisque celebratissimi, *Quatuordecim quaestiones in Sphaeram Joannis de Sacro Bosco*, quaest. II: Utrum sint praecise 9 sphareae caeleste et non plures nec pauciores [fol. 118, col. b].

Chapter 5

1. H. Denifle and A. Chatelain, *Chartularium Universitatis Parisiensis*, vol. 1 (1200-85), piece no. 473, March 7, 1277.

2. *Clarissimi theologi* Magistri Ricardi de Mediavilla Seraphici ord. min. convent. *Super quatuor libros Sententiarum Petri Lombardi quaestiones subtilissimae*, tomus secundus, lib. II, dist. XIV, art. 3, quaest. III: Utrum Deus potest movere ultimum caelum motu recto [p. 186 col. a].

3. [Ibid.]

4. R. F. P. Joannis Duns Scoti, Doctoris Subtilis, Ordinis Minorum, *Quaestiones Quodlibetales*, quaest. XI: Utrum Deus possit facere quod, manente corpore et loco, corpus non habeat *ubi* sive *esse*, in loco? [XII, pp. 266-67].

5. Ibid., [p. 262].

6. Duns Scoti, Doctoris Subtilis, Ordinis Minorum, *Quaestiones in librum IV Sententiarum*, dist. X, quaest. II: Utrum idem corpus possit esse localiter simul in diversis locis? [VIII, p. 513.]

7. Ibid., quaest. I.

8. Duns Scotus, *Quaestiones Quodlibetales*, quaest. XI [XII, pp. 266-67].

9. Duns Scoti, Doctoris Subtilis, Ordinis Minorum, *Quaestiones in librum II Sententiarum*, dist. II, quaest. VI: An locus angeli sit determinatus, punctualis, maximus et minimus? [VI, pt. 1, p. 193.]

10. Duns Scoti, Doctoris Subtilis, Ordinis Minorum, *Quaestiones in librum IV Sententiarum*, dist. X, quaest. I: Utrum possibile sit corpus Christi sub specie panis et vini realiter contineri? [VIII, p. 501].

11. Duns Scoti *Quaestiones quodlibetales*, quaest. XI [XII, p. 263].

12. [Ibid.]

13. The questions of John of Jandun on this treatise are to be found among questions on the *Parva Naturalia* which have had no other edition than the following: Ioan. Gandavensis *Philosophi acutissimi Quaestiones, Super Parvis Naturalibus, cum Marci Antonii Zimarae De Movente et Moto, ad Aristotelis et Averrois intentionem, absolutissima quaestione, ac variis margineis scholiis hinc inde ornatae* (1557).

14. Joannis de Janduno *Quaestiones super octo libros Aristotelis de Physico auditu*, lib. IV, quaest. IV: An locus sit ultimum continentis?

15. Ibid., quaest. V: Locus in quo nam genere sit?

16. Ibid., quaest. VI: An locus sit immobilis?

17. Ibid., quaest. IX: An ultima sphaera sit in loco? [Fol. 59, col. a.]

18. [Ibid.]

19. Joannis de Janduno *Quaestiones super parvis naturalibus, quaestiones de motibus animalium*, quaest. V: Num in motus progressivo ipsius animalis requiratur extra ipsum aliquod fixum? [P. 114, col. b.]

20. Ibid., quaest. VI: Num Caelum in motu suo indigeat aliquo corpore quiescente? Quaest. X: Utrum inanimata requirunt aliquod fixum in motu locali?

21. Joannis de Janduno *Quaestiones super octo libros Aristotelis de Physica auditu*, lib. IV, quaest. IX: Utrum ultima sphaera sit in loco? Joannis de Janduno *Quaestiones de Caelo et Mundo*, lib. II, quaest. VI: An Terra propter Caeli motum necessaria sit?

22. Joannis de Janduno *Quaestiones de motibus animalium*, quaest. X: Utrum inanimata requirunt aliquod fixum in motu locali? [P. 122, col. b.]

23. Ibid., quaest. VI: Num Caelum in motu suo indigeat aliquo corpore quiescente? [P. 115, cols. a, b.]

24. [Ibid., p. 115, col. b.]

25. Ibid., quaest. VII: Utrum fixio Caeli sit causaliter ex fixione Terrae? [P. 116, col. b; p. 117, col. a.]

26. Burleus *Super octo libros Physicorum*, lib. IV, tract. I, cap. VI, fol. 92, col. d.

27. Joannis de Janduno, *Quaestiones de motibus animalium* [p. 117, col. b].

28. Ibid. and *Quaestiones super octo libros Aristotelis de Physica auditu*, lib. IV, quaest. IX.

29. [Joannis de Janduno, *Quaestiones de motibus animalium*, p. 117, col. a.]

30. [Ibid.]

31. Aristotle *Metereologica* I, c. 2.

32. Joannis de Janduno *Quaestiones in libros de Caelo et Mundo*, lib. II, quaest. IV: An Terra propter Caeli motum necessaria sit?

33. Joannis de Janduno *Quaestiones super octo libros Aristotelis de Physica auditu*, lib. IV, quaest. IX: An ultima sphaera sit in loco?

34. Joannis de Janduno *Quaestiones de motibus animalium*, quaest. VI: Num Caelum in motu suo indigeat aliquo corpore quiescente.

35. Ibid., quaest. IX: Utrum Caeli motor sit majoris virtutis in movendo, quam Terra in quiescendo?

36. [Ibid., p. 123, col. a.]

37. Petri Aureoli Verberii Ordinis Minorum Archiepiscopi Aquensis S. R. E. Cardinalis. *Commentarium in secundum librum Sententiarum*, tomus secundus, lib. II, dist. II, quaest. III: De loco angelorum, art. 1: Utrum locus sit superficies corporis continentis immobiles primum, pp. 49 et seq.

38. We ought to note that in various passages Giles of Rome expresses himself, carelessly no doubt, as if the formal place were an attribute of the contained thing, instead of the container.

39. Joannis Canonici *Quaestiones super VIII libros Physicorum*, lib. IV, quaest. I.

40. Ibid.

41. Ibid., quaest. II.

42. Ibid., quaest. I.

43. Ibid., quaest. I (his visis . . .) and quaest. II (sed hic sunt duo dubia; primum . . .).

44. *Secundus liber Sententiarum Magistri* Francisci de Marchia, dist. III, quaest. IV: Utrum primum mobile sive ultima spera sit in loco, fol. 107, col. c to fol. 108, col. a.

45. Ibid., fol. 107, cols. c, d.

46. Ibid., col. d.

47. [Ibid.]

48. Ibid.

49. Joannis Canonici *Quaestiones super VIII libros Physicorum*, lib. IV, quaest. II.

50. Ibid., quaest. I.

51. [Ibid., fol. 50, col. c.]

52. *Questiones* Scoti *Super Universalia Porphyrii; necnon Aristotelis predicamenta ac Peryarmenias. Item super libros Elenchorum.* Et Antonio Andree *Super libros sex principiorum. Item questiones* Joannis Angelici (sic) *super questiones universales ejusdem Scoti,* Venetiis (1512).

53. *Questiones clarissimi doctoris* Antonii Andree *super sex principiis Gilberti Porretani,* quaest. VIII: Utrum caelum sit in loco, fol. 60, col. d [MDCXXII ed., quaest. XIV, p. 280, col. b to p. 281, col. a].

54. Ibid., quaests. XII, XIII, XIV, fol. 80.

55. Ant. Andree, *Conventualis Franciscani, ex Aragoniae provincia ac Ioannis Scoti doctoris subtilis discipuli celeberrimi, in quatuor Sententiarum Libros opus longe absolutissimum,* lib. II, dist. II, quaest. V: Utrum angelus sit in loco, fol. 53, col. c.

56. Ibid.

57. [Ibid., cols. c, d.]

58. *Profundissimi Sacre theologie professoris* F. Joannis de Bassolis *minorite in secundum sententiarum Questiones ingeniosissime et sane quam utiles,* dist. II, quaest. III, art. 4 [fol. 39, cols. c, d].

59. [Ibid., fol. 39, col. d.]

60. [Ibid.]

61. Bibliothèque Nationale, fonds latin, ms. no. 16130, fol. 131, col. b, from: "Quia communis opinio est quod motus, tempus, et locus quedam res alie a mobilis et locato . . ."; fol. 140, col. c: " . . . et sic contingit successive verificari contradictoria." Explicit *Tractatus de successivis editus a* Guillelmo Ockam. Someone else from the fourteenth century added Okam [p. 32].

62. Ibid. This part begins at fol. 134, col. d, as follows: "Consequenter videndum est de loco quod Philosophus 4 Physicorum diffinit sic: Locus est ultimum corporis continentis contigui immobile." It ends at fol. 137, col. d, thus: "Et haec dicta de loco et ejus definitione sufficiant gratia veritatis, et que modo minus diffuse dicta sunt, alias, si necesse fuerit, diffusius tractabuntur." It makes up chapters 20, 21, and 22 of the fourth part of the *Summulae in libros Physicorum* [p. 104, col. a to p. 112, col. b].

63. Gulielmi de Ockam *Summulae in libros Physicorum,* lib. IV, c. 20, 21.

64. *Quodlibeta septem* Venerabilis Inceptoris Fratris Guilhelmi de Ockham, quodlib. I, quaest. IV [Franciscan Institute ed., pp. 23-28].

65. Gulielmi de Ockam *Summulae in libros Physicorum,* lib. IV, c. 22 [p. 109, col. b].

66. Ibid.

67. [Ibid., p. 109, cols. a, b.]

68. [Ibid., p. 110, col. b; p. 111, col. a.]

69. [Ibid., p. 111, col. b.]

70. *Questiones magistri* Guilelmi de Ockam *super librum Phisicorum*, quaest. LXXVII: Utrum sit idem locus numero corporis continue quiescentis quando corpus circumstans continue movetur circa illud, fol. 14, col. c.

71. Ibid., quaest. LXXVIII: Utrum locus sit immobilis.

72. [Ibid.]

73. [Ibid.]

74. Guilhelmi de Ockham *Quodlibeta septem*, quodlib. VII, quaest. XI [Franciscan Institute ed., pp. 738-45].

75. [Ibid.]

76. [Ibid.]

77. Ibid., quaest. XIII [pp. 749-52].

78. [Ibid.]

79. *Questiones* magistri Guilelmi de Ockam *super librum Phisicorum*, quaest. LXXX: Utrum octava spera moveatur per se, fol. 14, col. d.

80. [Ibid.]

81. Ms.: *tunc.*

82. Ms.: *talis.*

83. Ms.: *partis.*

84. Ms.: illegible word.

85. Ms.: *anexas.*

86. Ms.: *acquirit.*

87. Ms.: *illas.*

88. Ms.: *ut.*

89. Ms.: *si.*

90. [Ockham, *Super libros Physicorum.*]

91. Burleus *Super octo libros Physicorum.* The pages have no pagination, although the *tabula dubiorum* indicates one. The theory of place forms the *Tractatus primus quarti libri*, whose seven chapters occupy fols. 76v-95r.

92. Burleus *Super octo libros Physicorum*, lib. IV, tract. I, cap. V, fols. 86v, 87r.

93. Ibid., fol. 87, col. d.

94. Ibid., fol. 88v; fol. 89, col. a.

95. [Ibid., fol. 88, col. c.]

96. This proposition is not correct; Burley must have supposed that the movement of the universe as a whole is oriented northward or southward.

97. [Burleus *Super octo libros Physicorum*, lib. IV, tract. I, cap. V, fol. 88, col. d.]

98. Ibid., fol. 87, col. c; fol 89, col. a.

99. Ibid., fol. 88, col. a.

100. Ibid., fol. 89r.

101. Ibid., fol. 89, cols. b, c.

102. Ibid., fol. 88, col. d.

103. Ibid., fol. 89, col. c.

104. [Ibid.]

105. Ibid., lib. I, tract. I, cap. VI, fol. 91, col. b.

106. Ibid., lib. IV, tract. I, cap. V, fol. 89, col. d.

107. Ibid., cap VI, fol. 91, col. b.

108. Ibid., fol. 92, col. d.

109. [Ibid.]

110. [Ibid.]

111. [Ibid.]
112. Ibid., fol. 79, col. c.
113. Nicolai Boneti *Physica*, lib. VIII, cap. X, fol. 143, cols. a, b.
114. Ibid., cap. XI, fol. 143, cols. b, c, d.
115. Ibid., cap. XII, fol. 143, col. d; fol. 144, cols. a, b, c, d.
116. [Ibid., fol. 144, col. a.]
117. [Ibid.]
118. [Ibid., fol. 143, col. d.]
119. [Should be "the concavity of the sphere of air."]
120. [Boneti *Physica*, lib. VIII, cap. XII, fol. 144, cols. a, b, c, d.]
121. [Ibid., col. d.]
122. Ibid., cap. XIII, fol. 144, col. d; fol. 145, col. a.
123. [Ibid.]
124. Let us recall that *objective* takes, according to Scholastic terminology, the meaning that the word *subjective* takes in modern philosophical usage.
125. Bonet restricts his thesis to natural means because in the consecrated host, accidents, separated from the substance which carried them, persist, even when the substance ceases to exist.
126. Nicolai Boneti *Metaphysica*, lib. VIII, cap. VI, fol. 76, cols. c, d.
127. Johannis Buridani *Subtilissime quaestiones super octo Physicorum libros*, lib. IV, quaest. I: Utrum sit aequalis suo locato, fol. 67, col. a.
128. Ibid., quaest. II: Utrum locus sit terminus corporis continentis, fol. 68, cols. b, c.
129. Ibid., quaest. IV: Utrum diffinitio loci sit bona, in qua dicitur: locus est ultimum corporis continentis immobile primum.
130. Ibid., quaest. III: Utrum locus sit immobilis.
131. [Ibid.]
132. [Ibid.]
133. [Ibid.]
134. Ibid., fol. 69, col. b.
135. Ibid., fol. 69, col. d; fol. 70, col. a.
136. Ibid., quaest. VI: Utrum ultima sphaera seu suprema sit in loco.
137. Ibid., fol. 72, cols. b, c.
138. Ibid., lib. III, quaest. VII: Utrum motus localis est res distincta a loco et ab eo quod localiter movetur.
139. Ibid., fol. 50, col. c.
140. *Quaestiones super libris De Caelo et Mundo magistri* Johannis Buridani, rectoris Parisius, lib. I, quaest. XV: Utrum possible est corpus recte motum esse infinitum, fol. 77, col. c.
141. Buridani *Quaestiones super octo Physicorum libros*, lib. III, quaest. VII, fol. 50, col. c.
142. Ibid., col. d.
143. Buridani *Quaestiones super libris De Caelo et Mundo*, lib. II, quaest. VI: Utrum sit ponendum coelum quiescens supra coelos motos, fol. 86, col. b.
144. [Ibid.]
145. [This conclusion does not follow. There is a problem with the text here. The next three paragraphs, which I have deleted, do not belong in this discussion. Someone has inserted materials belonging to a discussion of John Buridan II's *Questions on the Meterology* into this discussion. The latter material is inconsistent with the claims made by Buridan.]

146. Alberti de Saxonia *Quaestiones super libros de Physica auscultatione Aristotelis*, lib. IV, quaest. I: Utrum locus sit superficies? [Fol. 43, col. c.]

147. Ibid., lib. I, quaest. VI: Utrum omnis res extensa sit quantitas? [Fol. 5, col. b.]

148. Ibid., lib. IV, quaest. I [fol. 43, col. c].

149. Ibid., lib. IV, quaest. III: Utrum locus sit immobilis?

150. [Ibid, fol. 44, col. d.]

151. [Ibid.]

152. [Ibid.]

153. [Ibid., fol. 45, col. a.]

154. [Ibid.]

155. Ibid., lib. IV, quaest. VII: Utrum omne ens sit in loco? [Fol. 47, col. d.]

156. *Quaestiones subtilissimae* Alberti de Saxonia *in libros De Caelo et Mundo*, lib. I, quaest. I: Utrum cuilibet corpori simplici insit naturaliter tantum unus motus simplex?

157. Alberti de Saxonia *Quaestiones super libros de Physica auscultatione*, lib. IV, quaest. VII.

158. *Logica* Albertucili. *Perutilis logica* excellentissimi sacre theologie professoris magistri Alberti de Saxonia ordinis eremitarum Divi Augustini: *per reverendum sacre pagine doctorem magistrum Petrum Aurelium Sanutum Venetum ejusdem ordinis professum: quam diligentissime castigata: nuperrimeque impressa,* tract. primi, cap. XXV: De predicamento quando et aliis sex predicamentis, fol. 10, col. d.

159. Alberti de Saxonia *Quaestiones super libros de Physica auscultatione*, lib. IV, quaest. VIII [quaest. VII, fol. 48, col. a].

160. [Ibid.]

161. Ibid., lib. IV, quaest. VII [fol. 47, col. a]; Alberti de Saxonia *Quaestiones in libros De Caelo et Mundo*, lib. I, quaest. I; lib. II, quaest. VIII: Utrum omne caelum sit mobile?

162. Alberti de Saxonia *Quaestiones super libros de Physica auscultatione*, lib. IV, quaest. VII [fol. 47, col. d].

163. Ibid.

164. Alberti de Saxonia *Quaestiones in libros De Caelo et Mundo*, lib. II, quaest. X: Utrum illa consequentia sit bona: Caelum movetur, ergo necesse est Terram quiescere?

165. Ibid., lib. IV, quaest. X. Cf. ibid., quaest. VII.

166. Ibid., quaest. X.

167. [Ibid.]

168. [Ibid.]

169. [Ibid.]

170. [Ibid.]

171. Alberti de Saxonia *Tractatus proportionum*, Du motu circulari, 7a conclusio [p. 70].

172. Marsilii Inguen *Abbreviationes libri Physicorum*, fol. d3, cols. c, d.

173. Johannis Marcilii Inguen *Quaestiones super octo libros Physicorum*, lib. IV, quaest. III: Utrum locus sit ultima superficies corporis continentis?

174. Ibid., lib. IV, quaest. III.

175. Ibid., quaest. VII: Utrum omne ens sit in loco?

176. We consulted these *Questions* in the following text: *Questiones super tres primos libros Metheororum et super majorem partem quarti a Magistro* Jo. Buridan,

Bibliothèque Nationale, fonds latin, ms. no. 14723 (Ancien fonds Saint-Victor, ms. no. 712).

177. Ibid., lib. I, quaest. XXI, fol. 202, col. b. [Should be quaest. XX: Quaeritur consequenter 20 de permutatione marium ad aridam et econverso, fol. 200, col. c. to fol. 202, col. b.]

178. Ibid., fol. 204, col. a. [Quaest. XXI: Consequenter quaeritur 21 et ultimo circa primum metheororum, utrum possible est naturaliter tantos montes quanti maximi apparent nobis destrui, et reverti ibi terra ad planitiem, fol. 202, col. c to fol. 204, col. a.]

179. Joannis Canonici *Quaestiones super VIII libros Physicorum*, lib. IV, quaest. I.

180. [Ibid., fol. 50, col. b.]

181. Francisci de Mayronis *Scripta in quatuor Sententiarum*, lib. II, dist. XIV, quaest. IX (in the 1521) Venice edition, this question occupies fol. 151, col. d to fol. 152, col. 2; the question is not numbered, so that quaest. X is quaest. IX) [fol. 18, col. b].

182. [Ibid., fol. 18, col. c.]

183. Nicolai Boneti *Physica*, lib. VIII, cap. II [fol. 138, cols. c, d; fol. 139, col. a].

184. [Ibid., fol. 138, col. c.]

185. [Ibid., col. d.]

186. [Ibid., fol. 138, col. d; fol. 139, col. a.]

187. Ibid., cap. VII [fol. 141, col. c].

188. Ibid., cap. VIII [fol. 142, cols. a, b].

189. Ibid., cap. VI, [fol. 141, cols. a, b].

190. [Ibid., fol. 141, col. b.]

191. Francisci de Mayronis *Scripta in quatuor Sententiarum*, lib. II, dist. XIV, quaest. IX [fol. 18, col. c].

192. [Ibid., fol. 18, col. b.]

193. Nicole Oresme, *Traité du Ciel et du Monde*, livre I, c. XXIV, fol. 21, col. d; fol 22, col. a [p. 177].

194. Ibid., liv. IV: "Apres sunt trois chapitres du translateur, et sunt comment les chose dehors ce monde sunt en lieu, et comme elles sont meues, et est le premier chapitre des chose incorporelles, et le disieme chapitre," fol. 120, col. c [pp. 721-23].

195. Ibid., le XIe chapitre est quant a ce des choses corporelles, fol. 121, col. a [p. 725].

196. Ibid., liv. II, c. VIII, fol. 56, col. a to fol. 57, col. c [pp. 363-73].

197. See *Le Système du monde* 4:229.

198. See P. Duhem, *Le Mouvement absolu et le mouvement relatif*, conclusion (*Revue de Philosophie*).

Chapter 6

1. Nicolai de Orbellis *Physicorum*, in *Curcus librorum philosophiae naturalis*, lib. IV, cap. I [fol. 94, cols. b, c].

2. [Ibid., fol. 95, col. a.]

3. [Ibid., col. b.]

4. Magistri Georgii Bruxellensis *Physicorum*, in *Cursus optimarum questionum super Philosophiam Aristotelis*, lib. IV, quaeritur utrum diffinitio loci data a Philosopho sit bene assignata, fol. sign. Ff 3, col. d; fol. following, cols. a, b, c [fol. 72, col. d to fol. 74, col. b].

5. Ibid., quaeritur utrum ultima sphaera sit in loco, fol. following fol. sign. Ff 3, col. d [fol. 74, col. a to fol. 75, col. c].

6. Nicolai de Orbellis *Physicorum*, lib. IV, cap. I [fol. 95, col. b].

7. Georgii Bruxellensis *De Caelo et Mundo*, in *Curcus optimarum questionum super Philosophiam Aristotelis*, lib. II, dubitatur utrum sint octo sphaerae celestes, fol. sign. p, col. c, d [fol. 173, col. c].

8. [Ibid.]

9. Georgii Bruxellensis *Physicorum*, lib. IV, quaeritur utrum ultima sphaera sit in loco, fol. following fol. sign. Ff 3, col. d [fol. 73, cols. b, c].

10. George of Brussels also says in another place: "Every mobile heaven has a place by which it is contained. In fact, one must assume an immobile sphere which contains all the mobile spheres above the mobile spheres." *Physicorum*, lib. IV, quaeritur utrum tempus sit motus localis caeli, fol. sign. Hh 2, col. a [fol. 92, col. a].

11. Anonymous *Sententiae uberiores* . . ., fol. immediately preceding fol. sign. B recto.

12. [Ibid.]

13. Lamberti de Monte *Prohemium Phisicorum*, quaest: quaeritur ultima sphera sit in loco [fol. 79, cols. a, b, c].

14. [Ibid., col. c.]

15. Gabrielis Biel *Collectiorum . . . super quattuor libros sententiarum*, lib. II, dist. II, quaests. II, III.

16. Conradi Summenhart *Commentaria in Summam physice Alberti Magni*, tract. I, cap. X, prima difficultas.

17. Ibid., tertia difficultas.

18. Gregorii Reisch *Margarita philosophica*, lib. II: De principiis logicis, tract. II: De praedicamentis, cap. XII: De ubi [p. 143].

19. Ibid., lib. VIII: De principiis rerum naturalium, cap. XL: De loco et ejus speciebus [p. 758].

20. [Ibid.]

21. *Collecta et exercitata* Friderici Sunczel Mosellani, lib. IV, quaest. II: Utrum locus sit terminus sive ultimum corporis continentis.

22. Ibid., quaest. I: Utrum quilibet locus sit equalis suo locato; quaest. IV: Utrum diffinitio loci Aristotelis sit sufficiens.

23. [Ibid.]

24. Ibid., quaest. III: Utrum locus sit immobilis.

25. Ibid., quaest. VI: Utrum ultima sphera sit in loco.

26. Jodoci Isennachensis *Summa in totam Physicen*, lib. I., cap. IV: De loco, fol. sign. b i verso.

27. Ibid., fol. sign. b i verso; fol. sign. b ij recto and verso.

28. [Ibid.]

29. [Ibid.]

30. [Ibid.]

31. Ibid., fol. following fol. sign. b iij recto.

32. Pauli de Veneti *Summa totius philosophiae*, pars prima, cap. XIX [fol. 10, col. b].

33. See *Le Système du monde* 10:205.

34. Ibid., cap. XXL [fol. 11, col. b].

35. [Ibid.]

36. [Ibid., fol. 11, col. d.]

37. Ibid., pars II, cap. XIV [fol. 30, col. d].

38. See *Le Système de monde* 8:284-85.

39. Pauli Veneti *Expositio super libros Physicorum*, lib. IV, tract. I, cap. III, pars II, notandum sextum.

40. This interpretation of the meaning one ought to give to the words *simple place* and *composite place* in no way agrees with what the author of the *Six Principles* asserted about it.

41. Paul of Venice, *Expositio super libros Physicorum*, lib. IV, tract. I, cap. III, pars II, notandum septimum.

42. Ibid., sub prima rub.: contra.

43. Ibid., sub secunda rub.: contra.

44. As we have already observed, Paul of Venice attributes to Gilbertus Porretanus an opinion different from the one he professed.

45. [Pauli Veneti *Expositio super libros Physicorum*, lib. IV, tract., I, cap. III, pars II, sub secunda rub.: contra.]

46. [Ibid.]

47. Ibid., notandum sextum.

48. [Ibid.]

49. [Ibid.]

50. Ibid., cap. IV, notandum sextum.

51. Ibid., notandum quartum.

52. Ibid., lib. VI, tract. II, cap. III, pars II, in fine.

53. Ibid., sub rub.: quarto sequitur.

54. [Ibid.]

55. Ibid.

56. Ibid., lib. IV, tract I, cap. IV, notandum octavum.

57. [Ibid.]

58. Ibid., lib. VIII, tract. IV, cap. I, propositio quarta, notandum tertium.

59. Averrois Cordubensis *Commentarii in Aristotelis libros de Physico auditu*, lib. VIII, comm. 76.

60. Pauli Veneti *Universalia sexque principia*, Expositio praedicamentorum Aristotelis, capitulum: de ubi, primum notandum, fol. 115, cols. b, c.

61. Ibid., dubium tertium, fol. 116, col. c.

62. This argument can be seen, in almost the same words, in the *Summa totius philosophiae*, pars VI, cap. XXXVII.

Chapter 7

1. Joannis Duns Scoti *Scriptum Oxoniensis*, lib. II, dist. II, quaest. XI: Dico ergo ad quaestionem . . . *[Opera Omnia*, VI, pt. 1, p. 324].

2. The text has *actualis et positivi.*

3. Joannis Duns Scoti *Quaestiones Quodlibetales*, quaest. XI: Utrum Deus possit facere quod manente corpore et loco, corpus non habeat ubi, sive esse in loco [XII, p. 266].

4. See *Le Système du Monde* 3:291-98.

5. Ibid., 2:471-77.

6. Joannis Duns Scoti *Scriptum Oxoniensis*, lib. IV, dist. XLVIII, quaest. II: Ad questionem potest dici [X, pp. 315-18].

7. [Ibid., p. 317.]

8. [Ibid.]

9. [Ibid., pp. 317-18.]

10. [Ibid., p. 318.]

11. The text reads *Deus* instead of *tempus*, which makes no sense.

12. Petri Aureoli Verberii Ordinis minorum Archiepiscopi Aquensis S. R. E. Cardinalis. *Commentariorum in secundum librum Sententiarum Pars Secundus*, dist. II, quaest. I, art. 1: Utrum tempus sit duratio vel successio, sive quantitas continua, vel discreta, p. 33, col. a.

13. [Ibid.]

14. Ibid., p. 34, col. b.

15. Ibid., p. 35, col. b.

16. Ibid., art. 2: Utrum tempus secundum suum formale sit ens in anima vel extra animam, pp. 37-38.

17. Ibid., p. 38, col. b.

18. Averrois Cordubensis *Commentarii in Aristotelis libros de Physico auditu*, lib. IV, comm. 88.

19. Aureoli *Commentariorum in secundum librum Sententiarum*, dist. II, quaest. I, art. 2, p. 38, col. b; p. 39, col. a.

20. Ibid., art. 4: Utrum tempus sit passio primi motus, p. 41, col. b.

21. Ibid., art. 3: Utrum repugnet formali rationi temporis plurificari, p. 40, col. a.

22. Ibid., p. 39, col. b; p. 40, col. a.

23. We would say subjective.

24. Aureoli *Commentariorum in secundum librum Sententiarum*, dist. II, quaest. I, art. 5: Utrum sit idem nunc in toto tempore, p. 44, cols. a, b.

25. See *Le Système du monde* 2:472.

26. Aureoli *Commentariorum in secundum librum Sententiarum*, dist. II, quaest. I, art. 4: Utrum tempus sit passio primi mobilis, p. 41, cols. a, b.

27. Ibid., p. 42, col. a.

28. [Ibid.]

29. [Ibid.]

30. Gulielmi de Villa Hoccham *Summulae in libros Physicorum*, pars IV, cap. III, fol. 24, col. b [p. 87, cols. a, b].

31. *Tractatus de successivis* edita a Guillelmo de Ocham, cap. II, fol. 139, col. b [pp. 110-12].

32. Ibid., fol. 138, col. a [pp. 98-99].

33. *Questiones* magistri Guilelmi de Ockam *super librum Phisicorum*, quaest. XL: Utrum tempus sit motus, fol. 9, col. a.

34. *Tractatus de successivis* edita a Guillelmo de Ocham, cap. II, fol. 140, col. b [p. 120].

35. Gulielmi de Villa Hoccham *Summulae in libros Physicorum*, pars IV, cap. VIII, fol. 25, col. b [p. 91, col. a].

36. *Tractatus de successivis* edita a Guillelmo de Ocham, cap. II, fol. 140, col. c [pp. 121-22].

37. The manuscript has *majori* instead of *minori*.

38. *Questiones* magistri Guilelmi de Ockam *super librum Phisicorum*, quaest. XL: Utrum tempus sit motus, fol. 9, col. a.

39. Ibid., quaest. XLIII: Utrum aliquis motus inferior sit tempus, fol. 9, cols. b, c.

40. [Ibid.]

41. Averrois Cordubensis *Commentaria magna in libros Physicorum Aristotelis*, lib. IV, summa III: De tempore, cap. III, comm. 98.

42. [Ibid.]

43. [Ibid.]

44. [Ibid.]

45. [Ibid.]

46. Gulielmi de Villa Hoccham *Summulae in libros Physicorum*, pars. IV, cap. XI, fol. 26, cols. b, c [p. 94, col. a to p. 97, col. a].

47. *Tractatus de successivis editus a* Guillelmo de Ocham, cap. II, fol. 138, cols. c, d [pp. 105-7].

48. [Ibid., p. 105.]

49. [Ockham, *Summulae in libros Physicorum*, p. 94, col. b to p. 95, col. a.]

50. [Ibid., p. 95, col. a.]

51. [Ockham, *Tractatus de successivis*, pp. 105-7.]

52. The manuscript has *non* in most places where it ought to have *nos*.

53. *Questiones magistri* Guilelmi de Ockam *super librum Phisicorum*, quaest. XLV: Utrum secundum intentionem Philosophi, quilibet percipiens tempus percipit motum celi, fol. 9, col. d.

54. [Ibid.]

55. Gulielmi de Villa Hoccham *Summulae in libros Physicorum*, pars IV, cap. XI, fol. 26, col. d [p. 95, col. b].

56. *Liber Secundus Sententiarum* Magistri Francisci de Marchia, quaest. IV: Utrum tempus differat a motu, fol. 84, col. a.

57. [Ibid.]

58. [Ibid.]

59. The ms. has *tardus*.

60. The ms. has *quin*.

61. *[Liber Secundus Sententiarum* Magistri Francisci de Marchia, quaest. IV, fol. 84, col. a.]

62. Francisci de Mayronis *Scripta in quatuor libros Sententiarum*, lib. II, dist. XIV, quaest. XI, fol. 152, col. b [fol. 19, col. a].

63. Joannes Canonicus, *Quaestiones in libros Physicorum Aristotelis*, lib. IV, quaest. V, fol. 44, col. a [fol. 154, col. c].

64. [Ibid.]

65. Burleus *Super octo libros Physicorum*, lib. IV, tract. III, cap. II, 116th fol., cols. b, c.

66. Ibid., 117th fol., col. a.

67. Johannis Buridani *Quaestiones super octo Physicorum libros*, lib. IV, quaest. XII, fol. 78, col. d; fol. 79, col. a.

68. Alberti de Saxonia *Quaestiones super libros de Physica auscultatione*, lib. IV, quaest. XIV, quantum ad tertium [fol. 53, col. b].

69. [Ibid.]

70. [Ibid., fol. 53, cols. b, c.]

71. *Abbreviationes libri Physicorum* edite a Marsilio Inguen, fol. sign. f, cols. c, d.

72. [Ibid.]

73. *Quaestiones* Johannis Marcilii Inguen *Super octo libros Physicorum secundum nominaliam viam*, lib. IV, quaest. XVI.

74. [Ibid.]

75. [Ibid.]

76. Johannis Buridani *Quaestiones super octo Physicorum libros*, lib. IV, quaest. XII, fol. 88, col. d.

77. Ibid., quaest. XIV, fol. 81, col. a.

78. Alberti de Saxonia *Quaestiones super libros de Physica*, lib. IV, quaest. XIV, ad rationes [fol. 53, col. c].

79. [Ibid.]

80. Nicole Oresme *Traité du Ciel et du Monde*, liv. II, chap. XIII: il montre que le movement du ciel est regulier par trois autres raisons, fol. 68, cols. a, b [pp. 421, 423].

81. Joannis Canonici *Quaestiones super VIII libros Physicorum Aristotelis*, lib. VI, quaest. I, fol. 60, cols. a, b.

82. [Ibid.: However, Brother Gerard makes an effort to break down this agreed upon theory; he said that, in the definition of continua, in which continua are things whose extremities are one, we ought not accept extremity as standing for a part connected to another but [as standing] for the distinct discreteness of places such that the before of the one and the after of the other, or the above of the first and the below of the second, make up whole unity. And in this way it is possible for points, surfaces, lines (on both sides), and also instants in time to be continued. Granted that these things are indivisible with respect to quantitative parts, they are however divisible with respect to differences of place or time. Just as it is evident that a surface is distinguished by an inside and an outside. We say that insofar as a body is inside a surface, it is not outside; in fact, a tangent body is outside, and not inside. In this way, a surface, which exists indivisibly, is divided by an inside and an outside. In the same way, a point at the center of a circle can terminate a radius coming from the parts that are right, but not by terminating [the radius coming] from the parts that are left.

By this premise, he can reply to the Philosopher's reasoning about continua, in which it is said that in things whose extremities are one, extremity is not taken from some part by which it is connected to another, but for the respective differences of place, in such a way that the before of the one and the after of the other, or the above of the first and the below of the second, make up a whole unity; and in this way it is possible for points to be continued. Fol. 74, col. d.]

83. Nicolai Boneti *Tractatus de Praedicamentis*, lib.: de quantitate, cap. VII, fol. 155, col. d.

84. Ibid., cap. VIII, fol. 156, cols. b, c, d.

85. Nicolai Boneti *Physica*, lib. VI, cap. II, fol. 126, col. d.

86. Ibid., lib. IV, cap. IX, fol. 111, col. d.

87. Ibid., lib. VI, cap. II, fol. 126, col. d.

88. Ibid., lib. IV, cap. IX, fol. 111, col. d.

89. Nicolai Boneti *Tractatus de Praedicamentis*, lib.: de quantitate, cap. VI, fol. 155, col. a.

90. Ibid., cap. XI, fol. 158, col. d.

91. Ibid., cap. XII, fol. 159, col. d.

92. Ibid., cap. XI, fol. 158, col. d; fol. 159, cols. a, b.

93. [Ibid.]

94. In a passage of his *Natural Theology*, Bonet, recalling briefly what we have just detailed, writes: "You ought to remember what we have stated in our *Physics* [by which he means his *Treatise on Predicaments*] about the continuum. There we have stated that every magnitude has as foundation the agreement of two realities, of which neither is a magnitude, but of which both is some thing of magnitude. One does not say of such magnitude that it is divisible into several magnitudes, but that it is divisible into several realities. A line is constituted by such finite and indivisible magnitudes. The line can therefore be resolved into indivisible minima. I do not call these indivisibles points. Moreover, between two points, there is a magnitude which is not divisible into several magnitudes, although it is divisible into several realities. That should be enough about the division of the continuum, for it is treated sufficiently in the *Physics*; seek there for it." Nicolai Boneti *Theologia Naturalis*, lib. VII, cap. XVIII, fol. 293, col. b.

95. Nicolai Boneti *Tractatus de Praedicamentis*, lib.: de quantitate, cap. XII, fol. 159, cols. b, c, d.

96. Joannis Canonici *Quaestiones super VIII libros Physicorum Aristotelis*, lib. VI, quaest. I, fol. 60, col. d.

97. Moses Maimonides *The Guide of the Perplexed*, first part, c. LXXIII [I, pp. 195-97].

98. Nicolai Boneti *Physica*, lib. IV, cap. IX, fol. 111, col. b.

99. Ibid., cap. VIII, fol. 110, col. d; fol. 111, cols. a, b.

100. Ibid., cap. IX, fol. 111, cols. b, c, d.

101. Ibid., lib. VI, cap. I, fol. 126, cols. a, b.

102. Ibid., cap. II, fol. 126, col. c.

103. Ibid., cap. III, fol. 127, col. d; fol. 128, col. a.

104. Francisci de Mayronis *Scripta in quatuor libros Sententiarum*, lib. II, dist. XIV, quaest. XI, fol. 152, col. b [fol. 19, col. a].

105. Nicolai Boneti *Physica*, lib. VIII, cap. VIII, fol. 142, cols. b, c.

106. Ibid., lib. VI, cap. II, fol. 126, cols. c, d.

107. Joannis Canonici *Quaestiones super VIII libros Physicorum Aristotelis*, lib. VI, quaest. I, fol. 60, col. d [fol. 75, col. c].

108. Nicolai Boneti *Physica*, lib. IV, cap. XI, fol. 112, cols. b, c.

109. [Ibid.]

110. Ibid., cap. XIV, fol. 113, col. b.

111. Ibid., col. d.

112. Ibid., cap. XVII, fol. 115, cols. a, b.

113. Ibid., cap. XV, fol. 114, cols. b, c.

114. Ibid., lib. VI, cap. V, fol. 128, cols. c, d.

115. [Ibid., fol. 129, col. a.]

116. Ibid., lib. IV, cap. XVII, fol. 115, col. c.

117. Ibid., lib. VI, cap. II, fol. 126, col. d; fol. 127, cols. a, b.

118. Ibid., cap. IV, fol. 128, cols. b, c.

119. Ibid., cap. V, fol. 128, cols. c, d.

120. [Ibid., fol. 128, col. d.]

121. See also G. Milhaud, "Le Traité de la méthode d'Archimede," *Revue Scientifique* (October, 1908); and G. Milhaud, *Nouvelles etudes sur l'histoire de la Penseé scientifique* (Paris, 1911), pp. 144-47.

122. [Nicolai Boneti *Physica*, lib. VI, cap. V, fol. 128, col. d; fol. 129, col. a.]

123. Ibid., cap. VI, fol. 129, cols. a, b, c.

124. *Quaestiones* fratris Gratiadei de Esculo . . . *per ipsum in florentissimo studio patavino disputatae a feliciter*, quaest. XII, ad cujus intelligentiam, fol. 122, cols. a, b.

125. *Preclarissime quaestiones litterales* edite a fratre Gratia dei Esculano . . . *super libros Aristo. de Physico auditu*, lib. IV, lect. XXII, quaest. III, fol. 57, cols. c, d.

126. Instead of *mathematice*, the text has *metaphysice*, which is clearly erroneous.

127. Nicolai Boneti *Tractatus de Praedicamentis*, lib.: de quantitate, cap. XIII, fol. 159, col. d.

Chapter 8

1. Pauli Veneti *Expositio super octo libros Physicorum*, lib. IV, tract. III, cap. II, second fol. after the fol. sign. y iiij, cols. c, d.

2. Ibid., dubium primum, col. d, and the following folio, cols. a, b.

3. [Ibid.]

4. Ibid., fol. sign. z, col. b.

Chapter 9

1. The first number indicates the order of the proposition within Etienne Tempier's decree; the second number between [ ], the order of the proposition in Mandonnet's classification [pp. 340, 343].

2. Alfarabi, *Die Hauptfragen von* Abu Nasr Alfarabi, in *Philosophische Abhandlungen*, p. 100.

3. Friedrich Dieterici, *Die Philosophie der Araber in IX und X* . . . vol. II, pp. 28-29.

4. *Logica et Philosophia* Algazelis Arabis, *Incipit liber philosophie* Algazalis, lib. II, tract. I, cap. V: Quod non datur vacuum, fol. sign. g2, recto.

5. Maimonides *The Guide of the Perplexed* I, p. 379 [I, pp. 196-97].

6. Ibid., p. 383 [I, p. 198].

7. Averrois Cordubensis *In Aristotelis de Physico auditu libros VIII commentaria magna*, lib. IV, summa II, cap. III, comm. 71 [Minerva facsimile, p. 257].

8. [Ibid., p. 257.]

9. [Ibid., pp. 259, 260, 262.]

10. [Ibid., pp. 260, 261.]

11. [Ibid., p. 261.]

12. Aristotelis *De Caelo cum* Averrois Cordubensis *variis in eumdem commentariis, Commentaria magna*, lib. IV, summa II, cap. unicum, comm. 22.

13. Aristotelis *De Physico auditu libri VIII cum* Averroes Cordubensis *variis in eosdem commentariis, Commentaria magna*, lib. IV, comm. 71 [p. 262].

14. Divi Roberti Lincolniensis *Super octo libris Physicorum brevis et utilitis summa*, fol. sign. Q2, col. c [fol. 281, col. a].

15. Beati Alberti Magni *De physico auditu*, libri VIII, lib. IV, tract. II: De vacuo, cap. VII: Et est digressio declarans solutiones contradictionum Avicenna et Avempace contra inductas demonstratione.

16. Ulrici Engelberti *Liber de Summo Bono*, libri quarti, tract. 2, cap. 19, de loco . . ., fol. 271, col. d.

17. Sancti Thomae Aquinatis *In libros Physicorum Aristotelis expositio*, lib. IV, lect. XIII [p. 239].

18. [Ibid.]

19. [Ibid., p. 240.]

20. [Ibid.]

21. Egidii Romani *In libros de Physico auditu Aristotelis commentaria.* . . . *Eiusdem questio de gradibus formarum*, lib. IV, lect. XIII, dubium 3m, fol. 81, col. b [fol. 91, col. b].

22. Ibid., dubium 2m, fol. 80, col. d; fol. 81, col. a [fol. 91, col. a].

23. *Primus* D. Egidii Ro. Columne *Fundamentari doc. Theologorum principis*, dist. XXXCII, pars II, principalis 2a, quaest. II: Utrum angelus cum movetur, moveatur in tempore vel in instanti, fol. 198, cols. a, b, c.

24. *Questiones supra librum Phisicorum a magistro dicto* Bacuun, fol. 48, col. d [XIII, pp. 234-35].

25. Rogeri Bacon *Questiones naturales et primo questiones libri Phisicorum*, lib. IV, in fine, fol. 24, cols. b, c.

26. Rogeri Baconis *Opus tertium*, cap. XLII, p. 150.

27. Ibid., p. 153.

28. *Opera hactenus inedita* Rogeri Bacon, *Liber primus communium naturalium* Fratris Rogeri, pars. III, dist. II, cap. IV, pp. 208-10.

29. *Quaestiones* fratris Gratiadei de Esculo *excellentissimi sacre pagine doctoris predicamentorum ordinis per ipsum in florentissimo studio patavino disputatae feliciter*, quaest. XI: utrum supposito vacuo in ipso possit esse motus, et supposito, quod in vacuo aliquod grave descenderet utrum descenderet in instanti, fol. 119, col. d, fols. 120-21, cols. a, b [fols. 97-103].

30. *Preclarissime quaestiones litterales edite a fratre* Gratia deo Esculano *sacri ordinis predicamentorum super libros Aristo. de Physico auditu: secundum, ordinem lectionum Divi Thome Aquinatis*, lib. IV, lect. XI, fol. 46, col. c to fol. 49, col. c.

31. Gratiadei *Quaestiones in libros Physicorum . . . disputatae*, quaest. XI, fol. 120, cols. a, b; *Quaestiones litterales super libros de Physico auditu*, lib. IV, lect. XI, quaest. II, fol. 46, col. d; fol. 47, col. a.

32. Gratiadei *Quaestiones litterales super libros de Physico auditu*, lib. VIII, lect. VI, quaest. IV, fol. 84, col. d; fol. 85, col. a; *Quaestiones in libros Physicorum . . . disputatae*, quaest. XIII, secunda autem, fol. 123, cols. b, c.

33. *Eximii atque excellentissimi physicorum motuum cursusque siderei indagatoris* Michaelis Scotis *super auctore Sperae . . .*, quaest. II [p. 252].

34. Aristotle *De Caelo* I, c. IX [279a].

35. Guillelmi Parisiensis *De Universo*, *Opera Omnia*, vol. II, secundus tractatus, cap. III, fol. 98, col. a.

36. [Ibid.]

37. Ibid., cap. IV, fol. 98, cols. a, b.

38. *Questiones supra librum Phisicorum a magistro dicto* Bacuun, Queritur utrum sit ponere vacuum in natura, fol. 48, col. d. [XIII, p. 230].

39. [Ibid.]

40. Ibid., Dubitatur utrum sit ponere vacuum infra celum, fol. 47, col. c [p. 223].

41. [Ibid., p. 224.]

42. [Ibid., pp. 224-25.]

43. Rogeri Baconis *Opus tertium*, cap. XLIII, p. 154. Idem, *Liber primus communium naturalium* Fratris Rogeri, pars III, De existencia vacui secundum se, p. 217.

44. Fratris Rogeri Bacon Ordinis Minorum, *Opus majus ad Clementem quartum, Pontificem Romanum*, pars quarta, dist. IV, cap. XII: An possit esse plures mundi, et an materia sit extensa in infinitum, p. 102.

45. Fratris Rogeri Bacon *Opera quaedam hactenus inedita, Opus tertium*, cap. XLI, pp. 140-41.

46. Rogeri Baconis *Incipit secundum liber communium naturalium qui est de Caelestibus, vel de caelo et mundo*, pars III, cap. II, fol. 108, cols. a, b [Steele ed., pp. 374-75].

47. Ibid., cap. III, fol. 109, col. c [p. 379].

48. *Les Quatres premiers quodlibets de Godefroid de Fontaines*, quodlib. IV, quaest., VI, p. 255.

49. *Quodlibeta* Magistri Henrici Goethals a Gandavo Doctoris Solemnis, quodlib. XIII, quaest. III: Utrum Deus possit facere corpus aliquod extra caelum quod non tangat caelum, fol. 524v.

50. [Ibid.]

51. [Ibid.]

52. [Etienne Tempier, edited by Mandonnet, p. 352.]

53. *[Quodlibeta* Magistri Henrici Goethals a Gandavo, quodlib. XIII, quaest. III, fol. 525r.]

54. [Ibid.]

55. Ibid., quodlib. XV, quaest. I: Utrum Deus possit facere quod vacuum esset, fol. 475.

56. Ibid., quodlib. XIII, quaest. III, [fol. 525r].

57. [Ibid.]

58. [Ibid.]

59. *Clarissimi theologi* Magistri Ricardi de Mediavilla *Seraphici ord. min. convent. super quatuor libros Sententiarum Petri Lombardi quaestiones subtilissimae*, lib. I, dist. LXIV, art. 1, quaest. IV: Utrum Deus posset facere aliud universum, p. 392, col. b.

60. Ibid., lib. II, dist. XIC, art. 3, quaest. III: Utrum Deus posset movere ultimum caelum motu recto, vol. II, p. 186 [col. a].

61. [Ibid.]

62. [Ibid., cols. a, b.]

63. [Ibid., col. b.]

64. [Ibid.]

65. *Declaratio* Raymundi *per modum dialogi edita contra aliquorum philosophorum et eorum sequacium opiniones erroneas et damnatas a venerabili Padre Domino Episcopo Parisiensi*, cap. XLIV, in Otto Keitcher, *Raymundus Lullus und seine Stellung zur arabischen Philosophie*, in *Beitrage zur Geschichte, der Philosophie des Mittelalters*, vol. VV, pt. 4-5, p. 143.

66. Gulielmi Varonis *Quaestiones super libros Sententiarum*, lib. II, quaest. VIII: Queritur utrum Deus posset facere alium mundum simul cum isto, fol. 96, cols. c, d; fol. 97, cols. a, b, c.

67. [Ibid.]

68. [Ibid.]

69. Joannis Duns Scoti *Quodlibeta*, quaest. XI: Utrum Deus possit facere quod manente corpore et loco, corpus non habet ubi, sive esse in loco; De secundo *[Opera Omnia*, XII, p. 265].

70. [Ibid.]

71. [Ibid., pp. 265-66.]

72. [Ibid., p. 266.]

73. Joannis Canonici *Quaestiones super VIII libros Physicorum Aristotelis*, lib. IV, quaest. IV, fols. 42v, 43r.

74. Petri Aquilani *Cognomento Scotelli ex Ord. Min. in doctrina Ioan. Duns Scoti spectatissimi, Quaestiones in quatuor Sententiarum libros, ad eiusdem Doctrinam multum conferentes*, lib. II, dist. I, quaest. II: An aliquid a Deo possit esse ab eo sine fine durationis, p. 185, col. b; p. 186, col. a [fol. 76, col. b].

75. [Ibid.]

76. [Ibid., fol. 76, col. c.]

77. Magistri Roberti Holkot *Super quatuor libros Sententiarum questiones*, lib. II, quaest. II, art. 1: Utrum Deus ab aeterno sciverit se producturum mundum, 3m principale.

78. Ibid., art. 6: Deus potest facere quicquid non includit contraditionem.

79. Ibid., quaest. III: Utrum daemones libere peccaverunt; ad rationes Hibernici.

80. Burleus *Super octo libros Physicorum*, lib. IV, tract. I, cap. I, fol. immediately preceding fol. sign. 1, col. c [which is the same as fol. 78, col. c].

81. Joannis Duns Scoti *Quodlibeta*, quaest. XI, quantum ad primum *[Opera Omnia*, XIII, p. 263].

82. Burleus *Super octo libros Physicorum*, lib. IV, tract. I, cap. I [fol. 78, col. c]. One sees that Walter Burley had written commentaries on the *De Caelo et Mundo*. We have not been able to find any index of the actual existence of these commentaries as printed works; however they can be found as manuscripts with Burley's commentaries on the *Meterology* at the Oxford University Library; Cf. J. C. Houzeau and A. Lancaster, *Bibliographie générale de l'Astronomie*, vol. 1, nos. 1741, 1742.

83. Burleus *Super octo libros Physicorum*, lib. IV, tract, II, cap. IV, fol. immediately preceding fol. sign. o, col. c [which is the same as fol. 102, col. c].

84. Johannis Buridani *Subtilissime quaestiones super octo libros Physicorum*, lib. IV, quaest. VII: Utrum possible vacuum esse, fol. 72, col. d.

85. [Ibid.]

86. [Ibid., fol. 73.]

87. Ibid., lib. IV, quaest. VIII, fol. 73, col. d; fol. 74, cols. a, b.

88. Alberti de Saxonia *Quaestiones super libros de Physica*, lib. IV, quaest. VIII [fol. 48, col. c].

89. *Tractatus de sacramento altaris* Venerabilis Inceptoris Fratris Guilhelmi de Ockam *Anglici*, cap. XII [pp. 218-21].

90. [Alberti de Saxonia *Quaestiones super libros Physicorum*, lib. IV, quaest. VIII.]

91. [Ibid.]

92. *Subtiles doctrinaque plene abbreviationes libri Physicorum edite a prestantissimo philosopho* Marsilio Inguen *doctore parisiensi*, fol. sign. e 2, col. d.

93. Nicole Oresme *Traité du Ciel et du Monde*, fol. 21, cols. c, d; fol. 22, col. a [p. 175].

94. [Ibid., pp. 175-79.]

95. *Preclarissime quaestiones litterales edite a fratre* Gratia deo Esculano *sacri ordinis predicamentorum super libros Aristo. de Physico auditu* . . ., lib. IV, lect. XII, quaest. III, fol. 50, col. b.

96. [Ibid.]

97. Ibid., quaest. IV, fol. 50, col. c.

Chapter 10

1. Nicolai de Orbellis *Physicorum*, in *Curcus librorum philosophiae naturalis*, lib. IV, cap. IIII [fol. 96, col. d].

2. [Ibid., fol. 97, cols. a, b.]

3. Johannis Hennon *Physicorum*, lib. IV, quaest. I, dubium 3m, dictum 2m, fol. 84, cols. b, c.

4. Ibid., quaest. III, fol. 90, col. b.

5. Ibid., quaest. I, dubium 3m, dictum 2m, fol. 84, col. c.

6. Ibid., quaest. III, 2m dictum, fol. 90, col. a.

7. [Ibid.]

8. [Ibid., fol. 90, col. b.]

9. Georgii Bruxellensis *Physicorum*, lib. IV, quaeritur utrum possibile sit vacuum esse per aliquem potentiam, fol. sign. Gg, col. c [fol. 180, incorrectly marked 179, col. b].

10. Ibid., dubitatur utrum in motibus gravium et levium tota successio proveniat ex resistentia medii [fol. 180, incorrectly marked 179, col. b, to fol. 181, col. b].

11. Ibid., fol. sign. Gg, col. d [fol. 181, incorrectly marked 179, col. c].

12. Ibid., col. b [fol. 181, cols. b, c].

13. Ibid., col. c [fol. 181, col. c].

14. Ibid., lib. IV, dubitatur utrum grave positum in vacuo moveretur in instanti, fol. Hh 2, col. d [fol. 92, col. d].

15. Ibid., col. c [fol. 92, col. d].

16. Ibid., col. d [fol. 92, col. d].

17. *Collecta et exercitata* Friderici Sunczel, lib. IV, quaest. VII.

18. [Ibid.]

19. [Ibid.]

20. Conradi Summenhart *Commentaria in Summam physice Alberti Magni*, tract. I, cap. VIII, septima difficultas, fol. following the fol. sign. c 5, col. b.

21. [Ibid.]

22. Gregorii Reisch *Margarita philosophica*, lib. VIII: De principiis rerum naturalium, cap. XL [pp. 758-59].

23. Judoci Isennachensis *Summa in totam Physicen*, lib. I, cap. IV, fol. preceding fol. sign. i i verso.

24. [Ibid.]

25. Ibid., cap. III, fol. sign. g ij verso.

26. Pauli Veneti *Expositio super octo libros Physicorum*, lib. IV, tract. II, cap. III, pars II, fol. sign. xxij, cols. a, b, c, d; fol. sign. xxiij, cols. a, b, c.

27. Ibid., fol. sign. xxiij, col. d.

28. Ibid., pars I, second fol. after fol. sign. xiiij, col. a.

29. Ibid., cols. b, c.

30. [Ibid.]

31. Ibid., cap. I, fol. sign. xij, col. b.

32. Ibid., lib. III, tract. II, cap. IV, pars II, fol. sign. S, cols. b, c.

33. Pauli Veneti *Summa totius philosophiae*, pars prima, cap. XXI [fol. 11, col. d; fol. 12, col. a].

Chapter 11

1. Aristotle *De Caelo et Mundo* I, c. 9 (278b 22-24).

2. Ibid., c. 8 (276b 18-20).

3. [Ibid. (276b 25-26).]

4. Ibid., c. 9 (279a).

5. [Ibid. (279a 8).]

6. [Ibid. (279a 11).]

7. Joannis Stobaei *Eclogarum Physicarum*, cap. XXIV, vol. 1, p. 140.

8. Simplicii *In Aristotelis libros de Caelo commentarii*, lib. I, cap. VIII, p. 115, cols. a, b [comm. 79, p. 55].

9. Averrois Cordubensis *Commentaria in Aristotelis libros de Physico auditu*, lib. I, comm. 81.

10. Ibid., lib. VII, comm. 10.

11. [Ibid.]

12. Ibid., lib. VIII, comm. 35.

Chapter 12

1. See *Le Système du monde* 8:28-36.

2. *Eximii atque excellentissimi physicorum motuum cursusque siderei indagatoris* Michaelis Scoti *super auctore Sperae, cum quaestionibus diligiter emendatis, expositio confecta Illustrissimi Imperatoris Domini D. Frederici praecibus*, vol. 2, p. 146, note [p. 253].

3. [Ibid.]

4. Ernest Renan, *Averroes et Averroisme, essai historique*, p. 165.

5. Guillelmi Parisiensis *De Universo*, in *Opera*, prima partis principalis, pars I, cap. XIV, fol. 98, col. c.

6. Ibid., cap. XV, fol. 100, col. a.

7. Ibid., cap. XVI, fol. 100, cols. a, b.

8. Fratris Rogeri Bacon *Opus majus*, pars IV, dist. IV, cap. XII: An possint esse plures mundi et an materia mundi sit extensa in infinitum, p. 102 (labeled in error, p. 102).

9. [Aristotle *De Caelo* I, c. 9 (279 a).]

10. Fratris Rogeri Bacon *Opus tertium*, cap. XLI, pp. 140-41.

11. *Incipit secundus liber communium naturalium* [fratris Rogeri Bacon], *qui est de caelestibus, vel de Caelo et Mundo*, pars III, fol. 108, cols. a, b [pp. 374-75].

12. Alberti Magni *De Caelo et Mundo*, liber primus, tract. I, in quo subtilissime habetur utrum mundus sit unus vel plures, capitulum secundum, de contradictio eorum qui dicunt elementa diversorum mundorum moveri ad eundum mundum [I, tract. 3, c. 2, pp. 57-58].

13. Sancti Thomae Aquinatis *Expositio super libros de Caelo et Mundo Aristotelis*, lib. I, lect. XVI.

14. See *Le Système du monde* 8:231-319.

15. Aquinas *Expositio . . . de Caelo et Mundo*, lib. I, lect. XIX.

16. H. Denifle, and A. Chatelain, *Chartularium Universitatis Parisiensis*, vol. I, art. 473, p. 543; this proposition is classified as the twenty-seventh in Mandonnet's ranking.

17. *Les Quatres premiers Quodlibets de Godefroid de Fontaines*, quodlib. IV, quaest. VI: Utrum posset extra mundum istum fieri altera terra ejusdem speciei cum terra hujus mundi, quaestio longa, pp. 254-55, epitome, pp. 331-32.

18. [Ibid., pp. 254-55.]

19. [Ibid., p. 255.]

20. [Ibid.]

21. *Quodlibeta Magistri* Henrici Goethals a Gandavo *Doctoris Solemnis*, quodlib. XI, quaest. I: Utrum Deus possit facere, sub una specie specialissima angeli, aliquem alium angelum aequale in natura et essentia speciei angelo jam facto sub illa, fol. 418v; fol. 419r [fol. 439, incorrectly marked].

22. Ibid., quodlib. XIII, quaest. III: Utrum Deus possit facere corpus aliquod extra caelum quod non tangat caelum, fol. 524v.

23. See *Le Système du monde* 8:36-40.

24. *Clarissimi theologi Magistri* Ricardi de Mediavilla *Seraphici ord. min. convent. super quatuor libros Sententiarum Petri Lombardi quaestiones subtilissimae*, lib. I, dist. XLIII, art. I, quaest. IV, p. 392 [col. b].

25. [Ibid., p. 392, col. b; p. 393, col. a.]

26. See *Le Système du monde* 8:42.

27. D. Aegidii Romani *Ordinis Fratrum Eremitarum Sancti Augustini Archiepiscopi Bituriensis Opus Hexaemeron*, pars I, cap. VIII, fol. 7, col. b.

28. [Ibid.]

29. [Ibid.]

30. *Quodlibet domini* Egidii Romani, quodlib. VI, quaest. VIII: Utrum via nature possit generari vel fieri nova species que nunquam fuerit facta, fol. 77, cols. b, c, d.

31. Gulielmi Varonis *Quaestiones super libros Sententiarum*, lib. II, quaest. VIII: Quaeritur utrum Deus posset facere alium mundum simul cum isto, fol. 96, col. c.

32. [Ibid.]

33. See *Le Système du monde* 8:44-45.

34. Gulielmi Varonis *Quaestiones super libros Sententiarum*, lib II, quaest. VIII, fol. 96, col. b.

35. Ibid., fol. 97, col. b.

36. Ibid., col. c.

37. [Ibid.]

38. [Ibid.]

39. *Opera* Joannis de Bassolis *Doctoris Subtilis Scoti (sua tempestate) fidelis Discipuli Philosophi ac Theologi profundissimi, in quatuor Sententiarum libros*, lib. I, dist. XLIV, quaest. unica, fol. 214, col. b.

40. Ibid., col. d.

41. Ibid., col. b.

42. [Ibid.]

43. Thomae ab Argentina *Eremitarum Divi Augustini Prioris Generalis, Commentaria in quatuor libros Sententiarum*, lib. I, dist. XLIV, quaest. I: An Deus universum potuit facere melius, art. 4: Utrum Deus possit facere aliud universalium illo universum manente, fol. 120, cols. b, c, d.

44. [Ibid.]

45. [Ibid.]

46. [Ibid.]

47. [Ibid.]

48. Joannis de Janduno *Quaestiones in libros Aristotelis de Caelo et Mundo*, lib. I, quaest. XXIV: An sit possibile esse plures mundos [fol. 16, col. c].

49. [Ibid., fol. 16, col. d.]

50. [Ibid.]

51. Ibid., lib. IV, quaest. XIX.

52. Ibid., lib. I, quaest. XXIV.

53. See *Le Système du monde* 6:560-64.

54. Magistri Guilhelmi de Ockam *Super quatuor libros Sententiarum annotationes*, libri primi sententiarum, dist. XLIV, quaest. unica: Utrum Deus posset facere mundum meliorem isto mundo [Franciscan Institute ed., IV, p. 655].

55. [Ibid., pp. 657-58.]

56. [Ibid., p. 659.]

57. [Ibid., p. 660.]

58. Roberti Holkot *Quaestiones super libros Sententiarum*, lib. II, quaest. II, art. 1: Utrum Deus ab aeterno sciverit se producturum mundum.

59. Ibid., art. 6: Deus potest facere quicquid non includit contradictionem.

60. [Ibid.]

61. [Ibid.]

62. [Ibid.]

63. [Ibid.]

64. *Quaestiones super libros de Caelo et Mundo magistri* Johannis Buridani rectoris Parisius, lib. I, quaest. XVII: Decimoseptimo queritur utrum si essent plures mundi, terra unius moveretur naturaliter ad terram alterius, fol. 78, col. a [p. 84].

65. Ibid., col. b [pp. 84-85].

66. See *Le Système du monde* 7:282-87.

67. *Quaestiones . . . de Caelo et Mundo* Johannis Buridani, lib. I, quaest. XVII, fol. 78, cols. b, c [pp. 86-87].

68. *Quaestiones subtilissimae* Alberti de Saxonia *in libros de Caelo et Mundo*, lib. I, quaest. XIII: Utrum sint vel possint esse plures mundi [quaest. XI, fol. 95, col. a].

69. [Ibid., col. b.]

70. Ioannis de Ianduno *In libros Aristotelis de Caelo et Mundo quaestiones subtilissimae*, lib. I, quaest. XXIV: An sit possibile esse plures mundos?

71. Alberti de Saxonia *In libros de Caelo et Mundo*, lib. I, quaest. XII: Utrum supposito quod essent plures mundi, terra unius mundi moveretur ad terram alterius mundi [quaest X, fol. 94, col. d].

72. [Ibid.]

73. [Ibid.]

74. [Ibid., col. c.]

75. See *Le Système du monde* 6:216.

76. Ibid., 6:210.

77. Alberti de Saxonia *In libros de Caelo et Mundo*, lib. I, quaest. XIII [quaest. XI, fol. 95, cols. a, b].

78. Ibid., quaest. XII [quaest. X, fol. 94, col. d].

79. Ibid., quaest. XIII [quaest. XI, fol. 95, col. b].

80. Ibid., quaest. XII [quaest. X, fol. 94, col. d to fol. 95, col. a].

81. [Ibid.]

82. Bibliothèque Nationale, fonds latin, ms. no. 16621, fol. 213v.

83. Joannis de Dumbleton *Summa*, pars VI, cap. X, fol. 65, col. c.

84. See *Le Système du monde* 2:361-63.

85. Nicole Oresme *Traité du Ciel et du Monde*, liv. I, chap. 16, fol. 14, col. d [p. 131].

86. Oresme, fol. 15, col. c [p. 135].

87. [Ibid.]

88. [Ibid., p. 137.]

89. Ibid., chap. 19, fol. 17, col. a [p. 145].

90. Ibid., fol. 20, col. a [p. 167].

91. Ibid., col. d [p. 171].

92. Ibid., fol. 20, col. d; fol. 21, cols. a, b, c [pp. 171-75].

93. Ibid., chap. 4, fol. 5, col. d [p. 71].

94. Ibid., chap. 25, fol. 83, cols. b, c [pp. 527-29].

95. Joannis Stobaei *Eclogarum physicarum*, lib. I, cap. XXVI, p. 151.

96. Aristotle *The Generation of Animals* III, c. XI (761b).

97. Averrois Cordubensis *In libros Aristotelis de Caelo et Mundo commentarii*, lib. I, summa IV, comm. 16; lib. II, quaest. III, comm. 32; summa III, cap. I, comm. 42; summa III, cap. II, comm. 49; idem, *Sermo de substantia orbis*, cap. II.

98. Averrois Cordubensis *In libros de Caelo et Mundo commentarii*, lib. II, summa II, quaest. III, comm. 32.

99. Ibid., summa III, cap. I, comm. 42.

100. Ibid., cap. II, comm. 49.

101. [Ibid.]

102. Averrois Cordubensis *Sermo de substantia orbis*, cap. II.

103. Bedae Venerabilis *De constitutione mundi caelestis terrestrisque liber*, cap.: De furvo Lunae, vol. XC, col. 888.

104. Honorius Solitarius *De imagine mundi*, cap. XXIII: De luna, vol. CLXII, cap. LXIX , col. 138.

105. *Introductoire d'Astronomie*, De chascun planete par soi, Bibliothèque Nationale fonds français, ms. no. 1353, fol. 28, col. d; fol. 29, col. a.

106. Alberti Magni *De Caelo et Mundo*, lib. II, tract. III: De natura et figura et motibus stellarum, cap. VI: Et est digressio declarans qualiter stellae omnes illuminatur a Sole.

107. Ibid., cap. VIII: De motibus duobus scintil lationis et titubationis, utrum conveniant stellis; in quo est digressio declarans causam et figuram umbrae quae videtus in Luna; et de altercatione Averrois contra Avicennam.

108. Sancti Thomae Aquinatis *Expositio super libros de Caelo et Mundo Aristotelis*, lib. II, lect. XII.

109. *Tractatus de spera* Jo. de Sacro Bosco *cum glosis* R. Anglici, cap. IV, glosa II, fol. 42, cols. c, d; fol. 43, col. a [pp. 245-46].

110. [Ibid.]

111. Bibliothèque Nationale, fonds latin, ms. no. 7333, fol. 19, col. c; see *Le Système du monde* 3:456.

112. Petri Padubaensis *Lucidator Astronomiae*, fol. 118, col. b.
113. [Ibid.]
114. [Ibid.]
115. D. Aegidii Romani *Opus hexaemeron*, pars II, cap. X: Quod ex comparatione ipsius lucis ostenditur quod neque lumen neque splendor sunt forma substantialis corporis lucidi.
116. Ibid., cap. XXXV: Declarans unde habent esse diaphaneitates et tot modi coloris, et tot modi lucis, quot videmus in caelis.
117. Averrois Cordubensis *Sermo de substantia orbis cum* Joannis de Janduno *Expositione*, cap. II.
118. *Quaestiones super libros de Caelo et Mundo* magistri Johannis Buridani rectoris Parisius, lib. II, quaest. XIX: Utrum macula apparens in Luna proveniat ex diversitate partium Lunae vel ab aliquo extrinseco, fol. 96, col. d; fol. 97, col. a [pp. 214-15].
119. Ibid., fol. 97, col. b [p. 217].
120. Nicole Oresme *Traité du Ciel et du Monde*, liv. II, chap. XVI: il monstre que les estoils sont meues aux mouvemens des cieulx ou elles sont, et non autrement, fol. 75, cols. b, c, d; fol. 76, col. a [pp. 455-59].
121. Alberti de Saxonia *Quaestiones subtilissimae in libros de Caelo et Mundo*, lib. II, quaest. XXII: Utrum omnia astra alia a sole habeant lumen suum a Sole [quantum ad secundum].
122. [Ibid., quantum ad tertium.]
123. [Ibid.]
124. [Ibid.]
125. [Ibid.]
126. Ibid., quaest. XXIV: Utrum macula illa quae apparet in Luna causetur ex diversitate partium Lunae vel ab aliquo extrinseco.

Chapter 13

1. Johannis Hennon *De Caelo et Mundo*, in *Commentarii in Aristotelis libros Physicorum*, lib. I, quaest. II, dubium lm, fol. 156, col. d.
2. Pierre Mandonnet, *Siger de Brabant*, p. 237.
3. [See, for example, *Le Système du monde* 10:245-46.]
4. Johannis Hennon *De Caelo et Mundo*, lib. I, quaest. II, fol. 153, col. d.
5. Georgii *De Caelo et Mundo*, in *Curcus optimarum super Philosophiam Aristotelis*, lib. I, Quaeritur utrum possibile est esse plures mundos, 3d folio after folio sign. Nn. 3, cols. a, b [fol. 155, col. d].
6. [Ibid., fol. 156, col. a.]
7. Pauli de Veneti *Summa totius philosophiae*, pars secunda, cap. IV [fol. 25, col. a].
8. Pauli Veneti *Liber de compositione mundi*, cap. XXIX.
9. Ristoro d'Arezzo *Della composizione del mondo*, lib. VIII, cap. XXIV: Di conoscere se il Mondo e solo, o e piu mondo di fuori daquesto, pp. 323-24.
10. [Pauli Veneti *Liber de compositione mundi*, cap. XXIX.]
11. [Ibid.]
12. Ristoro d'Arezzo *Della composizione del mondo*, p. 324.
13. Johannes Majoris Scotus, *De infinito* [pp. 56, 58].
14. [Ibid., p. 62.]

15. [Ibid., p. 60.]
16. [Ibid., p. 62.]
17. [Ibid.]
18. *Recollectae* Gaietani *super octo libros Physicorum cum annotationibus textuum*, lib. IV, in principio, fol. 28, col. d.
19. [Nicole Oresme *Traité du Ciel et du Monde*, p. 173.]
20. Nicolai de Cusa *De docta ignorantia*, lib. II, cap. XI, p. 38 [pp. 109-10].
21. [Ibid., p. 109.]
22. Ibid., cap. XII, pp. 39-40 [pp. 112-13].
23. Ibid., p. 39 [pp. 111-12].
24. Ibid., cap. XIII, p. 42 [pp. 118-20].
25. Ibid., cap. XII, pp. 40-41 [p. 116].

Bibliography of Works Cited by Duhem

(with References to Modern Editions and English Translations)

Aegidius Romanus, Archbishop of Bourges, ca. 1243-1316

Egidii Romani *In libros de Physico auditu Aristotelis commentaria accuratissime emendata: et in marginibus ornata quotationibus textuum et commentorum. Ac aliis quamplurimis annotationibus. Cum tabula questionum in fine. Eiusdem questio de gradibus formarum. Cum privilegio.* Colophon: Preclarissimi summique philosophi Egidii Romani. De gradibus formarum tractatus Venetiis impresse mandato et expensis Heredum Nobilis viri domini Octaviani Scoti civis Modoetiensis. Per Bonetum Locatellum presbyterum 12 Kal. Octobr. 1502. [*Commentary on the Physics.* Facsimile. Frankfurt: Minerva, 1968.]

Egidius *cum* Marsilio *et* Alberto *de generatione. Commentaria fidelissimi expositoris.* D. Egidii Romani *In libros de generatione et corruptione* Aristotelis *cum textu intercluso singulis locis. Quaestiones item subtilissime eiusdem doctoris super primo libro de generatione: nunc quidem primum in publicum prodeuntes. Quaestiones quoque clarissimi doctoris* Marsilij Inguem (sic) *in prefatos libros de generatione. Item quaestiones subtilissime magistri* Alberti de Saxonia *in esodem libros de gene. Nusquam alias imprese. Omnia accuratissime revisa: atque castigata: ac quantum ars eniti potuit fideliter impressa.* Colophon: Impressum Venetiis mandato et expensis nobilis viri luce Antonii de Giunta florentini. Anno domini 1518 die 12 mensis Februarii.

D. Egidii Romani *Fidelissimi expositionis in Arist. libros de Gener. Commentaria, & subtilissimae quaest. super primo clarrismique doctoris Marsilii Inguen, et Magistri* Alberti de Saxonia *in eosdem accuratissimae quaestiones* . . . Venetiis, apud Hieronimum Scotum, MDLXVII.

Primus D. Egidii Ro. Columne *Fundamentari doc. Theologorum principis.* Bituricensis archiepi. S.R.E. Cardinalis ordinis Eremi. sancti Augu. Primus Sententiarum: Correctus a reverendo magistro Augustino Monifalconio euisdem ordinis. Colophon: Venetiis Impressus sumptibus et expensis heredum quondam Domini Octaviani Scoti civis Modoetiensis: ac sociorum. Die Martii 1521. [Microfilm available.]

Quodlibet domini Egidii Romani *Theoremata eiusdem de Corpore christi.* Guiliermus de Ocham *De sacramento altaris.* Cum Privilegio. Colophon: Impressum Venetiis per Simonem de Luere: Impensis domini Andree Torresani de Asula. 18 Januarii 1502. [Microfilm available. *Quodlibet* and *Theoremata de Corpore Christi.* Facsimile of 1646 edition. Frankfurt: Minerva, 1966.]

[*Opera exegetica.* Facsimile of 1554-55 edition. Frankfurt: Minerva, 1968.]

D. Aegidii Romani *Ordinis Fratrum Eremitarum Sancti Augustini Archiepiscopi Bituricensis Opus Hexaemeron sive de Mundo sex diebus constituto.* [Microfilm available.]

d'Ailly, Pierre, Cardinal, 1330-1420

Reverendissimi Domini Petri de Aliaco, Cardinalis et Episcopi. Cameracensis, Doctorisque celebratissimi. *Uberrimum Sphera Mundi comentum intersertis etiam questionibus domini* Petri de Aliaco. Iehan Petit. Colophon: Et sic est finis hujus egregii tractatus de Sphera Mundi Johannis de Sacro Busto Anglici et doctoris parisiensis. Una cum textualibus optimisque additionibus ac uberrimo commentario Petri Ciruelli Darocensis ex ea parte Terraconensis Hispaniae quam Aragoniam et Celtiberiam dicunt oriundi. Atque insertis persubtilibus quaestionibus reverendissimi domini cardinalis Petri de Aliaco ingeniosissimi doctoris quoque parisiensis. Impressem est hoc opusculum anno dominice nativitatis 1498 in mense februarii Parisius in camp Gallardo opera atque imprensis magistri Guidonis mercatoris. [For further bibliographical information, see *Le Système du monde* 4:169.] [*Sphera Mundi novit recognita cum commentariis & authoribus in hoc volumnie contentis vz* . . . Petri de Aliaco *Cardinalis Questiones,* fol. 116-31. Venetiis impensis nobilis viri divini luce Antonii de Giunta florentini. Die ultimo junii 1518.]

Albert of Saxony, see Albertus de Saxonia

Alberti, Leandro, 1479-1552

De viris illustribus Ordinis Praedicatorum libri sex in unum congesti autore Leandro Alberto Conomiensi *viro clarissimo.* Bologna, Hieronymus de Benedictis, for Johannes Baptista de Lopis, 1517.

Albertus de Saxonia, d. 1390

Logica Albertucili. *Perutilis logica excellentissimi sacre theologie professoris magistri Alberti de Saxonia ordinis eremitarum Divi*

Augustini: per reverendum sacre pagine doctorem magistrum Petrum Aurelium Sanutum Venetum ejusdem ordinis professum: quam diligentissime castigata: nuperrimeque impressa. Colophon: Explicit perutilis logica . . . impressa Venetiis ere et sollertia Heredum Domini Octaviani Scoti civis Modoetiensis et sociorum. Anno a Christi ortu MDXXII. [*Perutilis logica.* Facsimile. Hildesheim, N.Y.: Olms, 1974.]

Quaestiones subtilissimae Alberti de Saxonia *in libros De Caelo et Mundo.* Colophon: Expliciunt questiones . . . Impresse autem Venetiis Arte Boneti de locatellis Bergomensis. Impensa vero nobilis viri Octaviani scoti civis modoetiensis Anno salutis nostre 1492 nono kalendas novembris Ducante inclito principe Augustino barbadico. [Microfilm available.]

Alberti de Saxonia *Acutissimae Quaestiones super libros de Physica auscultatione Aristotelis.* [Microfilm available with the *De Caelo* in a 1516 edition.]

Alberti de Saxonia *Tractatus proportionum.* Bibliothèque Nationale, fonds latin, ms. no. 7368, fols. 14r-26v. In fol. 26v is: Expliciunt proportiones motuum. Deo gratias. [Busard, H. *Tractatus proportionum von Albert von Sachsen.* Wien: Springer in Komm., 1971]

Albertus Magnus, Saint, Bishop of Ratisbon, 1193(?)-1280

[*Sancti Doctoris Ecclesiae* Alberti Magni *Ordinis Fratrum Praedicatorum Episcopi Opera Omnia ad fidem codicum manuscriptorum edenda apparatu critico notis prolegomenis indicibus instruenda curavit institutum Alberti Magni Coloniense Bernhardo Geyer Praeside.* Monasterii Westfalorum in Aedibus Aschendorff, 1971.]

De Caelo et Mundo.

De Physico auditu.

Tabula Tractatuum Parvorum naturalium Alberti Magni Episcopi Ratisbo. de ordine Predicamentorum. *De sensu et Sensato. De Memoria et Reminiscentia. De Somno et Vigilia. De Motibus animalium. De etate sive de Juventute et Senectute. De Spiritu et Respiratione. De Morte et Vita. De Nutrimento et Nutribili. De Natura et Origine anime. De Unitate intellectus contra Averroem. De intellectu et intelligibili. De Natura locorum. De Causis et proprietatibus Elementorum. De Passionibus Aeris. De Vegetabilibus et Plantis. De principiis motus processivi. De Causis et processu universitatis a Causa prima. Speculum Astronomicum de Libris licitis et illicitis.* Colophon: Venetiis impensa heredum quondam domini Octaviani Scoti civis Modoentiensis: ac sociorum. Die 10 Martii 1517.

Alfarabi, see al-Farabi, Muhammad ibn Muhammad

Algazali, see al-Ghazzali

Andreae, Antonius, d. ca. 1320

Questiones Scoti *super universalia Porphyrii. Necnon Aristotelis predicamenta ac Peryarmenias. Item super libros Elenchorum.* Et Antonii Andree *Super libros sex principiorum. Item questiones* Joannis Angelici (sic) *super questiones universales eiusdem Scoti.* Colophon: Expliciunt questiones Doctoris subtilis Joannis Scoti super universalia Porphyrii: et Aristotelis predicamenta; et peryarmenias: ac elenchorum necnon discipuli ejus Antonii Andree super libro sex principiorum Gilberti Porretani: studiosissime correcte per Reverendissimum patrem magistrum Mauritium de portu Hibernicum archiepiscopum Tumanensem ordinis minorum. Impresse Venetiis per Philippum pincium Mantuanum. Anno domini 1512. die 9 Augusti.

Conventualis Franciscani, ex Aragoniae provincia ac Ioannis Scoti doctoris subtilis discipuli celeberrimi, in quatuor Sententiarum libros opus longe absolutissimum: Quod, cum diu latuerit: a F. Constantio A. Sarnano ejusdem ordinis, e tenebris jam nunc vindicatum . . . ; felicio auspicio prodit . . . Venetiis, Apud Damianum Zenarum. MDLXXVII. [Ionnis Duns Scoti, subtilis, *ac celeberrimi doctoris in universam* Aristotelis *logicam exactissimae Quaestiones. Quibus singulis perutiles quaedam adjecta sunt dubitationes cum earum solutionibus; necnon Tractatus de secundis intentionibus; simul cum quaestionibus* Antonii Andreae *super sex principiis* . . . Urbellis in Archiepisc. Moguntion, sumptibus Antonii Hierati, Bibliopolae Coloniensis, anno MDCXXII.]

Anonymous (student of Dumbleton)

Bibliothèque Nationale, fonds latin, ms. no. 16621.

Anonymous (student of Ockham)

Bibliothèque Nationale, fonds latin, ms. no. 16130.

Anonymous (astrologer of Baudoin de Courtenay)

Introductoire d'astronomie. Bibliothèque Nationale, fonds français, ms. no. 1353.

Anonymous

Sententie uberiores ex scriptis beati Thome et venerabilis Alberti super octo libros Phisicorum Aristotelis in studio Coloniensi Summatim congeste. This title is located on the recto of the first folio. The following title is located on the recto of the second folio (sign Aij): *Summa Sententiarum principaliorum primi libri Phisicorum.*

Antonius Andres, see Andreae, Antonius

Aquinas, Saint Thomas, see Thomas Aquinas, Saint

Aristotle, see Aristoteles

Aristoteles, 384-322

Aristotelis *Opera*. 5 vols. Paris: A. Firmin Didot, 1848-78.

Aristotelis *Groece*. Ex recensione Immanuelis Bekkeri edidit Academia Regia Borussica. Berolini, 1931.

[*The Works of Aristotle Translated into English*, edited by W. D. Ross. Oxford: Clarendon, 1908-52.]

Aureol, Peter, see Aureoli, Petrus

Aureoli, Petrus, Archbishop of Aix, 1280-1322

Vol. 1: *Commentariorum in primum librum Sententiarum Pars Prima*. Auctore Petro Aureolo Verberio Ordinis minorum Archiepiscopi Aquensis S.R.E. Cardinalis. Ad Clementem VIII. Pont. Opt. Max. Romae Ex Typographia Vaticana. MDXCVI, cum privilegio, et Superiorum permissu. Finis primae partis Commentariorum Petri.

Vol. 2: *Commentariorum in secundum librum Sententiarum Pars Secunda* . . . *Scriptum super primum Sententiarum*, editum a Petro Aureoli, Ordinis Fratrum Minorum, exit feliciter. Finis. On back of the last page: Romae, Ex Typographia Vaticana, MDXCVI. [Microfilm available. For further bibliographical information, see *Le Système du monde* 6:392-93.] [*Scriptum Super Primum Sententiarum*. Edited by E. M. Buytaert. St. Bonaventure, N.Y.: Franciscan Institute, 1952-56.]

Averroes Cordubensis, 1126-1198

[*Aristotelis opera cum Averrois commentariis*. 9 vols. Facsimile of Venice: Juncta, 1562-74 edition. Frankfurt: Minerva, 1962.]

Averrois Cordubensis *Commentarii in quatuor libros Aristotelis de Caelo et Mundo*.

Magni Commentatoris Averrois Cordubensis *Disputationes-queo Destructio destructionem philosophiae* Algazelis *dicuntur*. [For further bibliographical information, see *Le Système du monde* 6:497.] [*Averroes' Tahafut al-Tahafut (The Incoherence of the Incoherence)*. 2 vols. Translated from the Arabic by Simon van den Bergh, Oxford: Oxford University Press, 1954.] [*Averroes' "Destructio destructionem philosophiae Algazelis" in the Latin Version of Calo Calonymos*. Edited by B. H. Zeller. Milwaukee: Marquette University Press, 1961.]

Aristotelis Stagiritae *Metaphysicorum libri XIIII cum* Averrois Cordubensis *id eosdem Commentariis et Epitome*. Venetiis, apud Juntas, MDLIII.

Averrois Cordubensis *Commentarii in Aristotelis libros de Physico auditu*.

Sermo de substantia orbis. [For further bibliographical information, see *Le Système du monde* 4:534-35.] [*Sermo de substantia orbis*. Hebrew text and English translation by A. Hyman, Ph. D. diss., Harvard University, 1953.]

Avicenna, 980-1037

Metaphysica Avicenne *sive eius prima philosophia.* Colophon: Explicit metaphysica Avicenne sive eius prima philosophia optime Castigata per Reverendum sacre theologie bachalarium fratrem Franciscum de Macerata ordinis minorum et per excellentissimum artium doctorem dominum Antonium frachantianum vicentinum philosophiam legentem in gymnasio pativino. Impressa Venetiis per Bernardinum Benetum Expensis viri Jeronymi duranti anno domini 1495 die 26 martii. [*Metaphysica.* Facsimile of Venice 1495 edition. Frankfurt: Minerva, 1966.]

Bacon, Roger, 1215-ca. 1292

Communia naturalium, Opera hactenus inedita Rogeri Bacon *Liber secundus communium naturalium* Fratris Rogeri. *De celestibus.* Partes quinque edidit Robert Steele. Oxonii, MCMXIII. Bibliothèque Mazarine ms. no 3576. [The *De celestibus* is also in *Opera hactenus inedita,* ed. Steele.]

Fratris Rogeri Bacon Ordinis Minorum, Opus majus ad Clementem quartum, *Pontificem Romanum,* M.S. Codice Dubliniensi cum aliis quibusdam collato nunc primum edidit S. Jebb, M.C. Londini, typis Gulielmi Bowyer, MDCCXXXIII. *The Opus Majus of Roger Bacon.* Edited by John Henry Bridges. London, Edinburgh, and Oxford, 1900. [*The Opus Majus of Roger Bacon.* Edited and translated by R. B. Burke. Philadelphia: University of Pennsylvania Press, 1928.]

Rogeri Bacon *Opus tertium*: Fratris Rogeri Bacon *Opera quaedam hactenus inedita;* eds. Brewer, Greenman, Green, Congnan, and Roberts, London, 1859. [For more bibliographical information, see *Le Système du monde* 3:420n.]

Questiones naturales et primo questiones libri phisicorum. Bibliothèque Municipale d'Amiens, ms. no. 406, fol. 24. [In *Opera hactenus inedita,* ed. Steele.]

Questiones supra librum phisicorum a magistro dicto Bacuun. Bibliothèque Municipale d'Amiens, ms. no. 406, fol. 48. [In *Opera hactenus inedita,* ed. Steele.]

Baire, René Louis

Bulletin de la Société Mathématique de France, 1905.

Bassolis, Joannes de, d. ca. 1347

Vol. 1: *Opera* Joannis de Bassolis *Doctoris Subtilis Scoti (sua tempestate) fidelis Discipuli, Philosophi, ac Theologi profundissimi, in quatuor Sententiarum libros (credite) Aurea. Quae nuperrime Impensis non minimis, Curaque, et emendatione non mediocri, ab debitae integritatis sanitatem revocata, Decoramentisque marginalibus, ac Indicibus, ad notata: Opera denique, et Arte Impressionis mirifica Dextris Syderibus elaborata fuere. Venundantur a Francisco Regnault:*

et Ioanne Frellon. Parisiis. Cum gratia et privilegio. Colophon: Hic finem accipiunt subtilissime: et sane quam utiles questiones R. P. Fratris Jo. de Bassolis Minorite, ac Theologi profundissimi in primum Sententiarum. Niper ab Orontio Fine Delphinate (etsi sorruptum et maculatissimum exemplar nactus extiterit) priori integritati quam integerrime et emendatissime valuit, diligenter restitute. Ac marginariis adnotamentis haud parum conducentibus, cum earum indicibus studiose ab eodem decorate. Sumptibus autem non modicis Fidelium Bibliopolarum Alme universitatis Parisiensis Francisci Regnault et Joannis Frellon Typis mandate. In Aedibus scilicet Nycolai de Pratis chalcographi probatissimi. Anno Jesu Aeterni Regis sesquimillesim decimo septimo Nono Idus Septembres, Sole sub XXV parte Virginis gradiente in hemispherio Parisiensi. Leonis Pape X pontificatus Anno Quinto. [Microfilm available.]

Vol. 2: *Profundissimi Sacre theologie professoris F.* Joannis de Bassolis *minorite in secundum sententiarum Questiones ingeniosissime et sane quam utiles: vigilanti cura: ac improbo labore revise, et emendate: iuoctaque scientie exigentiam marginaris annotamentis decorate: Nuperrime autem nitidissimis caracteribus (sed nusquam antea) impresse: felici sydere in lucem prodeunt. Quibus premittitur Tabula questionum Articulorum, et omnium in hoc claro opere contentorum studiose et artificialiter collecta. Venundantur in vico Maturinorum apud Joannem Frellon fidelissimum Bibliopolam sub signo Avicludij commorantem Parhisius.* Colophon: Expliciunt preclarissime et sane quam utiles questiones super secundum sententiarum: a profundissimo et ingeniose theologo Fratre Joanne de Bassolis studiose composite et discusse. Impresse noviter in alma Parhisiorum Lutecia (Previa tamen diligenti examinatione, seu correctione, et debita coordinatione ipsarum) Sumptibus honestorum bibliopolarum Francisci Regnault et Joannis Frellon. Arte vero et nitidissimis caracteribus Nicolai de Pratis Calcographi probatissimi. Anno ab orbe redempto millesimo quingentesimo decimo sexto, die ultimo mensis Octobris. Laus Jesu. [Microfilm available. For further bibliographical information, see *Le Système du monde* 6:438-39.]

Baur, Ludwig

Die philosophischen Werke Robert Grosseteste, Bischof von Lincoln. In *Beitrage zur Geschichte der philosophie des Mittelalders,* vol. IX, Munster, 1912.

Beda Venerabilis, 673-735

Bedae Venerabilis *De constitutione mundi caelestis terrestrisque liber* in Bedae Venerabilis *Operum,* vol. i, in *Patrologiae latinae,* accurante J. P. Migne, vol. XC.

Bede, the Venerable, see Beda Venerabilis

Bernard de Verdun, fl. late thirteenth century

 Tractatus super totam Astrologiam, Bibliothèque Nationale, fonds latin, ms. no. 7333. [*Tractatus super totam Astrologiam*. Edited with a commentary by Polykarp Hartmann. Dietrich: Lollde-Verlag, 1961.]

Biel, Gabriel, d. 1495

 Inventarium seu repertorium generale: tametsi compendiorum et succinctum: verumtamen valde utile atque necessarium: contentorum in quattuor collectoriis profundissimi ac diligentissimi theologi Gabrielis Biel. *Super quattuor libro sententiarum*. Colophon (at the end of the fourth book): Impressit hoc opus probus vir Joannes Clein Allemannus chalcographus et bibliopolo in famatissimo Lugdunensi emporio: Anno dominice incarnationis Mccccxix Die xxiiii mensis Septembris. [*Collectiorum in iv libros Sententiarum Gulielmi Occam.* Facsimile. Hildesheim, N.Y.: Olms, 1977.]

Bonaventura, Saint, Cardinal, 1221-74

 Celebratissimi Patris Domini Bonaventurae Doctoris Seraphici *In secundum librum Sententiarum disputata*. [*The Works of Saint Bonaventure, Cardinal, Seraphic Doctor, translated from the Latin.* Paterson, N.J.: St. Anthony Guild Press, 1966.]

Bonaventure, Saint, see Bonaventura, Saint

Bonet, Nicholas, see Bonetus, Nicolaus

Bonetus, Nicolaus, Bishop of Malta, d. ca. 1343

 Fol. 2r (in a handwriting that is more recent than the text's and is almost erased, is this title): *Philosophia naturalis magistri* Nicolai Boneti. Bibliothèque Nationale, fonds latin, ms. no. 14716. Omnes homines, ymo omnes nature intellectuales cum universaliter scire desiderant, a primis scibilibus secundum naturam est incipiendum.

 Fol. 106r: Imponamus ergo finem dictis in ista metaphysica, que novem libris parcialibus est contenta in quorum primo terminatur et stabilitur subiectum quodest ens in quantum ens. et. cetera. Deo gratias. *Explicit metaphysica* Boneti.

 Fol. 107r: *Incipit Physica* fratris Nicolai Boneti ordinis fratrum minorum, etc. Quoniam autem sciendi et intelligendi desiderium [in] intellectu est inventum. . . .

 Fol. 186r: . . . ergo plura locata non possunt habere idem ubi et in hoc terminatur octavus liber philosophie naturalis. (There follows a summary of the contents of each of the eight books.)

 Fol. 188r: . . . et quomodo sunt et quomodo non. Et sic terminantur capitula et per consequens summa octavi libri philosophie naturalis. Deo gratias. *Explicit philosophia naturalis Magistri* Nicholai Boneti ordinis minorum.

 Tractatus de Praedicamentis and *Theologia Naturalis* Bibliothèque Nationale, fonds latin, ms. no. 16132. [Available microfilms of 1505 editions contain this manuscript and the *Philosophia naturalis.*]

Bricot, Thomas, see Georges de Bruxelles

Buridan, Jean, fl. 1328-58

    Expliciunt quaestiones super libros de Caelo et Mundo magistri Johannis Buridani rectoris Parisius. Royal Library of Munich, cod. lat. 19551. [*Iohannis Buridani Quaestiones super libri quattuor De Caelo et Mundo.* Edited by E. A. Moody. Cambridge, Mass.: Mediaeval Academy of America, 1942.]

    In Metaphysicam Aristotelis. Quaestiones argutissimae Magistri Joannis Buridani *in ultima praelectione ab ipso recognitae et emissae: ac ad archetypon diligenter repositae: cum duplice indicio: materiarum videlicet in fronte: et quaestionum in operis calce.* Vaenundantur Badio. Colophon: Hic terminantur Metaphysicales quaestiones breves et utiles super libros Metaphysice Aristotelis quae ab excellentissimo magistro Ioanne Buridano diligentissima cura et correctione ac emendatione in formam redactae fuerunt in ultima praelectione ipsius Recognite cursus accuratione et impensis Iodoci Badii Ascensii ad quartum idus Octobris MDXVIII. Deo gratias. [*Commentary on the Metaphysics.* Facsimile of 1518 edition. Frankfurt: Minerva, 1964.]

    Acutissimi philosophi reverendi Magistri Johannis Buridani *Subtilissime quaestiones super octo libros Physicorum Aristotelis diligenter recognite et revise a magistro* Johanne Dullart De Gandavo *antea nusquam impresse.* Venum exponuntur in edibus dionisi roce parisius in vico divi Jacobi sub divi martini intersignio. Colophon: Hic finem accipiunt questiones reverendi magistri Johannis Buridani super octo phisicorum libros impresse parhisiis opera ac industria Magistri Petri le dru Impensis vero honesti bibliopole Dionisii roce sub divo martino in via ad divum Jacobum Anno millesimo quingentesimo nono octavo calendas novembres. [*Commentary on the Physics.* Facsimile of 1509 edition. Frankfurt: Minerva, 1964.]

Buridan, Jean II

    Questiones super tres primos libros Metheororum et super majorem partem quarti a Magistro Jo. Buridan. Bibliothèque Nationale, fonds latin, ms. no. 14723.

Buridan, John, see Buridan, Jean

Burley, Walter, 1275-1345(?)

    Preclarissimi viri Gualtery Burlei *anglici sacre pagine professoris excellentissimi super artem veterem Porphyrii et Aristotelis expositio sive scriptum feliciter incipit* . . . Colophon: Explicit scriptum preclarissimi viri Gualtery Burlei Anglici sacre pagine professoris eximii. In artem veterem Porphyrii et Aristotelis arte ac diligentia

Boneti de locatellis sumptibus vero D. Octaviani Scoti Impressum Venetiis Anno 1488 Octavo idus Julii. [For other works contained in this work, see *Le Système du monde* 6:673.] [*Super artem veterem Porphyrii*. Facsimile of 1497 edition. Frankfurt: Minerva, 1967.]

Burleus *super octo libros Physicorum*. Colophon: Et in hoc finitur expositio excellentissimi philosophi Gualtery de Burley anglici in libros octo de Physico auditu. Aristo. stragerite (sic.) *emendata diligentissime*. Impressa arte et diligentia Boneti locatelli bergomensis. Sumptibus vero et expensis Nobilis viri Octaviani scoti modoetiensis. Et humato Jesu eiusque genitrici supra millesimum et quadringentesimum. Quarto nonas decembris. [*In Physicam Aristotelis expositio et quaestiones*. Facsimile. Hildesheim, N.Y.: Olms, 1972.]

Campanus of Novara, ca. 1205-96
Explicit *theorica planetarum* Campani. Bibliothèque Nationale, ms. no. 7401. [For further bibliographical information, see *Le Système du monde* 3:322-23.] [*Campanus of Novara and Medieval Planetary Theory. Theorica Planetarum*. Edited and translated by F. S. Benjamin and G. J. Toomer. Madison: University of Wisconsin Press, 1971.]

Carra de Vaux, Bernard
Avicenne. Paris: F. Alcan, 1900.

Couturat, Louis
De l'infini mathématique. Paris: F. Alcan, 1896.

Curtze, Ernst Ludwig Wilhelm Maximillian
Ueber die Handschrift R. 4 2. *Problematum Euclidis explicatio der Konigl. Gymnasialbibliothek zu Thorn*, in *Zeitschrift fur Mathematik und Physik*, vol. 13, supplement, 1868.

Denifle, Heinrich and Chatelain, Aemilio
Chartularium Universitatis Parisiensis. Vol. 1 (1200-85). Paris, 1889.

Dieterici, Friedrich
Die Philosophie der Araber in IX und X. Jahrhundert n Chr. aus der Theologie des Aristoteles, *den Abhandlungen* Alfarabi *und den Schriften der* Lautern Bruder *Herausgegeben und ubersetzt von* Dr. Friederich Dieterici. Vol. 5, *Die Naturanschauung und Naturphilosophie*. 2d ed., Leipzig: J.C. Henrichs, 1876.

Duns Scotus, Joannes, 1265-1308
R. F. P. Joannis Duns Scoti, Doctoris Subtilis, Ordinis Minorum, *Opera omnia quae hucusque reperiri potuerunt, collecta, recognita, notis, scholiis, et commentariis illustrata, a. P. P. Hibernis, Collegii Romani S. Isidori professoribus, jussu et auspiciis R.mi. P. F. Joannis Baptistae a Campanea, ministri generalis*. Lugduni, sumptibus

Laurentii Durand, MDCXXXIX. [*Opera Omnia*, 12 vols. Facsimile of 1639 edition. Hildesheim, N.Y.: Olms, 1968.]

Quaestiones Quodlibetales Joannis Duns Scoti Ordinis Minorum. Colophon: Expliciunt quaestiones Quodlibetales aeditae a fratre Joanne Duns ordinis fratrum minorum doctore subtilissimo: ac omnium theologorum principe. Per excellentissimum sacre theologiae doctorem magistrum Mauritium de Portu Hibernicum eiusdem ordinis fratrum minorum: in gymnasio Patavino ordinarie legentem maxima cum diligentia emendatae. Impressae Venetijs (mandato domini Andreae Torresani de Asula) per Simonem de Luere, 28 Julii 1506. Second Edition de Paris, 1513. [For further bibliographical information, see *Le Système du monde* 6:362-63.]

Secundus scripti Oxoniensis Doctoris Subtilis Fratris Joannis Duns Scoti ordinis Minorum *super Sententias*. Colophon: Venetiis per Simonem de Luere pro domino Andrea de Torresanis de Asula, 22 Octobris 1506. [For more bibliographical information, see *Le Système du monde* 3:493; 6:362.]

Durandus de Sancto Porciano, d. 1334

Durandi A Sancto Portiano *Super sententias theologicas Petri Lombardi commentariorum libri quatuor*. Per fratrem Iacobum Albertum Castrensene ad fidem veterem exemplarium diligenter recogniti. Venundatur Parisiis apud Ioannem Roigny sub basilio, et quatuor elementis, via ad divum Iacobum. 1539. [*Commentary on the Sentences*. Facsimile of 1579 edition. Ridegwood, N.J.: Gregg Press, 1964.]

Al-Farabi, Muhammad ibn Muhammad, 872-950

Alfarabi *Philosophische Abhandlungen aus Londoner, Leidener und Berliner-Handschriften herausgegeben von* Dr. Friedrich Dieterici. Leiden: E. J. Brill, 1890. Translated into German in al-farabi's *Philosophische Abhandlungen, aus dem Arabischen ubersetzt von* Dr. Fr. Dieterici. Leiden: Brill, 1892. [For further bibliographical information, see *Le Système du monde* 4:404.]

Francis of Mayronnes, see Franciscus de Mayronis

Franciscus de Marchia

Reportatio 4i libri Sententiarum Magistri Francisci de Marchia fratris minoris et sacre theologie doctoris. Bibliothèque Nationale, fonds latin, ms. no. 15852.

Franciscus de Mayronis, ca. 1285-1328

Fol. non sign. occupying the place of fol. A.I: Illuminati doctoris fratris Francisci de Mayronis. *In Primum Sententiarum foecundissimum scriptum sum Conflatus nominatum*. In back is a dedication to Maurice du Port, dated in Padua, 1504.

Fol. 136, col. d, colophon: Illuminati doctoris fratris Francisci

Maronis: Ordinis Minorum: provincie sancti Ludovici: scriptum preclarissimum super Primum sententiarum: Conflatus nuncupatum: fine felici terminatur: Correctum atque decoratum summa cura atqua sollertia sacre Theologie doctoris eximii patris fratris Mauritii de Hybernia: eiusdem religionis: dum actu theologiam publice in alma universitate Pativina legebat. [For further information about all four books see *Le Système du monde* 6:452-54.] [*Commentary on the Sentences.* Facsimile of 1520 Venice edition. Frankfurt: Minerva, 1966.]

Gaetano of Thienis (or Cajetanus de Thienis, canon of Padua), see Thiene, Gaetano

George of Brussels (or Georgius Bruxellensis), see Georges de Bruxelles

Georges de Bruxelles, and Bricot, Thomas, fl. 1486

Cursus optimarum questionum super Philosophiam Aristotelis cum interpretatione textus secundum viam Modernorum: ac secundum cursum magistri Georgii: *Per magistram* Thomas Bricot: *sacre Theologie professorum emendate. Incipiunt questiones super Philosophiam Aristotelis cum interpretatione textus eiusdem edite a magistro* Georgio *et per magistrum* Thomas Bricot *emendate.* Sine loco, anno. Typographo: Sed Basiliae, per Johannem Amerbach. [*Incip. textus abbreviatus Aristotelis super octo libris phisicorum & tota naturali philosopha nuper a magistro* Thoma Bricot . . . *una cum continuatione textus magistri* Georgii . . . a Jacobo Mailleri lugdini impresse anno salutis Mcccccii xvi Kal octobris.]

Al-Ghazzali, 1058-1111

Logica et Philosophia Algazelis Arabis. *Petrus Lichenstein Coloniensis Germanus: ex oris Erweruelde oriundus Ad laudem et honorem dei summi tonantis; et ad commune bonum seu utilitatem summis cum vigiliis laboribusque hoc preclarum in lucem opus prodire fecit Anno Virginei partus* MDVI *Idibus Februariis sub hemispherio Veneto. [For further bibliographical information, see Le Système du monde 4:403.] [Logica et Philosophia.* Facsimile of 1506 edition. Frankfurt: Minerva, 1969.]

Giles of Rome, see Aegidius Romanus

Godfrey of Fontaines, see Godefroid de Fontaines

Godefroid de Fontaines, ca. 1205-1309

Les Quatres premiers quodlibets de Godefroid de Fontaines. Edited by M. De Wulf and A. Pelzer. Louvain, 1904.

Les Quodlibets cinq, six et sept de Godefroid de Fontaines. Edited by M. De Wulf and J. Hoffmans. Louvain, 1914.

Graziadei of Ascoli, see Graziadei, Giovanni of Ascoli

Graziadei, Giovanni of Ascoli, fl. 1341

Incipiunt preclarissime quaestiones litterales edite a fratre Gratia Deo Esculano *sacri ordinis predicatorum super libros Aristo. de Physico auditu: secundum ordinem lectionum Divi Thome Aquinatis. Incipiunt quaestiones fratris* Gratiadei de Esculo *excellentissimi sacre pagine doctoris predicatorum ordinis per ipsum in florentissimo studio patavino disputate feliciter.* Colophon: Hic lector suavissime divina ope preclarissime questiones de physico auditu fratris Gratiadei Esculani sacri ordinis fratrum predicatorum: finem accipiunt unperque reperte: ac ex archetypo imprese: a Reverendoque in christo patre fratre Nicolae methonensi eiusdem ordinis maxima cum diligentia emendate: studio vero et impensa nobilis viri domini Alexandri Calcedonii civis Pisaurensis: arte vero et industria magistri Petri de quarengiis civis Bergomensis: Impresse: anno a nativitate domini Millesimo quingentesimo tertio Idibus Decembris: Venetijs Leonardo Lauretano principe.

Gratia Dei Esculani, seu ab Esculo. *Quaestiones in libros Physicorum Aristotelis,* in *Studio Patavino disputatae.* Colophon: Ad instantiam Antonii de regio. Anna incartationis christi. MCCCCLXXXIII pridie calendas Maias. Feliciter Venetiis impresse: ibidemque Ioanne Mocenigo principe illustrissimo regnante (Hain, *Repertorium bibliographicum,* no. 7877). [Microfilm available.]

Gregorius de Arimino, d. 1358

Gregorius de Arimino *In primum Sententiarum nuperrime impressus. Et quamdiligentissime sue integritati restitutus. Per doctissimum Sacre pagine professorem Fratrem Garamanta doctorem Parrhisiensem Augustinianum. Venundantur Parrhisijs a Claudio Chevallon in vico Jacobeo sub intersignio Solis aurei: et in vico divi Joannis Lateranensis sub intersignio divi Christofori.* Colophon: Explicit lectura primi sententiarum fratris Gregorii de Arimino: sacri ordinis heremitarum sancti Augusti. Theologie professoris precellentissimi prioris generalis quondam prefati ordinis. Qui legit Parisius anno domini 1344. Per venerabilem sacre Theologie professorem fratrem Petrum de Garanta (sic) quamdiligentissime castigata et sue pristine integritati restituta. Gregorius de Arimino *In secundum Sententiarum . . . sub signo Sancti Christofori.* Colophon: Explicit lectura Secundi sententiarum Fratris Gregorii de Arimino: . . . professoris excellentissimi . . . qui legit Parisius anno domini 1344. . . . (This edition, printed by Claude Chevallon, bears no date.) [*Super Primum et Secundum Sententiarum.* Reprint of 1522 edition. St. Bonaventure, N.Y.: Franciscan Institute, 1955.]

Gregory of Rimini, see Gregorius de Arimino

Grosseteste, Robert, Bishop of Lincoln, 1175(?)-1253

Divi Roberti Lincolniensis *Super octo libris Physicorum brevis et utilitis summa feliciter incipit.* This *Summa is to be found at the end of the following work: Emptor et lector Aveto. Divi Thome Aquinatis In libros Physicorum Aristotelis interpretatio sum et expositio* . . . Colophon: . . . Impressa in inclyta Venetiarum urbe per Bonetum Locatellum Bergomensem presbyterum mandato et sumptibus heredum nobilis viri domini Octaviani Scoti civis Modoetiensis Anno a nativitate Domini quarto supra millesimum quinquiesque centesimum, sexto Idus Aprilis [Venetiis, MDCVII apud Hieronymi Scoti]. [See also Baur, Ludwig. *Die philosophischen Werke Robert Grosseteste, Bischof von Lincoln.* Munster, 1912.]

Guilelmus Avernus, d. 1249

Guillelmi Parisiensis Episcopi doctoris eximii *Operum summa divinarum humanarumve rerum difficultates profundissime resolvens* . . . Venales habentur in via Jacobea in officina Francisci Regnault sub divi Caludii intersignio. The second volume has the title: *Pars secunda operum* Guillelmi Parisiensis Episcopi *Morales, theologas, atque philosophicas difficultates dubiaque inaudita dilucide aperiens* . . . Colophon: Summa hic finem capiunt Guillelmi Parisiensis episcopi . . . non modicis sumptibus honestissimi atuque probi viri Francisci regnault librarii jurati universitatis Parisiane vigilantissimi: solis luce V. Julii. Ab incarnato domino Anno XVI supra millesimum quingentesimum. [For more information about various editions of this work see *Le Système du monde* 3:250n.] [*Opera.* Facsimile of 1674 edition. Frankfurt: Minerva, 1963.]

Gulielmus Varonis, fl. end of thirteenth century

Gulielmi Varonis *Quaestiones super libros Sententiarum.* Bibliothèque municipale de Bordeaux, ms. no. 163.

Hennon, John, fl. 1463

Bibliothèque Nationale, fonds latin, ms. no. 6529. In fol. 4r, someone during the seventeenth century wrote: *Magistri* Johannis Hennon *Commentarii in Aristotelis libros Physicorum, parva naturalis et metaphysicam, completi die prima octobris anno 1473, ut habetur in ultima pagina hujus libri.*

Henricus Gandavensis, 1217-93

Quodlibeta Magistri Henrici Goethals a Gandavo Doctoris Solemnis Socii Sorbonici: et archidianconi Tornacensis, cum duplici tabella. Vaenundantur ab Iodoco Badio Ascensio, sub gratia et privilegio ad finem explicandis. Colophon: In chalcographia Iodoci Badii Ascensii. Cui Christianissimus Francorum rex concessit de singulari gratia privilegium et auctoritatem imprimendi et vendendi in regno suo hec et alia Magistri nostri Henrici de Gandavo opera:

Cum definisione ne alius quispiam audeat eadem imprimere aut impressa aluibi vaenundare sub poena confiscationis sic impressorum intro triennium ab undecimo Kalendas Septemb. Anni domini MDXVIII. [For further bibliographical information, see *Le Système du monde* 6:124.] [*Quodlibets*. Reprint of 1518 edition. Louvain, 1961.]

Henry of Ghent, see Henricus Gandavensis

Hentisbery, William, d. ca. 1373

Tractatus Gulielmi Hentisberi *de sensu composito et diviso.— Regule eiusdem cum sophismatibus.—Declaratio* Gaetani *supra easdem.—Expositio litteralis supra tractatum de tribus.—Questio Messini de motu locali cum expletione* Gaetani.—*Scriptum supro eodem* Angeli de Fosambruno.—Bernardi Torni *Annotata supra eodem.*—Simon de Lendenaria *Supra sex sophismata.—Tractatus* Hentisberi *de veritate et falsitate propositionis.— Conclusiones eiusdem*. Colophon: Impressa venetiis per Bonetum locatellum bergomensem: sumptibus Nobilis viri Octaviani scoti Modoetiensis. Millesimo quadringentesimo nonagesimo quarto sexto Kalendas iunias.

Holkot, Robert, see Holkot, Robertus

Holkot, Robertus, d. 1349

Magistri Roberti Holkot *Super quatuor libros Sententiarum questiones. Quaedam conferentie. De imputabilitate peccati questio longa. Determinationes quarundam aliarum questionum. Tabule duplices omnium predictorum*. Colophon: Hujus operis diligenter impressi Lugdunia magistro Johanne Trechsel alemanno, anno salutis nostre MCCCCSCVII. Ad nonas Aprilis. registrum . . . [*Commentary on the Sentences*. Facsimile of 1518 edition. Frankfurt: Minerva, 1967. Microfilm of 1497 edition available.]

Honorius Augustodunensis, fl. 1106-35

Honorius Solitarius *De imagine mundi* in Beati Anselmi *Opuscula*, Basileae(?) 1497(?) and Honorii Augustodunensis *Opera* in *Patrologiae latinae*, accurante J. P. Migne, vol. CLXXII. Paris: Carnier & Freres, 1895.

Honorius Inclusus, see Honorius Augustodunensis

Houzeau, Jean Charles, and Lancaster, A.

Bibliographie générale de l'Astronomie. Bruxelles, 1887.

Jesuits of the University of Coimbra

Commentarii Colegii Coninbricensis, Societatis Jesu, in octo libros Physicorum Aristotelis Stagiritae. Lyon, sumptibus Horatii Cardon, 1602.

Joannes XXI, Pope, d. 1277

Petri Hispani *Summulae logicales cum* Versorii Parisiensis *Clarissima expositione. Parvorum logicalium eidem* Petro Hispano *Ascriptum opus nuper in partes ac capita distinctum.* Venetiis, Apud Haeredes Melchioris Sessae. MDLXXXIII. [*The Summulae Logicales of Peter of Spain.* Translated by Joseph P. Mullaly. Notre Dame, Ind.: University of Notre Dame Press, 1945.] [*Tractatus* (called afterward *Summulae Logicales*). Edited by L. M. de Rijk. Assen: Van Gorcum & Company, 1972.]

Joannes Canonicus Anglus, fl. 1329

Joannis Canonici *Quaestiones super VIII libros Physicorum Aristotelis perutiles: nuperrime correcte et emendate: additis textibus Commentorum in margine: una cum utili Repertorio cunctorum auctoris notabilium indice.* Colophon: . . . Venetiis mandato heredum q. domini Octaviani Scoti civis ac patricij Modoetiensis: et sociorum. Anno a dominica incarnatione 1520 die 8 Maij. [Microfilm of 1520 edition available.]

Joannes de Dumbleton, fl. mid-fourteenth century

Joannis de Dumbleton *Summa.* Bibliothèque Nationale, fonds latin, ms. no. 16146.

Joannes de Janduno, d. 1328

Ioannis de Ianduno *In libros Aristotelis de Caelo et Mundo quae extant quaestiones subtilissimae: quibus nuper consulto a diecimus* Averrois *Sermonem de substantia orbis cum* Ioannis *Commentario ac quaestionibus.* Venetiis apud Iuntas. Anno MDLII. [Microfilm available. For a listing of other editions, see *Le Système du monde* 6:542.]

Joannis de Janduno *Quaestiones de motibus animalium.* These questions of John of Jandun are located among the questions on the *Parva naturalia* which have had no edition other than the following:

Ioan. Gandavensis *Philosophi acutissimi Quaestiones, Super Parvis Naturalibus, cum* Marci Antonii Zimarae *De Movente et Moto, ad Aristotelis et Averrois intentionem, absolutissima quaestione, ac variis margineis scholiis hinc inde ornatae.* Nunc denno per Albratium Apulum, in Gymnasio Patavino Philosophiam publice profitentem quam diligentissime emendatae . . . Venetiis, apud Hieronymum Scotum, MDLVII [also MDLXX].

Joannis de Janduno *Philosophi acutissimi super octo libros Aristotelis de Physica auditu subtilissimae quaestiones* Eliae *etiam Hebraei Cretensis quaestiones* . . . Venetiis, apud Iuntas Anno MDLI. [For a brief list of other editions and dates, see *Le Système du monde,* 6:542.] [*Commentary on the Physics.* Facsimile of 1551 edition. Frankfurt: Minerva, 1969.]

Ioannis de Ianduno Expositio super libro de substantia orbis. Item Questiones super eodem libro. Vincentiae impensa ingenique Henrici de Sancto Viso (no date) (Hain, *Repertorium bibliographicum*, no. 7464). [For further bibliographical information, see *Le Système du monde* 6:444-45.]

Jodocus of Eisenach (or Judocus von Eisenach), see Trutfetter, Jodocus

John of Baconthorpe, d. 1346

Edicitur per gratiam et Regium Privilegium sub pena in eo contenta ne quis aeditionem hanc iterum attentare ausit in toto hoc Mediolanensi Ducatu—En Lector Doctoris Resoluti Ioannis Bacconis *Anglici Carmelitae radiantissimum opus super quatuor sententiarum libris. In cuius Fonte lotius sapientiae uberrimus invenies latices. Nam si dei opt. maximi penetralia adire suadet animus: nemo melius: nemo accurabilis (sic) essentiam mandavit litteris. Si rerum causas: si naturae effectus: si caeli varios motus ac elementorum contrarias qualitates discere exoptas: una se hic offert officina: ubi omnia cuduntur in qua animam averrois: Mentem aristotelis intus et incute apertissime intueri licet. Christianae Religionis arma vulcaniis munitiora contra Indeos solus hic doctor in iii. et iiii. libro ministranda tradidit: Mesiae super quatuor sententiarum libris . . . adventum dilucidat Antichristi aperit venturam Fallaciam: Quem multi errore ducti venisse opinatur. Manmethi secutam posternit, scripturae nodos solvit. Enigmata cuncta serenat. Heus quisquis es. Id omne referas Petro terassae. Qui theologorum omnium clariss. Oratorum Facundis Carmelitanae religionis et partis opt. & Generalis meritis. Hoc opus aureum propiris impensis iussit (quae est sui animi aput) (sic) omnes graditudo prodire in lucem.* Colophon of the fourth volume: Theologi excellentissimi Ioannis Bacconis Anglici Carmelite Questiones disputate in quartum sententiarum. Explicit Mediolani. In officina libraria Leonard Vegii anno MDX die xxii Aprilis. Duce Mediolani VII Ludovico: ac Faelicissimo Francorum Rege. Gubernante Carolo de Ambosya omnium mortalium iustissimo, et gloria rei militaris illustrissimo: Praeside iafredo Caelo virorum omnium sapientissimo. (Vols. I, II, and III have the following dates: *Super quatuor Sententiarum libris . . .* Anno MDX, die xxiii Aprilis—Anno MDXI, Die xvii Februarii.— Mediolani Mccccx die xxv mensis Februarii.)

John of Bassols (or Joannes de Bassolis), see Bassolis, Joannes de

John of Jandun, see Joannes de Janduno

Lambertus de Monte, see Monte, Lambertus de

Lee, Sir Sidney

    Dictionary of National Biography. 63 vols. London: Smith, Elder & Co., 1885-1901.

Lull, Ramon, d. 1315

    Declaratio Raymundi *per modum dialogi edita contra aliquorum philosophorum et eorum sequacium opiniones erroneas et damnatas a venerabili Padre Domino episcopo Parisiensi;* in Otto Keitcher, *Raymundus Lullus und seine Stellung zur arabischen Philosophie,* in *Beitrage zur Geschichte, der Philosophie des Mittelalders,* vol. 7, parts 4-5. Munster, 1909.

Lull, Raymond, see Lull, Ramon

Maimonides, Moses (or Maimon, Moses ben),
see Moses ben Maimon

Major, John, 1469-1550

    Magister Johannes Majoris Scotus *Omnia Opera in Artes quas Liberales Vocant a perspicacissimo et fantassimo uno sanctarum litteratum professore profundissimo Magistro Johanne Majoris majori accuratione elaborata atque castigata quam antehac in lucem prodita sint majorique precio comparanda quam quispiam persolvere possit si ea ab equo judice pensiculantor.* Colophon: Impressum Cadomi per Larrentium Hostingue impensis vivorum industriosorum Michaelis Augier prope pontem ejusdem Cadomi commorantis et Johannis Mace e regione Sancti Salvatoris Redonis residentis (no date). [*Le traité "De L'infini" de Jean Mair.* Edited and translated by Hubert Elie. Paris: Vrin, 1937.]

Mandonnet, Pierre

    Les écrits authentiques de Saint Thomas d'Aquin (Revue Thomiste, 1909-10).

    Siger de Brabant (Etude critique), Les Philosophes Belges. Textes et Etudes. 2 vols. Louvain, 1908, 1911. [For further bibliographical information, see *Le Système du monde* 4:310.]

Marsilius of Inghen, see Marsilius von Inghen

Marsilius von Inghen, d. 1396

    Incipiunt subtiles doctrinaque plene abbreviationes libri Physicorum edite a prestantissimo philosopho Marsilio Inguen doctore parisiensi. Pavia, Antonius de Carcano, ca. 1490. [Microfilm available.]

    Quaestiones subtilissimae Johannis Marcilii Inguen *Super octo libros Physicorum secundum nominalium viam.* Colophon: Impresse Lugduni per honestum virum Johannem Marion, anno Domini MCCCCCXVIII. [*Quaestiones super octo libros Physicorum.* Facsimile of 1518 edition. Frankfurt: Minerva, 1964.]

Michael Scot, see Scott, Michael

Milhaud, G.

"Le Traité de la méthode d'Archimede." *Revue Scientifique.* October 1908. *Nouvelles etudes sur l'histoire de la Pensée scientifique.* Paris, 1911.

Monte, Lambertus de, d. 1499

Prohemium Phisicorum. Colophon: Copulata prediligenti studio correcta circa octo phisicorum Aresto tilis Lamberti de Monte artium ac sacre theologie professoris iuxta doctrinam excellentissimi doctoris sancti Thome de Aquino ordinis predicamentorum hic felciter finem habent.

Moses ben Maimon, 1135-1204

Moise ben Maimoun dit Maimonide. *Le guide des égares, traité de Théologie et de Philosophie.* 3 vols. Translated by S. Munk. Paris, 1856-66. [*The Guide of the Perplexed.* Translated by Shlomo Pines. Chicago: University of Chicago Press, 1963.]

Nicholas of Cusa, see Nicolaus Cusanus,

Nicholas of Orbellis, see Orbellis, Nicolaus de

Nicolaus Cusanus, Cardinal, 1401-65

D. Nicolai de Cusa *Cardinalis, utriusque juris doctoris, in omnique philosophia incomparabilis viri Opera . . . Librorum catalogum versa pagina indicabti. Cum privilegio Caes. Majest. Basileae, ex. officina Henricpetrina.* In fine: Basileae, ex officina Henricpetrina. Mense Augusto, Anno MDLXV. [*Nicolai de Cusa Opera omnia iussu et auctoritate.* Heidelberger Akademie. Leipzig: Felix Meiner, 1944.]

Nifo, Agostino, ca. 1473-1545

Aristotelis Stagiritae *De Caelo et Mundo libri quatuor, e graeco in latinum ab* Augustino Nipho *Philosopho Suessano conversi, et ab eodem etiam . . . aucti expositione . . .* Venetiis, apud Hieronymum Scotum, 1550.

Nipho, Agostino, see Nifo, Agostino

Ockham, William, d. ca. 1349

[Guilhelmi de Ockham *Opera Philosophica et Theologica ad fidem codicum manuscriptorum edita.* St. Bonaventure, N.Y.: Franciscan Institute, 1977-].

Quodlibeta (sic) *Septem una cum tractatus de sacramento altaris* Venerabilis inceptoris fratris Guilhelmi de Ockham anglici, sacre theologie magistri, de ordine fratrum minorum. Colophon of the *Quodlibeta*: Expliciunt quodlibeta septum venerabilis inceptoris magistri Wilhelmi de Ockam anglici, veritatum speculatoris acerrimi, fratris ordinis minorum post ejus lecturam oxoniensem (super sententias) edita. Impressa Argentine Anno domini Mccccxci. Finita post Epiphanie (sic) domini.—Colophon of the *Tractatus de Sacramento Altari*: Explicit tractatus gloriosus de corpore christi et in primis de puncti linee superficiei corporis quantitatis et substantie distinctione. Venerabilis inceptoris magistri Guilhelmi de Ockam anglici. Veritatis indagatoris profundissimi sacre theologie professoris doctissimi, ed ordine fratrum minorum post lecturam oxoniensem, catholice editus. Impressus Argentine anno domini Mccccxci. Finitus post festum Epiphanie domini. [Microfilm available. For further bibliographical information, see *Le Système du monde* 6:578-79.] [*The De Sacramento Altaris of Ockham*. Edited and translated by T. Bruce Birch. Burlington, Ia.: The Lutheran Literary Board, 1930.]

Tabule ad diversa hujus operis Magistri Guilhelmi de Ockam *Super quatuor libros Sententiarum annotationes et ad centilogii theologici ejusdem conclusiones facile reperiendas apprime conducibiles.* Colophon (at the end of the *Questions on the Sentences*): Impressum est autem hoc opus Lugduni per M. Johannem Trechsel Alemannum: virum hujus artis solertissimum. Anno domini nostri MCCCCXCV. [Microfilm available. For further bibliographical information, see *Le Système du monde* 6:577-78.]

Bibliothèque Nationale, fonds latin, ms. no. 16130. . . . Explicit *Tractatus de successivis editus a* Guillelmo de Ocham. [*The tractatus de successivis, attributed to William Ockham*. Edited by Philotheus Boehner. St. Bonaventure, N.Y.: The Franciscan Institute, 1944.]

Venerabilis inceptoris fratris Guilielmi de Villa Hoccham Anglie: *Academie Nominalium Principis Summulae in libros Physicorum adsunt.* Cum gratia un patet in suis privilegiis. Colophon: Expliciunt auree summule in lib. physicorum Fratris Guilielmi de Villa Hoccham Anglie: Academie Nominalium principis: Sacrarum litterarum professoris; ex ordine minorum: Correcte vigili studio ac labore venerabilis patris Fratris Augustini de Filizano ordinis sancti Augustini Sacre Theologie professoris. Impresseque Venetiis per Lazarum de Soardis. Anno 1506. Die 17 Augusti. (1st ed.: Bologne, 1494 by Benedictus Hectoris Bononiensis.) [Microfilm available.] [*Philosophia Naturalis*. Facsimile of 1637 edition. Ridgewood, N.J.: Gregg Press, 1963.]

Quaestiones magistri Guilelmi de Okam *super librum Phisicorum.* Bibliothèque Nationale, fonds latin, ms. no. 17841. [For further bibliographical information, see *Le Système du monde* 6:579-80.]

[*Ockham's Theory of Terms, Part I of the Summa Logicae.* Translated by M. Loux. Notre Dame, Ind.: University of Notre Dame Press, 1974.]

[*Ockham's Theory of Propositions: Part II of the Summa Logicae.* Translated by A. Freddoso and H. Schuurman. Notre Dame, Ind.: University of Notre Dame Press, 1979.]

Orbellis, Nicolaus de, d. 1475

Curcus librorum philosophiae naturalis venerabilis magistri Nicolai de Orbelli ordinis minorum secundum viam doctoris subtilis Scoti. Colophon: Expliciunt libri Ethicorum Basilee impressi: Anno incarnationis domini Mccccciii.

Oresme, Nicolas, d. 1382

Traité du Ciel et du Monde. Bibliothèque Nationale, fonds français, ms. no. 1083. [Le livre du ciel et du monde. Edited by A.D. Menut, and A. J. Denomy. Translated by A. D. Menut. Madison: University of Wisconsin Press, 1968.]

Tractatus de difformitate qualitatum. Bibliothèque Nationale, fonds latin, ms. no. 7371.

Oresme, Nicole, see Oresme, Nicolas

Paul of Venice (or Paul Nicoletti), see Paulus Venetus

Paulus Venetus, d. 1429

Expositio Magistri Pauli Veneti super libros de generatione et corruptione Aristotelis. Liber de compositione mundi cum figuris . . . Divi Pauli Veneti Theologi clarissimi: philosophi summi: ac astronomi maximi Augustiniani libellus quem inscripsit de compositione mundi Aureus incipit. Colophon: Pauli Veneti Theologi clarissimi: ac philosophi summi liber aureus quem de compositione mundi edidit, Feliciter explicit. Correctus a proprio originali per venerabilem virum fratrem Jacobum Baptisam Aloyxium de Ravenna lectorem in conventu Venetiarum sancti Stephani. Impressus Venetiis mandato et expensis nobilis Viri Domini Octaviani Scoti Civil Modoetiensis duodecimo kalendas Junias 1498. Per Bonetum Locatellum Bergomensem. Finis. [Microfilm available. For further bibliographical information, see Le Système du monde 4:208.]

Pauli de Venetiis Summa totius philosophiae. Expositio librorum naturalium Aristotelis. Colophon: Explicit sexta et ultima pars summae naturalium acta et compilata per reverendum artium et theologie doctorem magistrum Paulum de Venetija ordinis fratrum heremitarum sancti Augustini transumpta ex proprio originali manu propria prefati magistri confecta Venetijs impressionem habuit impensis Iohannis de Colonia sociique ejus manthen de Gherretzem. Anno a natali christiano. MCCCClxxvi. [Microfilm available. For further bibliographical information, see Le Système du monde, 4:283-84.] [Summa philosophiae naturalis magistri Pauli Veneti noviter recognita per ditiis purgata ac pristine integritati restituta cùm privilegio . . . Venetiis per Bonetum locatellum bergomensem. Sumptibus nobilis viri divi Octaviani Scoti civis Modoetiensis. Anno a saluti fera incarnatione tertio et quingentesimo supra millesimum. 7 die Martys.] [Summa philosophie naturalis. Facsimile of Venice 1503 edition. Hildesheim, N.Y.: Olms, 1974.]

Expositio Pauli Veneti *super octo libros Physicorum Aristotelis necnon super comento Averrois cum dubiis eiusdem.* Colophon: Explicit liber Phisicorum Aristotelis: expositum per me fratrem Paulum de Venetiis: artium liberalium et sacre theologie doctorem: ordinis fratrum heremitarum beatissimi Augustini. Anno domini MCCCCIX die ultima mensis Junii: qua festum celebratur commemorationis doctoris gentium et christianorum Apostoli Pauli. Impressum Venetiis per providum virum dominum Gregorium de Gregoriis. Anna nativitatis domini MCCCCXCIX. Die XXIII mensis Aprilis. [Microfilm available.]

Pauli Veneti *Universalia sexque principia.* Colophon: Expliciunt predicamenta Aristot. exposita per me fratrem Paulum de Venetijs artium liberalium et sacre theologie doctorem ordinis heremetiarum beatissimi augustini, etc. Anno domini Mcccxxviii die martij . . . Impressa Venetiis per Bonetum Locatellum bergomensem. Sumptibus nobilis viri d. Octaviani Scoti civis Modoetiensis. Anno ab incarnatione Jesu Christi Domini Nostri nonagesimo quarto supra millesimum et quadrigentesimum. Nono calendas octobres.

Peter of Spain (or Petrus Hispanus), see Joannes XXI

Petrus Aquilanus, see Pietre dell' Aquila

Petrus de Alvernia, see Petrus of Auvergne

Petrus of Auvergne, Bishop of Clermont, d. 1304

In presenti volumine infrascripta invenies *opuscula* Aristotelis *cum expositionibus* Sancti Thome: *ac* Petri de Alvernia. Perquam diligenter visa recognita: erroribusque innumeris purgata. Sanctus Thomas *De sensu et sensato. De memoria et reminiscentia. De somno et vigilia. Ultimo altissimi* Proculi (sic) *de causis cum* ejusdem Sancti Thome *commentationibus.* Petrus de Alvernia *De motibus animalium. De longitudine et brevitate vite. De juventute et senectute. De respiratione et inspiratione. De morte et vita.* Egidius Romanus *De bona fortuna.* Colophon: . . . Impressa vero Venetiis mandato sumptibusque Heredum nobilis viri domini Octaviani Scoti civis Modoentiensis, per Bonatum Locatellus presbyterum Bergomensem. Anno a partu virgineo saluberrimo septimo supra millesimum quinquiesque centisimum quinto Idus Novembris.

[S. Thomae Aquinatis *Commentaria quae extant in eos, qui parva naturalia Aristotelis dicuntur libros, diligentissime castigata, duplici nuper textus tralatione, antiqua videlicet recognita & nova* Nicol Leonici *opposita* Pietri Item de Alvernia, *Ordinis Praedicatorum in quosdam huius operis a divo Thoma in expositos libros refertissima expositio, ab innumeris erroribus dermo expurgata* . . . Venetiis, apud Lucam Antonium Iuntam. Anno MDLXVI.]

Petrus Lombardus, Bishop of Paris, 12th century

Petri Lombardi *Episcopi Parisiensis, Sententiarum libri quatuor.* Edited by J.P. Migne. Paris, 1853.

Petrus Padubaensis

Petri Padubaensis *Lucidator Astronomiae.* Bibliothèque Nationale, fonds latin, ms. no. 2598.

Pierre d'Ailly (or Petrus de Aliaco), see d'Ailly, Pierre

Pietre dell' Aquila, d. 1361

Petri Aquilani *Cognomento Scotelli ex Ord. Min. in doctrina Ioan. Duns Scoti spectatissimi, Quaestiones in quatuor Sententiarum libros, ad eiusdem Doctrinam multum conferentes.* Venetiis, MDLXXXIIII, Apud Hieronymum Zenarium et Fratres.

Prantl, Carl von

Geschichte der logik im Abendlande. 4 vols. Leipzig, 1855-70.

Quétif, Jacobus, and Echard, Jacobus

Scriptores Ordinis Praedicamentorum. Paris, 1719.

Reisch, Gregor, d. 1525

[Gregorii Reisch] *Margarita philosophica nova cui insunt sequentia. Epigrammata in commentationem operis. Institutio Grammaticae Latinae. Praecepta Logices. Rhetoricae informatio. Ars Memorandi* Ravennatis. Beroaldi *Modus componendi Epi. Arithmetica. Musica plana. Geometrie Principia. Astronomia cum quibusdam de Astrologia. Philosophia Naturalis, Moralis Philosophia cum figuris.* Argentinae. 1512. Colophon: Accipe candide lector Margaritam philosophicam . . . Cum quo te bene valere industrius vir Joannes gruningerius operis excussor et optat et praecatur. Ex Argentorato veteri Pridie Kalendas Junii. Anno redemptionis nostrae duodecimo supra mille quingentos. [*Margarita philosophica, Rationalis, Moralis philosophiae principia, duodecim libris dialogice complectus, olim ab ipso autore recognita: nuper autem ab Orontio fineo Delphi nate castigata & aucta, una cum appendicibus . . .* Basileae, 1535.]

Reisch, Gregory, see Reisch, Gregor

Renan, Ernest

Averroes et Averroisme, essai historique. Paris: A. Durand, 1852.

Richard of Middleton, see Ricardus de Mediavilla

Ricardus de Mediavilla, ca. 1249-1308(?)

Quodlibeta Doctoris eximii Ricardi de Media Villa *Ordinis Minorum, quaestiones octuaginta continentia.* Brixiae, apud Vincentium Sabbium, MDXCI.

Clarissimi theologi Magistri Ricardi de Mediavilla Seraphici ord. min. convent. *super quatuor libros Sententiarum Petri Lombardi quaestiones subtilissimae, Nunc demum post alias editiones diligentius, ac laboriosius (quoad fieri potuit) recognitae, et ab erroribus innumeris castigatae* . . . Brixiae, apud Vincentium Sabium, MDXCI (4 vols.).

[*Commentary on the Sentences* and *Quodlibets.* Facsimiles of 1591 editions. Frankfurt: Minerva, 1963.]

Ristoro d'Arezzo, fl. 1282

Della composizione del Mondo Ristoro d'Arezzo. *Testo italiano del 1282.* Gia pubblicato da Enrico Narducci ed ora in piu comoda forma ridotto. Milano, 1864. [*Della composizione del mondo.* Facsimile of 1854 edition. Bologna: Formi, 1970.]

Robert Anglicus, fl. 1271

Tractatus de spera Jo. de Sacro Bosco *cum glosis* R. Anglici. Bibliothèque Nationale, fonds latin, ms. no. 7392. [See Lynn Thorndyke, *The Sphere of SacroBosco and Its Commentators* (Chicago, 1949) for Robertus Anglicus's *Tractatus de Spera.*]

Sacro Bosco, Joannes de, fl. 1230

Joannis de Sacro Bosco Anglici V. C. *Spaera* (sic) *mundi feliciter incipit.* Colophon: Impressi Andreas hoc opus cui Francia nomen tradidit: at civis Ferrariensis ego . . . MCCCLXXII. [See Lynn Thorndyke, *The Sphere of SacroBosco and Its Commentators* (Chicago, 1949) for Sacro Bosco's *Sphere.*]

Scott, Michael, fl. 1210

Eximii atque excellentissimi physicorum motuum cursusque siderei indagatoris Michaelis Scotis *super auctore Sperae cum quaestionibus diligiter emendatis, incipit expositio confecta Illustrissimi Imperatoris Domini D. Frederici praecibus.* In fine: Bononiae . . . per Iustnianum de Ruberia, MCCCCLXXV, die Septembris. [For further bibliographical information, see *Le Système du monde* 3:246n.] [See Lynn Thorndyke, *The Sphere of SacroBosco and Its Commentators* (Chicago, 1949) for Scot's *Commentary on the Sphere.*]

Scotus, John Duns, see Duns Scotus, Joannes

Simplicius of Cilicia, 6th century

Simplici *Commentarius in IV libros Aristotelis de Caelo.* Ex rec. Sim. Karsteni, Trajecti ad Rhenum, MDCCCLXV. Simplicii *In Aristotelis de Caelo commentaria.* Edidit. I. L. Heiberg, Berolini MDCCCLXXXXIV. [For further information about editions, see *Le Système du monde* 3:356.]

Stobaeus, Joannes

Joannis Stobaei *Eclogarum Physicarum*. In Joannis Stobaei *Eclogarum physicarum et ethicarum libri duo*. Leipzig: Augustus Meineke, 1860-64.

Summenhardt, Konrad, d. 1501

Conradi Summenhart *Commentaria in Summam physice Alberti Magni*. Colophon: Habes nunc Candidissime lector Conradi Summenhart Theologi eruditas commentationes in Albertum recognitas quam plenissime ex corrupto exemplari recognisco potuere. Que miro ingenio literis sunt excuse a solerti Henrico gran Chalcographo in Hagenaw. Hec tam magnum artificium tam aptissimum cultum redolent. Ut que ex aliis libris adhuc obscuriora videntur: hic in promptu patent ad nutem: et sine interprete (sed frequenti exercitatione) percipi possunt. Ocius eme: attentius legito. Ex istis enim totam et naturam et philosophiam consequere. Vale ex Hage cursim Anno 1507 septimo kal. maias.

Sunczel, Frederic, see Sunczel, Fridericus

Sunczel, Fridericus, fl. end of the fifteenth century

Collecta et exercitata Friderici Sunczel Mosellani *liberalium studiorum magistri in octo libros phisicorum Aristotelis: in almo studio Ingolstadiensi. Cum adiectione textus nove translationis Johannis Argiropoli bizatij circa questiones*. Colophon: Laus deo: finiunt collecta et exercitata Friderici Sunczel philosophie: magistri in octo libros phisicorum Aristotelis cum additione textus nove translationis Johannis Argiropolo civea questiones positi cum numero commentariorum opposito super quem mota est questio Impressa sub hemisperio veneto Impensis Leonardi Alantse Bibliopole viennensis Arte vero et ingenio Petri Lichtenstein Coloniensis anno M dvj die xxviii Mensis madij Maximilliano primo Romanorum Rege fautissime imperante, etc. [Microfilm of 1499 edition available.]

Swineshead, Richard, d. ca. 1365

Ro. (Richard) Swineshead *De primo motore*. Bibliothèque Nationale, fonds latin, ms. no. 16621.

Tannery, Jules

Introduction à la théorie des fonctions d'une variable. 1st ed. Paris, 1886.

Thomas ab Argentina, see Thomas of Strasburg

Thomas Aquinas, Saint, 1225-74

[*Opera Omnia*, 34 vols. Paris: Vives, 1871-82.]
Sancti Thomae Aquinatis *De aeternitate mundi contra*

murmurantes opusculum, in S. Thomae Aquinatis *Opuscula;* opusc. XV. [St. Thomas Aquinas, Siger of Brabant, and St. Bonaventure. *On the Eternity of the World (De Aeternitate Mundi).* Translated by C. Vollert, L. H. Kendzierski, and P. M. Byrne. Milwaukee: Marquette University Press, 1964.]

Sancti Thomae Aquinatis *Quaestiones disputatae de potentia Dei. [On the Power of God.* 3 vols. Translated by English Dominican Fathers. London: Burns, Oates & Washbourne, 1932-34.]

Sancti Thomae Aquinatis *Quaestiones disputatae de Scientia Dei.*

Sancti Thomae Aquinatis *Quodlibetales quaestiones* Sancti Thomae. [*Quaestiones de Quodlibet,* Torino-Rome: Casa Marietti, 1949.]

Sancti Thomae Aquinatis *Expositio super libro de Caelo et Mundo Aristotelis.* [*Sententia de Caelo et Mundo.* Torino-Rome: Casa Marietti, 1952.]

Sancti Thomae Aquinatis *Opuscula;* opusc. LII: De natura loci.

Divi Thomae Aquinatis *Totius logicae Aristotelis summa* in D. Thomae Aquinatis *Opuscula;* opusc. XLVIII. [*Opuscula Philosophica,* Torino-Rome: Casa Marietti, 1954].

Emptor et lector aveto. Sancti Thomae Aquinatis *In libros Physicorum Aristotelis interpretatio sum et expositio* . . . Colophon: Expliciunt preclarissima commentaria Divi Thomae Aquinatis . . . Impressa vero in inclita Venetiarum urbe per Bonetum Locatellum Bergomensem presbyterum mandata et sumptibus heredum Nobilis Viri domini Octaviani Scotis Civis. Modoetiensis Anno a nativitate domini quarto supra millesimum quinquiesque centisimum. Sexto Idus Apriles. [*Commentary on Aristotle's Physics.* Translated by R. J. Blackwell, et al. New Haven: Yale University Press, 1963.]

Sancti Thomae Aquinatis *Summa Theologica. [Summa Theologiae.* 60 vols. Translated by Blackfriars. New York: McGraw-Hill, 1964-].

[*On Truth.* 3 vols. Translated by R. Mulligan, J. McGlynn, and R. Schmidt. Chicago: Regnery, 1952-54.]

Thomas of Strasburg, d. 1357

Thomae ab Argentina Eremitarum Divi Augustini prioris generalis *Commentaria in quatuor libros Sententiarum.* Genuae. Apud Antonium Orerium, MDLXXXV. [*Commentary on the Sentences.* Facsimile of 1564 edition. Ridgewood, N.J.: Gregg Press, 1965].

[Thorndike, Lynn]

[*The Sphere of SacroBosco and Its Commentators.* Chicago: The University of Chicago Press, 1949.]

Tiene, Gaetano, 1387-1462

Recollectae Gaietani *super octo libros Physicorum cum annotationibus textuum.* Colophon: Impressum est hoc opus per

Bonetum Locatellum, jussu et expensis nobilis viri Domini Octaviani Scotis civis Modoetiensis. Anno Salutis 1496. [Microfilm available.]

Trutfetter, Jodocus, 1460-1519

Summa in totam Physicen: hoc est philosophiam naturalem conformiter.siquidem verae sophiae: que est Theologia per D. Jodocum Isennachensem *in gymnasio Erphordiensi elucubrata et edita.* Colophon: Impressum Erffordie per Mattheum Maler finitum Feria quinta post Dionisij Anno Millesimo Quingentesimo decimoquarto.

Ulrich von Strasburg, d. 1277

Ulrici Engelberti *Liber de Summo Bono.* Bibliothèque Nationale, fonds latin, ms. no. 15900.

Ulricus Engelberti de Argentina, see Ulrich von Strasburg

William of Auvergne, see Guilelmus Avernus

William of Ockham, see Ockham, William

William Varon, see Gulielmus Varonis

Selected Duhem Bibliography: Historical and Philosophical Works

Works Published Separately

Un Fragment inédit de l'Opus Tertium de Roger Bacon, précédé d'une étude sur ce fragment. Claras Aquas: Quaracchi, 1909.

Le Système du monde: Histoire des doctrines cosmologiques de Platon à Copernic. 10 vols. Paris: Hermann, 1913-59.

Collected Articles

L'Evolution de la Méchanique. Paris: A. Joanin, 1903. First published in *Revue Générales des Sciences pures et appliquées* 14 (1903). Translated by M. Cole as *The Evolution of Mechanics.* Alphen aan den Rijn: Sijthoff and Noordhoff, 1980.

Les Origines de la Statique. 2 vols. Paris: Hermann, 1905-06. First published in *Revue des Questions scientifiques*, 3d ser., vol. 4 (1903); 5´(1904); 6 (1904); 7 (1905); 9 (1906); 10 (1906).

La Théorie Physique, son objet et sa structure. Paris: Chevalier & Rivière, 1906. First published in *Revue de Philosophie* 4, no. 1 (1904); 4, no. 2 (1904); 5, no. 1 (1905). 2d ed., Paris: Rivière, 1914. Translated by P. Wiener as *The Aim and Structure of Physical Theory.* Princeton: Princeton University Press, 1954.

Etudes sur Léonard de Vinci, ceux qu'il a lus et ceux qui l'ont lu. 1st series. Paris: Hermann, 1906. Containing:
  "Albert de Saxe et Léonard de Vinci." *Bulletin Italien* 5 (1905).
  "Léonard de Vinci et Villalpand." Ibid.
  "Léonard de Vinci et Bernardino Baldi." Ibid.
  "Bernardino Baldi, Roberval et Descartes." Ibid., 6 (1906).
  "Thémon le fils de juif et Léonard de Vinci." Ibid.
  "Léonard de Vinci, Cardan et Bernard Palissy." Ibid.
  "La Scientia de ponderibus et Léonard de Vinci."
  "Albert de Saxe."
  (The last two articles have not been published separately.)
Etudes sur Léonard de Vinci. 2d ser. Paris: Hermann, 1909. Containing:

579

"Léonard de Vinci et les deux infinis."

"Léonard de Vinci et la pluralité des mondes."

(The last two articles have not been published separately.)

"Nicholas de Cues et Léonard de Vinci." *Bulletin Italien* 7 (1907).

"Léonard de Vinci et les Origines de la Géologie." Ibid. 8 (1908).

Etudes sur Léonard de Vinci. 3d ser. Paris: Hermann, 1913. Containing:

"Jean I Buridan (de Béthune) et Léonard de Vinci." *Bulletin Italien* 9 (1909).

"La tradition de Buridan et la Science italienne au XVIe siècle." Ibid.

"Dominique Soto et la Scholastique parisienne." *Bulletin Hispanique* 12 (1910).

"La Dialectique d'Oxford et la Scholastique italienne." *Bulletin Italien* 12 (1912).

SOZEIN TA PHAINOMENA: Essai sur la notion de théorie physique de Platon à Galilée. Hermann, 1908. First published in *Annales de Philosophie Chrétienne* 79 (1908). Translated by E. Doland and C. Maschler as *To Save the Phenomena.* Chicago: University of Chicago Press, 1969.

Le Mouvement absolu et le Mouvement relatif. Montligeon: Imprimerie-librairie de Montligeon, 1909. First published in *Revue de Philosophie* 7 (1907); 8 (1908).

La Science allemande. Paris: Hermann, 1915.

Articles and Notes

"Quelques réflexions au sujet des théories physiques." *Revue des Questions Scientifiques,* ser. 2, vol. 1 (1892).

"Notation atomique et hypothèses atomistiques." *Revue des Questions Scientifiques,* ser. 2, vol. 1 (1892).

"Physique et Métaphysique." *Revue des Questions Scientifiques,* ser. 2, vol. 2 (1893).

"L'Ecole anglaise et les théories physiques." *Revue des Questions Scientifiques,* ser. 2, vol. 2 (1893).

"Quelques réflexions au sujet de la Physique expérimentale." *Revue des Questions Scientifiques,* ser. 2, vol. 3 (1894).

"L'Evolution des théories physiques du XVII siècle jusqu'à nos jours." *Revue des Questions Scientifiques,* ser. 2, vol. 5 (1896).

"Physique de croyant." *Annales de Philosophie chrétienne* 77, ser. 4, no. 1 (1905-06). Translated as an appendix to *The Aim and Structure of Physical Theory.* Princeton: Princeton University Press, 1954.

"Sur l'histoire du principe employé en Statique par Torricelli." *Comptes Rendus des Séances de l'Académie des Sciences* 143 (1906).

"Sur quelques découvertes scientifiques de Léonard de Vinci." *Comptes Rendus des Séances de l'Académie des Sciences* 143 (1906).

"Le P. Marin Mersenne et la pésanteur d l'air—Première partie: Le P. Marsenne et le poids spécifique de l'air." *Revue Générale des Sciences Pures et Appliquées,* 15 September 1906.

"Le P. Marin Mersenne et la pésanteur de l'air—Seconde partie: Le P. Mersenne et l'expérience du Puy-de Dome." *Revue Générale des Sciences Pures et Appliquées*, 30 September 1906.

"De l'accélération exercée par une constante. Notes pur servir à l'histoire de la dynamique." *Comptes Rendus du Deuxième Congrès de Philosophie*. Geneva: Kundig et fils, 1906.

"Sur un fragment, inconnu jusqu'ici, de l'Opus Tertium de Roger Bacon." *Comptes Rendus des Séances de l'Académie des Sciences* 146 (1908).

"Sur la découverte de la loi de la chute des graves." *Comptes Rendus des Séances de l'Académie des Sciences* 146 (1908).

"La valeur de la théorie physique, à propos d'un livre récent." *Revue Générale des Sciences Pures et Appliquées*, 15 January 1908. Translated as an appendix to *The Aim and Structure of Physical Theory*. Princeton: Princeton University Press, 1954.

"Ce que l'on disait des Indes occidentales avant Christophe Colomb." *Revue Générale des Sciences Pures et Appliquées*, 30 May 1908.

"Un précurseur français de Copernic: Nicole Oresme." *Revue Générales des Sciences Pures et Appliquées*, 15 November 1909.

"Du temps ou la Scholastique latine a connu la Physique d'Aristote." *Revue de Philosophie* 9, no. 2 (1909).

"Thierry de Chartres et Nicolas de Cues." *Revue des Sciences Philosophiques et Théologiques* 3 (1909).

"La Physique neo-platonicienne au moyen age." *Revue des Questions Scientifiques*, ser. 3, vol. 18 (1910).

"Sur les *Meteorologicorum libri quatuor*, fausserhent attribués à Jean Duns Scot." *Archivium Franciscanum Historicum* 3 (1910).

"Le temps selon les philosophes hellènes." *Revue de Philosophie* 11, no. 2 (1911).

"Un document relatif à la réforme du calendrier." In *Hommage à Louis Olivier*. Paris: Imprimerie de la Court d'Appel, 1911.

"Physics," "Nemore, Lordanus de," "Saxony, Albert of," "Saxony, John of," and "Saxony, Thierry of." In *Catholic Encyclopaedia*. New York: R. Appleton, 1911.

"La précession des equinoxes selon les astromes grecs et arabes." *Revue des Questions Scientifiques*, ser. 3, vol. 21 (1912).

"La nature du raisonnement mathématique." *Revue de Philosophie* 12, no. 2 (1912).

Preface for: Albert Maire, *L'oeuvre scientifique de Blaise Pascal*. Paris: Hermann, 1912.

"Le temps et le mouvement selon les Scholastiques." *Revue de Philosophie* 13, no. 2 (1913); 14, no. 1, 2 (1914).

"François de Meyronnes O.F.M. et la question de la rotation de la terre." *Archivium Franciscanum Historicum* 6 (1913).

"Examen logique de la théorie physique." *Revue Scientifique* 51 (1913).

"Roger Bacon et l'horreur du vide." In *Seventh Centenary of the Birth of Roger Bacon*. Oxford: Clarendon Press, 1914.

"L'Optique de Malebranche." *Revue de Métaphysique et de Morale* 23 (1916).

Selected Bibliography of Works on Medieval Physical Science after Duhem

Aiton, E. J. *The Vortex Theory of Planetary Motions.* London: Macdonald; New York: American Elsevier, 1972.

Armstrong, A. H., ed. *The Cambridge History of Later Greek and Early Medieval Philosophy.* Cambridge: Cambridge University Press, 1970.

Baron, M. *Origin of the Infinitesimal Calculus.* Oxford: Clarendon Press, 1969.

Baudry, L. *Lexique philosophique de Guillaume d'Ockham.* Paris: Lethielex, 1958.

_____ . "Sur trois manuscrits occamistes." *Archives d'Histoire Doctrinale et Littéraire du Moyen Age* 10 (1936): 129-62.

Bottin, F. "Analisi linguistica e fisica Aristotelica nei 'Sophismata' di Richard Kilmington." In *Filosofia e politica, e altri saggi,* edited by C. Giacon. Padua: Antenore, 1973.

_____ . "L'Opinio de 'Insolubilibus' di Richard Kilmington." *Revista Critica di Storia della Filosofia* 28 (1973): 568-90.

_____ . "Un testo fondamentale nell'ambito della 'Nuova Fisica' di Oxford: I. Sophismata di Richard Kilmington." *Miscellanea Mediaevalia* 9 (1974): 201-5.

Bourbaki, N. *Elements d'histoire des mathématiques.* Paris: Hermann, 1960.

Boyer, C. B. *The Concepts of Calculus: A Critical and Historical Discussion of the Derivative and Integral.* New York: Hafner, 1949.

_____ . *A History of Mathematics.* New York: Wiley, 1968.

_____ . *The History of the Calculus and Its Conceptual Development.* New York: Dover, 1959.

Brampton, C. K. "Guillaume d'Ockham et la date probable de ses opuscules sur l'Eucharistie." *Etudes Franciscaines* 14 (1964): 77-79.

Breidert, W. *Das aristotelische Kontinuum in der Scholastik.* Beiträge zur Geschichte der Philosophie und Theologie des Mittelalters, new ser. 1. Munster: Aschendorff, 1970.

Brown, S. "Walter Burley's Treatise *De Suppositionibus* and Its Influence on William of Ockham." *Franciscan Studies* 32 (1972): 15-64.

Buescher, G. *The Eucharistic Teaching of William of Ockham.* Washington, D. C.: Catholic University of America Press, 1950.

Butterfield, H. *Origins of Modern Science: 1300-1800.* New York: The Free Press, 1965.

Cantor, M. *Vorlesungen uber Geschichte der Mathematik,* vols. 1-2. Leipzig: Teubner, 1880-92 and subsequent editions.

Carré, M. H. *Realists and Nominalists.* Oxford: Oxford University Press, 1946.

Carruccio, E. *Mathematics and Logic in History and in Contemporary Thought.* Translated by I. Quigly. Chicago: Aldine, 1964.

Clagett, M. *Archimedes in the Middle Ages.* 5 vols. Madison: The University of Wisconsin Press; Philadelphia: American Philosophical Society, 1964-84.

————. *Giovanni Marliani and Late Medieval Physics.* New York: Columbia University Press, 1941.

————. "The *Liber de motu* of Gerard of Brussels and the Origins of Kinematics in the West." *Osiris* 12 (1956): 73-175.

————. *The Science of Mechanics in the Middle Ages.* Madison: The University of Wisconsin Press, 1959.

Coleman, J. "Jean de Ripa O. F. M. and the Oxford Calculators." *Mediaeval Studies* 37 (1975): 130-89.

Cornford, F. M. "The Invention of Space." In *Essays in Honour of Gilbert Murray,* 215-35. London: Allen & Unwin, 1936. Reprinted in *Concepts of Space and Time,* edited by M. Capek, pp. 3-16.

Corvino, F. "Questioni inedite di Occam sul continuo." *Revista Critica di Storia della Filosofia* 13 (1958): 191-208.

Crombie, A. C. *Augustine to Galileo.* London: The Falcon Press, 1952. Reissued as *Medieval and Early Modern Sciences.* 2 vols. Garden City, N.Y.: Doubleday Anchor, 1959.

————. *Robert Grosseteste and the Origins of Experimental Science 1100-1700.* Oxford: Clarendon Press, 1953.

————. "Sources of Galileo's Early Natural Philosophy." In *Reason, Experiment, and Mysticism in the Scientific Revolution,* edited by M. L. R. Bonelli and W. R. Shea, pp. 157-75, 303-05. New York: Science History Publications, 1975.

Crosby, H. L., Jr. *Thomas of Bradwardine: His Tractatus de Proportionibus. Its Significance for the Development of Mathematical Physics.* Madison: The University of Wisconsin Press, 1961.

Dick, S. J. *Plurality of Worlds.* Cambridge: Cambridge University Press, 1982.

————. "Plurality of Worlds and Natural Philosophy: An Historical Study of the Origins of Belief in Other Worlds and Extraterrestrial Life." Ph.D. diss., Indiana University, 1977.

Dictionary of Scientific Biography. 15 vols. Edited by C. C. Gillespie. New York: Scribner, 1970- .

Dijksterhuis, E. J. *The Mechanization of the World Picture.* Translated by C. Dikshoorn. Oxford: Oxford University Press, 1961.

Dreyer, J. L. E. *A History of Astronomy from Thales to Kepler.* 2d ed. New York: Dover, 1953.

Easton, S. C. *Roger Bacon and His Search for a Universal Science.* London: Blackwell, 1952.

Effler, R. R. *John Duns Scotus and the Principle "Omne quod movetur ab alio movetur."* St. Bonaventure, N.Y.: The Franciscan Institute, 1962.

Eldredge, L. "Late Medieval Discussions of the Continuum and the Point of the Middle English Patience." *Vivarium* 17 (1979): 90-115.

Emden, A. B. *A Biographical Register of the University of Oxford to A.D. 1500.* Vol. 1. Oxford: Oxford University Press, 1957.

The Encyclopedia of Philosophy. Edited by Paul Edwards. New York: Macmillan, 1967.

Faral, E. "Jean Buridan: Notes sur les manuscrits, les editions et le contenu de ses ouvrages." *Archives d'Histoire Doctrinale et Littéraire du Moyen Age* 21 (1946): 1-53.

Feldman, S. "Platonic Themes in Gersonides' Cosmology." In *Salo Wittmayer Baron Jubilee Volume,* pp. 383-405. Jerusalem: American Academy for Jewish Research, 1975.

Franklin, A. *The Principle of Inertia in the Middle Ages.* Boulder, Colorado: Colorado Associated University Press, 1976.

Fuerst, A., O.S.B. *An Historical Study of the Doctrine of the Omnipresence of God in Selected Writings between 1220-1270.* Ph. D. diss. The Catholic University of America. In *Studies in Sacred Theology,* 2d ser., no. 62. Washington, D.C.: Catholic University of America Press, 1951.

Fumagalli, M. T. B. *The Logic of Abelard.* Synthese Historical Library, vol. 1. Dordrecht: Reidel, 1970.

Funkenstein, A. "The Dialectical Preparation for Scientific Revolutions: On the Role of Hypothetical Reasoning in the Emergence of Copernican Astronomy and Galilean Mechanics." In *The Copernican Achievement,* edited by R. Westman. Contributions of the UCLA Center for Medieval and Renaissance Studies, no. 7, pp. 165-203. Berkeley: University of California Press, 1975.

Ghisalberti, A. *Giovanni Buridano dalla metafisica alla fisica.* Publicazioni della Universita Catolica del Sacro Cuore. Milano: Vita e pensiero, 1975.

Gilson, E. *History of Christian Philosophy in the Middle Ages.* New York and London: Sheed and Ward, 1955.

————. *Reason and Revelation in the Middle Ages.* New York: Scribner's, 1955.

Gingerich, O. "The Mercury Theory from Antiquity to Kepler." In *Actes du XIIe congrès international d'histoire des sciences, Paris 1968,* vol. 3A, pp. 57-64. Paris: Albert Blanchard, 1971.

Glorieux, P. *La Littérature quodlibetique.* 2 vols. Paris: Vrin, 1925-35.

Grabmann, M. *Die Sophismataliteratur des 12 und 13. Jahrhunderts mit Textausgabe eines Sophisma des Boetius von Dacien.* Beitrage zur Geschichte der Philosophie und Theologie des Mittelalters, vol. 36, pt. 1. Munich: Aschendorff, 1940.

Grant, E. "The Arguments of Nicholas of Autrecourt for the Existence of Interparticulate Vacua." In *Actes du XIIe congrès international d'histoire des sciences, Paris 1968*, vol. 3A, pp. 65-68. Paris: Albert Blanchard, 1971.

————. "Aristotle, Philoponus, Avempace, and Galileo's Pisan Dynamics." *Centaurus* 2 (1965): 79-95.

————. "Bradwardine and Galileo: Equality of Velocities in the Void." *Archive for History of Exact Sciences* 2, no. 4 (1965): 344-64.

————. "The Concept of *Ubi* in Medieval and Renaissance Discussions of Place." In *Science, Medicine, and the University: 1200-1550, Essays in Honor of Pearl Kibre*, part I. Special eds. N. G. Siraisi and L. Demaitre. *Manuscripta* 20, no. 2 (1976): 71-80.

————. "The Condemnation of 1277." In *Cambridge History of Later Medieval Philosophy*, edited by N. Kretzmann, A. Kenny, and J. Pinborg. Cambridge: Cambridge University Press, 1982.

————. "The Condemnation of 1277, God's Absolute Power, and Physical Thought in the Late Middle Ages." *Viator* 10 (1979): 211-244.

————. "Henricus Aristippus, William of Moerbeke and Two Alleged Mediaeval Translations of Hero's *Pneumatica*." *Speculum* 46 (1971): 656-69.

————. *In Defense of the Earth's Centrality and Immobility: Scholastic Reaction to Copernicanism in the Seventeenth Century.* Transactions of the American Philosophical Society 74, pt. 4. Philadelphia: American Philosophical Society, 1984.

————. "Jean Buridan: A Fourteenth Century Cartesian." *Archives Internationales d'Histoire des Sciences* 16 (1963): 251-55.

————. "Medieval and Seventeenth Century Conceptions of an Infinite Void Space beyond the Cosmos." *Isis* 60 (1969): 57-59.

————. "The Medieval Doctrine of Place: Some Fundamental Problems and Solutions." In *Studii sul XIV secolo in memoria di Anneliese Maier*, edited by A. Maieru and A. P. Bagliani. Rome: Edizioni di Storia e Letteratura, 1981.

————. "Medieval Explanations and Interpretations of the Dictum that 'Nature Abhors a Vacuum.'" *Traditio* 29 (1973): 327-55.

————. "Motion in the Void and the Principle of Inertia in the Middle Ages." *Isis* 55 (1964): 265-92.

————. *Much Ado about Nothing: Theories of Space and Vacuum from the Middle Ages to the Scientific Revolution.* Cambridge: Cambridge University Press, 1981.

————. "On the Origin of the Medieval Version of Equality of Fall for Unequal Bodies in the Void: A Critique of Duhem's Explanation."

In *Actes du XIe congrès international d'histoire des sciences, Varsovie-Cracovie 24-31 Aout 1965*, vol. 3, pp. 19-23. Warsaw/Cracow: 1967.

———. *Physical Science in the Middle Ages*. New York: Wiley, 1971.

———. "Place and Space in Medieval Physical Thought." In *Motion and Time, Space and Matter, Interrelations in the History of Philosophy and Science*, edited by P. K. Machamer and R. G. Turnbull, pp. 137-67. Columbus: Ohio State University Press, 1976.

———. "The Principle of the Impenetrability of Bodies in the History of Concepts of Separate Space from the Middle Ages to the Seventeenth Century." *Isis* 69 (1978): 551-71.

———. "Scientific Thought in Fourteenth-Century Paris: Jean Buridan and Nicole Oresme." In *Machaut's World: Science and Art in the Fourteenth Century*, edited by M. P. Cosman and B. Chandler. Annals of the New York Academy of Sciences, vol. 314, pp. 105-24. New York: New York Academy of Sciences, 1978.

———, ed. *A Source Book in Medieval Science*. Cambridge, Mass.: Harvard University Press, 1974.

Guerlac, R. *Humanism and Medieval Logic: Juan Luis Vives on the Pseudo-dialecticians*. Dordrecht: Reidel, 1979.

Hall, A. R. *The Scientific Revolution: 1500-1800*. Boston: The Beacon Press, 1974. 1st ed., Longmans Green & Co., 1954.

Harries, K. "The Infinite Sphere: Comments on the History of a Metaphor." *Journal of the History of Philosophy* 13 (1975): 5-15.

Haskins, C. H. *The Renaissance of the Twelfth Century*. Cambridge, Mass.: Harvard University Press, 1928.

———. *Studies in the History of Mediaeval Science*. 2d ed. Cambridge, Mass.: Harvard University Press, 1927.

Hayes, Z., O.F.M. *The General Doctrine of Creation in the Thirteenth Century with Special Emphasis of Matthew of Aquasparta*. Munich: Schöningh, 1964.

Henry, J. "Francesco Patrizi da Cherso's Concept of Space and Its Later Influence." *Annals of Science* 36 (1979): 549-73.

Hintikka, J. "Aristotelian Infinity." *Philosophical Review* 75 (1966): 197-218.

Hisette, R. *Enquète sur les 219 articles condamnés à Paris le 7 mars 1277*. Philosophes Médiévaux, vol. 22. Louvain: Publications Universitaires; Paris: Vander-Oyez, 1977.

Hoskin, M. A., and Molland, A. G. "Swineshead on Falling Bodies: An Example of Fourteenth Century Physics." *British Journal for the History of Science* 3 (1966): 150-82.

Hossfeld, P. "Die naturwissenschaftlich/naturphilosophische Himmelslehre Alberts des Grossen." *Philosophia Naturalis* 11 (1969): 329-30.

Jaki, S. L. *Planets and Planetarians: A History of Theories of the Origins of Planetary Systems*, Edinburgh: Scottish Academy Press, 1978.

Johnson, F. R. *Astronomical Thought in Renaissance England, A Study of the English Scientific Writings from 1500 to 1645.* New York: Octagon Books, 1968. 1st ed., Baltimore: Johns Hopkins University Press, 1937.

Jolivet, J. *Arts du Language et Théologie chez Abelard.* Etudes de Philosophie Médiévale, no. 57. Paris: Vrin, 1969.

Junghaus, H. *Ockham im Lichte der neueren Forschung.* Berlin: Lutherisches Verlagshus, 1968.

Juschkewitsch, A. P. *Geschichte der Mathematik in Mittelalter.* Leipzig: Teubner, 1964.

Kenny, A. "Vacuum Theory and Technique in Greek Science." *Transactions, Newcomen Society* 37 (1964-65): 47-56.

Kline, M. *Mathematical Thought from Ancient to Modern Times.* Oxford: Clarendon Press, 1972.

Knuuttila, S., and Lehtinen, A. I. "Change and Contradiction: A Fourteenth Century Controversy." *Synthese* 40 (1979): 189-207.

————. *"Plato in infinitum remisse incipit esse albus*: New Texts on the Late Medieval Discussion on the Concept of Infinity in Sophismata Literature." In *Essays in Honour of Jaakko Hintikka*, edited by E. Saarinen, R. Hilpinen, I. Niinilauto, and M. Provence Hintikka. Dordrecht: Reidel, 1979.

Koyré, A. *From the Closed World to the Infinite Universe.* Baltimore: Johns Hopkins University Press, 1957.

————. "Le vide et l'espace infini au XIVe siècle." *Archives d'Histoire Doctrinale et Littéraire du Moyen Age* 24 (1949): 45-91.

Kren, C. "Homocentric Astronomy in the Latin West: The *De reprobatione ecentricorum et epiciclorum* of Henry of Hesse." *Isis* 59 (1968): 269-81.

Kretzmann, N. "Comments on Murdoch's the Analytic Character of Late Medieval Learning: Natural Philosophy without Nature." In *Nature in the Middle Ages*, edited by L. Roberts. Forthcoming.

————. "Incipit/Desinit." In *Motion and Time, Space, and Matter*, edited by P. K. Machamer and R. G. Turnbull, 101-36. Columbus: Ohio State University Press, 1976.

————. "Socrates Is Whiter Than Plato Begins To Be White." *Nous* 11 (1977): 3-15.

————. "Time Exists—But Hardly, or Obscurely (*Physics* IV, 10; 217b69-218a33)." *The Aristotelian Society*, supplementary vol. 50 (1976): 91-114.

————. ed. *Infinity and Continuity in Ancient and Medieval Philosophy.* Ithaca, N. Y.: Cornell University Press, 1981.

Kuhn, T. S. *The Copernican Revolution: Planetary Astronomy in the Development of Western Thought.* Cambridge, Mass.: Harvard University Press, 1957.

Lasswitz, K. *Geschichte der Atomistik von Mittelalter bis Newton.* 2 vols. 2d ed. Leipzig: L. Voss, 1926.

Leff, G. *The Dissolution of the Medieval Outlook: An Essay on the Intellectual and Spiritual Change in the Fourteenth Century.* New York: New York University Press, 1976.

_____. *Paris and Oxford Universities in the Thirteenth and Fourteenth Centuries.* New York: Wiley, 1968.

_____. *William of Ockham: The Metamorphosis of Scholastic Discourse.* Manchester: University of Manchester Press, 1975.

Lindberg, D., ed. *Science in the Middle Ages.* Chicago: University of Chicago Press, 1978.

Lindhagen, A. "Die Neumondtafel des Robertus Grosseteste." *Arkiv for Matematik, Astronomi och Fysik* 11, no. 2 (1916): 1-41.

Litt, T. *Les corps céleste dans l'univers de Saint Thomas d'Aquin.* Louvin and Paris: Publication Universitaire, 1963.

Little, A. G., ed. *Roger Bacon Essays.* Oxford: Oxford University Press, 1914. Reprinted, New York: Russell and Russell, 1972.

Lohr, C. H., S.J. "Medieval Latin Aristotle Commentaries." *Traditio* 23 (1967), 24 (1968), 26 (1970), 27 (1971), 28 (1972), 29 (1973), 30 (1974).

_____. "Renaissance Latin Aristotle Commentaries: Authors A-B." *Studies in the Renaissance* 21 (1974): 228-89; "Authors C." *Renaissance Quarterly* 28 (1975): 689-741; "Authors D-F." *Renaissance Quarterly* 29 (1976): 714-45; "Authors G-K." *Renaissance Quarterly* 30 (1977): 681-741; "Authors L-M." *Renaissance Quarterly* 31 (1978): 532-603.

MacDonald, D. B. "Continuous Re-creation and Atomic Time in Muslim Scholastic Theology." *Isis* 9 (1927): 326-44.

McVaugh, M. "Arnald of Villanova and Bradwardine's Law." *Isis* 58 (1966): 56-64.

Maier, A. *Ausgehendes Mittelalter: gesammelte Aufsatze zur Geistesgeschichte des 14. Jahrhunderts.* 3 vols. Rome: Edizioni di Storia e Letteratura, 1964-77.

_____. *An der Grenze von Scholastik und Naturwissenschaft. Die Struktur der materiellen Sub-stanz; Das Problem der Gravitation; Die Mathematik der Forlatituden.* 2d ed. *Studien zur Naturphilosophie der Spatscholastik,* vol. 3. Rome: Edizioni di Storia e Letteratura, 1952.

_____. *Metaphysische Hintergrunde der spatscholastischen Naturphilosophie. Studien zur Naturphilosophie der Spatscholastik,* vol. 4. Rome: Edizioni di Storia e Letteratura, 1955.

_____. *Die Vorlaufer Galileis im 14. Jahrhundert. Studien zur Naturphilosophie der Spatscholastik,* vol. 1. Rome: Edizioni di Storia e Letteratura, 1949.

_____. *Zwei Grundprobleme der scholastischen Naturphilosophie.* 2d ed. Rome: Edizioni di Storia e Letteratura, 1951. 3d ed. Rome: Edizioni di Storia e Letteratura, 1968.

_____. *Zwischen Philosophie und Mechanik.* Rome: Edizioni di Storia e Letteratura, 1958.

_____. *On the Threshold of Exact Science.* Translated by S. Sargent. Philadelphia: University of Pennsylvania Press, 1982.

Maieru, A. *Terminologia logica della tarda scholastica.* Rome: Edizioni dell'Ateneo, 1972.

————. "Il 'Tractatus de sensu composito et diviso' di Guglielmo Heytesbury." *Revista Critica di Storia della Filosofia* 21 (1966): 243-63.

Michel, P. H. *The Cosmology of Giordano Bruno.* Translated by Dr. R. E. W. Maddison. Paris: Hermann; London: Methuen; Ithaca, N.Y.: Cornell University Press, 1973. Original French edition, Paris: Hermann, 1962.

Molland, A. G. "The Geometrical Background to the 'Merton School.'" *British Journal for the History of Science* 4 (1968-69): 108-25.

Moody, E. A. "Galileo and Avempace: The Dynamics of the Leaning Tower Experiment." *Journal of the History of Ideas* 12 (1951): 163-93, 375-422.

————. "Galileo and His Precursors." In *Galileo Reappraised*, edited by C. Golino, pp. 23-43. Berkeley: University of California Press, 1966.

————. "Ockham and Aegidius of Rome." *Franciscan Studies* 9 (1949): 417-42.

————. *Studies in Medieval Philosophy, Science and Logic: Collected Papers, 1933-1969.* Berkeley: University of California Press, 1975.

Murdoch, J. E. "The 'Equality' of Infinites in the Middle Ages." *Actes du XIe Congrès International d'Histoire des Sciences* 3 (1965): 171-74.

————. "Geometry and the Continuum in the Fourteenth Century: A Philosophical Analysis of Thomas Bradwardine's *Tractatus de continuo.*" Ph.D. diss. University of Wisconsin-Madison, 1957.

————. "Infinity and Continuity." In *The Cambridge History of Later Medieval Philosophy*, edited by N. Kretzmann, A. Kenny, and J. Pinborg. Cambridge: Cambridge University Press, 1982.

————. "*Mathesis in Philosophiam Scholasticam Introducta*, The Rise and Development of the Application of Mathematics in Fourteenth Century Philosophy and Theology." Arts Liberaux et Philosophie au Moyen Age Actes du Quatrième Congrès International de Philosophie Médiévale, Montreal 27 aout-2 septembre 1967, pp. 215-54. Montreal: Institut d'Etudes Médiévales; Paris: Librairie Philosophique J. Vrin, 1969.

————. "The Medieval Language of Proportions: Elements of the Interaction with Greek Foundations and the Development of New Mathematical Techniques." In *Scientific Change*, edited by A. C. Crombie, pp. 261-65. London: 1963.

————. "Naissance et développement de l'atomisme au bas moyen age latin." In *La science de la nature: Théories et pratiques*, Cahiers d'Etudes Médiévales, vol. 2. Montreal: Bellarmin; Paris: Vrin, 1974.

————. "Propositional Analysis in Fourteenth-Century Natural Philosophy: A Case Study." *Synthese* 40 (1974): 117-46.

_____. *"Rationes mathematice:"* Un Aspect du rapport des mathématiques et de la philosophie au moyen age. Paris: Université de Paris, 1962.

_____. "Superposition, Congruence, and Continuity in the Middle Ages." In *Mélanges Alexandre Koyré*, vol. 1, pp. 416-41. Paris: Hermann, 1964.

Murdoch, J. E., and Sylla, E. D., eds. *The Cultural Context of Medieval Learning: Proceedings of the First International Colloquium on Philosophy, Science, and Theology in the Middle Ages—September 1973.* Dordrecht: Reidel, 1975.

Murdoch, J. E., and Synan, E. A. "Two Questions on the Continuum: Walter Chatton (?), O.F.M., and Adam Wodeham, O.F.M." *Franciscan Studies* 26 (1966): 212-88.

Normore, C. G. "The Logic of Time and Modality in the Later Middle Ages: The Contribution of William of Ockham." Ph.D. diss., University of Toronto, 1976.

Pedersen, O. "The Development of Natural Philosophy, 1250-1350." *Classica et Mediaevalia* 14 (1953): 86-155.

_____. "A Fifteenth Century Glossary of Astronomical Terms." In *Classica et mediaevalia Francisco Blatt septuagenario dedicata*, edited by O. S. Due, H. F. Johansen, and B. D. Larsen. Classica et mediaevalia, dissertationes, vol. 9, pp. 584-94. Copenhagen: 1973.

_____. "The Life and Work of Peter Nightingale." In *Vistas in Astronomy*. Vol. 9. *New Aspects in the History and Philosophy of Astronomy*, edited by A. Beer, pp. 3-10. Oxford: 1967.

_____. *Peter Nightingale: A Problem of Identity with a Survey of Manuscripts.* Cahiers de l'Institut du Moyen Age, no. 19. Copenhagen: 1977.

Pedersen, O., and Pihl, M. *Early Physics and Astronomy: A Historical Introduction.* New York: Science History Publications, 1974.

Peters, F. E. *Aristotle and the Arabs: The Aristotelian Tradition in Islam.* New York: New York University Press, 1968.

Pieper, J. *Scholasticism: Personalities and Problems of Medieval Philosophy.* Translated by R. Winston and C. Winston. New York: McGraw-Hill, 1960.

Pinborg, J., ed. *The Logic of John Buridan.* Opuscula graecolatina, no. 9. Copenhagen: Museum Tusculanum, 1976.

Pines, S. *Beitrage zur Islamischen Atomenlehre.* Inaugural dissertation. Berlin: A. Heine, 1936.

_____. "La dynamique d'Ibn Bajja." In *L'adventure de la science*, pp. 442-68. Vol. 1 of *Mélanges Alexandre Koyré publiés l'occasion de son soixante-dixième anniversaire.* Vols. 12-13 of *Histoire de la pensée.* Paris: Hermann, 1964.

_____. "Philosophy, Mathematics, and the Concepts of Space in the Middle Ages." In *The Interaction between Science and Philosophy,* edited by Y. Elkana, pp. 75-90. Atlantic Highlands, N.J.: Humanities Press, 1974.

————— . "Thabit ibn Qurra's Conception of Number and Theory of the Mathematical Infinite." *Actes du XIe Congrès International d'Histoire des Sciences* 3 (1965): 160-66.

Poulet, G. *The Metamorphoses of the Circle.* Translated by C. Dawson and E. Coleman in collaboration with the author. Baltimore: Johns Hopkins University Press, 1966.

Ritter, G. *Studien zur Spatscholastik: I. Marsilius von Inghen und die okkamistische Schule in Deutschland.* Sitzungsberichte der Heidelberger Akademie der Wissenschaften, no. 4. Heidelberg: C. Winter, 1921.

Saggi, L. M. "Commentariola et Textus: Ioannis Baconthorpe textus de Immaculata Conceptione." *Carmelus* 2 (1955): 216-303.

Sambursky, S. *The Physical World of Late Antiquity.* New York: Basic Books, 1962.

————— . "Place and Space in Late Neoplatonism." *Studies in History and Philosophy of Science* 8 (1977): 173-87.

Sarton, G. *Introduction to the History of Science.* 3 vols. in 5 parts. Baltimore: Williams & Wilkins, 1927-48.

Schmitt, C. B. *Gianfrancesco Pico della Mirandola (1469-1533) and His Critique of Aristotle.* The Hague: Martinus Nijhoff, 1967.

Scott, T. K., Jr. "John Buridan on the Objects of Demonstrative Science." *Speculum* 40 (1965), 654-73.

Shapiro, H. *Motion, Time and Place according to William Ockham.* St. Bonaventure, N.Y.: Franciscan Institute, 1957.

————— . "Walter Burley and Text 71." *Traditio* 16 (1960): 395-404.

Sola, F. de P. "*De Sacramento Altaris*: Dos manuscritos Ockhamistas . . . Introduccion, transcripcion, y notas." *Pensamiento* 22 (1966): 279-352.

Steenberghen, F. Van. *Aristotle in the West: The Origins of Latin Aristotelianism.* Translated by L. Johnston. Louvain: Nauwelaerts, 1955.

————— . *Maitre Siger de Brabant.* Philosophes Médiévaux, no. 21, Louvain: Nauwelaerts, 1977.

————— . *Siger de Brabant d'après ses oeuvres inédites*, vol. 2. Vol. 12 of *Les Philosophes Belges.* Louvain: Editions de l'Institut Supérieur de Philosophie, 1942.

Sweeney, L. J., S.J. "Divine Infinity: 1150-1250." *The Modern Schoolman* 35 (1957-58): 38-51.

————— . "John Damascene and Divine Infinity." *The New Scholasticism* 35 (1961): 76-106.

Sweeney, L. J., S.J., and Ermatinger, C. J. "Divine Infinity according to Richard Fishacre." *The Modern Schoolman* 35 (1957-58): 191-211.

Sylla, E. "Autonomous and Handmaiden Science: St. Thomas Aquinas and William of Ockham on the Physics of the Eucharist." In *The*

Cultural Context of Medieval Learning, edited by J. E. Murdoch and E. D. Sylla. Dordrecht: Reidel, 1975.

————— . "Godfrey of Fontaines on Motion with Respect to Quantity of the Eucharist." In *Studi sul XIV secolo in memoria di Anneliese Maier*, edited by A. Maieru and A. Paravicini Bagliani. Rome: Edizioni di Storia e Letteratura, 1981.

————— . "Medieval Concepts of the Latitude of Forms: The Oxford Calculators." *Archives d'Histoire Doctrinale et Littéraire du Moyen Age* 40 (1973): 223-83.

————— . "Medieval Quantifications of Qualities: The 'Merton School.'" *Archive for History of Exact Sciences* 8 (1971): 12-15.

————— . "The Oxford Calculators." In *The Cambridge History of Later Medieval Philosophy*, edited by N. Kretzmann, A. Kenny, and J. Pinborg. Cambridge, Cambridge University Press, 1982.

————— . "The Oxford Calculators and the Mathematics of Motion, 1320-50: Physics and Measurements by Latitudes." Ph.D. diss., Harvard University, 1970.

Taton, R., ed. *La science antique et medievale*. Paris: Presses Universitaires de France, 1957. Translated as *Ancient and Medieval Science*. New York: Basic Books, 1963.

Thomson, S. H. "Unnoticed *Questiones* of Walter Burley on the Physics." *Mitteilungen des Instituts fur osterreichische Geschichtsforschung* 62 (1954): 390-405.

Thorndike, L. *A History of Magic and Experimental Science*. 8 vols. New York: Columbia University Press, 1923-58.

————— . *The Sphere of Sacrobosco and Its Commentators*. Chicago: The University of Chicago Press, 1949.

Tigerstedt, E. N. *The Decline and Fall of the Neoplatonic Interpretation of Plato. An Outline and Some Observations*. Commentatines Humanarum Litterarum, vol. 52. Helsinki: Societas Scientiarum Fennica, 1974.

Tonquedec, J. de. *Questions de cosmologie et de physique chez Aristote et Saint Thomas*. Paris: Vrin, 1950.

Tuveson, E. "Space, Deity, and the 'Natural Sublime.'" *Modern Language Quarterly* 12 (1951): 20-38.

Tweedale, M. M. *Abailard on Universals*. Amsterdam and New York: North-Holland, 1976.

Van Ess, J. *Theology and Science: The Case of Abu Ashaq an Nazzam*. Second Annual United Arab Emirates Lecture in Islamic Studies. Ann Arbor: University of Michigan Press, 1978.

Vignaux, P. *Philosophy in the Middle Ages*. Translated by E. C. Hall. Cleveland and New York: World, 1959.

Waard, C. de. *L'Experience barometrique, ses antecedents et ses explications: Etude historique*. Thouars: Imprimerie Nouvelle, 1936.

Wallace, W. A., O.P. "Aquinas on the Temporal Relation between Cause and Effect." *Review of Metaphysics* 27 (1974): 569-84.

————. "Buridan, Ockham, Aquinas: Science in the Middle Ages." *The Thomist* 40 (1976): 475-83.

————. *Causality and Scientific Explanation.* Vol. 1, *Medieval and Early Classical Science.* Ann Arbor: University of Michigan Press, 1972.

————. "The Enigma of Domingo de Soto: *Uniformiter difformis* and Falling Bodies in Late Medieval Physics." *Isis* 59 (1968): 384-401.

————. *Galileo and His Sources.* Princeton: Princeton University Press, 1984.

————. "Galileo Galilei and the *Doctores Parisienses.*" In *New Perspectives on Galileo,* edited by R. E. Butts and J. C. Pitt. Dordrecht: Reidel, 1978.

Weinberg, J. R. *Nicolaus of Autrecourt: A Study in 14th Century Thought.* Princeton: Princeton University Press, 1948.

Weisheipl, J. A., O.P. "The Celestial Movers in Medieval Physics." *The Thomist* 24 (1961): 286-326.

————. "The Concept of Matter in Fourteenth-Century Science." In *The Concept of Matter in Greek and Medieval Philosophy,* edited by E. McMullin. Notre Dame, Ind.: University of Notre Dame Press, 1963.

————. *The Development of Physical Theory in the Middle Ages.* Ann Arbor: University of Michigan Press, 1971.

————. *Early 14th-Century Physics and the Merton "School" with Special Reference to Dumbleton and Heytesbury.* D. Phil. thesis, Oxford University, 1956.

————. *Friar Thomas d'Aquino, His Life, Thought, and Work.* Garden City, N.Y.: Doubleday, 1974.

————. "Motion in a Void: Aquinas and Averroes." In *St. Thomas Aquinas 1274-1974, Commemorative Studies,* 467-88. Foreword by E. Gilson. Toronto: Pontifical Institute of Mediaeval Studies, 1974.

————. "Ockham and Some Mertonians." *Mediaeval Studies* 30 (1968): 163-213.

————. "The Place of John Dumbleton in the Merton School." *Isis* 50 (1959): 439-54.

————. "The Principle *Omne quod movetur ab alio movetur* in Medieval Physics." *Isis* 56 (1965): 26-45.

————. "Repertorium Mertonense." *Mediaeval Studies* 31 (1969): 174-224.

————. "Roger Swyneshed, O.S.B., Logician, Natural Philosopher, and Theologian." In *Oxford Studies Presented to Daniel Callus.* Oxford: Clarendon Press, 1964.

Williams, D. C. "The Myth of Passage." *Journal of Philosophy* 48 (1951): 457-72.

Wilson, C. *William Heytesbury, Medieval Logic and the Rise of Mathematical Physics.* Madison: University of Wisconsin Press, 1956.

Wippel, J. F. "The Condemnations of 1270 and 1277 at Paris." *Journal of Medieval and Renaissance Studies* 7 (1977): 169-201.

Wippel, J. F., and Wolter, Allen, O.F.M., eds. *Medieval Philosophy from St. Augustine to Nicholas of Cusa.* New York: Freed Press; London: Collier Macmillan Publishers, 1969.

Wolf, A. *A History of Science, Technology and Philosophy.* 2d ed. London: George Allen and Unwin, 1950.

Wolff, M. *Fallgesetz und Massebegriff. Zwei wissenschaftshistorische Untersuchungen zur Kosmologie des Johannes Philoponus.* Quellen und Studien zur Philosophie, vol. 2. Edited by G. Patzig, E. Scheibe, and W. Wieland. Berlin: Walter de Gruyter, 1971.

Wolfson, H. A. "The Plurality of Immovable Movers in Aristotle, Averroes, and St. Thomas." In *Studies in the History of Philosophy and Religion: Harry Austryn Wolfson,* edited by I. Twersky and G. H. Williams. Cambridge, Mass.: Harvard University Press, 1973.

————. "The Problem of the Souls of the Spheres from the Byzantine Commentaries on Aristotle through the Arabs and St. Thomas to Kepler." In *Studies in the History of Philosophy and Religion: Harry Austryn Wolfson,* edited by I. Twersky and G. H. Williams, 22-59. Cambridge, Mass.: Harvard University Press, 1973.

Worthen, T. D. "Pneumatic Action in the Kepsydra and Empedocles' Account of Breathing." *Isis* 61 (1970): 520-30.

Zimmermann, A. *Verzeichnis ungedruckter Kommentare zur Metaphysik und Physik des Aristotles aus der Zeit von etwa 1250-1350.* Leiden and Cologne: Brill, 1971.

Zoubov, V. P. "Jean Buridan et les concepts du point au quatorzième siècle." *Mediaeval and Renaissance Studies* 5 (1961): 43-95.

————. "Walter Catton, Gérard d'Odon et Nicolas Bonet." *Physics* 1 (1959): 261-78.

Index